METHODS IN MOLECULAR BIOLOGY

Series Editor
John M. Walker
School of Life and Medical Sciences
University of Hertfordshire
Hatfield, Hertfordshire, UK

For further volumes:
http://www.springer.com/series/7651

For over 35 years, biological scientists have come to rely on the research protocols and methodologies in the critically acclaimed Methods in Molecular Biology series. The series was the first to introduce the step-by-step protocols approach that has become the standard in all biomedical protocol publishing. Each protocol is provided in readily-reproducible step-by step fashion, opening with an introductory overview, a list of the materials and reagents needed to complete the experiment, and followed by a detailed procedure that is supported with a helpful notes section offering tips and tricks of the trade as well as troubleshooting advice. These hallmark features were introduced by series editor Dr. John Walker and constitute the key ingredient in each and every volume of the Methods in Molecular Biology series. Tested and trusted, comprehensive and reliable, all protocols from the series are indexed in PubMed.

Manuscript Instructions https://drive.google.com/file/d/1K4Az5vpAFVzYadJ3jKx2-QSo7Dqa_SNA/view?usp=sharing

Ustilago maydis

Methods and Protocols

Edited by

An N. T. Phan

Institute of Applied Microbiology-iAMB, Aachen Biology and Biotechnology-ABBt, RWTH Aachen University, Aachen, Germany

Lars M. Blank

Institute of Applied Microbiology-iAMB, Aachen Biology and Biotechnology-ABBt, RWTH Aachen University, Aachen, Germany; WSS Research Centre "Catalaix", Aachen, Germany

Jan Schirawski

Matthias-Schleiden-Institut/Genetik, Friedrich-Schiller-Universität Jena, Jena, Germany

Humana Press

Editors
An N. T. Phan
Institute of Applied Microbiology-iAMB
Aachen Biology and Biotechnology-ABBt
RWTH Aachen University
Aachen, Germany

Lars M. Blank
Institute of Applied Microbiology-iAMB
Aachen Biology and Biotechnology-ABBt
RWTH Aachen University
Aachen, Germany

WSS Research Centre "Catalaix"
Aachen, Germany

Jan Schirawski
Matthias-Schleiden-Institut/Genetik
Friedrich-Schiller-Universität Jena
Jena, Germany

ISSN 1064-3745 ISSN 1940-6029 (electronic)
Methods in Molecular Biology
ISBN 978-1-0716-5363-0 ISBN 978-1-0716-5364-7 (eBook)
https://doi.org/10.1007/978-1-0716-5364-7

This Humana imprint is published by the registered company Springer Science+Business Media, LLC, part of Springer Nature.
The registered company address is: 1 New York Plaza, New York, NY 10004, U.S.A.

Preface

Ustilago maydis, a basidiomycete responsible for corn smut disease, is historically classified as a model to study fungal pathogenicity, cell biology, genetics, and host–microbe interactions. Its tractable genetics, well-annotated genome, and amenability to both classical and modern molecular techniques have enabled researchers to uncover fundamental principles that extend far beyond the Ustilaginaceae. Over the past two decades, *U. maydis* has also emerged as a versatile platform for biotechnology, helping bridge basic fungal biology with applied research in metabolic engineering and bioprocess development.

This book brings together an extensive collection of protocols contributed by experts across the *U. maydis* community. It serves as a reliable and practical resource that strengthens the collective toolkit of the *U. maydis* community and fosters further discoveries in fungal biology, plant pathology, applied microbiology, and beyond.

Aachen, Germany — *An N. T. Phan*
Aachen, Germany — *Lars M. Blank*
Jena, Germany — *Jan Schirawski*

Acknowledgments

This work was partially funded by:

(1) The Deutsche Forschungsgemeinschaft (DFG, German Research Foundation)—SFB1535—Mibinet—Project ID 458090666.
(2) Germany's Excellence Strategy—Cluster of Excellence 2186 "The Fuel Science Center"—Project ID: 390 919 832.
(3) The Deutsche Forschungsgemeinschaft (DFG, German Research Foundation)—Exzellenzcluster EXC 2186 "The Integrated Fuel & Chemical Science Center" (FSC^2).
(4) The Bundesministerium für Bildung und Forschung (BMBF), Germany, in the project "UpRePP—Upcycling of Regional Waste Streams for the Production of Platform Chemicals" in the project cluster "Modellregion, BioRevierPLUS: InBio, Innovationscluster Integrierte Bioraffinerie, TP2" (Grant Number 031B1135B/031B1135BX).
(5) European Union's Horizon 2020 research and innovation programme under the Marie Skłodowska-Curie grant agreement No. 101034323 (Talent4BBI) and co-funded by BiOrbic Research Ireland Centre for Bioeconomy.
(6) The project SPaRe (Steigerung der Energieeffizienz des Wertstoffkreislaufs Papier durch Erschließung anfallender Rejektströme) was funded by the BMWE (Bundesministerium für Wirtschaft und Energie): Project number: 03EN2131A.
(7) The Werner Siemens Foundation in the frame of the WSS Research Centre "Catalaix".

Contents

Contributors

FLORIAN ALTEGOER • *Institute of Microbiology, Cluster of Excellence on Plant Sciences, Heinrich Heine University, Düsseldorf, Germany*

NILS BÄCKER • *Center for Synthetic Microbiology (SYNMIKRO), Marburg University, Marburg, Germany; Department of Biology, Marburg University, Marburg, Germany*

GERT BANGE • *Center for Synthetic Microbiology (SYNMIKRO), Marburg University, Marburg, Germany; Department of Chemistry, Marburg University, Marburg, Germany; Max-Planck Institute for Terrestrial Microbiology, Marburg, Germany*

RAMÓN RAMOS BARRALES • *Centro Andaluz de Biología del Desarrollo (CABD), Universidad Pablo de Olavide-CSIC-Junta de Andalucía, Seville, Spain*

SHIKHI BARURI • *Department of Biology, Program on Disease Evolution, University of Louisville, Louisville, KY, USA*

CHIBBHI BHASKAR • *Institute of Plant and Microbial Biology, Academia Sinica, Taipei, Taiwan*

BASTIEN BISSARO • *Biodiversité et Biotechnologie Fongiques (BBF), INRAE, Aix Marseille University, Marseille, France*

LARS M. BLANK • *Institute of Applied Microbiology-iAMB, Aachen Biology and Biotechnology-ABBt, RWTH Aachen University, Aachen, Germany; WSS Research Centre "Catalaix", Aachen, Germany*

JOSÉ ANTONIO CERVANTES-CHÁVEZ • *Facultad de Ciencias Naturales, Unidad de Microbiología Básica y Aplicada, Universidad Autónoma de Querétaro. Campus Aeropuerto, Carretera a Chichimequillas, Querétaro, México*

MINH-QUANG CHAU • *Institute of Plant and Microbial Biology, Academia Sinica, Taipei, Taiwan*

SHIVAM CHAUDHARY • *Matthias Schleiden Institute/Genetics, Friedrich Schiller University Jena, Jena, Germany*

ZIYANG CHEN • *Department of Plant Pathology, Institute of Crop Science and Resource Conservation (INRES), University of Bonn, Bonn, Germany*

HARIHARAN V CHINNASAMY • *Microbial Adaptation and Chromatin Remodeling Laboratory, Department of Biological Sciences and Bioengineering, Environmental Microbiology Laboratory, Center for Environmental Sciences and Engineering, Indian Institute of Technology, Kanpur, Uttar Pradesh, India*

JOSÉ IGNACIO IBEAS CORCELLES • *Centro Andaluz de Biología del Desarrollo (CABD), Universidad Pablo de Olavide-CSIC-Junta de Andalucía, Seville, Spain*

LUKAS DORIAN DITTIGER • *Matthias Schleiden Institute/Genetics, Friedrich Schiller University Jena, Jena, Germany*

ARMIN DJAMEI • *Department of Plant Pathology, Institute of Crop Science and Resource Conservation (INRES), University of Bonn, Nussallee, Bonn, Germany*

GUNTHER DOEHLEMANN • *Institute for Plant Sciences and Cluster of Excellence on Plant Sciences (CEPLAS), University of Cologne, Cologne, Germany*

THOMAS DREPPER • *Bioeconomy Science Center (BioSC), Jülich, Germany; Institute of Molecular Enzyme Technology, Heinrich-Heine-University Düsseldorf, Jülich, Germany*

MICHAEL FELDBRÜGGE • *Institute for Microbiology, Heinrich Heine University, Düsseldorf, Germany*
FLORIAN FINKERNAGEL • *Genomics Core Facility, Center for Tumor Biology and Immunology (ZTI), Philipps University, Marburg, Germany*
DAMON FREITAG • *Center for Synthetic Microbiology (SYNMIKRO), Marburg University, Marburg, Germany; Department of Chemistry, Marburg University, Marburg, Germany; Microcosm Earth Center, Marburg University & Max Planck Institute for Terrestrial Microbiology, Marburg, Germany*
JOHANNES FREITAG • *Center for Synthetic Microbiology (SYNMIKRO), Marburg University, Marburg, Germany; Department of Biology, Marburg University, Marburg, Germany*
LORENZO GARCÍA TEJADA • *Institute of Applied Microbiology-iAMB, Aachen Biology and Biotechnology-ABBt, RWTH Aachen University, Aachen, Germany*
SELINA GRANDERATH-MIEGEL • *Institute of Applied Microbiology-iAMB, Aachen Biology and Biotechnology-ABBt, RWTH Aachen University, Aachen, Germany*
LUCA ANTONIA GREBE • *AVT-BioVT – Biochemical Engineering, RWTH Aachen University, Aachen, Germany*
GUADALUPE GUERRA SÁNCHEZ • *Departamento de Microbiología, Escuela Nacional de Ciencias Biológicas, Instituto Politécnico Nacional, Carpio y Plan de Ayala S/N, Santo Tomás, Miguel Hidalgo, Ciudad de México, México*
XIAOWEI HAN • *National Key Laboratory of Agricultural Microbiology, Huazhong Agricultural University, Wuhan, China; Hubei Hongshan Laboratory, Wuhan, China; Hubei Key Laboratory of Plant Pathology, College of Plant Science and Technology, Huazhong Agricultural University, Wuhan, China*
KAI HEIMEL • *Georg-August-University Goettingen, Institute of Microbiology and Genetics, Department of Microbial Cell Biology Grisebachstr, Goettingen, Germany*
MAX HEINEN • *Institute of Microbiology, Cluster of Excellence on Plant Sciences, Heinrich Heine University, Düsseldorf, Germany*
WILLIAM K. HOLLOMAN • *Department of Microbiology and Immunology, Weill Cornell Medicine, New York, NY, USA*
EVAN JOHN • *Department of Plant Pathology, Institute of Crop Science and Resource Conservation (INRES), University of Bonn, Nussallee, Bonn, Germany*
JUDITH M. KLATT • *Center for Synthetic Microbiology (SYNMIKRO), Marburg University, Marburg, Germany; Department of Chemistry, Marburg University, Marburg, Germany; Microcosm Earth Center, Marburg University & Max Planck Institute for Terrestrial Microbiology, Marburg, Germany*
MILORAD KOJIC • *Institute of Molecular Genetics and Genetic Engineering, University of Belgrade, Belgrade, Republic of Serbia*
MATTHIAS KRETSCHMER • *Michael Smith Laboratories, The University of British Columbia, East Mall, Vancouver, Canada*
JAMES W. KRONSTAD • *Michael Smith Laboratories, The University of British Columbia, East Mall, Vancouver, Canada*
STEFAN KRÜGER • *Metabolomics and Metabolism Laboratory, Heinrich Heine University, Düsseldorf, Germany; Cluster of Excellences in Plant Science (CEPLAS), Heinrich Heine University, Düsseldorf, Germany*
JIA-CHI KU • *Plant Cell Biology, Biochemistry, and Biotechnology, University of Regensburg, Regensburg, Germany*

Jung-En Kuan • *Department of Plant Pathology, Institute of Crop Science and Resource Conservation (INRES), University of Bonn, Nussallee, Bonn, Germany*
Juliane Kühne • *Matthias Schleiden Institute/Genetics, Friedrich Schiller University Jena, Jena, Germany*
Nora Marie Kühne • *Georg-August-University Goettingen, Institute of Microbiology and Genetics, Department of Microbial Cell Biology Grisebachstr, Goettingen, Germany*
Pascal Künzel • *Institute of Bio- and Geosciences Biotechnology (IBG-1), Forschungszentrum Jülich, Jülich, Germany*
Venkata S. Kuppireddy • *Department of Biology, Program on Disease Evolution, University of Louisville, Louisville, KY, USA*
Wolfgang Laudensack • *University of Kaiserslautern Landau, Institute of Bioprocess Engineering, Kaiserslautern, Germany*
Claudia Geraldine Leon-Ramírez • *Departamentos de Ingeniería Genética, Centro de Investigación y de Estudios Avanzados del IPN (CINVESTAV), Unidad Irapuato, Gto, México*
Xuepei Li • *National Key Laboratory of Agricultural Microbiology, Huazhong Agricultural University, Wuhan, China; Hubei Hongshan Laboratory, Wuhan, China; Hubei Key Laboratory of Plant Pathology, College of Plant Science and Technology, Huazhong Agricultural University, Wuhan, China*
Yu-Han Lin • *Institute of Plant and Microbial Biology, Academia Sinica, Taipei, Taiwan*
Lay-Sun Ma • *Institute of Plant and Microbial Biology, Academia Sinica, Taipei, Taiwan*
Jørgen Barsett Magnus • *AVT-BioVT – Biochemical Engineering, RWTH Aachen University, Aachen, Germany*
Domingo Martínez-Soto • *Departament of Microbiology, Centro de Investigación Científica y de Educación Superior de Ensenada (CICESE), Ensenada, Código, México*
Saravanan Matheshwaran • *Microbial Adaptation and Chromatin Remodeling Laboratory, Department of Biological Sciences and Bioengineering, Environmental Microbiology Laboratory, Center for Environmental Sciences and Engineering, Indian Institute of Technology, Kanpur, Uttar Pradesh, India*
Mira Milisavljevic • *Institute of Molecular Genetics and Genetic Engineering, University of Belgrade, Belgrade, Republic of Serbia*
Christian Müller • *Matthias Schleiden Institute/Genetics, Friedrich Schiller University Jena, Jena, Germany*
Andreas Müsgens • *Institute of Applied Microbiology-iAMB, Aachen Biology and Biotechnology-ABBt, RWTH Aachen University, Aachen, Germany*
Gouri S. Nair • *Microbial Adaptation and Chromatin Remodeling Laboratory, Department of Biological Sciences and Bioengineering, Environmental Microbiology Laboratory, Center for Environmental Sciences and Engineering, Indian Institute of Technology, Kanpur, Uttar Pradesh, India*
Hirofumi Nakagami • *Protein Mass Spectrometry, Max-Planck Institute for Plant Breeding Research, Cologne, Germany*
María Dolores Pejenaute Ochoa • *Centro Andaluz de Biología del Desarrollo (CABD), Universidad Pablo de Olavide-CSIC-Junta de Andalucía, Seville, Spain; School of Life Sciences, University of Dundee, Dundee, UK*
Juan Pablo Pardo • *Departamento de Bioquímica, Facultad de Medicina, Universidad Nacional Autónoma de México, Circuito Interior, Ciudad Universitaria, Coyoacán, Ciudad de México, México*

JOSÉ PÉREZ-MARTÍN • *IBV, CSIC, Valencia, Spain*
FERNANDO PÉREZ-RODRÍGUEZ • *Departamento de Ciencias Biomédicas, Escuela de Medicina, Universidad Quetzalcóatl Irapuato (UQI), Irapuato, Código, México*
MICHAEL H. PERLIN • *Department of Biology, Program on Disease Evolution, University of Louisville, Louisville, KY, USA*
KOMAL PERVAIZ • *Department of Plant Pathology, Institute of Crop Science and Resource Conservation (INRES), University of Bonn, Bonn, Germany*
AN N. T. PHAN • *Institute of Applied Microbiology-iAMB, Aachen Biology and Biotechnology-ABBt, RWTH Aachen University, Aachen, Germany*
LESLEY PLÜCKER • *Institute for Microbiology, Heinrich Heine University Düsseldorf, Düsseldorf, Germany; Bioeconomy Science Center (BioSC), Jülich, Germany*
JOHANNES POSTMA • *Institute for Microbiology, Heinrich Heine University, Düsseldorf, Germany*
PAUL RICHTER • *AVT-BioVT – Biochemical Engineering, RWTH Aachen University, Aachen, Germany*
LUCERO ROMERO-AGUILAR • *Departamento de Bioquímica, Facultad de Medicina, Universidad Nacional Autónoma de México, Circuito Interior, Ciudad Universitaria, Coyoacán, Ciudad de México, México*
JOSÉ RUIZ-HERRERA • *Departamentos de Ingeniería Genética, Centro de Investigación y de Estudios Avanzados del IPN (CINVESTAV), Unidad Irapuato, Gto, México*
BJÖRN SANDROCK • *Department of Biology - Genetics, Philipps-Universität Marburg, Marburg, Germany*
ISABEL M. L. SAUR • *Institute for Plant Sciences and Cluster of Excellence on Plant Sciences (CEPLAS), University of Cologne, Cologne, Germany*
BARRY J. SAVILLE • *Environmental and Life Sciences Graduate Program, Trent University, Peterborough, ON, Canada; Department of Forensic Science, Trent University, Peterborough, ON, Canada*
KERSTIN SCHIPPER • *Institute of Microbiology, Heinrich Heine University Düsseldorf, Düsseldorf, Germany; Bioeconomy Science Center (BioSC), Jülich, Germany*
JAN SCHIRAWSKI • *Matthias Schleiden Institute/Genetics, Friedrich Schiller University Jena, Jena, Germany*
AMANDA M. SETO • *Environmental and Life Sciences Graduate Program, Trent University, Peterborough, ON, Canada*
SAINATH SHANMUGASUNDARAM • *Institute of Microbiology, Heinrich Heine University Düsseldorf, Düsseldorf, Germany*
WEI SHI • *Institute for Plant Sciences and Cluster of Excellence on Plant Sciences (CEPLAS), University of Cologne, Cologne, Germany*
JEANETTE H. SUTHERLAND • *Department of Biology, Farmingdale State College, Farmingdale, NY, USA*
RIAZ TABASSUM • *Department of Plant Pathology, Institute of Crop Science and Resource Conservation (INRES), University of Bonn, Bonn, Germany*
SHIGEYUKI TANAKA • *Faculty of Agriculture, Setsunan University, Hirakata, Osaka, Japan*
EDA P. TENORIO • *Departamento de Bioquímica, Facultad de Medicina, Universidad Nacional Autónoma de México, Circuito Interior, Ciudad Universitaria, Coyoacán, Ciudad de México, México*
JESSICA TIEFENBACHER • *Department of Biology - Genetics, Philipps-Universität Marburg, Marburg, Germany*
ANTONIO DE LA TORRE • *CIALE, Universidad de Salamanca, Salamanca, Spain*

ROLAND ULBER • *University of Kaiserslautern Landau, Institute of Bioprocess Engineering, Kaiserslautern, Germany*
LAURA VALDÉS-SANTIAGO • *SECIHTI-Tecnológico Nacional de México/Instituto Tecnológico Superior de Irapuato, Guanajuato, México*
KARINA VAN DER LINDE • *Plant Cell Biology, Biochemistry, and Biotechnology, University of Regensburg, Regensburg, Germany*
HÉCTOR VÁZQUEZ MEZA • *Departamento de Bioquímica, Facultad de Medicina, Universidad Nacional Autónoma de México, Circuito Interior, Ciudad Universitaria, Coyoacán, Ciudad de México, México*
SIMON WEGMANN • *Institute for Microbiology, Heinrich Heine University Düsseldorf, Düsseldorf, Germany; Bioeconomy Science Center (BioSC), Jülich, Germany*
JENS WEIERMÜLLER • *Kronos International Inc., Peschstraße 5, Leverkusen, Germany*
PHILIPP WESTHOFF • *Metabolomics and Metabolism Laboratory, Heinrich Heine University, Düsseldorf, Germany; Cluster of Excellences in Plant Science (CEPLAS), Heinrich Heine University, Düsseldorf, Germany*
NICK WIERCKX • *Institute of Bio- and Geosciences IBG-1: Biotechnology, Forschungszentrum Jülich, Jülich, Germany*
LASSE VAN WIJLICK • *Institute of Microbiology, Heinrich Heine University Düsseldorf, Düsseldorf, Germany*
BING YANG • *National Key Laboratory of Agricultural Microbiology, Huazhong Agricultural University, Wuhan, China; Hubei Hongshan Laboratory, Wuhan, China; Hubei Key Laboratory of Plant Pathology, College of Plant Science and Technology, Huazhong Agricultural University, Wuhan, China*
ROSELINE A. YAO • *Biodiversité et Biotechnologie Fongiques (BBF), INRAE, Aix Marseille University, Marseille, France*
FELICIA ZLATI • *Institute of Bio- and Geosciences IBG-1: Biotechnology, Forschungszentrum Jülich, Jülich, Germany*
PATRICE JULIAN ZSOLDOS • *Department of Plant Pathology, Institute of Crop Science and Resource Conservation (INRES), University of Bonn, Bonn, Germany*
MATIAS D. ZURBRIGGEN • *Institute of Synthetic Biology and CEPLAS-Cluster of Excellence on Plant Sciences, Heinrich Heine University Düsseldorf, Düsseldorf, Germany*

Check for updates

Chapter 1

Induction Medium to Analyze In Vitro Morphological Changes and Biotrophic Features in Sporidia of *Ustilago maydis*

Matthias Kretschmer and James W. Kronstad

Abstract

Biotrophic fungal pathogens such as *Ustilago maydis* rely on a host to complete their lifecycle. *U. maydis* has the advantage, in contrast to obligate biotrophs such as rust fungi, that the saprophytic yeast-like sporidia can be cultivated in artificial media, while the dikaryotic filamentous and infectious stage depends on the host to complete the sexual phase of the lifecycle. Key phenotypes during plant infection include filamentous growth after mating, an increase of fungal biomass in developing tumors, production of an extracellular matrix (ECM) around fungal cells, and melanin formation during sporulation. Furthermore, effector proteins are only highly expressed and secreted during plant infection and function to manipulate plant immunity and metabolism to enable colonization of plant tissue. In general, the display of virulence traits only during in planta growth hampers research progress, and existing artificial media containing glucose don't mimic the plant environment. We noted that other carbon sources (e.g., amino acids, fatty acids, organic acids, secondary plant metabolites, vitamins, and plant hormones) are all present in planta in combination with preferred carbon sources such as glucose or sucrose. Considering these other carbon sources, we developed an artificial medium that can be used to phenocopy in planta development of the fungus. The use of this specialized medium reveals that *U. maydis* can exploit a variety of different host carbon sources as signals and to support proliferation during pathogenic development.

Keywords Corn smut, Organic acids, Carbon source, Pathogenicity, Nutritional virulence

1 Introduction

Pathogenic microorganisms must obtain all of the micronutrients and macronutrients needed for their survival and reproductive development during colonization of their respective hosts. In response, the host is known to use the limitation of essential micronutrients such as iron to reduce pathogen development and growth, leading to so-called nutritional immunity [1]. Successful pathogens, therefore, have developed and employ a variety of sophisticated mechanisms and strategies to acquire essential nutrients from the host, thus displaying so-called nutritional virulence [2]. As described below, nutritional

An N. T. Phan et al. (eds.), *Ustilago maydis*, Methods in Molecular Biology, vol. 3050,
https://doi.org/10.1007/978-1-0716-5364-7_1,

considerations of carbon sources are important for both beneficial and pathogenic associations of plants and microbes.

In a mutualistic, symbiotic situation, both organisms benefit from the interaction and the exchange of nutrients. For example, during mycorrhizal associations or bacterial nitrogen fixation, phosphate or biologically fixed nitrogen, respectively, are exchanged for a variety of carbon sources such as sugars, lipids/fatty acids, or organic acids such as malate [3–5]. The nutrient exchange often takes place at a defined plant-microorganismal interface in specialized structures or tissues. For example, biological nitrogen fixation takes place in nodules containing differentiated bacterial cells in symbiosomes, organelle-like structures with limited oxygen availability to facilitate the enzymology of nitrogen fixation. In this situation, the legume hosts provide the bacteria with carbon in the form of organic acids (e.g., malate) produced by plant cells from long-distance transported sugars such as sucrose [5]. Nutrients such as sucrose, lipids, and organic acids within the plant tissue are either transported symplastically and/or apoplastically to reach the plant–microbe interface. The individual contributions of symplastic vs apoplastic transport are not well-established. However, it is known, e.g., that apoplastic concentrations of organic acids such as citrate and malate in sugar beet leaves can reach the low millimolar range (4.3 mM) and environmental conditions such as iron deficiency can further increase their apoplastic concentrations up to 12.2 mM [6]. Organic acid concentrations in other nonsymplastic structures such as the xylem can reach even higher values, especially during iron scarcity. Thus, the above example and others illustrate the importance and use of nonpreferred carbon sources during beneficial microbial–plant interactions.

In a plant disease setting, only the pathogen benefits from the interaction with a host. Necrotrophic plant pathogens access all available nutrients after the killing of the host cells or tissues, and thus don't require special adaptation to nutrient sources. In contrast, the more intricate and intimately associated biotrophic pathogens manipulate the host to acquire all necessary nutrients for reproductive growth by: (1) inhibiting plant immune responses to supress plant cell death (i.e., via the hypersensitive response) and; (2) altering plant metabolism to benefit the pathogen. *Ustilago maydis* is a biotrophic fungal pathogen of corn and teosinte, which is able to infect all meristematic plant tissues above soil level [7]. The fungus incites dramatic tumors that are chlorotic and show a disruption of the sink to source transition in corn leaves [8]. This change leads to the infected area being in a sink state for carbon and nitrogen rather than developing as a source tissue [9]. During tumor development, the fungal biomass constantly increases and fungal cells are embedded in an extracellular matrix (ECM; [10]). Sporulation is observed at late stages of tumor development, and spores are covered in a layer of black melanin. Mating and pathogenicity in *U. maydis* are governed by the mating pheromone pathway and a signaling cascade, which leads to the infectious filamentous dikaryon and to the transcriptional changes required for biotrophy including expression of effector genes [10]. Interestingly, *U. maydis* preferentially colonizes the leaf tissue in the region between the mesophyll and the bundle sheath cells [11]. Corn plants have apoplastic organic acids, as mentioned above, and furthermore, use organic acids such as malate and pyruvate in a shuttle during C-4 photosynthesis for CO_2 fixation. Initially, CO_2 is fixed in malate in the mesophyll cells and then transported to the bundle sheath cells for formation of sugars [12]. Pyruvate or other organic acid derivatives are shuttled back to the mesophyll cells to replenish the organic acid pool needed for initial CO_2 fixation. Thus, organic acids are abundant and fulfill an important physiological function in corn hosts.

There is ample evidence that pathogenic microorganisms are able to use a variety of nonpreferred or unusual carbon sources to support their pathogenic development. These sources range from organic acids and glucose derivatives in the bipartite metabolism of intracellular bacterial pathogens, to the use of malate in symbiotic nitrogen fixing bacteria and sugars and lipids in mycorrhizal fungi [2–5]. Similar to the bipartite metabolism of intracellular bacterial pathogens, *U. maydis* as a biotroph reacts to a mixture of glucose plus organic acids with improved growth, formation of a glucose-derived extracellular matrix, and the oxygen dependent formation of melanin [13]. Notably, the combination of glucose plus organic acids leads to in vitro activation of mating pathway components and effector gene expression [13]. Previously, it was known that activation of the mating pathway in *U. maydis* governs the expression of effector genes in planta [10, 14].

Here, we describe a simple defined and adjustable artificial medium adapted from previous media described by Holliday [15] to test the impact of nutrients and mixtures of preferred and nonpreferred nutrients on virulence traits and phenotypic development of *U. maydis* in vitro. This medium can be adapted to assess the use of uncommon carbon sources such as fatty acids, organic acids, or even vitamins (e.g., ascorbate), which are able to support growth of *U. maydis* as sole sources of carbon (Figs. 1 and 2; [13, 16, 17]). The medium can further be employed to analyze the effects of combinations of carbon sources on growth and virulence traits of *U. maydis*. For example, a combination of hexose or pentose sugars plus tricarboxylic acid cycle (TCA)-related organic acids can be tested to simulate in planta conditions (Figs. 1 and 2). Thus, the medium described here is a new tool to facilitate the investigation of nutrient dependent virulence traits of *U. maydis* in vitro. In the future, we anticipate that the spatial and temporal manipulation of factors such as media nutrient composition of preferred and nonpreferred carbon sources and growth parameters such as oxygen availability, growth space (fluidic chambers), pH, and temperature would further stimulate pathogenic-like development of fungal sporidia of *U. maydis* in an in vitro setting. It is interesting to point out that the optimization of growth parameters is similar to approaches used to achieve stem cell differentiation. The ultimate end-goal is to achieve completion of the entire life cycle in vitro for haploid or diploid fungal cells without the need for plant infections. Achieving this goal would enable additional genetic approaches to identify mechanisms of virulence.

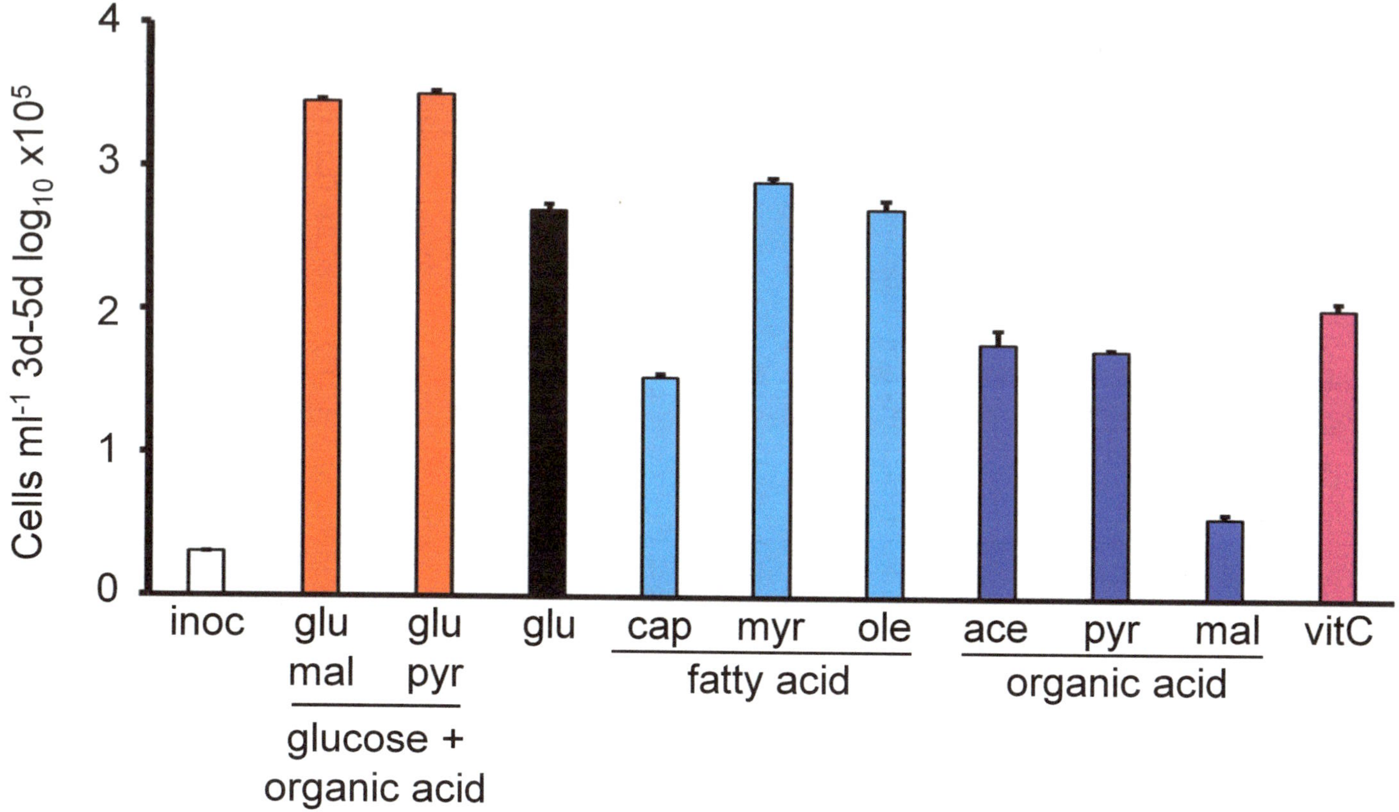

Fig. 1 Growth potential of *U. maydis* on uncommon carbon sources and mixtures of preferred and nonpreferred carbon sources. The cell numbers of the wild-type *U. maydis* strain 002 in liquid minimal medium with 1% glucose were compared to growth on fatty acids after 5 days for caproic acid (cap; 0.01%), myristic acid (myr; 1%), and oleic acid (ole; 1%), or for growth on organic acids after 3 days for acetate (ace; 1%), pyruvate (pyr; 1%), malate (mal; 1%), or ascorbate (vitC; 0.1%). Furthermore, growth on a combination of 1% glucose plus 0.5% of the organic acids pyruvate or malate after 3 days is shown. The level of the initial inoculum is also depicted. Cultures were incubated in a shaker at 30°C with 180 rpm for 3 or 5 days. The cell numbers were determined with a Neubauer cell counter, and the averages plus standard deviation log_{10} transformed are shown. (Adapted from Kretschmer et al. [13, 16, 17])

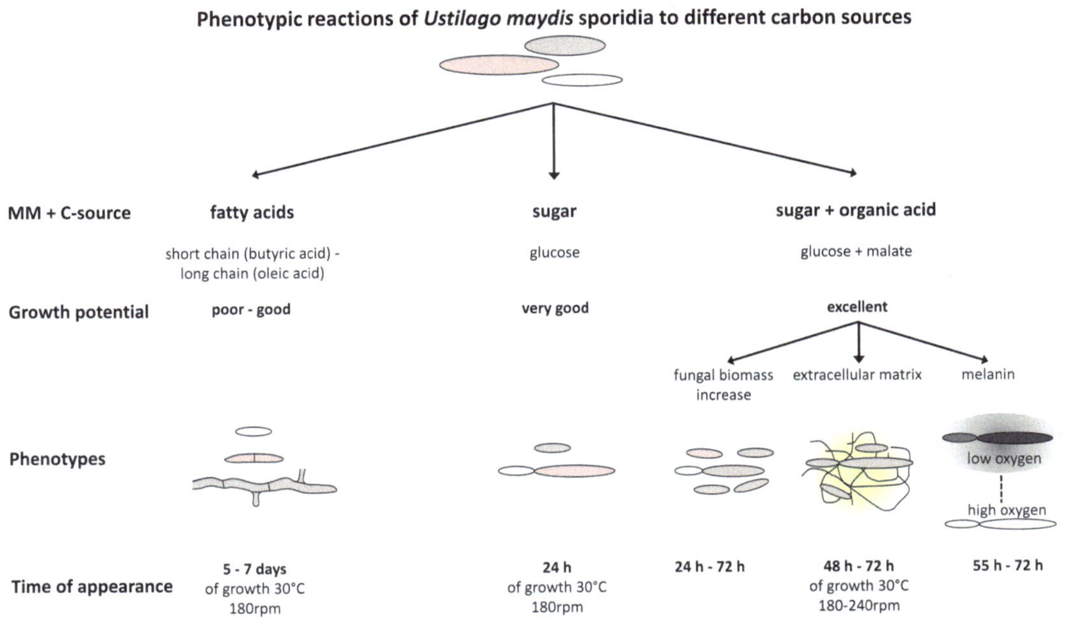

Fig. 2 Summary of carbon source-dependent phenotypes of sporidia of *U. maydis* in vitro in a salt-based minimal medium in liquid culture. Fatty acids can serve as carbon sources for *U. maydis* in minimal medium and induce morphological changes including filamentous growth in liquid cultures after extended growth periods. A mixture of sugar and organic acid leads to a time dependent appearance of three distinct phenotypes: (1) increased cellular numbers; (2) the development of a glucose-based extracellular matrix and; (3) the appearance of the black pigment melanin. Melanin formation is dependent on oxygen availability, and the pigment can be cell associated and shed into culture media. Oxygen levels in liquid cultures of minimal medium can either be influenced by additional stirring during incubation or, alternatively, an increase in shaking in the incubator to increase aeration. Without additional stirring, melanin will appear at ~55 h of incubation at 180–200 rpm depending on the shaker model. (Adapted from Kretschmer et al [13, 16, 17])

2 Materials

Prepare all solutions using ultrapure water such as milliQ and use highest purity available for each reagent. All steps can be carried out at room temperature unless indicated otherwise. Storage of the media and components is at room temperature unless indicated otherwise. Keep good sterile technique.

2.1 Individual Media Components to Prepare Ahead of Time

1. Glucose: 50% w/v glucose in milliQ water, filter sterilized (*see* **Note 1**).
2. Trace elements solution needed for salt solution: 30 mg H_3BO_3, 70 mg $MnCl_2$, 200 mg $ZnCl_2$, 20 mg $Na_2MO_4 \times 2H_2O$, 50 mg $FeCl_3$, 200 mg $CuSO_4$. Make up to 500 mL in water. Do not autoclave and store at 4°C (*see* **Note 2**).

3. *U. maydis* salt solution: 16 g KH_2PO_4, 4 g Na_2SO_4, 8 g KCl, 2 g $MgSO_4 \times 7H_2O$, 1 g $CaCl_2 \times 2H_2O$, 8 mL trace elements solution. Make up to 1 L in water. Do not autoclave and store at 4°C (*see* **Note 3**).

2.2 Popular All-Purpose Culture Media for U. maydis

1. PDB/PDA: Dissolve 24 g **P**otato **D**extrose **B**roth powder, add 20 g agar per L for plates only (PDA), make up to 1 L in milliQ water and autoclave.
2. YEPS-light: 4 g yeast extract, 4 g peptone, 20 g sucrose, add 20 g agar per L for plates only, make up to 1 L in water and autoclave.
3. Complete medium: 2.5 g casamino acids, 1.5 g ammonium nitrate, 10 g yeast extract, 62.5 mL salt solution. Add water to a volume of 980 mL. Adjust pH to 7.0 and autoclave. After cooling, add 20 mL of 50% glucose per 1 L to reach a 1% final glucose concentration. Add 20 g agar per L for plates only.
4. Double complete medium for mating assessment of *U. maydis* and for biolistic transformation: 10 g casamino acids, 3 g ammonium nitrate, 20 g yeast extract, 125 mL salt solution, add 20 g agar per L for plates only, add water to a volume of 980 mL, adjust pH to 7.0, and autoclave. After cooling, add 20 mL of 50% glucose per 1 L to reach a 1% final glucose concentration.

Variations: (a) To test mating on plates, add 10 gw/v activated charcoal per L before autoclaving and add glucose after autoclaving; (b) For biolistic transformation on plates, add sorbitol to 0.9 M before autoclaving and omit glucose; (c) For mutant selection on plates, add filter sterilized hygromycin B (250 µg per mL) or nourseothricin (125 µg per mL) together with glucose after cooling.

2.3 U. maydis Minimal Medium as Base for the Biotrophy Medium

1. Minimal medium without a carbon source (adapted from Holliday [15]): 3 g ammonium nitrate, 62.5 mL *U. maydis* salt solution. Add water to 800 mL and adjust pH to 7.0 with NaOH. Make up to 1 L with water and filter sterilize or autoclave. Add the desired carbon source, when preparing the assay (*see* **Note 4**).
2. Fatty acid-based minimal medium to induce a morphological switch of *U. maydis*: 3 g ammonium nitrate, 62.5 mL *U. maydis* salt solution. Dissolve in 800 mL of water, adjust pH with NaOH to 7.0, bring to 1 L and filter sterilize or autoclave. Add 10 gw/v or v/v of fatty acids. Add the fatty acids directly and individually in each test tube as they will not mix or dissolve in a water-based medium (*see* **Note 5**).
3. Organic acid-based minimal medium to induce biotrophic features in *U. maydis*: 3 g ammonium nitrate, 62.5 mL *U. maydis* salt solution, 10 gw/v organic acid. Dissolve in 800 mL of water, adjust pH to 7.0 with NaOH, bring to 1 L and filter sterilize (*see* **Note 6**).

2.4 Equipment

1. Microcentrifuge tubes, 1.5 or 2 mL, sterile.
2. Tips, sterile.
3. Vortex.
4. Centrifuge.
5. Biosafety cabinet.
6. Sterile glass test tubes for cell culture (18 x 150 mm Borosilicate glass).
7. Syringes (10 mL).
8. Sterile filter units.
   - for small volumes: 10 mL syringe-based units.
   - for larger volumes of media: up to 250 mL vacuum filter units.
9. Microscope.
10. Neubauer cell counter.
11. Incubator. Different models require adaptation in rpm settings (*see* **Note 7**).
    - Barnstead Lab-Line MaxQ 5000 Incubator Shaker 180 rpm.
    - New Brunswick Excella E24 Incubator Shaker 240 rpm.
12. pH-meter.
13. Autoclave.
14. Parafilm.
15. Petri dishes.
16. Pipettors.
17. Viscometer type Ubbelohde size 1 constant 0.00957.
18. Techan® plate reader (405 nm for melanin).
19. Heating plate/stirring plate.
20. 30°C incubator nonshaking.

3 Methods

All steps were carried out at room temperature under sterile conditions.

3.1 General Setup and Tips to Consider

A salt-based, minimal medium adapted from Holliday [15] was used for *U. maydis* strains FB1, 001, 002, SG200, and D132 to induce biotrophic features in sporidia in vitro in 5 mL liquid cultures (in test tubes). All strains showed similar results regarding increases in cell number, culture viscosity, and melanin formation. Strain FB2 was an exception and did not behave like the other strains (in our hands). We tested three different isolates of the FB2 strain from different laboratories with similar results. We note that strain FB2 also has other phenotypic differences compared with strain FB1 (e.g., in iron sensing; [18] and unpublished). The salt-based minimal medium described here was adapted for use with *U. maydis*, but shows similar results for other smut and hemibiotrophic fungi [13]. However, the appearance of the phenotypes was variable for the different fungi suggesting that further adaptation for each individual fungus is needed. The organic acids in their biological active form or, respectively, their salts were purchased from Sigma Aldrich® or Alfa Aesar® with the highest available purity. The organic acids used were: pyruvate, citric and isocitric acid, ketoglutarate, succinate, fumarate, and malic (not maleic) acid. All of the acids provoked partial or complete phenotypes after 3 days of growth. The organic acids with the most prominent expression of the phenotypes in *U. maydis* were pyruvate and malate suggesting a possible connection to the C4-photosynthetic metabolism of the corn host. There are several factors that impact the expression of the observed phenotypes including the use of clean, residue-free test tubes, water quality (only milliQ), temperature, iron content, oxygen availability, and pH. Changes in these parameters and medium composition will influence the phenotypic response of *U. maydis*.

3.2 Cell Inoculum Preparation

1. Streak *U. maydis* sporidia on a PDA plate from a −80°C glycerol freezer stock.
2. Seal the petri dish with parafilm.
3. Incubate at 30°C for 2–4 days until colonies appear.
4. Pipette 5 mL of PDB in a sterile glass test tube.
5. Use a yellow 200 μL pipette tip to pick-up a colony of *U. maydis* and inoculate the 5 mL PDB.
6. Incubate at 180 rpm at 30°C in a shaker with the test tubes angled at ~45° for overnight (*see* **Note 8**).

3.3 Harvesting the Cells

1. Harvest 2 mL of the overnight culture in a sterile 2 mL Eppendorf tube by centrifugation at 10,000 rpm for 1.5 min.
2. Discard the supernatant.
3. Wash the cells with 1 mL of sterile minimal medium without a carbon source.
4. Vortex the cells for 1 min.
5. Centrifuge at 10,000 rpm for 1.5 min.
6. Wash the cells as described above for a total of three times to remove any residual carbon source from the overnight culture.
7. Determine the cell number by cell counting with a Neubauer cell counter on a microscope.

8. Use 10^6 cells as inoculum for each 5 mL test medium in a sterile test tube.

3.4 To Test if Unusual Compounds are a Carbon Source for U. maydis

1. Prepare solutions of 0.1%, 0.5%, and 1% of the to be tested compound in minimal medium without another carbon source (*see* **Note 9**).
2. a. Prepare a sterile stock solution, and add the desired amount of compound to a test tube if a 100–1000× stock solution is possible. However, make sure the compound does not change the pH of the medium.
   b. If the compound does not change the pH, but a 100–1000× stock solution is not possible, it can be directly dissolved in the needed volume of media without a carbon source and filter sterilized.
   c. If the pH is changed, add the compound during media preparation and only prepare 50–100 mL minimal medium for each dilution as most of the compounds are not stable in liquid media for an extended period of time.
3. Depending on the preparation of the media pipette 5 mL of the medium with the different compound concentrations in a sterile test tube. Make sure to include the compound as final concentrations of 0.1%, 0.5%, and 1% if a stock was prepared (*see* **Note 10**).
4. Include one control. Include one repeat of 5 mL minimal medium without carbon source, and add 100 μL of 50% glucose for a final concentration of 1% glucose. If other glucose concentrations are desired, increase or decrease the glucose amount.
5. Add 10^6 *U. maydis* cells to each test tube.
6. Incubate at 180 rpm at 30°C in a shaker for 3 days with the test tubes angled at ~45°.
7. Anticipated results: Count the cell numbers microscopically every 24 h during the incubation time (*see* **Note 11**). Observe morphological changes if present.

3.5 To Test Morphological Changes Induced by Fatty Acids in U. maydis

1. a. For solid fatty acids. Weigh in to a sterile test tube 50 mg w/v, which equals 1% of the solid fatty acid. Pipette 5 mL minimal medium without carbon source into the test tube (*see* **Note 12**).
   b. For liquid fatty acids. Filter sterilize the fatty acids with a 10 mL syringe and a syringe filter. Pipette 5 mL minimal medium without carbon source into the sterile test tube.

   c. Add 50 µL v/v of the fatty acid in each test tube to reach 1% final concentration (*see* **Note 13**). Short chain fatty acids are toxic and should be used at 0.01%–0.1% instead.

2. Include one control. Include one repeat of 5 mL minimal medium without carbon source and add 100 µL of 50% glucose for a final concentration of 1% glucose. If other glucose concentrations are desired, increase or decrease the glucose amount.
3. Add 10^6 *U. maydis* cells to each test tube.
4. Incubate at 180 rpm at 30°C in a shaker for 5 days with the test tubes angled at ~45°.
5. Anticipated results: Count the cell numbers with a Neubauer cell counter under a microscope after 5 days incubation. Investigate morphological changes by microscopy and staining (e.g., cell wall staining with calcofluor white; *see* **Note 14**).

3.6 Induction of Biotrophic Features in U. maydis *by a Combination of Glucose and Organic Acids*

1. Glucose only control (G; *see* **Note 15**): Pipette 5 mL minimal medium without a carbon source into the sterile test tube. Add 100 µL v/v of 50% glucose to reach a final glucose concentration of 1%. The glucose concentration can be decreased to 0.5% or increased to 2% without significant changes in the growth of *U. maydis.*
2. Minimal medium 1% glucose plus 0.5% organic acid [e.g., Glucose plus Malate (GM medium)]: Add 2.5 mL minimal medium without a carbon source into the sterile test tube. Add 2.5 mL organic acid 1% minimal medium (*see* **Note 16**). Add 100 µL v/v 50% glucose.
3. Minimal medium 1% glucose plus 0.5% organic acid plus extra aeration [e.g., Glucose plus Malate plus Aeration ((GMA medium)]: Add 2.5 mL minimal medium without a carbon source into the sterile test tube. Add 2.5 mL organic acid 1% minimal medium. Add 100 µL v/v 50% glucose. Add a sterile yellow 200 µL pipette tip to enhance media stirring (see **Note 17**).
4. Add 10^6 *U. maydis* cells to each test tube.
5. Incubate at 180 rpm at 30°C in a shaker for 3 days with the test tubes angled at ~45°.
6. Anticipated results: Count the cell numbers with a Neubauer cell counter under a microscope every 24 h. Investigate culture viscosity with a viscometer (Ubbelohde size 1) after 72 h. Investigate melanin formation at 72 h. Effector gene expression can be assessed at 72 h (*see* **Note 18**).

4 Notes

1. Carbon sources should not be autoclaved. However, mild heating on a hot plate with stirring is needed to reach a 50% glucose solution.
2. In this solution, iron chloride will not completely dissolve, and the solution should be shaken before use.
3. The solution should be shaken before use.
4. Not all carbon sources can be autoclaved, are easily dissolved, and are pH neutral.
5. Fatty acids can induce filamentous growth in *U. maydis* sporidia without the need of a mating partner. Fatty acids are nonpreferred carbon sources and need a longer culture time of 5–7 days until growth is evaluated and morphological changes can be observed (Table 1). Fatty acids can be solid or liquid at room temperature. Liquid fatty acids were filter sterilized, solid fatty acids were used as supplied by the manufacturer. Tested fatty acids were short chain, medium, and long chain fatty acids usually at 1%. Short chain fatty acids are highly toxic in higher concentrations and thus were used at lower concentrations ranging from 0.01% to 0.1%. The use of detergents such as Tween® to increase solubility of the fatty acids is not advised as they contain fatty acids, which can be broken-down by *U. maydis.*
6. A combination of the preferred carbon source glucose plus TCA cycle-related organic acids is able to induce biotrophic features in *U. maydis* sporidia without the need of a mating partner. The observed phenotypes include increased growth rate, expression of an extracellular matrix, and melanin formation (Table 2). Effector gene expression is also induced. For the media, 10 g of organic acids per L equals 1%. In general, organic acids were used as 0.5% by mixing the 1% solution with an equal volume of minimal medium without a carbon source. This approach was also employed to be able to use other concentration ranges of organic acids by changing the volume of minimal medium without carbon source. The organic acids related to the TCA cycle were pyruvate, citrate, isocitrate, ketoglutarate, succinate, fumarate, malate, and oxaloacetate. Organic acids were used in the biological active form. Organic acids are heat sensitive, and thus cannot be autoclaved. Glucose was added to each test tube individually just before setting up the study. Organic acids in minimal media can be stored at room temperature for up to 1–2 month. However, the activity as carbon sources will decrease over time. Best results are achieved with freshly prepared media.

Table. 1

Summary of fatty acid use by the wild-type *U. maydis* strain 002 in minimal medium as sole carbon sources and as inducers of morphological changes. (Adapted from Kretschmer [16])

	Sugar control	Short chain fatty acids		Medium chain fatty acid	Saturated long-chain fatty acids		Unsaturated long-chain fatty acids	
	Glucose	Butyric	Caproic	Lauric	Myristic	Palmitic	Oleic	Linoleic
	$C_6H_{12}O_6$	$C_4H_8O_2$	$C_6H_{12}O_2$	$C_{12}H_{24}O_2$	$C_{14}H_{28}O_2$	$C_{16}H_{32}O_2$	$C_{18}H_{34}O_2$	$C_{18}H_{32}O_2$
Concentration of carbon source in minimal medium in %	1	0.01	0.01	0.1	1	1	1	1
Use as sole carbon source by *U. maydis* after 5 days of growth	Very good	Poor	Poor	Medium	Good	Medium	Good	Medium
Inducer of filamentation in *U. maydis*	**No**	Poor	No	Yes	Yes	Yes	Yes	Yes
Toxicity for *U. maydis*	None	Toxic	Toxic	None	None	None	None	None

Table. 2

A combination of glucose and tricarboxylic acid cycle (TCA)-related organic acids leads to an increased cellular growth rate and variable levels of melanin induction. (Adapted from Kretschmer [13])

	Sugar control	Mixed carbon sources containing combinations of sugar and organic acids [1% glucose + tri carboxylic acid cycle (TCA) organic acids (*see* Note 19)]								
	Glucose	Pyruvate	Citrate	Aconitate	Isocitrate	Ketoglutarate	Succinate	Fumarate	Malate	Oxaloacetate
	$C_6H_{12}O_6$	$C_3H_3NaO_3$	$C_6H_5Na_3O_7$	$C_6H_3Na_3O_6$	$C_6H_5Na_3O_7$	$C_5H_4Na2O_5$	$C_4H_4Na_2O_4$	$C_4H_2Na2O_4$	$C_4H_4Na_2O_5$	$C_4H2Na_2O_5$
Concentrations in minimal medium in %	1 (0.5 to 2)	0.5	0.5	0.5	0.5	0.5	0.5	0.5	0.5 (0.005–0.5)	0.5
Cell number of *U. maydis* after 3 days of growth	Normal	High	High	High	High	High	High	High	High	Reduced (toxic)
Melanin induction in *U. maydis* after 3 days of growth	None	High	High	Low	Low	Medium	Low	Medium	High	None

7. In our experience, shaking incubators from different manufacturers are not equal in shaking speeds. Thus, it has to be experimentally evaluated, which speed setting is successful in initiating the phenotypes. In general, if cells settle in the GM medium the speed is too low, if melanin is not induced the speed is too high.
8. We have wire test tube racks (4×10) installed in the shakers (Sigma Aldrich® HS120766). Test tubes are inserted in the second row at the top of the rack and the bottom of the test tube ends in row one of the rack. This leads to an angle of the test tube of around 45°.
9. Different dilutions should be tested for each compound as some will support growth at lower concentrations but are toxic in higher concentrations. For example, ascorbate supports growth at 0.1% but is toxic at 0.5%. Further, most of the unusual carbon sources are not heat stable and thus should not be autoclaved.
10. If glass test tubes are reused, make sure they don't contain any residue from former growth conditions or any other compounds such as soap from the glass washing facility. *U. maydis* is very sensitive in minimal medium to any form of residue, which will impact its growth.
11. Some compounds will change the cell shape or induce color changes in the medium or the cells of *U. maydis*. Thus, OD_{600} readings or 96 well plate-based growth curves (e.g., in a Tecan® plate reader) should be used with caution.
12. Solid fatty acids were weight on a fine balance and added directly to the sterile test tubes. This leads to a not 100% sterile environment. Regardless, contaminations were only rarely observed.
13. For solid fatty acids, it is best to weigh in the fatty acids before adding media to the test tube. While for liquid fatty acids, it is best to add the media before adding the fatty acids.
14. Fatty acids that are liquid at room temperature will form an emulsion, and thus, OD_{600} readings cannot be used to determine the cell number. Depending on the fatty acid used, filamentation will occur in sporidia. Oleic acid is one of the best inducers of filamentation compared to other fatty acids [19]. However, the cultures will be a mixture of yeast-like sporidia and multicellular filamentous hyphae, which complicates the determination of cell numbers. Detergent-based compounds such as Tween® have been used to dissolve fatty acids in media or plates, but as they are fatty acid based and can be degraded by *U. maydis*. Their use should be avoided.
15. G, GM, and GMA media from the publication of Kretschmer [13] are the media described here: G is 1% glucose, GM is 1% glucose plus 0.5% malate, and GMA is 1% glucose plus 0.5% malate plus a yellow (200 μL) pipette tip acting as a stir bar.

16. Other percentages or combinations can be achieved by adjusting the ratio of minimal medium without carbon source plus organic acid minimal medium, if desired.
17. A higher shaker speed may also be used. However, this would impede melanin formation in the GM condition if tested at the same time.
18. The increase in cell number in GM will be apparent after 24 h but more pronounced at 48 h and 72 h. The increase in viscosity of the medium will start to appear at 48 h and can be measured at 72 h as the flow through time of a defined volume over time in a Ubbelohde size 1 viscometer. Three 5 mL cultures will be needed for one measurement in the instrument. Melanin formation first appears at ~55 h and can be measured in the supernatant with absorbance at 405_{nm} after brief centrifugation to remove cells. Melanin can also be extracted from the wall of cells and further analyzed. If other fungi are investigated, the phenotypes may appear at times different from those identified for *U. maydis.* Thus, incubation times may need to be adjusted accordingly. In our experience, different sized incubators (small table top vs big floor shakers) are not identical in shaker speed even when the same rotation speed is programmed. As the melanin phenotype is oxygen dependent, the shaker speed has to be adjusted depending on the shaker model used. If effector expression is desired, then higher aeration is advantageous, because we found that the expression of effectors was highest in GMA vs GM or G at 72 h.
19. Some organic acids can be obtained as sodium salts and others as the acid form. Each of the organic acids will dissociate to the salt form upon pH adjustment in the minimal medium with sodium hydroxide (NaOH).

Acknowledgments

This work was supported by a discovery grant from the Natural Sciences and Engineering Research Council of Canada (NSERC) and by the NSERC-CREATE Program: Plant Responses To Eliminate Critical Threats (PRoTECT). J.W.K. is the Power Corporation Fellow of the Canadian Institute for Advanced Research (CIFAR) program on the Fungal Kingdom: Threats & Opportunities.

References

1. Murdoch CC, Skaar EP (2022) Nutritional immunity: the battle for nutrient metals at the host–pathogen interface. Nat Rev Microbiol 20:657–670
2. Best A, Abu Kwaik Y (2019) Nutrition and bipartite metabolism of intracellular pathogens. Trends Microbiol 27:550–561

3. Bhalla K, Qu X, Kretschmer M et al (2022) The phosphate language of fungi. Trends Microbiol 30:338–349
4. Keymer A, Pimprikar P, Wewer V et al (2017) Lipid transfer from plants to arbuscular mycorrhiza fungi. eLife 6:e29107
5. Booth NJ, Smith PMC, Ramesh SA et al (2021) Malate transport and metabolism in nitrogen-fixing legume nodules. Molecules 26:6876
6. López-Millán AF, Morales F, Abadía A et al (2000) Effects of iron deficiency on the composition of the leaf apoplastic fluid and xylem sap in sugar beet. Implications for iron and carbon transport. Plant Physiol 124:873–884
7. Kämper J, Kahmann R, Bölker M et al (2006) Insights from the genome of the biotrophic fungal plant pathogen *Ustilago maydis*. Nature 444:97–101
8. Horst RJ, Engelsdorf T, Sonnewald U et al (2008) Infection of maize leaves with *Ustilago maydis* prevents establishment of C4 photosynthesis. J Plant Physiol 165:19–28
9. Horst RJ, Doehlemann G, Wahl R et al (2010) *Ustilago maydis* infection strongly alters organic nitrogen allocation in maize and stimulates productivity of systemic source leaves. Plant Physiol 152:293–308
10. Lanver D, Müller AN, Happel P et al (2018) The Biotrophic Development of *Ustilago maydis* studied by RNA-Seq analysis. Plant Cell 30:300–323
11. Villajuana-Bonequi M, Matei A, Ernst C et al (2019) Cell type specific transcriptional reprogramming of maize leaves during *Ustilago maydis* induced tumor formation. Sci Rep 9: 10227
12. Arrivault S, Obata T, Szecówka M et al (2017) Metabolite pools and carbon flow during C4 photosynthesis in maize: 13CO_2 labeling kinetics and cell type fractionation. J Exp Bot 68:283–298
13. Kretschmer M, Damoo D, Sun S et al (2022) Organic acids and glucose prime late-stage fungal biotrophy in maize. Science 376:1187–1191
14. Matei A, Doehlemann G (2016) Cell biology of corn smut disease-*Ustilago maydis* as a model for biotrophic interactions. Curr Opin Microbiol 34:60–66
15. Holliday R (1974) Ustilago maydis. In: King RC (ed.) Handbook of genetics, vol. 1. Plenum Press, New York, pp 575–595
16. Kretschmer M, Klose J, Kronstad JW (2012) Defects in mitochondrial and peroxisomal β-oxidation influence virulence in the maize pathogen *Ustilago maydis*. Eukaryot Cell 11:1055–1066
17. Kretschmer M, Lambie S, Croll D et al (2018) Acetate provokes mitochondrial stress and cell death in *Ustilago maydis*. Mol Microbiol 107:488–507
18. McCotter SW, Kretschmer M, Lee CWJ et al (2023) The monothiol glutaredoxin Grx4 influences iron homeostasis and virulence in *Ustilago maydis*. J Fungi (Basel) 9:1112
19. Klose J, de Sá MM, Kronstad JW (2004) Lipid-induced filamentous growth in *Ustilago maydis*. Mol Microbiol 52:823–835

Chapter 2

Induction of *Ustilago maydis* Appressorium and Filamentation

Evan John, Jung-En Kuan and Armin Djamei

Abstract

Ustilago maydis is a model smut-causing fungus infecting maize. Two compatible mating-type isolates of the fungus must form the infectious dikaryotic filaments that form infection. However, several laboratory techniques and strains have been developed that can induce this phenotype under controlled conditions, in order to study the pathogen outside the complexities of the host and microbiome. Three such methods are described here, which cover the use of charcoal-supplemented agar to induce filamentation, appressorium formation on hydrophic surfaces with 16-hydroxyhexadecanoic acid, as well as inducing b-dependent filamentation in liquid culture with engineered strains.

Keywords *Ustilago maydis*, SG200, FB1, FB2, AB33, AB31, bE, bW, Mating type, Filamentation, Charcoal, Appressorium, 16-hydroxyhexadecanoic acid, Parafilm, Arabinose, NO_3^-

1 Introduction

Ustilago maydis, became, in the past decades, a well-studied model system among the smut fungi of the Ustilaginales order with a large variety of different biological questions to be answered [1–6]. Whereas on the plant it is biotrophic, it can also grow saprophytically as haploid yeast-like "sporidia"; a form of asexual propagation in the absence of the maize host. On the other hand, the pathogenic morphotype follows sporidial fusion of compatible mating-types, and is characterized by filamentous growth, appressoria (plant penetration structures) formation, and hyphal spread during the early-mid stages of the infection cycle [7]. Mating and sexual development are genomically encoded through a tetrapolar system (Fig. 1A, B). The *a*-locus encodes a pheromone/receptor system responsible for sporidial fusion with two opposite compatible alleles having been described. The *b*-locus encodes for two homeodomain proteins, termed bE and bW, which form a heterodimeric transcription factor after mating, which drives filamentation/pathogenic development.

An N. T. Phan et al. (eds.), *Ustilago maydis*, Methods in Molecular Biology, vol. 3050,
https://doi.org/10.1007/978-1-0716-5364-7_2,

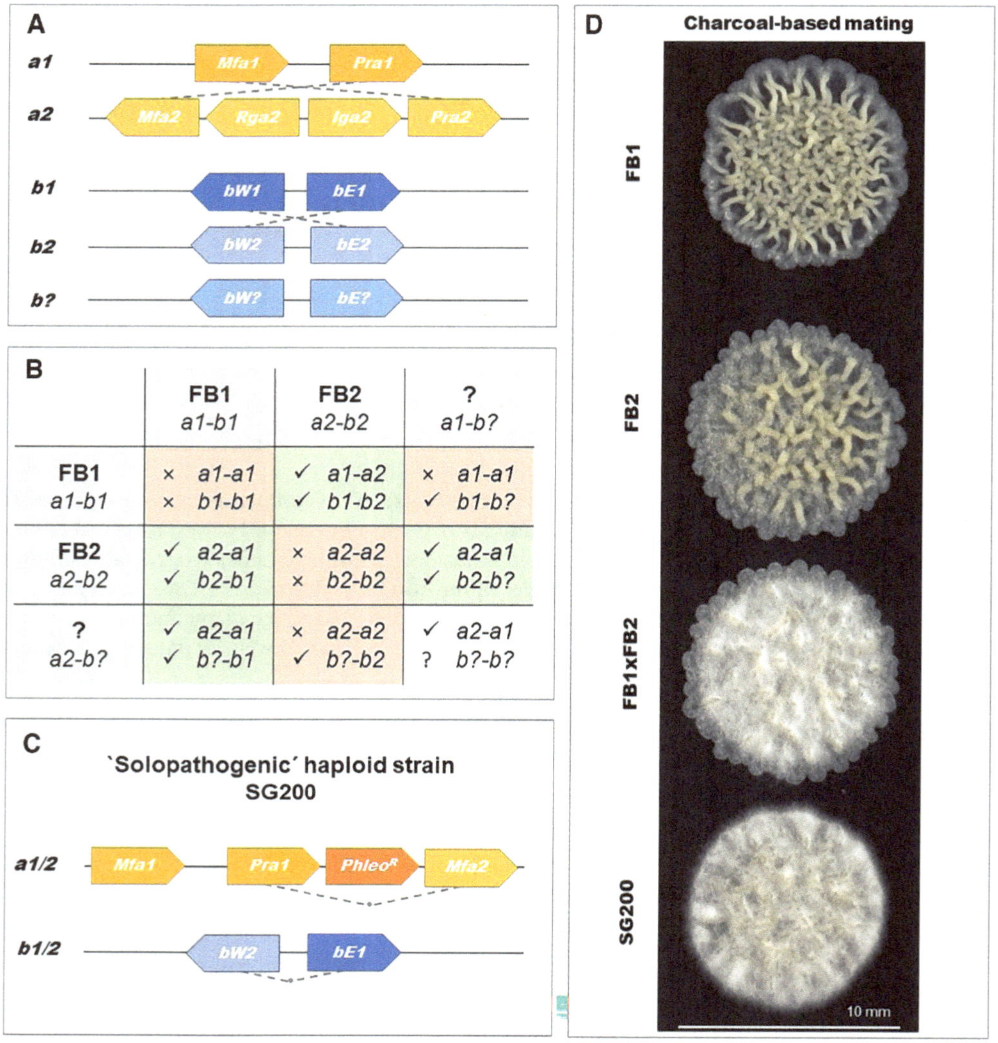

Fig. 1 (**A**) Schematic overview of allelic variants at the *a*- locus and *b*-locus in *Ustilago maydis*. Dashed lines represent the compatible gene combinations required for sporidial fusion (*a*-locus, two known alleles) and filamentation/pathogenic development (*b*-locus, multiple alleles described). (**B**) A compatibility table for *U. maydis*. Green boxes indicate compatible mating-type strains as a result of two different alleles at the *a* and *b* loci. The widely employed FB1 and FB2 strains are indicated. The *b*-locus also has many other natural variants that have been described [10] that are marked as *b*?. (**C**) A schematic overview of the engineered *a*- and *b*- loci of the "solopathogenic" haploid strain SG200 [14, 16]. (**D**) Example results of the charcoal-based filamentation-compatibility assay described in this protocol using the strains FB1, FB2, and SG200

The *b*-locus is multiallelic and typically any two alternate alleles are compatible [8–10].

In order to facilitate controlled investigation and isolate the fungus from the complexities of its host and microbiome, researchers have developed several methods to artificially induce the early pathogenic state. For example, activated charcoal is employed as a supplement in solid-agar media, on which coinoculated sporidial suspensions harboring compatible mating types grow filamentously, and are able to mate in the absence of the plant host [11–13]. Another important development was engineering of the haploid/"solopathogenic" strain "SG200" (*mfal-pral-mfa2-bW2-bE1*), which simultaneously expresses both *bE1* and *bW2* that encode compatible homeodomain transcription factors of the *b*-locus. A complete pheromone-receptor system was also engineered in SG200 from the compatible *a*-locus variants *a1* and *a2* (Fig. 1C). These modifications together enable pathogenic development to proceed without the dependency of an additional mating partner, and the strain is able to complete the infection cycle in maize [14–16]. Prerequisite plant chemical and physical cues have also been identified as triggers for early-infection development including hydroxy-fatty acids and hydrophobicity, respectively [17]. When provided artificially, these induce reliably both filamentation and appressoria in *U. maydis*. Furthermore, inducible promoters have been used to conditionally express a *bE1/2* chimer alongside *bW2*, which forms a compatible heterodimer in a haploid-strain background. In the engineered strains "AB33" and "AB31," filamentation is driven by the *nia1* (nitrate-reductase encoding) and *crg1* (arabinase encoding) promoters, respectively [15]. Cultivation in minimal media restricted to the appropriate nitrogen/carbon sources rapidly induces filamentation in liquid culture.

All the described strains and techniques above have been and remain in continual use to study aspects of the pathogenic development of *U. maydis* under controlled conditions [18–24]. As such, three protocols covering filamentation and mating in vitro, *b*-filament formation in axenic culture and appressoria formation on hydrophobic surfaces are described here for ease of application and reference.

2 Materials

2.1 Ustilago maydis *Axenic Filamentation on Charcoal-Supplemented Agar*

1. Sterile potato dextrose agar (PDA): 2.4% w/v potato dextrose broth, 2% w/v agar.
2. *U. maydis* compatible mating-type strains (e.g., FB1 and FB2) or solopathogenic haploid strains (e.g., SG200) grown on PDA at 28°C up to 2 days. Can be stored at 4°C up to 1 week.
3. Sterile YEPS-light (YEPS-L) liquid media: 0.4% w/v yeast extract, 0.4% w/v peptone, 2% w/v sucrose.
4. Sterile PDA supplemented with 1% w/v activated charcoal.
5. Sterile diH_2O.
6. Sterile elongated (15–20 cm) test tubes compatible with orbital shaker.

2.2 Triggering Ustilago maydis Appressorium Formation on Hydrophic Surface with 16-Hydroxyhexadecanoic Acid

1. Sterile potato dextrose agar (PDA): 2.4% w/v potato dextrose broth, 2% w/v agar.
2. *U. maydis* compatible mating-type strains (e.g., FB1 and FB2) or solopathogenic haploid strains (e.g., SG200) grown on PDA at 28°C up to 2 days. Can be stored at 4°C up to 1 week.
3. Sterile YEPS-light (YEPS-L) liquid media: 0.4% w/v yeast extract, 0.4% w/v peptone, and 2% w/v sucrose.
4. Sterile diH_2O.
5. Sterile elongated (15–20 cm) test tubes compatible with orbital shaker.
6. 16-hydroxyhexadecanoic acid16-hydroxyhexadecanoic acid (16-FA, purity grade >95%) stock solution (2.5 mg/mL in absolute ethanol—freshly prepared optimal or store max 1 month at −20°C).
7. Parafilm. Pre-mount ~2 × 3 cm on microscope slides using sticky-tape.
8. Clean petri-dish.
9. Lint-free clean tissue.
10. Light microscope for observation with epifluorescence capacity if using derivative strains expressing fluorescence-reporter markers or if manual staining is to be undertaken.

2.3 Inducing Ustilago maydis b-Dependent Filamentation in Liquid Culture

1. Sterile potato dextrose agar (PDA): 2.4% w/v potato dextrose broth, 2% w/v agar.
2. *U. maydis* haploids strain harboring an inducible-promoter driven compatible *bW/bE* heterodimer (e.g., AB33—nitrate, AB31—arabinose) [15] grown on potato dextrose agar 28°C up to 2 days. Can be stored at 4°C up to 1 week.
3. Sterile YEPS-light (YEPS-L) liquid media (0.4% w/v yeast extract, 0.4% w/v peptone, and 2% w/v sucrose).
4. Sterile diH_2O.
5. Salt solution: 16% w/v KH_2PO_4, 4% w/v Na_2SO_4, 8.0% w/v KCl, 1.3% w/v $CaCl_2$-$2H_2O$, 8.0% v/v trace elements, 2.0% w/v $MgSO_4$. Filter-sterilized and stored 4°C up to 6 months.
6. Trace elements: 0.06% w/v H_3BO_3, 0.14% w/v MnCl-$4H_2O$, 0.4% w/v $ZnCl_2$, 0.4% w/v Na_2MoO_4-$2H_2O$, 0.1% w/v $FeCl_3$-$6H_2O$, and 0.04% w/v $CuSO_4$-$5H_2O$. Filter-sterilized and stored 4°C up to 6 months.
7. Minimal Medium (MM): 6.25% v/v Salt solution, 0.3% w/v KNO_3, 1% w/v glucose* in diH_2O. Autoclaved to sterilize minimum 5 min 121°C. *For carbon source induction replace glucose with e.g., arabinose (for AB31 strain). Prolonged autoclaving can lead to caramelization, which can be mitigated by combining a 2% w/v sterile carbon solution with 2× concentration MM v/v in equal ratio.

8. Sterile elongated (15–20 cm) test tubes compatible with orbital shaker.
9. Microcentrifuge.
10. Light microscope for observation with epifluorescence capacity if using derivative strains expressing fluorescence-reporter markers or if manual staining is to be undertaken.

3 Methods

3.1 Ustilago maydis *Axenic Filamentation on Charcoal-Supplemented Agar*

This method is originally described in "Corn smut dikaryon in culture" (Day and Anagnostakis [25]).

*3.1.1 Day 1—*U. maydis *Inoculation*

1. Inoculate *U. maydis* strain(s) in ~2–5 mL YEPS-L in elongated test tubes and shake on a slanted angle (~45°, *see* **Note 1**) at 28°C, 180 rpm, O.N.

3.1.2 Day 2—Inoculation on Charcoal Medium Surface

1. Measure the OD_{600} (*see* **Note 2**). If OD_{600} is 0.6–0.8, it can be directly used for the next step. Otherwise dilute the culture to OD_{600}=0.2 in 5 mL YEPS-L, 28°C and shake at 180 rpm. Avoid letting cultures stand for extended periods (*see* **Note 3**).
2. Take 1 mL of OD_{600}=0.5 of cells (*see* **Note 4**), centrifuge by 4000 *g*, 5 min, remove supernatant.
3. Wash by 1 mL sterile diH_2O, 4000 g, 5 min, remove supernatant.
4. Resuspend in 1 mL sterile diH_2O (*see* **Note 5**).
5. Take 20 μL for each strain alone also mixtures of both compatible strains to drop on the PD-Charcoal agar. Leave ~2 cm between droplets.
6. Incubate at 28°C, O.N.

3.1.3 Day 3–4—Observation

1. Successful mating filament formation will lead to white fuzzy colonies (e.g., see Fig. 1 and Regenfelder et al. [26]) in 24–72 h (*see* **Note 6**).

3.2 Triggering Ustilagomaydis *Appressorium Formation on Hydrophic Surface with 16-Hydroxy-hexadecanoic Acid*

This method is originally described in "Physical-chemical plant-derived signals induce differentiation in *Ustilago maydis*" (Mendoza-Mendoza et al. [17]).

3.2.1 Day 1—*U. maydis* Inoculation

1. Inoculate *U. maydis* strain(s) in ~2–5 mL YEPS-L in elongated test tubes and shake on a slanted angle (~45°, *see* **Note 1**) at 28°C, 180 rpm, O.N.

3.2.2 Day 2—Appressorium Formation on Artificial Surface

1. Measure the OD_{600} (*see* **Note 2**). If OD_{600} is 0.6–0.8, it can be directly used for the next step. Otherwise dilute the culture to OD_{600}=0.2 in 5 mL YEPS-L, 28°C and shake them at 180 rpm. Avoid letting cultures stand longer time (*see* **Note 3**).
2. Measure the OD_{600} after 2 h and afterwards until the cultures reach an OD_{600} between 0.6 and 0.8.
3. Take 0.1 OD_{600} of cells (*see* **Note 7**), centrifuge by 4000*g*, 5 min, remove supernatant.
4. Wash by 1 mL sterile diH_2O, 4000 g, 5 min, remove supernatant.
5. Resuspend the cell pellet in 1 mL 2% v/v YEPS-L diluted in sterile diH_2O and containing 10 μL 16-FA stock solution (*see* **Note 8**). Incubate at 28°C, 1 h, inverting occasionally.
6. Invert to resuspend, then pipette small drops of 1 μL on hydrophobic surfaces such as the premounted Parafilm (*see* **Note 9**).
7. Place the inoculated Parafilm slide in a clean petri dish with a moist paper towel under the slide to promote humidity. Seal it with Parafilm.
8. Incubate away from direct light at 28°C, O.N. Appressorium will appear ~20 h.

3.2.3 Day 3—Check for Filamentation and Appressoria Formation by Microscopic Analysis

1. Use sterile diH_2O to wash off nonattached cells.
2. Screen by light microscopy (*see* **Note 10**). Successful induction will be evident by long filaments across the hydrophobic surface. Appressoria will be observable at a subset (generally between 2% and 5%) of filament tips, distinguishable by bulbous endings. For a diagrammatic overview and expected results, see Fig. 2.

3.3 Inducing *Ustilago maydis* b-Dependent Filamentation in Liquid Culture

This method describes strains and techniques originally established in "Identification of genes in the bW/bE regulatory cascade in *Ustilago maydis*" (Brachmann et al. 2001) [15].

3.3.1 Day 1—*U. maydis* Inoculation

1. Inoculate *U. maydis* strain(s) such as AB33 or AB31 (NO_3^- or arabinose inducible, respectively) in ~2–5 mL YEPS-L and shake on a slanted angle (~45°, *see* **Note 1**) at 28°C, 180 rpm, O.N.

3.3.2 Day 2—Filament Induction in Liquid Culture

1. Measure the OD_{600} (*see* **Note 2**). If OD_{600} is 0.6–0.8, it can be directly used for the next step. Otherwise dilute the culture to OD_{600}=0.2 in 5 mL YEPS-L, 28°C and shake at 180 rpm. Avoid letting cultures stand longer time (*see* **Note 3**).

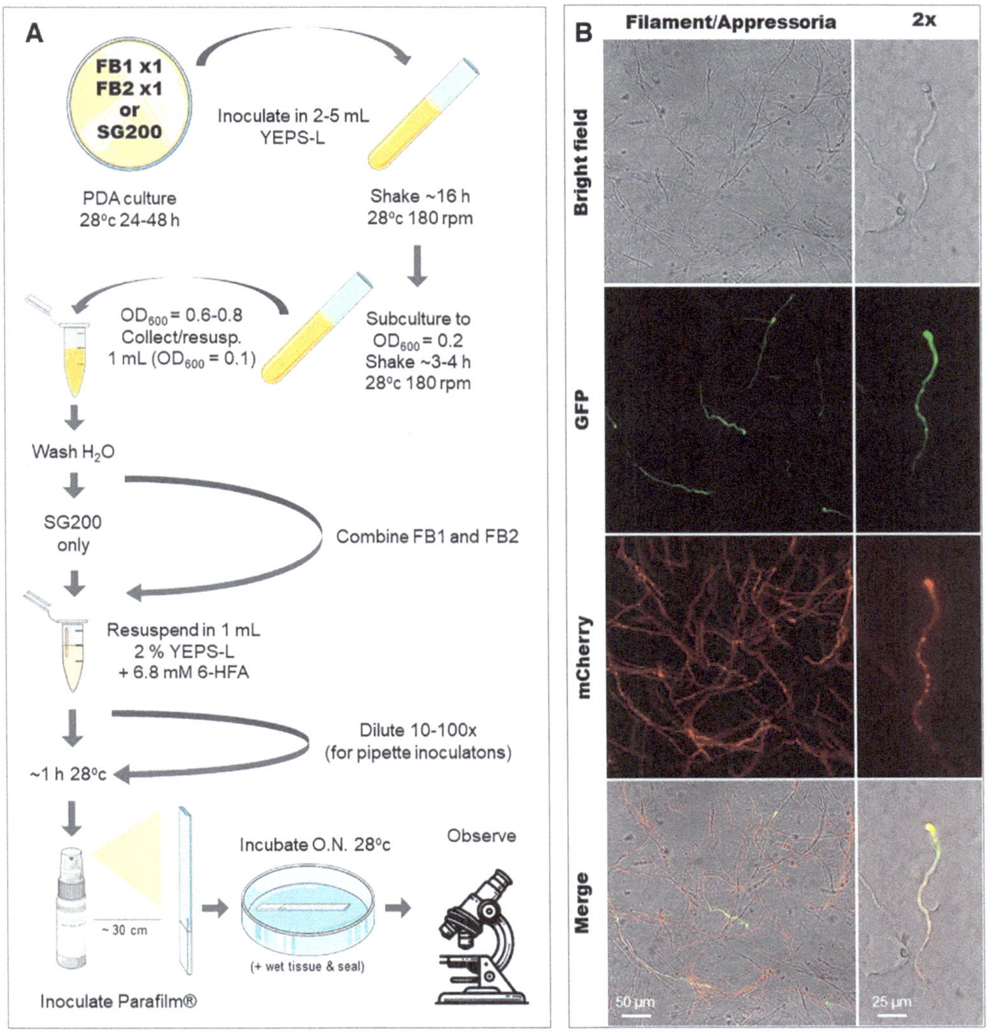

Fig. 2 (**A**) Flow diagram of the steps undertaken to induce appressorium formation on a hydrophobic surface, as described in this protocol, using the combination of FB1 and FB2 or the "solopathogenic" strain SG200. Artwork images were adapted from Servier (https://smart.servier.com) under CC-BY 3.0 licence. (**B**) An example of the expected results following microscopic examination. Here, an SG200-derived strain that constitutively expresses cytoplasmic mCherry, and which also expresses cytoplasmic GFP under an appressorium-specific promoter, is used to assist with visualization. The appressoria are characterized by the bulbous head at the tip of the respective GFP-fluorescent filaments.

2. Take the equivalent of 1 mL culture (OD_{600} = 0.8–1), centrifuge by 4000*g*, 2 min, remove supernatant.
3. Wash cells twice by resuspension in 1 mL sterile diH_2O, then 4000 g 2 min, remove supernatant. It is important that YEPS-L is thoroughly removed in order to reduce contaminating nitrogen/carbon sources.
4. Resuspend cells in 5 mL MM (NO_3^- or arabinose inducing) in a sterile growth tube, return to shake 28°C, 180 rpm.

5. Elongation of sporidial cells will be observable beginning ~ 3–4 h post transfer. Extended filamentation can be observable ~ 8 h.
6. Filaments can be captured at these immature phases by collecting 1 mL from the culture (centrifuge by 4000*g*, 2 min, remove supernatant), washing once (as described above), and storing 4°C overnight.

3.3.3 Day 3—Check Filaments by (Fluorescent) Microscopic Analysis

1. After 20–24 h shaking, fully mature filaments are observable and typically form mycelial aggregates. After collecting 1 mL and washing (as described above), immature vs mature filaments can be observed.
2. For direct observation, resuspend in sterile diH_2O to a desired density and directly mount 1 μL on slides for microscopy (for best results, ensure coverslip covers the entire circumference of the droplet to avoid liquid/filament movement across the slide during observation). For a diagrammatic overview and expected results, see Fig. 3.
3. Alternatively, induced filaments can be conveniently used for comparative (immuno)staining, gene expression and other downstream assessment of fungal physiology in this derived state.

4 Notes

1. The angle promotes aeration and is important as *U. maydis* prefers aerobic conditions.
2. OD measurements >1.0 are inaccurate, dilute 1/10 before measuring if necessary.
3. Better not more than 5 min without shaking to maintain aerobic conditions. Measure the OD_{600} after 2 h and afterwards until the cultures reach an OD_{600} between 0.6 and 0.8.
4. At this stage compatible mating-type cultures, if used, are mixed. The final OD_{600} for each strain should be equal.
5. Here, it is also useful to also make a 1/10 dilution in sterile diH_2O.
6. Prolonged incubation will lead to dense hyphal agglomeration and subsequent curling and detachment of the colony from the agar surface. This can hamper microscopic analysis but can be delayed if diluted inoculations used.
7. At this stage compatible mating-type cultures, if used, are mixed.
8. It is better to use freshly prepared 16-FA. Final working concentration will be 0.025 mg/mL (=6.8 mM).

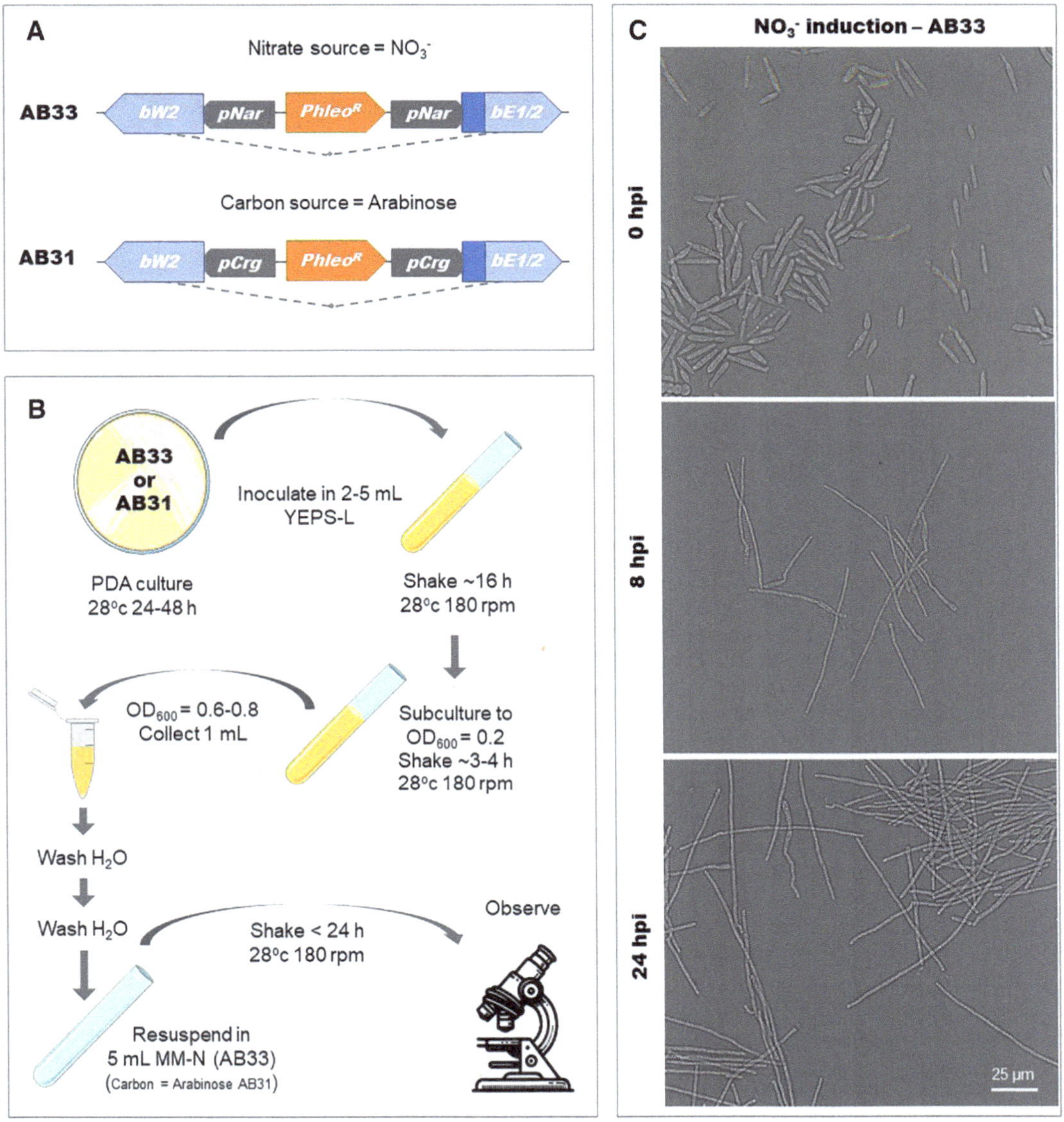

Fig. 3 (**A**) A schematic overview of the engineered and *b*- locus, which drives filamentation in the AB33 and AB31 strains [15] described in this protocol. The engineered *bE1/2* chimer, as well as *bW*, are conditionally expressed by the *pNar* and *pCrg* respective promoters in response to NO_3^- as the sole nitrogen source or arabinose as the sole carbon source. (**B**) Flow diagram of the steps undertaken in this protocol. Artwork images were adapted from Servier (https://smart.servier.com) under CC-BY 3.0 licence. (**C**) Representative microscopic images of AB33 at 0, 8, and 24 h post-induction (hpi) in NO_3^- minimal medium (MM-N). At 0 hpi, sporidia are the dominant form, while at 8 hpi, extended filaments are forming. At 24 hpi, most filaments have reached maximum length and begin to aggregate into clumps

9. Spraying on Parafilm with fine mist will also work. A 10× and 100× dilution in the 2% YEPS-L+16-FA can improve separation of subsequent filaments.
10. Light microscopy is sufficient to observe filamentation and appressoria. The attached filaments are also amenable to

staining and subsequent epifluorescence microscopy. For example, Calcoflour-White staining will highlight the fungal cell-wall while immunostaining can be used to highlight specific cell-surface components. Alternatively, *U. maydis* strains expressing fluorescent reporters (such as GFP or RFP) can also be substituted and observed using epifluorescence microscopy.

References

1. Tanaka S, Djamei A, Presti LL et al (2015) Experimental approaches to investigate effector translocation into host cells in the *Ustilago maydis*/maize pathosystem. Eur J Cell Biol 94:349–358. https://doi.org/10.1016/j.ejcb.2015.06.007
2. Rabe F, Seitner D, Bauer L et al (2016) Phytohormone sensing in the biotrophic fungus *Ustilago maydis*—the dual role of the transcription factor Rss1. Mol Microbiol 102:290–305. https://doi.org/10.1111/mmi.13460
3. Alcântara A, Bosch J, Nazari F et al (2019) Systematic Y2H screening reveals extensive effector-complex formation. Front Plant Sci 10:1437. https://doi.org/10.3389/fpls.2019.01437
4. Saado I, Chia K-S, Betz R et al (2022) Effector-mediated relocalization of a maize lipoxygenase protein triggers susceptibility to *Ustilago maydis*. Plant Cell 34:2785–2805. https://doi.org/10.1093/plcell/koac105
5. Bindics J, Khan M, Uhse S et al (2022) Many ways to TOPLESS—manipulation of plant auxin signalling by a cluster of fungal effectors. New Phytol 236:1455–1470. https://doi.org/10.1111/nph.18315
6. Khan M, Uhse S, Bindics J et al (2024) Tip of the iceberg? Three novel TOPLESS-interacting effectors of the gall-inducing fungus *Ustilago maydis*. New Phytol 244:949–961. https://doi.org/10.1111/nph.19967
7. Kahmann R, Kämper J (2004) *Ustilago maydis*: how its biology relates to pathogenic development. New Phytol 164:31–42. https://doi.org/10.1111/j.1469-8137.2004.01156.x
8. Bolker M, Urban M, Kahmann R (1992) The a mating type locus of *Ustilago maydis* specifies cell signaling components. Cell 63:441–450. https://doi.org/10.1016/0092-8674(92)90182-c
9. Gillissen B, Bergemann J, Sandmann C et al (1992) A two-component regulatory system for self/non-self recognition in *Ustilago maydis*. Cell 68:647–657. https://doi.org/10.1016/0092-8674(92)90141-X
10. Kämper J, Friedrich MW, Kahmann R (2020) Creating novel specificities in a fungal nonself recognition system by single step homologous recombination events. New Phytol 228:1001–1010. https://doi.org/10.1111/nph.16755
11. Holliday R (1974) *Ustilago maydis*. In: King RC (ed) Bacteria, bacteriophages, and fungi, vol 1. Springer US, Boston, MA, pp 575–595
12. Banuett F, Herskowitz I (1989) Different a alleles of *Ustilago maydis* are necessary for maintenance of filamentous growth but not for meiosis. Proc Natl Acad Sci 86:5878–5882. https://doi.org/10.1073/pnas.86.15.5878
13. Uhse S, Pflug FG, Stirnberg A et al (2018) In vivo insertion pool sequencing identifies virulence factors in a complex fungal–host interaction. PLOS Biol 16:e2005129. https://doi.org/10.1371/journal.pbio.2005129
14. Bölker M, Genin S, Lehmler C et al (1995) Genetic regulation of mating and dimorphism in *Ustilago maydis*. Can J Bot 73:320–325. https://doi.org/10.1139/b95-262
15. Brachmann A, Weinzierl G, Kämper J et al (2001) Identification of genes in the bW/bE regulatory cascade in *Ustilago maydis*. Mol Microbiol 42:1047–1063. https://doi.org/10.1046/j.1365-2958.2001.02699.x
16. Kämper J, Kahmann R, Bölker M et al (2006) Insights from the genome of the biotrophic fungal plant pathogen *Ustilago maydis*. Nature 444:97–101. https://doi.org/10.1038/nature05248
17. Mendoza-Mendoza A, Berndt P, Djamei A et al (2009) Physical-chemical plant-derived signals induce differentiation in *Ustilago maydis*. Mol Microbiol 71:895–911. https://doi.org/10.1111/j.1365-2958.2008.06567.x
18. Fernández-Álvarez A, Marín-Menguiano M, Lanver D et al (2012) Identification of O-mannosylated virulence factors in *Ustilago maydis*. PLOS Pathog 8:e1002563. https://doi.org/10.1371/journal.ppat.1002563
19. Lanver D, Berndt P, Tollot M et al (2014) Plant surface cues prime *Ustilago maydis* for biotrophic development. PLOS Pathog

10:e1004272. https://doi.org/10.1371/journal.ppat.1004272
20. Ma L-S, Wang L, Trippel C et al (2018) The *Ustilago maydis* repetitive effector Rsp3 blocks the antifungal activity of mannose-binding maize proteins. Nat Commun 9: 1711. https://doi.org/10.1038/s41467-018-04149-0
21. Schmitz L, Schwier MA, Heimel K (2019) The unfolded protein response regulates pathogenic development of *Ustilago maydis* by Rok1-dependent inhibition of mating-type signaling. mBio 10:e02756–19. https://doi.org/10.1128/mBio.02756-19
22. Ludwig N, Reissmann S, Schipper K et al (2021) A cell surface-exposed protein complex with an essential virulence function in *Ustilago maydis*. Nat Microbiol 6:722–730. https://doi.org/10.1038/s41564-021-00896-x
23. Lin Y-H, Xu M-Y, Hsu C-C et al (2023) *Ustilago maydis* PR-1-like protein has evolved two distinct domains for dual virulence activities. Nat Commun 14: 5755
24. Darino M, Chia K-S, Marques J et al (2021) *Ustilago maydis* effector Jsi1 interacts with Topless corepressor, hijacking plant jasmonate/ethylene signaling. New Phytol 229:3393–3407. https://doi.org/10.1111/nph.17116
25. Day PR, Anagnostakis SL (1971) Corn smut dikaryon in culture. Nat New Biol 231:19–20. https://doi.org/10.1038/newbio231019a0
26. Regenfelder E, Spellig T, Hartmann A et al (1997) G proteins in *Ustilago maydis*: transmission of multiple signals?. EMBO J 16:1934–1942. https://doi.org/10.1093/emboj/16.8.1934

Check for updates

Chapter 3

Ustilago maydis Seedling Infection Assay and Scoring

Patrice Julian Zsoldos, Ziyang Chen and Armin Djamei

Abstract

This protocol outlines a reproducible method for assessing the virulence of *Ustilago maydis* strains in *Zea mays* seedlings through targeted syringe inoculation of the leaf whorl. Using seedlings at the V3–V4 growth stage ensures optimal susceptibility and developmental uniformity. The method can be used with both solopathogenic strains and compatible mating-type combinations, enabling comparative analyses of infection efficiency and symptom development. Fungal cultures are prepared under controlled conditions to maintain virulence, and post-infection humidity is regulated to enhance colonization success. Symptom progression is monitored over 10–12 days and quantified using an ordinal disease scoring system. This assay provides a robust tool for evaluating effector mutants, analyzing host–pathogen interactions, and comparing strain virulence under standardized conditions.

Keywords *Ustilago maydis*, Maize smut, Leaf whorl infection, Pathogenicity assay, Fungal virulence, plant–microbe interaction, Disease scoring, Seedling infection protocol

1 Introduction

Ustilago maydis is a biotrophic fungal pathogen causing smut disease in maize (*Zea mays*), characterized by the formation of tumor-like galls on aerial tissues. including leaves, stems, and tassels [1–6]. The infection process relies on the fusion of two compatible haploid sporidia (e.g., mating types FB1 and FB2), which form an infectious dikaryon required for penetration and colonization of maize tissue. Alternatively, solopathogenic strains such as SG200 bypass this mating requirement by harboring both mating-type loci in a single genome, allowing infection from a single strain [1].

To evaluate pathogenicity under controlled conditions, maize plants are typically infected at the seedling stage. The vegetative development of maize is classified using the leaf collar method, where the number of fully developed leaves with visible collars defines the V-stage. For instance, the V3 stage corresponds to the appearance of the third leaf collar, indicating that three leaves are fully expanded. The V4 stage similarly marks the development of four-leaf collars [7].

An N. T. Phan et al. (eds.), *Ustilago maydis*, Methods in Molecular Biology, vol. 3050,
https://doi.org/10.1007/978-1-0716-5364-7_3,

Under optimal growth chamber conditions, maize typically reaches the V3–V4 stage around 7 days after sowing. This stage is ideal for syringe-inoculation into the leaf whorl, as the meristematic tissue is actively dividing and accessible, enhancing fungal colonization and gall formation.

This protocol describes a standardized syringe-inoculation assay to evaluate *U. maydis* virulence in maize seedlings, focusing exclusively on leaf-whorl infections. Disease severity is monitored and quantified using an ordinal symptom scale.

2 Materials

- *U. maydis* compatible mating-type strains (e.g., FB1 & FB2) or solopathogenic haploid strains (e.g., SG200); grown on potato dextrose agar at 28°C up to 2 days. Can be stored at 4°C up to 1 week.
- Potato dextrose agar (PD Agar): 2.4% w/v potato dextrose broth, 2% w/v agar, dissolved in H_2O.
- YEPS-light (YEPS-L): 0.4% w/v yeast extract, 0.4% w/v peptone, 2% w/v sucrose, dissolved in H_2O.
- Sterile deionized Water (diH_2O).
- Sterile Erlenmeyer flasks (100 or 250 mL).
- Maize seeds.
- Peat-based potting soil in round pots (20 cm diameter).
- Spectrophotometer (OD_{600}).
- Sterile 1 mL syringes with 26G needles.
- Sterile 50 mL Falcon tubes.
- Growth chamber (28°C, 14 h light/10 h dark cycle).

3 Methods

Carry out all procedures at room temperature, unless otherwise specified. The general workflow is described in Fig. 1, which shows the main steps involved.

3.1 Day 7 Before Infection: Maize Planting

1. Plant at least four kernels per 20 cm pot in soil and water them. The number of seeds should be chosen so that the outcome is at least four plants per pot, given the germination rate.
2. Grow in a growth chamber at 28°C Day/22°C Night (14 h/10 h light/dark) for 7 days until the three-leaf stage.

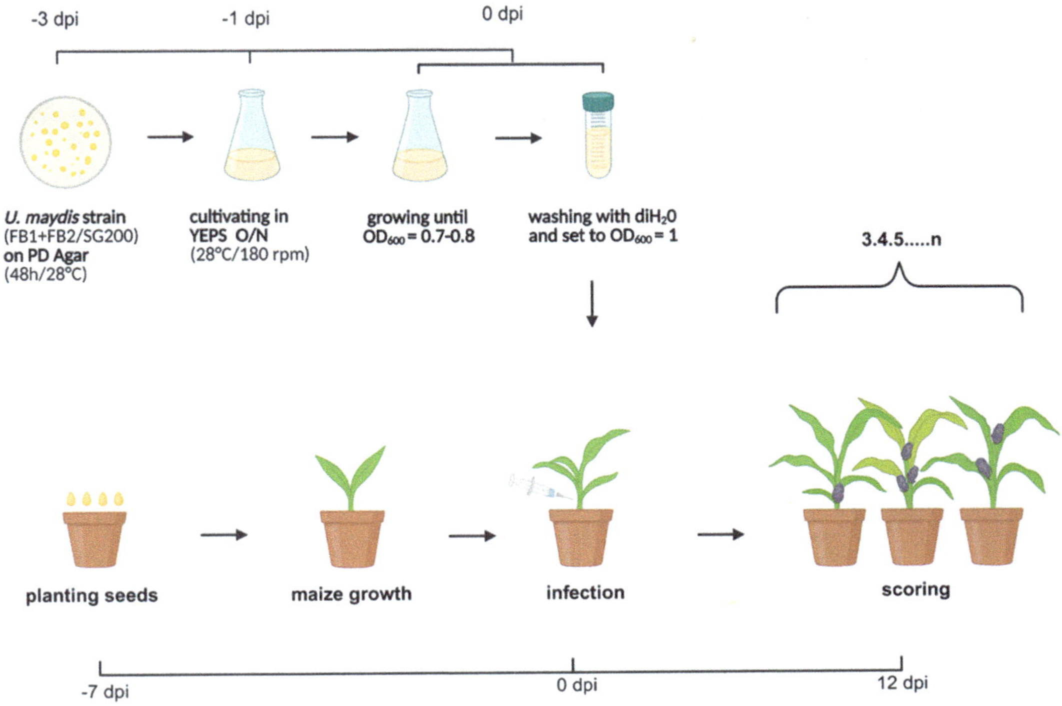

Fig. 1 Workflow of the *Ustilago maydis* seedling infection assay. Timeline of key steps in the standardized infection assay for evaluating *U. maydis* virulence on maize seedlings. Maize is grown to the V3–V4 stage under controlled conditions. In parallel, fungal strains are revived on PDA, cultured in liquid medium, and prepared to an OD_{600} of 1.0. On day 0 post infection, 7-day-old seedlings are syringe-inoculated with 300 μL of fungal suspension into the stem below the leaf whorl. Plants are kept at high humidity post-infection, and symptoms are scored at 12 days post-inoculation

3.2 Day 3 Before Infection: U. maydis Inoculation

1. Retrieve *U. maydis* strains from −80°C glycerol stocks and streak on PD Agar.
2. Incubate for 48 h at 28°C (*see* **Note 1**).

3.3 Day 1 Before Infection: Pre-Culturing

1. Inoculate 20 mL YEPS light in 100 mL Erlenmeyer flasks with fresh colonies (not older than 1 week).
2. Shake overnight (ON) at 28°C in an incubator shaker.

3.4 Day 0: Inoculum Preparation and Infection

1. Dilute ON cultures to $OD_{600} = 0.2$ in 60 mL fresh YEPS light (in 250 mL Erlenmeyer flasks). Measure OD_{600} of 1:10 diluted culture if OD > 0.8 to ensure accuracy.
2. Incubate at 28°C shaking until $OD_{600} = 0.7–0.8$ (approx. 4–6 h).
3. Transfer culture to 50 mL Falcon tubes and centrifuge at 3000 × *g* for 5 min (*see* **Note 2**).

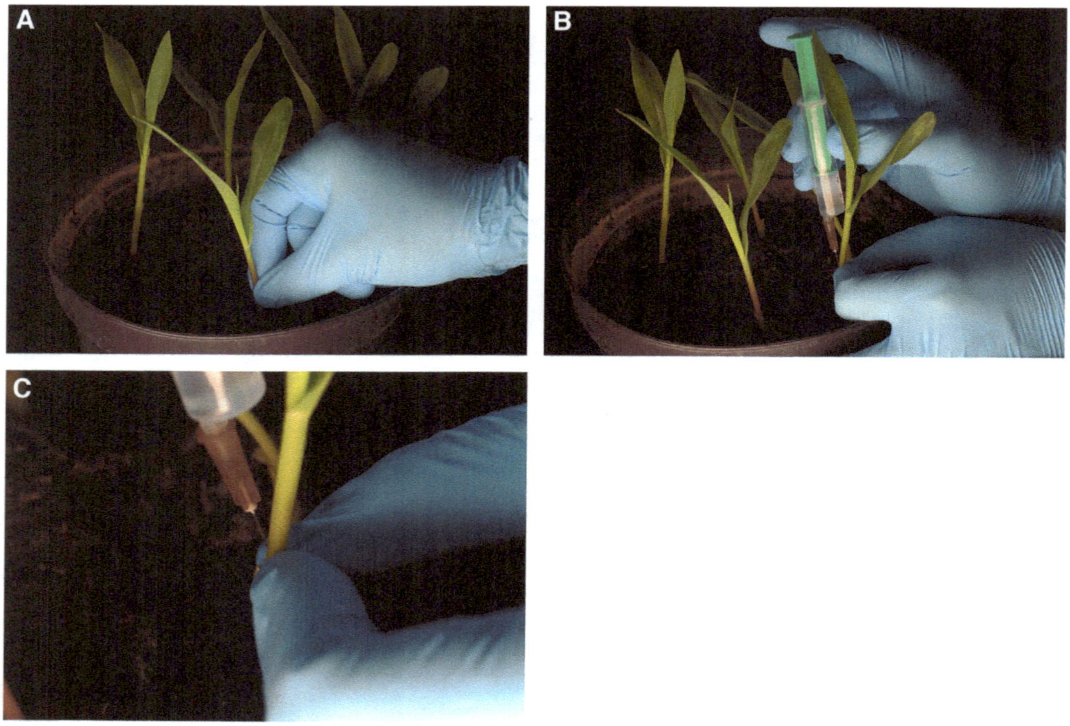

Fig. 2 Syringe-based inoculation of maize seedlings with *Ustilago maydis*. (**A**) Maize seedlings at the V3–V4 stage suitable for infection. Plants should exhibit 3–4 fully expanded leaf collars and be grown under uniform, controlled conditions. (**B**) Positioning the syringe for whorl injection. The seedling stem is stabilized by holding it firmly at the base with two fingers while a 1 mL syringe with a 26G needle is positioned parallel to the stem. The needle is then inserted into the base of the leaf whorl at a shallow angle, approximately 1 cm above the soil level. (**C**) Delivery of fungal inoculum. Approximately 300 μL of *U. maydis* cell suspension ($OD_{600} = 1.0$) is gently injected into the stem. Liquid should emerge from the whorl, confirming delivery to the inner tissue. Stabilizing the stem reduces movement and prevents mechanical damage during injection

4. Discard supernatant and wash pellet with diH2O (room temperature). Ensure that all media is fully removed during centrifugation to avoid any residual sugars that could interfere with the infection process.
5. Repeat centrifugation, resuspend pellet in diH_2O, and adjust OD_{600} to 1.0 using cuvettes and spectrophotometer.[1]
6. Fill 1 mL syringes with inoculum. Inject ~300 μL per plant ~1 cm above the soil line into the stem opposite the coleoptile (see Fig. 2) until inoculum exits into the central leaf whorl (*see* **Notes 3 and 4**).
7. Avoid overhead watering post-injection for the first 24 h (*see* **Note 5**).

[1] For diploid infection: mix FB1 and FB2 cultures at 1:1 ratio in a fresh Falcon tube. For solopathogenic strains, use a single culture.

3.5 Day 1 to Day 2: Post-Infection Care

1. Maintain growth at 28°C Day/22°C Night (14 h/10 h light/dark).
2. Resume regular watering after 24 h.

3.6 Day 12: Symptom Scoring

1. Score disease symptoms 10–12 days post-infection. Assign each plant the highest severity score observed, following the provided ordinal scale (see Table 1). Carefully inspect the stems, leaves, and nodes, as galls can vary in size and location.

3.7 Statistical Analysis of Disease Scores

Disease categories represent ordinal, non-numeric data and must be analyzed with methods that respect the inherent ranking. To ensure statistical robustness, a minimum of three biological replicates per strain or treatment group is recommended. The following statistical approaches can be used to perform a reliable analysis (*see* **Note 6**). The scoring system and categories were adapted from the work of Schilling et al. and Redkar et al. [6, 8]:

1. Ordinal Logistic Regression (OLR): Suitable for modeling ordered categorical response variables. Use polr() from the MASS package in R. Assumes proportional odds between levels.
2. Cumulative Link Mixed Models (CLMM): Extend OLR by accounting for random effects such as replicate or block. Use the clmm() function from the ordinal package in R to accommodate experimental design structure.
3. Non-parametric alternatives should be used if model assumptions are violated, or sample size is small:

Table 1
Criteria to score visible symptoms at 12 days post-infection [3, 4]

Symptom category	Description
No symptoms	No visible changes
Chlorosis	Yellowing of infected leaves
Ligula swelling	Swelling at the leaf base
Small galls	Galls <1.5 mm
Normal galls	Galls 2–4 mm
Heavy galls	Large galls causing visible growth distortion
Stunted growth	Overall reduction in plant height and vigor
Dead	Plant necrosis

Scoring rule: Always record the most severe symptom observed per individual plant. If a plant displays multiple symptoms, assign it to the category reflecting the greatest severity (e.g., a plant with both chlorosis and heavy galls is scored as "Heavy galls"). This approach ensures a conservative assessment of virulence

- Kruskal–Wallis test Detects differences in disease score distributions among multiple groups.
- Dunn'sposthoctest:Performspairwisecomparisonsfollowing significant Kruskal–Wallis; apply Benjamini–Hochberg correction to control false discovery rate.

3.8 Visualization

- Stacked bar plots: Show proportional distribution of symptom categories across genotypes.
- Ordinal severity index: Optionally convert categories to ranked integers (e.g., 0 = no symptoms to 7 = dead) for exploratory boxplots or linear modeling.
- Always include error bars (e.g., standard error or confidence intervals) and replicate identifiers when possible.

4 Notes

1. Use freshly streaked *U. maydis* colonies (≤5–7 days old) to inoculate liquid YEPS light cultures. Incubate cultures in 100–250 mL Erlenmeyer flasks with continuous shaking at 28°C to ensure sufficient aeration. Avoid letting cultures stand without shaking, as *U. maydis* is sensitive to low oxygen conditions. Use cultures in exponential phase ($OD_{600} \approx 0.8$–1.0) and proceed directly to washing and OD adjustment. Do not store cultures at room temperature after reaching the target OD.
2. During centrifugation (e.g., 3000 × *g*, 5 min), use swing-out rotors to ensure even pelleting. Fully decant or aspirate the supernatant to remove all residual medium. Traces of YEPS (especially sugars) may interfere with infection efficiency. Wash pellets twice with sterile distilled water before OD normalization.
3. Infect seedlings into the lower stem or leaf whorl, near the meristematic zone. Use 1 mL syringes with 26G needles inserted at a shallow angle. Consistent pressure and insertion depth are key to avoiding leakage or plant damage. For reproducibility, train with test plants prior to large experiments.
4. Inject ~300 μL per plant. Avoid over-injection and maintain consistent technique across all plants and treatments. Mark plants post-injection to avoid confusion during scoring.
5. Immediately after inoculation, maintain high humidity (≥90%) for at least 12–24 h. Cover with transparent foil, domes, or use a mist system inside the growth chamber. Avoid overhead watering during this period. Sudden drops in humidity drastically reduce infection success.

6. Perform at least three biological replicates, each from independently prepared fungal cultures and plant cohorts. Consider staggering infections across days to detect day-to-day variability. Maintain precise documentation of each replicate.

References

1. Kämper J et al (2006) Insights from the genome of the biotrophic fungal plant pathogen *Ustilago maydis*. Nature 444(7115):97–101
2. Bindics J et al (2022). Many ways to TOPLESS—manipulation of plant auxin signalling by a cluster of fungal effectors. New Phytol 236(4):1455
3. Saado I et al (2022) Effector-mediated relocalization of a maize lipoxygenase protein triggers susceptibility to *Ustilago maydis*. Plant Cell 34(7):2785–2805. pp
4. Khan M et al (2024) Tip of the iceberg? Three novel TOPLESS-interacting effectors of the gall-inducing fungus *Ustilago maydis*. New Phytol 244:949–961
5. Uhse S et al (2018) In vivo insertion pool sequencing identifies virulence factors in a complex fungal–host interaction. PLoS Biol 16(4):e2005129
6. Redkar A et al (2015) A secreted effector protein of *U. maydis* guides maize leaf cells to form tumors. Plant Cell 27(4):1332–1351
7. Abendroth LJ et al (2011) Corn growth and development. Iowa State University Extension and Outreach. https://store.extension.iastate.edu/product/6434
8. Schilling L et al (2014) Virulence of the maize smut *U. maydis* is shaped by organ-specific effectors. Mol Plant Pathol 15(8):780–789

Chapter 4

Apoplastic Fluid Collection/Analysis of *Ustilago maydis*-Infected Maize Leaves

Yu-Han Lin, Chibbhi Bhaskar and Lay-Sun Ma

Abstract

This chapter presents a detailed protocol for collecting and analyzing apoplastic-localized proteins from *Ustilago maydis*-infected maize leaves, using vacuum infiltration followed by centrifugation. It is suitable for protein profiling and the immunoblot detection of secreted effectors, providing a valuable tool for investigating host-pathogen interactions in the apoplast and offering insights into how *U. maydis* modulates this extracellular space during infection.

Keywords *U. maydis*-infected maize, Apoplastic fluid extraction, Apoplastic-localized effectors, Vacuum infiltration–centrifugation, Trichloroacetic acid precipitation

1 Introduction

In plants, the space outside the plasma membrane, known as the apoplast, comprises the porous cell wall, extracellular spaces, and xylem, all of which are immersed in apoplastic fluid [1]. The apoplast functions as a dynamic interface where plants and microbes directly interact. It represents the first battleground in plant–microbe interactions, safeguarded by a diverse repertoire of defense proteins, including hydrolytic enzymes, protease inhibitors, and reactive oxygen species-producing enzymes, that accumulate in this compartment to restrict pathogen invasion [2, 3]. These proteins play a pivotal role in early immune responses and strongly influence the outcome of infection. However, this defense barrier can be undermined by pathogen-secreted effectors. For example, the *U. maydis* effectors Pit2 and Pep1 inhibit the activity of maize apoplastic papain-like cysteine proteases (PLCPs) and peroxidase POX12, respectively, thereby suppressing host immunity [4, 5]. Given this central role, functional analysis of apoplastic proteins allows the identification of defense-associated proteins, pathogen effectors, and their host

An N. T. Phan et al. (eds.), *Ustilago maydis*, Methods in Molecular Biology, vol. 3050,
https://doi.org/10.1007/978-1-0716-5364-7_4,

targets, providing insight into both pathogen effector strategies and the molecular arms race between plants and pathogens.

The vacuum infiltration–centrifugation technique is a simple, rapid, and widely used method for isolating leaf apoplastic fluid from a variety of plant species [6–8]. In this approach, leaves are first infiltrated with a defined buffer and then subjected to gentle centrifugation. This process releases extracellular fluid while minimizing contamination from ruptured cells, ensuring that the recovered proteins accurately represent the apoplastic environment.

This chapter details a modified protocol for isolating apoplastic fluid from *U. maydis*-infected maize leaves. The protocol is structured in a step-by-step format and includes two main parts: (1) extraction of apoplastic fluid and (2) protein precipitation using trichloroacetic acid (TCA). The resulting protein samples are suitable for downstream applications such as immunoblot detection.

2 Materials

The following materials are provided as a reference and should be adjusted according to the experimental requirements.

2.1 Maize Growth Conditions

1. Grow maize (*Zea mays*) cultivar Honey 236 (Taiwan) until seedlings are 7 days old or at the three-leaf stage.
2. Growth chamber or greenhouse with controlled conditions (25–28°C, 15/9 h light/dark photoperiod, 55%–60% relative humidity).

2.2 U. maydis Infection of Maize Leaves

1. *U. maydis* haploid solopathogenic strain SG200.
2. YEPSL liquid medium: Preparing a medium containing 1% yeast extract, 2% Bacto tryptone, and 2% sucrose in Milli-Q water. Sterilize the medium by autoclaving.
3. Centrifuge.
4. Sterile distilled water.
5. Seven-day-old maize seedlings.

2.3 Vacuum-Infiltration Preparation

1. Clean scissors for cutting infected maize leaves.
2. 250-mL beakers.
3. Sterile Milli-Q water.
4. $CaCl_2$-NaAcetate buffer: Preparing a solution containing 5 mM NaAcetate, pH 4.3, and 0.2 M $CaCl_2$ in Milli-Q water. Sterilize the solution by autoclaving.
5. Styrofoam discs (cut to fit the beaker, with small holes punched for air escape).
6. Vacuum machine (e.g., Eppendorf Vacufuge Plus).
7. Paper towels.

8. 20-mL syringes (without plunger and needle).
9. 50-mL Falcon tubes.
10. Refrigerated centrifuge with swinging-bucket rotor.
11. 1.5-mL microcentrifuge tubes.

2.4 Protein Precipitation Preparation

1. 100% Trichloroacetic acid (TCA): Dissolve 100 g of TCA in Milli-Q water and adjust the final volume to 100 mL.
2. Acetone, precooled to 4°C.
3. 4× SDS sample buffer (SSB): Preparing a buffer containing 0.25 M Tris-HCl (pH 6.8), 8% SDS, 40% glycerol, and 0.02% bromophenol blue in Milli-Q water. Dilute to 1× SSB as needed.
4. 1 M Dithiothreitol (DTT): Preparing a solution containing 1 M of DTT in Milli-Q water.
5. Vortex mixer (optional).
6. Heating block set at: 60°C for acetone evaporation or 100°C for protein denaturation.

3 Methods

3.1 Sample Preparation

3.1.1 U. maydis Infection of Maize Seedlings

1. Sow maize seeds and grow them under the conditions described in Sect. 2.1 until the third leaf stage, which usually takes about 7 days (*see* **Note 1**).
2. To prepare *U. maydis* SG200 for infection, subculture *U. maydis* cells into 50 mL of YEPSL liquid medium in a 500-mL flask at an initial OD_{600} of 0.2. Grown the culture at 28°C with shaking until the OD_{600} reaches 0.8.
3. Harvest cells by centrifugation at 1300 ×*g* for 5 min at 25°C.
4. Discard the supernatant and resuspend the cell pellet in approximately 30 mL of sterile distilled water.
5. Centrifuge the suspension again to pellet the cells, then discard the supernatant.
6. Resuspend the washed cells in sterile distilled water and adjust the suspension to a final OD_{600} of 1.0.
7. Use the prepared *U. maydis* suspension (OD_{600} = 1.0) to inoculate the stems of 7-day-old maize seedlings.

3.1.2 Collection of Infected Maize Leaf Samples

1. Harvest infected leaves at desired time points by cutting them into 4–6 cm segments below infection holes using clean scissors. Segments of this length will lie flat in a 250-mL beaker without bending.

2. Gently rinse leaf segments with sterile Milli-Q water to remove any surface-attached soil particles.
3. Place the segments flat in a 250-mL beaker without bending them (*see* **Note 2**).
4. Place a perforated Styrofoam piece on top of the leaf segments to keep them from floating, ensuring efficient infiltration (*see* **Note 3**; Fig. 1).
5. Add ~100 mL of precooled sterile Milli-Q water or $CaCl_2$-NaAcetate buffer in the beaker (*see* **Note 4**).

3.2 Vacuum Infiltration

1. Place the beaker containing leaf segments inside a vacuum machine.
2. Apply vacuum at room temperature for 15 min, until the final pressure reaches 60 mm Hg (*see* **Note 5**).
3. Release the vacuum and allow the leaves to remain under atmospheric pressure for an additional 5 min. The leaves are slowly infiltrated with the surrounding solution, and become water-soaked and turn dark green (*see* **Note 6**).
4. Repeat the vacuum treatment once more for optimal infiltration if the leaves do not turn dark green.

Fig. 1 A 250-mL transparent beaker filled with 100 mL ice-cold water, containing infected leaf segments submerged in the water. A white, porous Styrofoam floats on the surface to keep the segments fully submerged

3.3 Apoplastic Fluid Extraction

1. Remove the infiltrated leaves and place them between 1 and 2 sheets of paper towel to absorb excess surface water (Fig. 2).
2. Gently press the leaves to remove any remaining surface moisture (*see* **Note 7**).
3. Stack the dried leaf segments in a uniform orientation and insert them vertically into a 25-mL syringe without a needle.
4. Place the syringe into a 50-mL Falcon tube to collect the apoplastic fluid (Fig. 3).
5. Centrifuge the Falcon tubes at 2000 × g for 15 min at 4°C using a swinging-bucket rotor.
6. Apoplastic fluid will collect at the bottom of the Falcon tube (Fig. 4).
7. Carefully transfer the apoplastic fluid into 1.5-mL microcentrifuge tubes and keep them on ice.
8. Repeat Step 5 to recover any remaining fluid.
9. Centrifuge the apoplastic fluid-containing microcentrifuge tubes at 14,000 × g for 5 min at 4°C to remove cell debris or precipitate.

Fig. 2 After vacuum infiltration, infected leaf segments are arranged in rows on a white paper towel to remove excess surface liquid prior to centrifugation

Fig. 3 A gloved hand inserts a 25-mL syringe containing infected leaf segments into a 50-mL centrifuge tube, which is positioned in a metal rack prior to centrifugation

10. Transfer the supernatant to new, clean 1.5-mL microcentrifuge tubes. Keep them on ice for immediate use, or store at −80°C for long-term storage (*see* **Note 8**).

3.4 Trichloroacetic Acid (TCA) Precipitation (Optional)

TCA precipitation of apoplastic fluid is required to detect low-abundance proteins by immunoblotting. All steps should be performed on ice unless otherwise specified. Reagents, including TCA, acetone, and centrifuge tubes, must be precooled to 4°C.

1. Add TCA to the extracted apoplastic fluid to a final concentration of 20%. Mix briefly by vortexing for 3 s, and incubate the samples overnight at 4°C.
2. Centrifuge the samples at 14,000 × *g* for 45 min at 4°C.
3. After centrifugation, a protein pellet will form at the bottom of the 1.5-mL microcentrifuge tube. Carefully discard the supernatant without disturbing the pellet (*see* **Note 9**).
4. Add ~1 mL of precooled acetone to the pellet. Vortex vigorously to wash any protein adhering to the tube walls (*see* **Note 10**).
5. Centrifuge at 20,000 × *g* for 15 min at 4°C to collect the pellet.

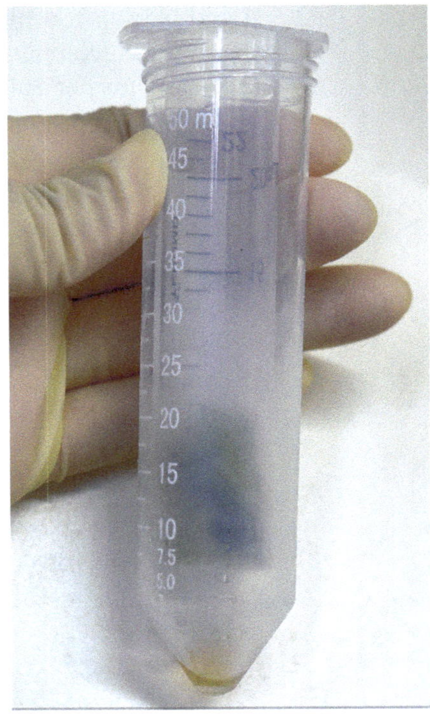

Fig. 4 A gloved hand holding a transparent plastic 50-mL centrifuge tube containing a 25-mL syringe packed with leaf segments. At the bottom of the tube, apoplastic fluid has accumulated as a light-yellow liquid

6. Carefully aspirate approximately 0.95 mL of the supernatant acetone, leaving ~50 μL in the tube to avoid disturbing the loose protein pellet (*see* **Note 11**).
7. Repeat Steps 4–6 a total of three times to ensure complete removal of TCA and acetone.
8. Open the tube cap and place the tube in a dry heating block at 65°C to evaporate the remaining acetone (*see* **Note 12**).
9. Add 50 μL of 2× SSB buffer containing 100 mM DTT. Mix thoroughly and heat at 100°C for 10 min to denature the proteins (*see* **Note 13**).
10. The protein sample is now ready for downstream immunoblot analysis.

4 Notes

1. Infection efficiency may vary depending on the age of the maize cultivar. Carefully monitor seedling development and ensure plants reach the three-leaf stage, which typically occurs 6–8 days after sowing, depending on temperature and light conditions.

2. Avoid placing too many leaf segments in a single beaker, as this can reduce infiltration efficiency. Use no more than 20 leaf segments per 250-mL beaker.
3. During vacuum treatment, air released from the leaves may form bubbles that can prevent complete soaking. The holes in the Styrofoam disc allow this air to escape quickly, promoting uniform infiltration.
4. Use $CaCl_2$-NaAcetate buffer for coimmunoprecipitation assays as described in [9]. For detecting effector presence, filtration with Milli-Q water is sufficient.
5. If an Eppendorf Vacufuge Plus machine is available, run it in *D-AQ* mode for 15 min. Successful infiltration is indicated when the leaves turn dark green and settle at the bottom of the beaker.
6. Do not apply excessive pressure, as it may damage leaf tissue and affect apoplastic fluid purity.
7. When handling multiple samples, keep them on ice until centrifugation.
8. When preparing apoplastic fluid for coimmunoprecipitation, remove $CaCl_2$ by dialysis or using a buffer-exchange device.
9. TCA-precipitated protein pellet may or may not be visible. So, exercise caution to avoid loss of sample.
10. Proteins may adhere not only to the bottom but also to the tube walls. Vortex the tube with acetone for ~10 s to improve protein recovery.
11. After acetone precipitation, the protein pellet may easily detach from the tube. Leave a small volume of acetone to prevent accidental loss of proteins.
12. Monitor the tube every 10 s to prevent over-drying of the pellet. Once acetone has evaporated, promptly close the cap and remove the tube from the 65°C incubator. Over-drying will reduce the solubility of proteins in SSB buffer.
13. If TCA residues remain in the samples, the solution will turn yellow in SSB buffer instead of blue. Add 1–2 μL of 1 M Tris (~pH 12) to adjust pH and the solution should return to blue.

References

1. Hoson T (1998) Apoplast as the site of response to environmental signals. J Plant Res 111(1101):167–177. https://doi.org/10.1007/BF02507163
2. Wang Y, Wang Y, Wang Y (2020) Apoplastic proteases: powerful weapons against pathogen infection in plants. Plant Commun 1(4):100085. https://doi.org/10.1016/j.xplc.2020.100085
3. Doehlemann G, Hemetsberger C (2013) Apoplastic immunity and its suppression by filamentous plant pathogens. New Phytologist 198(4):1001–1016

4. Misas Villamil JC, Mueller AN, Demir F et al (2019) A fungal substrate mimicking molecule suppresses plant immunity via an inter-kingdom conserved motif. Nat Commun 10(1): 1576. https://doi.org/10.1038/s41467-019-09472-8
5. Hemetsberger C, Herrberger C, Zechmann B et al (2012) The *Ustilago maydis* effector Pep1 suppresses plant immunity by inhibition of host peroxidase activity. PLOS Pathog 8(5):e1002684
6. Joosten MH (2012) Isolation of apoplastic fluid from leaf tissue by the vacuum infiltration-centrifugation technique. Methods Mol Biol 835:603–610. https://doi.org/10.1007/978-1-61779-501-5_38
7. Witzel K, Shahzad M, Matros A et al (2011) Comparative evaluation of extraction methods for apoplastic proteins from maize leaves. Plant Methods 7:48. https://doi.org/10.1186/1746-4811-7-48
8. Lohaus G, Pennewiss K, Sattelmacher B et al (2001) Is the infiltration-centrifugation technique appropriate for the isolation of apoplastic fluid? A critical evaluation with different plant species. Physiol Plant 111(4):457–465. https://doi.org/10.1034/j.1399-3054.2001.1110405.x
9. Ma LS, Wang L, Trippel C et al (2018) The *Ustilago maydis* repetitive effector Rsp3 blocks the antifungal activity of mannose-binding maize proteins. Nat Commun 9(1):1711. https://doi.org/10.1038/s41467-018-04149-0

Check for updates

Chapter 5

Chlorazol Black E Staining of Plant Tissue Infected with *Ustilago maydis* to Monitor Pathogenic Development On and *In Planta*

Nora Marie Kühne and Kai Heimel

Abstract

The routine analysis of mutant strains of *Ustilago maydis* includes monitoring of fungal pathogenic growth. Defined developmental stages including filamentous growth on the leaf surface, appressoria formation, and plant penetration, as well as hyphal proliferation *in planta* can be visualized using a straight forward and low-cost staining procedure using Chlorazol Black E staining.

Keywords *Ustilago maydis*, Biotrophic growth, Hyphal morphology, Fungal proliferation, Chlorazol Black E

1 Introduction

The analysis of fungal proliferation after plant infection is crucial to generate meaningful conclusions on gene functions of mutant strains of *Ustilago maydis*. To this end, a variety of staining procedures can be used to study discrete stages of pathogenic development. While Calcofluor White staining allows to specifically monitor pathogenic growth of fungal hyphae on the leaf surface, WGA (wheat germ agglutinin)-Alexafluor—propidium iodide staining allows to investigate *in planta* development in detail but requires costly equipment including a confocal laser scanning microscope and vacuum infiltration device. Alternatively, Chlorazol Black E staining can be used to investigate all stages of pathogenic development on and *in planta*, using a low cost and easy procedure using conventional light microscopy. Its high affinity for chitin over plant-derived polysaccharides allows a high contrast and robust visualization of fungal hyphae. Here, a detailed protocol is provided including plant infection, sample collection, and staining, highlighting best practices to generate high contrast samples with relatively low hands-on time and long-term storage possibility of generated samples.

An N. T. Phan et al. (eds.), *Ustilago maydis*, Methods in Molecular Biology, vol. 3050,
https://doi.org/10.1007/978-1-0716-5364-7_5,

2 Materials

2.1 Plant Infection

1. $YEPS_{light}$ liquid medium [1]: 10 g yeast extract, 4 g Bacto Peptone, 4 g sucrose in 1 L dH_2O. Autoclave at 121°C for 10 min. Store at RT.
2. *Zea mays* plants (e.g., Cultivar Early Golden Bantam), three-leaf stage.
3. 50 mL falcon tubes.
4. 1 mL syringes.

2.2 Sample Collection

1. Razor blades.
2. 15 mL falcon tubes.

2.3 Destaining Procedure

1. 100% ethanol (p.a.).

2.4 KOH Treatment and Staining with Chlorazol Black E [2]

1. 10% KOH solution: 10 g KOH in 100 mL dH_2O.
2. Chlorazol Black E staining solution: add 0.03% (w/v) Chlorazol Black E to a mix of dH_2O, lactic acid and glycerol in a 1:1:1 ratio.
3. Waste container.
4. Flat tweezers.

2.5 Microscopy

1. Microscope slides.
2. Cover glasses (24 mm × 40 mm).
3. 50% glycerol in dH_2O.
4. Flat and fine tweezers.
5. Microscope with bright field or Normarski optics with connected camera.

3 Methods

3.1 Maize Plant Infection with *U. maydis* [3]

1. Inoculate maize plants (e.g., *Zea mays* cultivar Early Golden Bantam) at three-leaf stage (7–9 days after sowing) with respective *U. maydis* strains (*see* **Note 1**). To this end, grow strains overnight in 20 mL $YEPS_{light}$ liquid medium at 28°C under constant shaking.
2. In the next morning, dilute cultures in 50 mL $YEPS_{light}$ to an $OD_{600} = 0.25$ and incubate for another 4 h at 28°C under constant shaking.
3. Harvest cells by centrifugation at 2300× g for 5 min in 50 mL falcon tubes. Resuspend the cell pellet in sterile water to an $OD_{600} = 1$ and keep the suspension on ice.

4. Inoculate 300–500 µL of the cell suspension into the leaf whorl of maize seedlings.

3.2 Collecting Infected Plant Material

1. After the desired incubation time (1 dpi: filament formation, appressoria, penetration, 2 dpi: early proliferation, 3–4 dpi: full proliferation, 5–8 dpi: early to full tumor development) collect leaf samples using a razor blade. To this end, cut samples of 2–3 cm beneath the inoculation points (samples need to fit in 2 mL microcentrifuge tubes). Usually, the third leaf is collected, at later stages also the fourth leaf can be sampled. For each strain and stage, collect at least five samples.
2. Collect all samples of a given stage and strain in a 15 mL falcon tube.

3.3 Destaining of Infected Plant Material

1. Add ethanol to the 15 mL falcon tubes to fully cover all samples.
2. Incubate samples at room temperature in the dark for 3 days or longer until no chlorophyll is visible in leaf samples (*see* **Note 2**).

3.4 Staining with KOH and Chlorazol Black E

1. Carefully decant the ethanol and wash the collected samples once with water. Place single samples in 2 mL microcentrifuge tubes using a flat tweezer (*see* **Note 3**).
2. To each sample, add 10% KOH so that all plant material is covered (*see* **Notes 4 and 5**).
3. Incubate samples at 90°C in a water bath or thermomixer for 3–4 h.
4. Carefully decant the KOH solution and collect it in a suitable glass container. Neutralize before disposal. After this step, samples have to be handled with maximum care to not disturb tissue integrity (*see* **Note 6**).
5. Wash samples with 2 mL deionized water.
6. Add CBE staining solution, so that leaf samples are fully covered. Incubate the samples in CBE staining solution at 60°C overnight.
7. In the next morning, carefully decant and collect the staining solution in an appropriate glass bottle.
8. Add 2 mL of 50% glycerol to destain the samples overnight or longer (*see* **Note 7**).

3.5 Microscopy of Chlorazol Black E Stained Leaf Samples

1. Carefully decant the glycerol solution until 100–200 µL are left. Use this liquid to transfer the sample on a microscope slide using a flat or fine tweezer (*see* **Note 8**).
2. Use a micropipette to aspirate excessive amounts of glycerol.
3. Cover the samples with cover glasses of 24 mm × 40 mm size.
4. OPTIONAL: Conserve samples using nail polish.

5. Proceed with microscopy using conventional bright field or Normarski (differential interference contrast, DIC) optics.

4 Notes

1. Prepare for each strain and developmental stage at least five to ten plants. Do not forget to include the WT control.
2. The longer the incubation time, the better the destaining process. Remaining chlorophyll results in background staining. Usually, destaining for one week or longer gives best results. Samples collected at later timepoints may require longer incubation and can be stored in ethanol for several weeks. Ethanol can be exchanged during extended incubation times.
3. For leaf samples from older plants or later developmental stages, it is advisable to cut out the midrib.
4. The KOH treatment destabilizes plant tissue, so samples must be handled with extreme care to keep them intact for microscopic analysis.
5. When working with KOH and Chlorazol staining solutions, use appropriate safety equipment including gloves and safety glasses.
6. After decanting the KOH solution, it helps if the leaf "sticks" to the side of the tube until it is ready to be removed.
7. Destain final samples for at least one day in 50% glycerol. Longer destaining gives better contrast.
8. When preparing samples for microscopy, it is advantageous to transfer the sample onto the slide with some residual liquid to prevent tissue disintegration.

References

1. Tsukuda T, Carleton S, Fotheringham S et al (1988) Isolation and characterization of an autonomously replicating sequence from *Ustilago maydis*. Mol Cell Biol 8(9):3703–3709
2. Brundrett M, Bougher N, Dell B et al (1996) Working with mycorrhizas in forestry and agriculture, vol. 32 (monograph). Australian Centre for International Agricultural Research, Canberra
3. Kämper J, Kahmann R, Bölker M et al (2006) Insights from the genome of the biotrophic fungal plant pathogen *Ustilago maydis*. Nature 444(7115):97–101. https://doi.org/10.1038/nature05248

Chapter 6

Wheat Germ Agglutinin-Alexa Fuor488/Propidium Iodine Costaining and Fluorescence Microscopic Imaging of *Ustilago maydis* Infected Maize Anthers

Jia-Chi Ku and Karina van der Linde

Abstract

Ustilago maydis infects all aerial maize tissues, forming tumors and altering host development. However, most published methods focus on seedling leaf infections in order to understand the interaction between maize and *U. maydis*. Cytological analysis, particularly fluorescence microscopy, is vital not only for examining fungal infection progression, but also for understanding the host's response. Thus, a detailed protocol is presented here for dissecting *U. maydis*-infected maize anthers, followed by costaining with wheat germ agglutinin-Alexa Fluor 488 (WGA-AF488) and propidium iodide (PI). WGA-AF488 specifically labels fungal cell walls, while PI stains all cells. This robust method enables clear cytological analysis of fungal structures and the status of the host cell, overcoming imaging challenges posed by thick maize tissues.

Keywords Anther, Alexa Fluor488 (AF488), Propidium iodine (PI), Trojan Horse approach, *Ustilago maydis* , Wheat germ agglutinin (WGA), *Zea mays* (maize)

1 Introduction

In angiosperms, the anther produces the male germline and, together with the supporting filament forms the basic male reproductive unit: the stamen. Consequently, anther development is a key part of plant reproduction and can directly impact agricultural yield. The multi-branched male inflorescence of maize, known as the tassel, contains hundreds of spikelets. Each spikelet contains two florets, each of which contains three anthers (see Fig. 1). Each anther typically consists of four lobes. In the middle of each lobe, the male germline matures and is finally released as pollen summarized in [1, 2]. During tassel development, the first anthers form in the spikelets located in the middle of the main spike. This is followed by anther initiation in the spikelets on the main spike base and side branches. Thus, a developing tassel contains anthers at various stages of development [3, 4].

An N. T. Phan et al. (eds.), *Ustilago maydis*, Methods in Molecular Biology, vol. 3050,
https://doi.org/10.1007/978-1-0716-5364-7_6,

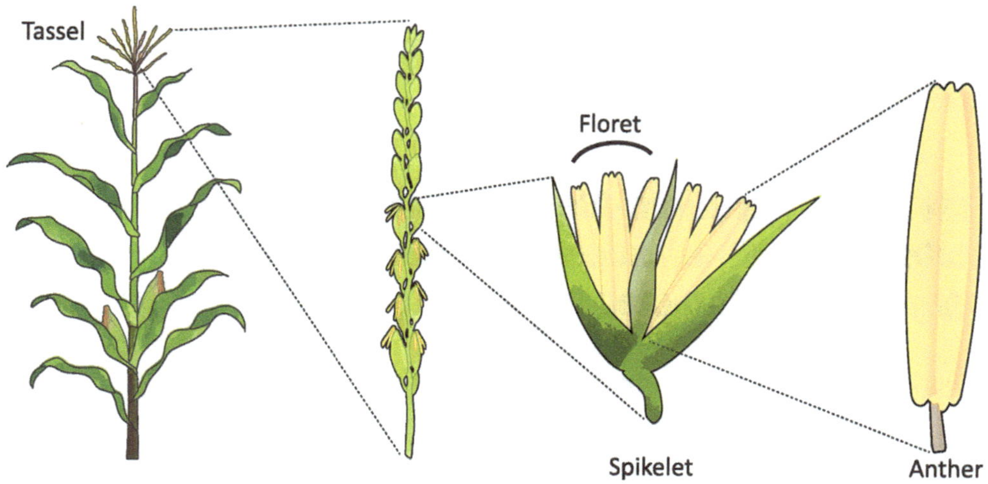

Fig. 1 Overview of the male inflorescence of maize (tassel). A typical maize tassel contains hundreds of spikelets that are arranged on the main spike and the tassel branches. Each spikelet consists of two florets, while each of them harbors three anthers

Ustilago maydis can infect all aerial organs of maize, including tassels and anthers, transforming the infected organs into nutrient-rich tumor tissue [5–7]. This is tightly linked to organ-specific transcriptional reprogramming on both sides of the interaction [8]. Further investigation subsequently identified sets of organ-specific fungal effectors, which collaborate with core, generally deployed effectors to tailor the infection program to the developmental status and cellular architecture of each maize organ [9–11]. The same mechanisms that *U. maydis* uses to deliver effectors into the plant apoplast can also be exploited in research to "microinject" maize proteins, thereby studying their function in situ while circumventing the use of transgenic maize lines. This *U. maydis*-based Trojan Horse approach has been successfully applied to investigate the function of ZIP1 (*Zea mays* IMMUNE PEPTIDE1) in plant pathology and of MAC1 (MULTIPLE ARCHESPORIAL CELLS1) in anther development [12, 13].

For Trojan horse assays and also for studies of *U. maydis* infection in anthers, cytological analysis by microscopic imaging are indispensable. To overcome common problems with microscopic imaging in maize, such as the lack of transgenic fluorescence maize reporter lines and tissue thickness, fluorescence dye-based staining techniques are a valuable tool. Among the most common stainings in maize-*U. maydis* research is the combination of wheat germ agglutinin (WGA)-Alexa Flour488 (AF488) with propidium iodine, and for leaf infections a detailed protocol of this staining techniques has been published [11, 12, 14, 15]. The lectin WGA binds to *N*-acetyl-glucosamines, the building blocks of chitin. When conjugated to the fluorophore

AF488, it enables highly specific fluorescent labeling of fungal cell walls with minimal background interference [16–18]. PI is a membrane-impermeable fluorescent dye that binds to nucleic acids. In plant or fungal tissues, intact plasma membranes exclude PI, therefore, it primarily stains cells with compromised membrane integrity, such as dead or damaged cells. In these cells, the dye can enter the cytoplasm and nucleus [19, 20]. In the laboratory, membrane permeability can also be induced by incubating in a KOH solution combined with heat treatment. Once inside, PI fluoresces strongly upon binding nucleic acids, making it a reliable general cell stain.

In-depth protocols on tassel infections with *U. maydis*, preparation of TH strains, and guidelines on the spatiotemporal progression of anther infections have already been published [21–23]. Moreover, previous studies mapped the developmental status of anthers based on their position on the tassel and to tassel length. This allows researchers to easily dissect anthers for experiments. Then, the dissected anthers can be precisely developmentally staged because anther development is highly correlated with anther length [3, 4]. The following sections provide a detailed, step-by-step protocol, including the necessary materials and equipment, for dissecting *U. maydis*-infected anthers, followed by WGA-AF488/PI costaining and microscopic imaging. Figure 2 briefly visualizes the workflow to provide an overview of the key steps.

2 Material

2.1 Anther Dissection

1. Maize plants with *U. maydis*-infected tassels.
2. Scalpel.
3. Humidity box: Petri dish (diameter: 90–100 mm) with a layer of wet paper towel.
4. 1.5 mL reaction tubes.
5. ≥96% ethanol.
6. Extra-fine pointed, stainless steel forceps (tip width 0.05 mm).
7. Micro ruler (microscopic slide with engraved micrometer, 0.1 mm markings).

2.2 Costaining

1. Single-channel micro pipettes (1000 μL and 200 μL) with fitting pipette tips.
2. ≥96% ethanol.
3. 10% KOH: 200 g/L 85% KOH pellets, in H_2O.
4. 10×PBS: 80 g/L NaCl, 14.2 g/L Na_2HPO_4, 2 g/L KCl, 2.4 g/L KH_2PO_4, in H_2O. If necessary, adjust pH to 7.4 with HCl.
5. WGA-AF488 stock solution: 4 mg/mL WGA-AF488, in H_2O.

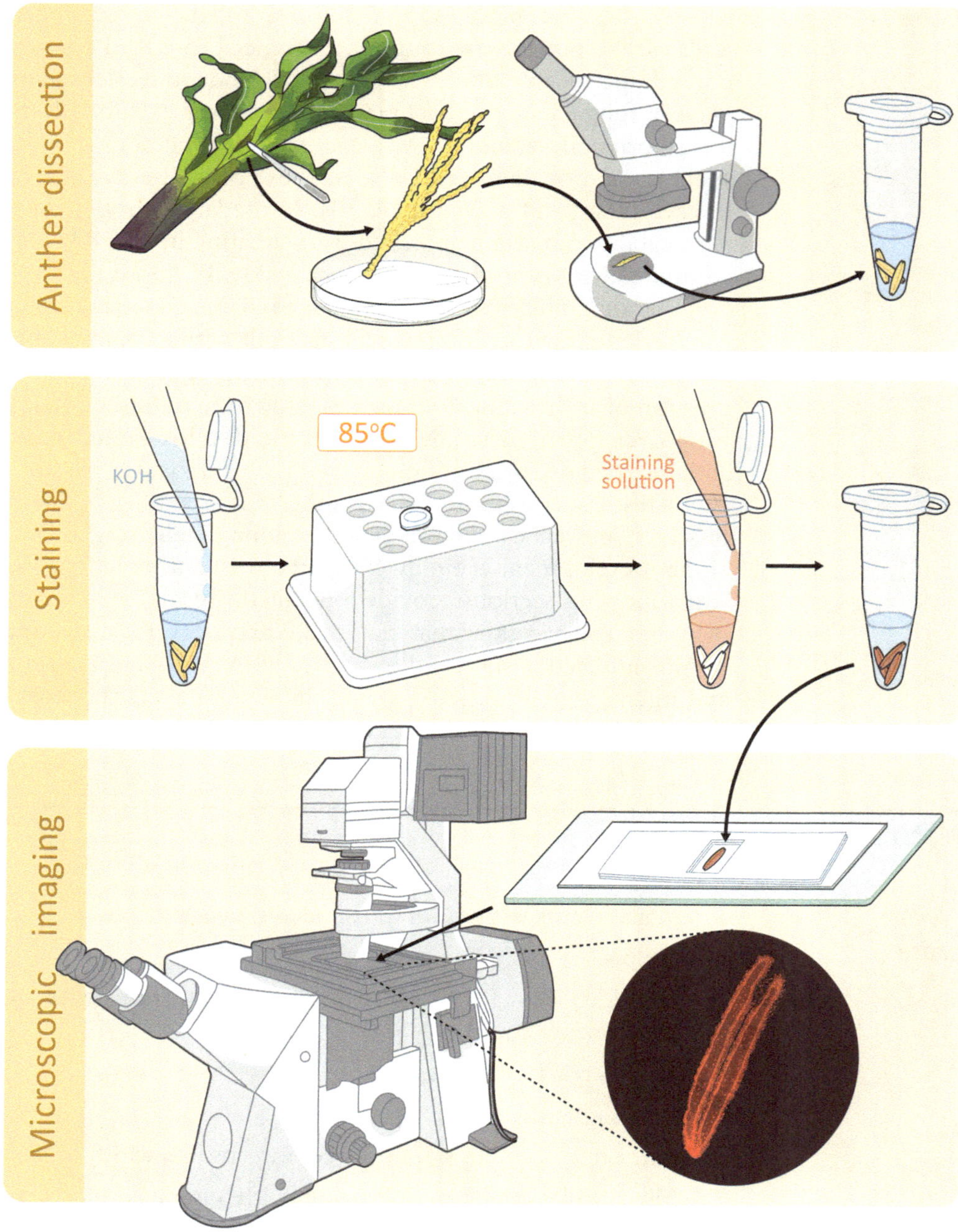

Fig. 2 Workflow for anther dissection, staining and microscopic imaging. For anther dissection first, the tassel is extracted from the maize plant and stored in a humidity box. Then, anthers can be dissected from the spikelets using a stereo microscope. For staining, the anther cell membranes are first permeabilized by KOH-heat treatment. Then, the staining solution is infiltrated. To image the stained anthers a microscopic slide with a sample chamber is prepared, and anthers are pipetted into the sample chamber before mounting the slide in a confocal laser scanning microscope

6. PI stock solution: 10 mg/mL PI, in 1× PBS.
7. WGA-AF488/PI staining solution: 20 μg/mL PI, 10 μg/mL WGA-AF488, 0.02% Tween20, in 1× PBS.

2.3 Microscopic Imaging

1. Double-sided tape, thickness ~75 μm.
2. Microscope slides.
3. Scalpel.
4. Stainless steel scissors.
5. 200 μL single-channel micro pipette with fitting pipette tips.
6. Coverslips, 22 × 22 mm, thickness 1.5 H.

2.4 Equipment

1. Stereo microscope.
2. Heating block.
3. Adjustable vacuum pump with desiccator.
4. Confocal laser scanning microscope.

3 Methods

3.1 Anther Dissection

1. Use a scalpel to cut the plant base under the tassel (*see* **Note 1**).
2. Dissect the tassel. Either peel off the individual leaves until you reach the tassel or slice the stem longitudinally with a scalpel to expose the tassel.
3. Place the tassel, still attached to approximate stem base, in the humidity box.
4. Take a few spikelets at a time from the tassel and place them on the micro ruler under the stereo microscope.
5. Dissect the anthers from the florets using the extra-fine pointed forceps. Measure anthers to ensure collecting the desired anther developmental stage (*see* **Note 2**).
6. Wet the forceps tip by dipping them into ≥96% ethanol to pick up the dissected anthers and transfer them in a reaction tube with ≥96% ethanol (*see* **Note 3**).
7. Store dissected anthers at 4°C in ≥96% ethanol before proceeding to the costaining (*see* **Note 4**).

3.2 Costaining

1. Remove the ethanol with a 1000 μL-pipette from the tube and add 1 mL of ≥96% ethanol (*see* **Note 5**).
2. Remove the ethanol completely using the 200 μL-pipette.
3. Add 500 μL of 10% KOH and mix by gently inverting the tube ten times (*see* **Note 6**).

4. Incubate at 85°C for 2 min in a heating block.
5. Remove the sample from the heating block and directly add 1×PBS to fill up the entire reaction tube (*see* **Note 7**).
6. Remove the supernatant and add 1 mL 1× PBS.
7. Remove the supernatant and again add 1 mL 1× PBS.
8. Remove the supernatant and add 1 mL WGA-AF488/PI staining solution.
9. Vacuum infiltrate the staining solution for 5 min at 400 mbar.
10. Release the vacuum slowly and invert the tube ten times.
11. Incubate for 10 min at RT.
12. Repeat Steps 9–11 four times.
13. Remove staining solution and add 1 mL 1×PBS (*see* **Note 8**).
14. Remove the supernatant and add 1 mL 1×PBS. Repeat this step.
15. Store stained anthers in 1×PBS at 4°C in the dark until microscopic imaging (*see* **Note 9**).

3.3 Microscopic Imaging

1. Prepare a sample chamber by stacking approximately 4 cm long strips of double-sided tape on a microscopic slide. With a scalpel, cut a square in the middle of the stack of tape and remove the tape square (*see* **Note 10** and Fig. 3).

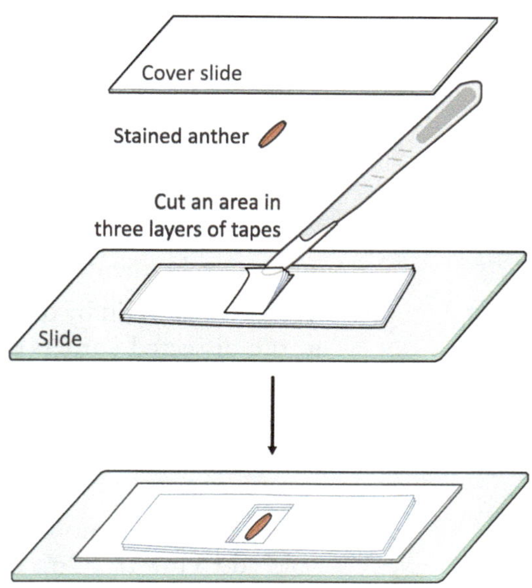

Fig. 3 Preparation of a microscopic slide for imaging. Layers of double-sided tape are stacked and pressed onto a microscopic slide. Then, a square incision is made with a scalpel. The stained anthers are pipetted into the resulting sample chamber and sealed with a cover slip

2. Cut off the of the opening of a 200 μL pipette tip with the scissors, so that anthers can be pipetted. Using the trimmed pipette tip, transfer the anthers in 1× PBS solution into the sample chamber on the microscopic slide.
3. Seal the sample chamber with a coverslip and position the microscopic slide with the sample under the confocal laser scanning microscope.
4. Measure anther length in the DIC channel and acquire Z-stacks of whole anther lobes. Excite PI with a laser at 561 nm and detect between 580 and 630 nm. Emission of WGA-AF488 (Ex: 488 nm) is collected between 500 and 540 nm.
5. Analyze the images using the software provided by the confocal laser scanning microscope platform.

4 Notes

1. The tassel can be located by pressing the stalk gently.
2. Under the stereo microscope, carefully clip off the spikelet base using the extra-fine pointed forceps. To dissect premeiotic anthers (<1000 μm), carefully peel off the glumes and the lemma to the expose the anthers. Then, cut individual anthers, at the filament, from the floret with the forceps. Bigger anthers (>1000 μm) can be "squeezed" out of the spikelet by gently pressing on the tip of the spikelet with the forceps once the base is removed.
3. Do not leave dissected anthers out in the air for more than 2 min, or else they will dry out.
4. Dissected anthers can be stored for up to 8 weeks in.
5. At all following steps, be careful not to suck anthers into the pipette tip, and wait to proceed to the next step of the protocol until anthers have settled on the bottom of the reaction tube.
6. Anthers should be completely covered with KOH solution. If needed, add more KOH solution to the reaction tube.
7. During the KOH-heat treatment, anthers become opaque and are hard to see. Give anthers extra time to settle to the bottom of the reaction tube.
8. The used staining solution can be stored at 4°C and reused.
9. Stained anthers can be stored for up to 3 months.
10. For premeiotic anthers, three layers of tape are used, and the square sample chamber should be approximately 5 × 5 mm.

Acknowledgments

The authors would like to thank Hsuan Pai (https://paiinthelab.blog/) for the graphic design of figures in this manuscript. This work was funded by the DFG Research Unit RU 5235 "Cereal Stem Cell Systems (CSCS)" TP 8 (LI 2437/3-1).

References

1. van der Linde K, Walbot V (2019) Pre-meiotic anther development. In: Grossniklaus U (ed) Current topics in developmental biology: plant development and evolution, vol 131. Academic Press (San Diego), pp 239–256. https://doi.org/10.1016/bs.ctdb.2018.11.001
2. Marchant DB, Walbot V (2022) Anther development–the long road to making pollen. Plant Cell 34(12):4677–4695. https://doi.org/10.1093/plcell/koac287
3. Kelliher T, Walbot V (2011) Emergence and patterning of the five cell types of the *Zea mays* anther locule. Dev Biol 350(1):32–49. https://doi.org/10.1016/j.ydbio.2010.11.005
4. Egger RL, Walbot V (2015) Quantifying *Zea mays* tassel development and correlation with anther developmental stages as a guide for experimental studies. Maydica 60(4):M34. https://doi.org/10.5555/20173230936
5. Sosso D, van der Linde K, Bezrutczyk M et al. (2019) Sugar partitioning between *Ustilago maydis* and its host *Zea mays* L during infection. Plant Physiol 179(4):1373–1385. https://doi.org/10.1104/pp.18.01435
6. Matei A, Ernst C, Gunl M et al. (2018) How to make a tumour: cell type specific dissection of *Ustilago maydis*-induced tumour development in maize leaves. New Phytol 217(4):1681–1695. https://doi.org/10.1111/nph.14960
7. Doehlemann G, Wahl R, Horst RJ et al. (2008) Reprogramming a maize plant: transcriptional and metabolic changes induced by the fungal biotroph *Ustilago maydis*. Plant J 56(2):181–195. https://doi.org/10.1111/j.1365-313X.2008.03590.x
8. Skibbe DS, Doehlemann G, Fernandes J et al. (2010) Maize tumors caused by *Ustilago maydis* require organ-specific genes in host and pathogen. Science 328(5974):89–92. https://doi.org/10.1126/science.1185775
9. Schilling L, Matei A, Redkar A et al. (2014) Virulence of the maize smut *Ustilago maydis* is shaped by organ-specific effectors. Mol Plant Pathol 15(8):780–789. https://doi.org/10.1111/mpp.12133
10. Seitner D, Uhse S, Gallei M et al. (2018) The core effector Cce1 is required for early infection of maize by *Ustilago maydis*. Mol Plant Pathol 19(10):2277–2287. https://doi.org/10.1111/mpp.12698
11. Redkar A, Hoser R, Schilling L et al. (2015) A secreted effector protein of *Ustilago maydis* guides maize leaf cells to form tumors. Plant Cell 27(4):1332–1351. https://doi.org/10.1105/tpc.114.131086
12. van der Linde K, Timofejeva L, Egger RL et al. (2018) Pathogen Trojan horse delivers bioactive host protein to alter maize (*Zea mays*) anther cell behavior in situ. Plant Cell 30(3):528–542. https://doi.org/10.1105/tpc.17.00238
13. Ziemann S, van der Linde K, Lahrmann U et al. (2018) An apoplastic peptide activates salicylic acid signalling in maize. Nat Plants 4(3):172–180. https://doi.org/10.1038/s41477-018-0116-y
14. Doehlemann G, Reissmann S, Aßmann D et al. (2011) Two linked genes encoding a secreted effector and a membrane protein are essential for *Ustilago maydis*-induced tumour formation. Mol Microbiol 81(3):751–766. https://doi.org/10.1111/j.1365-2958.2011.07728.x
15. Redkar A, Jaeger E, Doehlemann G (2018) Visualization of growth and morphology of fungal hyphae in planta using WGA-AF488 and propidium iodide co-staining. Bio Protoc 8(14):e2942
16. Figueroa-López AM, Cordero-Ramírez JD, Quiroz-Figueroa FR et al. (2014) A high-throughput screening assay to identify bacterial antagonists against *Fusarium verticillioides*. J Basic Microbiol 54(Suppl 1):S125–S133. https://doi.org/10.1002/jobm.201200594
17. Ayliffe M, Devilla R, Mago R et al. (2011) Nonhost resistance of rice to rust pathogens. Mol Plant Microbe Interact 24(10):1143–1155. https://doi.org/10.1094/mpmi-04-11-0100
18. Doehlemann G, Wahl R, Vraneš M et al. (2008) Establishment of compatibility in the *Ustilago maydis*/maize pathosystem. J Plant Phys 165(1):29–40. https://doi.org/10.1016/j.jplph.2007.05.016
19. Bureau C, Lanau N, Ingouff M et al. (2018) A protocol combining multiphoton microscopy

and propidium iodide for deep 3D root meristem imaging in rice: application for the screening and identification of tissue-specific enhancer trap lines. Plant Methods 14: 96. https://doi.org/10.1186/s13007-018-0364-x

20. Collins ASP, Kurt H, Duggan C et al. (2024) Parallel, continuous monitoring and quantification of programmed cell death in plant tissue. Adv Sci 11(23):e2400225. https://doi.org/10.1002/advs.202400225
21. Kutzner L, van der Linde K (2024) A Trojan horse approach using *Ustilago maydis* to study apoplastic maize (*Zea mays*) peptides in situ. Methods Mol Biol 2731:115–132. https://doi.org/10.1007/978-1-0716-3511-7_9
22. van der Linde K, Timofejeva L, Egger RL et al. (2018) Application of the pathogen Trojan horse approach in premeiotic maize (*Zea mays*) anther development. Plant Signal Behav 13(12):e1547575. https://doi.org/10.1080/15592324.2018.1547575
23. Fiedler I-C, Weiberg A, van der Linde K (2019) Using *Ustilago maydis* as a Trojan horse for in situ delivery of maize proteins. J Vis Exp 144:e58746

Chapter 7

A Genomic DNA-Based Real-Time PCR Assay for Quantification of *Ustilago maydis* Biomass in Plant Samples

Xuepei Li, Bing Yang and Xiaowei Han

Abstract

Ustilago maydis is a basidiomycete fungus that causes common smut in maize, serving as a model system for studying smut fungi and biotrophic pathogens. Reliable quantification of fungal biomass *in planta* facilitates tracking fungal proliferation and disease development, evaluating fungal mutant virulence, and assessing host resistance. Traditional methods such as symptom scoring or microscopy after staining are semiquantitative and prone to subjectivity. In this chapter, a DNA-based quantitative real-time PCR (qPCR) assay for sensitive and reproducible measurement of *U. maydis* biomass in infected maize tissues is described, based on the relative amplification ratio of a fungal gene to a maize reference gene. This assay provides a robust and quantitative tool for investigating fungal virulence and plant resistance.

Keywords *Ustilago maydis*, Maize, Fungal biomass, qPCR, Quantitative analysis

1 Introduction

1.1 *Ustilago maydis* is a Model System for Studying Smut Fungi

Smut fungi are a large group of fungal species mostly belonging to the Basidiomycota phylum, comprising more than 1600 species [1]. They infect a great diversity of plants and cause smut diseases of many agriculturally important crops, including maize (*Zea mays*), wheat, rice, sugarcane, sorghum, and millet [2–4]. Among them, the common smut, caused by the biotrophic fungus *Ustilago maydis*, is one of the common diseases that frequently occurs in global maize growing areas. Unlike most smut diseases that are only visible in floral organs, *U. maydis* infects all aerial parts of maize, such as ears, tassels, leaves, and stems, where tumors (also called galls) are often produced [5, 6]. Beyond its agricultural relevance, the *U. maydis*-maize pathosystem has become a popular model system for studying smut fungi and biotrophic plant pathogens in the past decades, due to its genetic tractability, well-defined infection cycle, and amenability to both molecular and genomic approaches [7]. Studies in this system have provided fundamental

An N. T. Phan et al. (eds.), *Ustilago maydis*, Methods in Molecular Biology, vol. 3050,
https://doi.org/10.1007/978-1-0716-5364-7_7,

insights into fungal genetics, lifestyle, virulence mechanisms, host resistance, and the molecular dialogs within plant–microbe interactions [3, 8–10].

1.2 Methods for Evaluating U. maydis Colonization in Infected Tissues

Quantification of fungal colonization *in planta* is a key requirement in various investigations in the *U. maydis*-maize pathosystem. Assessment of *U. maydis* colonization not only enables monitoring of disease progression but also provides a quantitative basis for evaluating virulence of fungal mutants and resistance of different maize genotypes [11, 12]. Traditional methods to evaluate fungal proliferation and disease severity include visual scoring of disease symptoms according to the number and size of tumors and microscopic observation of fungal hyphae after wheat germ agglutinin (WGA)-AF488 and propidium iodide (PI) co-staining [13, 14]. While informative, these approaches are labor-intensive, semiquantitative, and prone to subjectivity due to variation in scoring practices. In some cases, subtle differences in fungal growth and virulence between wild-type and mutant strains may not be discernible from visual symptom assessment alone but can be captured by other quantitative assays, such as quantitative real-time PCR (qPCR) assay for fungal biomass measurement.

1.3 qPCR for Sensitive and Quantitative Measurement of Fungal Biomass in Plant Tissues

qPCR is a DNA amplification technique that monitors product accumulation in real time, allowing sensitive and specific quantification of target DNA sequences [15, 16]. The underlying principle is that amplification curves directly correlate with the initial amount of target DNA. This method was subsequently adapted to evaluate gene expression by combining reverse transcription with the use of constitutively expressed reference genes for normalization [17–19]. Moreover, in the field of plant–pathogen interactions, plant pathologists rapidly optimized and applied the qPCR method to quantify pathogen load of infected plant tissues across a vast diversity of pathosystems, including *Magnaporthe oryzae*, *Verticillium dahliae*, *Botrytis cinerea*, *Rhizoctonia solani*, *Fusarium graminearum*, and *U. maydis* [20–26]. Using genomic DNA extracted from infected plant tissues as a template, a single-copy fungal specific gene is amplified and normalized to a plant reference gene. The resulting ratio provides a relative measure of fungal biomass within plant tissues [20, 27]. This approach offers high sensitivity and reproducibility, making it suitable for diverse experimental applications in dissecting host–pathogen interactions [27].

In this chapter, a detailed genomic DNA-based qPCR protocol for *U. maydis*-infected maize tissues is described, providing a quantitative and reproducible framework for analyzing infection dynamics, characterizing fungal virulence genes, and evaluating maize cultivar resistance.

2 Materials

2.1 Maize Seedling Inoculation

1. YESPL liquid medium: 0.4% yeast extract, 0.4% peptone, 2% sucrose.
2. 1 mL syringe.
3. Spectrophotometer.
4. Centrifuge.

2.2 Genomic DNA Extraction

1. Mortar and pestle.
2. Acid-washed glass beads (diameter 0.4–0.6 mm).
3. *Ustilago* lysis buffer: 10 mM Tris-HCl (pH = 8.0), 100 mM NaCl, 1 mM Na_2-EDTA, 1% SDS, 2% (v/v) Triton X-100.
4. Phenol and chloroform mix (1:1, v/v).
5. TE-RNase A buffer: 10 mM Tris-HCl, 1 mM EDTA, 12 μg/mL RNase A, pH = 8.0.
6. Tissue lyser.
7. Centrifuge.
8. Spectrophotometer.
9. Thermoshaker.

2.3 qPCR Analysis

1. 2× Q3 SYBR qPCRQuantitative polymerase chain reaction (qPCR) Master Mix (Universal).
2. Specific primers for maize *GAPDH* and *U. maydis PPI/RAB7* genes (Table 1; *see* **Note 1**).
3. iCycler.

Table 1
Primers used in qPCR for *U. maydis* biomass measurements

Primers	Sequence (5′-3′)	Target genes	References
ZmGAPDH-F	CTTCGGCATTG TTGAGGGTTTG	*GRMZM2G046804* (*ZmGAPDH*)	[25]
ZmGAPDH-R	TCCTTGGCTGA GGGTCCGTC		
UmPPI-F	ACATCGTCAA GGCTATCG	*UMAG_03726* (*UmPPI*)	[28]
UmPPI-R	AAAGAACACC GGACTTGG		
UmRAB7-F	GAGCAGGCTTT CCAGACCATCG	*UMAG_05511* (*UmRAB7*)	[12]
UmRAB7-R	CAATTGCATCCG CTCTGCTGG		

3 Methods

Unless otherwise specified, all operations are performed at room temperature.

3.1 Plant Growth Conditions and U. maydis Inoculation

1. *Z. mays* plants are grown in a temperature-controlled greenhouse (14 h/10 h light/dark cycle, 28°C/22°C) and used for infection.
2. *U. maydis* strains are grown on PDA plate at 28°C. *U. maydis* colonies are inoculated into YEPSL medium and incubated at 28°C, 200 rpm for overnight (12–16 h).
3. Transfer overnight culture into fresh YEPSL medium, adjust the culture to $OD_{600} \approx 0.2$, and incubate until $OD_{600} \approx 1.0$ (*see* **Note 2**).
4. Harvest cultures by centrifugation at 2000 × *g* for 5 min, resuspend cell pellets in autoclaved ddH_2O, and adjust the suspensions to $OD_{600} = 1.0$.
5. Inject cell suspensions into the stem of 7-day-old maize seedlings with a syringe [13] (Fig. 1a; *see* **Note 3**).

3.2 Sample Collection and Genomic DNA Preparation

1. Excise 2 cm to 5 cm leaf segments 1 cm below the injection holes at selected time points. Harvested leaf samples are immediately frozen in liquid nitrogen and stored at −70°C until further use (Fig. 1b; *see* **Note 4**).
2. Grind leaf tissues in a mortar with liquid nitrogen (*see* **Note 5**).
3. Transfer ~100 μL frozen powder to a 1.5 mL tube. Add 500 μL *Ustilago* lysis buffer and ~100 μL acid-washed glass beads. The samples are homogenized using a tissue lyser at 60 Hz for 180 s.
4. Add 500 μL phenol-chloroform (1:1) and mix thoroughly. Centrifuge at 15,000 × *g* for 15 min.
5. Transfer 400 μL supernatant to a new 1.5 mL tube, add 1 mL ethanol, and mix thoroughly. Centrifuge at 15,000 × *g* for 5–8 min and discard supernatant (*see* **Note 6**).
6. Add 1 mL 75% ethanol to briefly wash the pellet on the bottom, centrifuge at 15,000 × *g* for 2–3 min, and discard supernatant. Dry the DNA pellet for 3–4 min (*see* **Note 7**).
7. Dissolve the pellet in 50 μL TE buffer for 15 min at 60°C with gentle shaking (*see* **Note 8**).
8. Quantify genomic DNA concentration using spectrophotometer and store at −20°C until further use (Fig. 1c).

3.3 qPCR Reaction and Data Analysis

1. qPCR reaction setup: 1 μL of genomic DNA (100–150 ng/μL) is mixed with 0.4 μL gene specific primers (10 μM) and

5 µL 2× Q3 SYBR qPCR Master Mix. Add ddH_2O to a total reaction volume of 10 µL. Primers used for target gene amplification are listed in Table 1 (Fig. 1d; *see* **Note 9**).

2. Cycling conditions: 95°C for 1 min 30 s; 45 cycles of 95°C for 15 s, 60°C for 30 s, 72°C for 30 s [12]. The qPCR reaction is carried out using an iCycler.

3. Data analysis: relative fungal biomass is calculated using the $2^{-\Delta\Delta Ct}$ method [18]. For each biological replicate, use the mean Cq value of three technical replicates as the Ct. Calculate ΔCt as (Ct_{PPI}−Ct_{GAPDH}), then determine ΔΔCt relative to the control sample, and compute $2^{-\Delta\Delta Ct}$ for each replicate. Use the three biological replicates to calculate the mean and standard error of $2^{-\Delta\Delta Ct}$, representing relative fungal biomass. Perform statistical comparisons of fungal biomass between strains or treatments using Student's *t*-test (Fig. 1e; *see* **Note 10**).

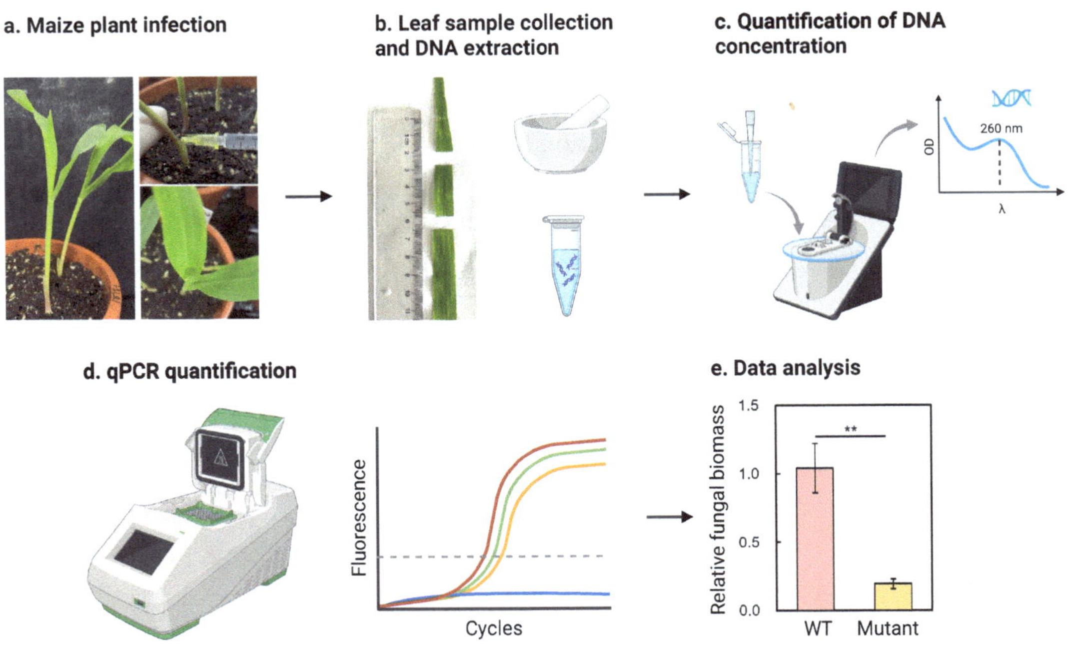

Fig. 1 The workflow depicting steps of the fungal biomass assay for *U. maydis* infection. (**a**) Seven-day-old maize seedlings are inoculated with *U. maydis* cell suspensions at OD_{600} of 1.0. (**b**) 2–5 cm leaf sections are excised 1 cm below the injection holes at desired time points, homogenized, and subjected to genomic DNA extraction. (**c**) Extracted genomic DNA samples are quantified using a spectrophotometer. (**d**) qPCR analysis is performed using an iCycler with defined cycling conditions. (**e**) Relative fungal biomass is calculated using the $2^{-\Delta\Delta Ct}$ method and statistically analyzed (e.g., wild-type (WT) vs. mutant, $**P<0.01$). This figure is created with BioRender.com

4 Notes

1. Two *U. maydis* housekeeping genes, which encode peptidyl-prolyl isomerase PPI (*UMAG_03726*) or small GTPase RAB7 (*UMAG_05511*), are recommended as fungal-specific genes for biomass quantification, and the maize *GAPDH* gene (*GRMZM2G046804*, encoding glyceraldehyde-3-phosphate dehydrogenase) is used as the internal reference gene for normalization.
2. For *U. maydis*, the doubling time is approximately 2 h at 28°C with shaking at 200 rpm. Thus, after adjusting the culture to OD_{600} 0.2 in fresh YEPSL medium, it typically requires about 5 h of incubation to reach OD_{600} 1.0.
3. It is recommended to grow more maize plants than needed. Choose plants with similar size and appearance for infection to minimize host variation on fungal biomass quantification.
4. Time points for sample collection are chosen according to the specific purposes of study. Harvest three biological replicates per strain or condition, and for each replicate collect eight to ten leaves representing the current disease development. At early time points [i.e., 1–6 days post inoculation (dpi)], 2 cm to 5 cm leaf blade sections are collected 1 cm below the infection holes (usually from the third and fourth leaves). At later time points (i.e., 7–12 dpi), all visibly infected tissues adjacent to the injection holes can be harvested.
5. Keep samples frozen in liquid nitrogen during grinding and extraction processes and avoid thawing at any stage.
6. If no DNA precipitation is observed, the samples can be incubated at room temperature or at −20°C for 10 min to facilitate DNA precipitation.
7. Carefully remove supernatant without disturbing the DNA pellet on the bottom. The pellet should not be over-dried, which might make it difficult to dissolve later.
8. The optimal temperature for RNase A enzymatic activity is 60°C. Incubation on a thermoshaker at 60°C, 750 rpm for 15 min is recommended for dissolving DNA pellet.
9. To prepare the qPCR reactions, samples for each strain at one time point should include three biological replicates and three technical replicates. Adjust the concentration of genomic DNA samples to 100–150 ng/μL used as templates. To avoid repeated freeze-thaw cycles of the SYBR qPCR Master Mix, aliquot 500 μL or even smaller volume of the Master Mix in 1.5 mL tubes. Take care to prevent contamination during aliquoting.

10. Normally, if the Cq value of the reference gene *GAPDH* falls within the range of 20–25, the DNA concentration of the template is considered appropriate. Relative fungal biomass of the control sample will be 1.0 in the calculation.

Acknowledgments

We thank Dr. Kenichi Tsuda for critical reading of the manuscript and members of the Plant Immunity Team at HZAU for helpful discussions. This work was supported by the National Natural Science Foundation of China (32372494 to X.H.), the National Key Research and Development Program of China (2022YFA1304400 to X.H.), and the Fundamental Research Funds for the Central Universities (2662025ZKPY008 to X.H.).

References

1. Vánky K (2008) Smut fungi (Basidiomycota p.p., Ascomycota p.p.) of the world. Novelties, selected examples, trends. Acta Microbiol Immunol Hung 55:91–109
2. Carris LM, Castlebury LA, Goates BJ (2006) Nonsystemic bunt fungi—*Tilletia indica* and *T. horrida*: a review of history, systematics, and biology. Annu Rev Phytopathol 44:113–133
3. Zuo W, Ökmen B, Depotter JRL et al (2019) Molecular interactions between smut fungi and their host plants. Annu Rev Phytopathol 57:411–430
4. Wu E, Liu L, Zhu M et al (2022) The life cycle and ultrastructure of the host response of the smut pathogen, *Anthracocystis destruens*, on broomcorn millet. Phytopathology 112:996–1002
5. Brefort T, Doehlemann G, Mendoza-Mendoza A et al (2009) *Ustilago maydis* as a pathogen. Annu Rev Phytopathol 47:423–445
6. Kahmann R, Steinberg G, Basse C et al (2000) *Ustilago maydis*, the causative agent of corn smut disease. In: Kronstad JW (Ed.)Eds. Fungal pathology. Springer, Netherlands, Dordrecht, 347–371
7. Dean R, Van Kan JAL, Pretorius ZA et al (2012) The top 10 fungal pathogens in molecular plant pathology: top 10 fungal pathogens. Mol Plant Pathol 13:414–430
8. Kahmann R (2024) *Ustilago maydis*: ein faszinierendes Modellsystem für pathogene Pilze. BIOspektrum 30:258–263
9. Ziemann S, van der Linde K, Lahrmann U et al (2018) An apoplastic peptide activates salicylic acid signalling in maize. Nat Plants 4:172–180
10. Han X, Altegoer F, Steinchen W et al (2019) A kiwellin disarms the metabolic activity of a secreted fungal virulence factor. Nature 565:650–653
11. Schurack S, Depotter JRL, Gupta D et al (2021) Comparative transcriptome profiling identifies maize line specificity of fungal effectors in the maize–*Ustilago maydis* interaction. Plant J 106:733–752
12. Lanver D, Müller AN, Happel P et al (2018) The biotrophic development of *Ustilago maydis* studied by RNA-seq analysis. Plant Cell 30:300–323
13. Kämper J, Kahmann R, Bölker M et al (2006) Insights from the genome of the biotrophic fungal plant pathogen *Ustilago maydis*. Nature 444:97–101
14. Doehlemann G, van der Linde K, Assmann D et al (2009) Pep1, a secreted effector protein of *Ustilago maydis*, is required for successful invasion of plant cells. PLOS Pathog 5:e1000290
15. Heid CA, Stevens J, Livak KJ et al (1996) Real time quantitative PCR. Genome Res 6:986–994
16. Higuchi R, Fockler C, Dollinger G et al (1993) Kinetic PCR analysis: real-time monitoring of DNA amplification reactions. Biotechnology 11:1026–1030
17. Bustin SA (2000) Absolute quantification of mRNA using real-time reverse transcription

polymerase chain reaction assays. J Mol Endocrinol 25:169–193

18. Livak KJ, Schmittgen TD (2001) Analysis of relative gene expression data using real-time quantitative PCR and the $2^{-\Delta\Delta CT}$ method. Methods 25:402–408
19. Rajeevan MS, Ranamukhaarachchi DG, Vernon SD et al (2001) Use of real-time quantitative PCR to validate the results of cDNA array and differential display PCR technologies. Methods 25:443–451
20. Qi M, Yang Y (2002) Quantification of *Magnaporthe grisea* during infection of rice plants using real-time polymerase chain reaction and northern blot/phosphoimaging analyses. Phytopathology 92:870–876
21. Naef A, Senatore M, Défago G (2006) A microsatellite based method for quantification of fungi in decomposing plant material elucidates the role of *Fusarium graminearum* DON production in the saprophytic competition with *Trichoderma atroviride* in maize tissue microcosms: quantitative microsatellite PCR assays. FEMS Microbiol Ecol 55:211–220
22. Duressa D, Rauscher G, Koike ST et al (2012) A real-time PCR assay for detection and quantification of *Verticillium dahliae* in spinach seed. Phytopathology 102:443–451
23. Horevaj P, Milus EA, Bluhm BH (2011) A real-time qPCR assay to quantify *Fusarium graminearum* biomass in wheat kernels: *Fusarium graminearum* biomass in wheat kernels. J Appl Microbiol 111:396–406
24. Lievens B, Brouwer M, Vanachter ACRC et al (2006) Real-time PCR for detection and quantification of fungal and oomycete tomato pathogens in plant and soil samples. Plant Sci 171:155–165
25. Di Stasio M, Brefort T, Mendoza-Mendoza A et al (2009) The dual specificity phosphatase Rok1 negatively regulates mating and pathogenicity in *Ustilago maydis*. Mol Microbiol 73:73–88
26. Brouwer M, Lievens B, Van Hemelrijck W et al (2003) Quantification of disease progression of several microbial pathogens on *Arabidopsis thaliana* using real-time fluorescence PCR. FEMS Microbiol Lett 228:241–248
27. Klosterman SJ (2012) Real-time PCR for the quantification of fungi in planta. In: Bolton MD, Thomma BPHJ (Eds.) Plant fungal pathogens. Humana Press, Totowa, NJ, 121–132
28. Brefort T, Tanaka S, Neidig N et al (2014) Characterization of the largest effector gene cluster of *Ustilago maydis*. PLOS Pathog 10:e1003866

Chapter 8

Random (UV) Mutagenesis, Unbiased Screening and Identification of DNA Repair Mutants by Complementation Cloning

Milorad Kojic and Mira Milisavljevic

Abstract

As a highly radiation-resistant microbial BRCA2 organism, *Ustilago maydis* is an attractive genetic system to study the mechanisms underlying genome protection. Despite the completion of its genome sequence nearly two decades ago, the functions of many of its 6500+ genes remain unknown. This gap underscores the critical need to identify novel genes, some of which may play vital roles in safeguarding the genome. One effective approach to discover these genes is to conduct forward genetic screens starting with the wild-type and looking for candidates that exhibit pronounced sensitivity to genotoxic agents. Therefore, a united protocol is hereby presented for efficient random mutagenesis, straightforward genetic screening, and the rapid detection of mutated genes by complementation cloning. As the quest for new DNA repair factors continues, it is anticipated that adherence to these procedures will facilitate the discovery of additional factors, thereby shedding light on the molecular networks that coordinate the maintenance of genome integrity in *U. maydis* and other BRCA2 organisms.

Keywords UV mutagenesis, DNA repair, Complementation cloning, Genetic screening

1 Introduction

Significant progress in our understanding of the eukaryotic molecular pathways involved in genome protection has resulted from studies on a few fungal species, including *Ustilago maydis*. Developed by Robin Holliday several decades ago as a model for studying recombinational repair [1, 2], *U. maydis* has contributed, and still contributes [3], greatly to our knowledge of this vital aspect of cellular biology.

While the premier fungal systems, *Saccharomyces cerevisiae* and *Schizosaccharomyces pombe*, have certainly driven much of the progress in genome integrity research, *U. maydis* offers unique advantages. Namely, unlike these well-studied yeasts, it possesses a BRCA2 ortholog known as Brh2 [4], along with its regulatory partner, Dss1 [5]. Indeed, this distinctive characteristic positions *U. maydis* as a

An N. T. Phan et al. (eds.), *Ustilago maydis*, Methods in Molecular Biology, vol. 3050,
https://doi.org/10.1007/978-1-0716-5364-7_8,

valuable resource for future BRCA2 research that goes beyond what traditional yeasts can offer.

In addition, *U. maydis* is also distinguished by its remarkable ability to survive exposure to extreme levels of UV and gamma radiation [6, 7]. Investigating this capacity may provide insights not only into the evolution of DNA repair mechanisms and *U. maydis'* ecological success but also into certain aspects of cancer biology, particularly in relation to the resistance to radiotherapy treatments.

Developing this argument further, it should also be noted that *U. maydis* has a significant number of predicted protein-encoding genes (approximately 30%) [8] that either lack significant matches in any database or exhibit similarities only to hypothetical proteins. This highlights the need to define novel genes, some of which may encode proteins underlying DNA repair and recombination in *U. maydis*, and by extension, in other BRCA2 organisms. Furthermore, this finding clearly demonstrates that identifying new DNA-repair factors in *U. maydis* cannot be accomplished merely by extrapolating from well-established yeast models.

While various methods can be used to uncover such genes, it is believed that the most effective strategy to fully explore the genetic potential of *U. maydis* lies in the implementation of comprehensive and systematic genetic screens. Unfortunately, previous screenings for DNA-repair genes have been limited in scope, resulting in a relatively slow and fragmented understanding of the DNA repair networks within this organism. This limitation largely stems from the fact that only a few research groups have dedicated their efforts to unraveling the complexities of the DNA damage response in this promising system. Nevertheless, for the reasons outlined above, it may be reasonably assumed that *U. maydis* will continue to serve as a significant contributor to the collective understanding of eukaryotic mechanisms underlying genome maintenance.

In recognition of this, methods are described below for:

1. Generating mutants of *U. maydis* using UV radiation.
2. Screening those mutants for putative DNA-repair defects.
3. Identification of the mutated genes by complementation cloning.

The core idea of this approach involves mutating haploid cells to generate various mutations, followed by screening for those mutants that exhibit a DNA-damage sensitive phenotype. The strategy is founded on the principle that the (hyper)sensitivity to DNA-damaging agents is a reliable indicator of defects in DNA repair mechanisms. Isolating mutants that show significant sensitivity to genotoxic agents enables the identification of the mutated genes through complementation with a genomic DNA library.

Finally, the approach is facilitated by the fact *U. maydis* can grow and divide as haploids. This allows for the straightforward identification and isolation of recessive mutations, which are frequently loss-of-function alleles and as such particularly useful in determining the normal function of the gene responsible. In addition, a particular advantage of this method is the potential to reveal hypomorphic alleles of essential genes [9], which would be missed by other approaches like screening deletion collections. So, there are strong grounds to believe that an unbiased mutant screen would provide an unparalleled approach to survey the considerably unexplored *U. maydis* genome for genes involved in DNA repair.

2 Materials

2.1 Strains of U. maydis and Growth Media

1. Frozen glycerol stock of *U. maydis* haploid wild-type strain.
2. YEPS medium: 10 g/L yeast extract, 20 g/L peptone, 20 g/L sucrose.
3. YEPS agar medium (YEPS-A): YEPS plus 20 g/L agar. Weigh and mix yeast extract, peptone, and sucrose in H_2O to a large flask. Add agar after complete dissolution. Autoclave (120°C, 20 min). Allow medium to cool to approximately 50°C before adding the antibiotics or the DNA-damaging chemicals, at the appropriate concentration (*see* **Note 1**).
4. YEPS-A 0.5 M sorbitol medium: YEPS-A plus 91 g/L D-sorbitol.

2.2 Additional Reagents and Equipment

1. Basic equipment for *U. maydis* culture, such as Erlenmeyer flask, bottles, orbital shakers, inoculating loops, plate spreaders, flame for sterilizing the tools, and so forth.
2. Incubators set at 30°C.
3. Centrifuges: Centrifuge (benchtop) and Microcentrifuge.
4. Light microscope equipped with a 40× long-distance working objective.
5. Neubauer hemocytometer.
6. Vortex mixer.
7. 254-nm UV source (e.g., germicidal lamp).
8. UV light dosimeter.
9. Replica plating block-a wooden or metal cylinder that fits closely inside a petri plate.
10. Replica-plating velvets, sterile. These velvets can be washed, autoclaved, and reused.
11. DNA-damage agents such as Methyl methanesulfonate (MMS), Diepoxybutane (DEB), etc.
12. The replication stressor hydroxyurea (HU).

13. Glycerol 50%: prepare 50% solution of glycerol and autoclave (120°C, 20 min). Glycerol is used for cryopreservation of cells (final concentration 25%, freezing at −80°C).
14. Solutions, enzymes, and buffers required for *U. maydis* protoplast preparation and transformation: SCS: 20 mM Sodium citrate pH 5.8, 1 M Sorbitol STC: 10 mM Tris-Cl pH 7.5, 0.1 M $CaCl_2$, 1 M Sorbitol SPTC: 40% PEG3350, 0.1 M $CaCl_2$, 10 mM Tris-Cl pH 7.5, 1 M Sorbitol. Dissolve PEG before adding sorbitol, and heparin (10 mg/mL in H_2O). After filter sterilization, SCS and SCT buffers are stored at 4°C, while SPTC is kept at room temperature.
15. Tubes: microcentrifuge, 1.5 mL, 15 and 50 mL screw cap tubes.
16. 15 × 100 mm plastic Petri dishes.
17. Square petri dishes with grid.

3 Methods

3.1 UV Mutagenesis

3.1.1 Principal Considerations

Most mutagenesis studies in *U. maydis* have historically relied on UV irradiation. The process involves exposing haploid cells to UV radiation, which induces (pre)mutagenic DNA lesions (primarily cyclobutane pyrimidine dimers and 6–4 photoproducts) that can result in mutations [10]. These mutations are generally random.

The technique provides a reasonably simple way to generate mutants, but there are several important methodological considerations to make. First, with the advancements in techniques that permit the identification of genes relevant to specific phenotypes (e.g., cloning, whole-genome sequencing) and the precise manipulation of target genes (e.g., gene replacement strategies and CRISPR–Cas technology), it has become standard practice in the lab to employ doses of UV radiation considerably higher than those traditionally used (*see* **Note 2**). Within the mutagenesis procedure, doses of UV radiation yielding approximately 0.5%–1% survival of *U. maydis* cells inoculated on Petri plates are routinely applied. It is important to determine the UV dose that achieves this survival rate empirically for each particular strain employed in mutagenesis, prior to irradiating a large number of Petri plates. Secondly, if the intention is to subsequently screen for mutants by replica plating, the system should be calibrated to produce around 400–500 survivors per plate. Therefore, based on these two methodological premises, the appropriate number of cells to inoculate per Petri plate is between 5×10^4 and 10^5.

3.1.2 Procedure

1. Starting from a frozen stock, streak the wild-type strain on a YEPS agar plate and allow it to grow for 2 days at 30°C. The streaking should be for single colony isolation.

2. Inoculate 5 mL YEPS with a single *U. maydis* colony; use an Erlenmeyer whose volume is at least ten times the volume of the medium. Incubate the inoculum at 30°C with vigorous agitation (~200 rpm) until the culture is freshly saturated (early stationary phase at a density of $1–2 \times 10^8$, which typically takes at least 16 h).
3. Determine the concentration of suspension by counting cells on a Neubauer hemocytometer. Adjust cell suspension to a final concentration of 10^6 cells/ mL by dilution in sterile H_2O. Following the cell dilution, it is a good practice to verify the cell concentration using a hemocytometer.
4. Plate 100 μL of the 1×10^6 cells/mL water cell suspension on 6 YEPS plates. Use a sterile spreader to evenly distribute the inoculum across the agar surface. It is important to continue spreading until the liquid is completely absorbed and to avoid spreading the inoculum all the way to the edge of the plate. Allow the spread plates to sit for 10–20 min so that the added moisture can dry (*see* **Note 3**).
5. Irradiate each of the plates with different UV doses (150, 200, 250, 300, 350, and 400 J/m^2) (*see* **Note 4**).
6. Incubate plates from Step 5 at 30°C until colonies form (~3–4 days). Observe at which dose such a survival is obtained that about 400–500 colonies are formed per plate.
7. Once the appropriate dose of UV light is determined, prepare 20–30 plates following the steps described above (1–4) and irradiate all of them with the identified dose.
8. Incubate plates at 30°C until colonies form (~3–4 days).
9. Screen colonies for DNA-repair mutants.

3.2 Screening for Mutants by Replica Plating

3.2.1 Principle

Colonies growing on the original (master) plate are transferred by a sterile velvet to one or more secondary (test) plates containing different selective DNA damaging agents. The technique preserves the original spatial pattern of colonies, and the mutants are identified by their failing to appear on test plates. Since a screen typically requires examination of large numbers of candidates, this method developed by the Lederbergs [11] is, indeed, a very convenient way to find among them the few with the desired phenotype. Yet, as noted above, the technique requires master plates containing well-separated colonies (*see* **Note 5**).

3.2.2 Procedure

1. Prepare the appropriate number of test plates, i.e., YEPS plates containing selective DNA-damaging agent (MMS, DEB) or HU. For testing UV-sensitive mutants, regular YEPS plates

are used, which are irradiated with the desired dose of UV immediately after colony transfer.

2. Label the bottom side of all the plates to enable accurate alignment of the master plates with the secondary plates (replicas), and exact identification of colonies after replica plating.
3. Mount a piece of sterile velvet by stretching it on a replica block and secure the velvet to the block by a rubber bend.
4. Invert the master plate and press it down gently onto the velvet; pressing to hard will cause the colonies to run together on the velvet.
5. Remove and save the master plate.
6. Take new plate with genotoxic compound, place it inverted on the velvet, oriented like the master plate, press it lightly and then remove it. As many as three different testing plates per velvet can often be replica plated.
7. Repeat the same procedure for all the plates to be replica plated.
8. Remove excess cells from the replica plates using a toothpick. Always use a new, sterile toothpick for each new spot.
9. Incubate all the plates at 30°C for 2–3 days.
10. Examine the location of each growing colony and compare them to the colonies on the master plate; colonies that appears on the master but disappear from the test plates represent DNA repair mutants. Mutants that exhibit hypersensitivity after 2 days and confirmed after 3 days could be considered as good (putative) candidates.
11. Using a sterile loop or toothpick, gently touch the candidate colony on the original plate and streak the cells onto a fresh YEPS agar plate (*see* **Note 6**).
12. The candidates detected by the screen and purified by single-colony isolation need to be further tested (verified) by standardized semiquantitative pronging survival assays. So, for each candidate start 2 or 3 YEPS liquid cultures of single colony from the same mutant (meaning you pick individual, distinct colonies from the same streak) and incubate them at 30°C. Make sure that the positive and negative controls are included. Determine the concentration of cell suspension of each culture by counting on a Neubauer hemocytometer and adjust cell suspensions to a final concentration of 4×10^7 cells/mL. Prepare a four-step tenfold serial dilution for each culture and spot 10 μL of each dilution onto YEPSA and YEPSA+DNA-damaging compound square plates. Allow the spotted agar plates to dry on the bench for about 10 min before transferring them to the 30°C incubator. Growth is

evaluated at 24, 48, and 72 h. Candidates that are genuine should show a reproducible DNA-damage sensitivity phenotype (*see* **Note 7**).

13. At this stage, create frozen stocks of newly identified/confirmed mutants, ensuring the long-term preservation and future usage. Prepare a YEPS culture for each mutant by picking isolated colonies from the fresh streak plate. Combine equal volumes of the cell culture and 50% glycerol solution in a sterile cryovial, mix well and freeze the mixture at −80°C. Discard the candidates that do not show a reproducible phenotype.

3.3 Identification of the Mutated Genes by Complementation Cloning

3.3.1 Principle

Complementation cloning is a technique that relies on the principle that introducing a specific wild-type gene into a particular/corresponding (recessive) mutant can restore the wild-type phenotype. This restoration indicates that the introduced gene is likely the one that was mutated in the original mutant. For this method to be effective, a well-established protocol for efficient transformation is necessary, along with a comprehensive library of wild-type genes in a self-replicating vector. Importantly, the mutants of interest must exhibit sensitivity to a DNA-damaging agent with a differential from wild-type large enough to enable successful selection for complementation. After cloning the genomic DNA fragment responsible for restoring the wild-type phenotype, this fragment is isolated and sequenced, facilitating the identification of the mutated gene.

3.3.2 Procedure

1. Prepare an overnight logarithmic phase culture by incubation at 30°C on an orbital shaker with an agitation of ~200 rpm. A low-density culture can be used directly for protoplast preparation. If the culture is in the stationary phase, inoculate 30 mL of fresh YEPS medium with 1/50 of the overnight culture, then incubate at 30°C for 3 h before making protoplasts.
2. Transform the protoplasts (~10^6–10^7) with 0.5–1 µg of the DNA library and plate them on selective agar medium containing an antibiotic (*see* **Note 8**).
3. Incubate the plates at 30°C for 2–3 days.
4. Once colonies become visible, replica-plate them onto agar medium containing plasmid-selective antibiotic, along with the appropriate selective agent such as MMS, HU, DEB. The concentration of the selective agent should be determined

based on the sensitivity of the mutated strain and the expected resistance of the transformants carrying the complementing genomic fragment. Incubate the Petri dishes at 30°C for 3–5 days, regularly checking for the growth of resistant clones.

5. Carefully transfer the transformant colonies to a new agar plate supplemented with the appropriate plasmid-selective antibiotic. Use a sterile toothpick or loop to streak the resistant clones for single colony isolation.
6. For an assessment of acquired resistance through complementation, a semiquantitative assay (as described in Step 12 of Sect. 3.2) or, alternatively, a "quick" test may be conducted (*see* **Note 9**). Make sure to include both positive and negative controls in the testing.
7. Isolate the total DNA from DNA-damage resistant clones, from 5–10 mL of overnight cultures. Transform electrocompetent *E. coli* cells using this DNA, and then plate the transformation mix onto LA plates containing plasmid-selective antibiotic (*see* **Note 10**).
8. Isolate plasmids from ten *E. coli* transformants and analyze their restriction patterns. By comparing the shared restriction fragments, one can determine whether the plasmids from different clones of the same DNA-repair mutant contain overlapping genomic segments. This information can help in the further identification of the complementing gene.
9. To further validate that a cloned genomic fragment is indeed complementary, reintroduce the isolated plasmid back into the original mutant strain. If the cloned fragment is indeed complementary, the mutant phenotype should be corrected, and all (re)transformants should be DNA-damage resistant.
10. Using Sanger sequencing with vector primers flanking the cloned genomic fragment, its boundaries and size can be determined via the Mycocosm JGI platform, which hosts the assembled and annotated genome of *U. maydis* (https://mycocosm.jgi.doe.gov/Ustma2_2/Ustma2_2.home.html). The simplest approach is to use the BLAST option: copy each sequence obtained with the flanking primers, and the program will indicate their positions on the chromosome. By identifying the smallest position (start) and largest position (end), one can ascertain the fragment's boundaries, and the program will display all genes located within that genomic fragment.
11. If genomic fragments contain two or more genes, and if the restriction patterns of the plasmids from different clones do not clearly indicate a single-gene overlap then it is necessary to subclone individual genes using suitable restriction digestions (*see* **Note 11**).

12. Once the complementing gene within the fragment is identified, it is necessary to sequence the identified gene from the genomic DNA of the original strain to confirm the existence of the inactivating mutation in the genome. For this purpose, one need to design several pairs of primers that cover the entire open reading frame (ORF), generating amplicons of approximately 0.7 kb each, having overlapping end regions of about 100 bp. This step rules out the possibility that the genomic fragment carries a suppressor gene rather than the complementing one.
13. The final confirmation that the identified mutated gene is responsible for the mutant phenotype involves reconstituting the mutation in a wild-type strain. It is also a common practice to generate a deletion of the gene (using the gene replacement technique) and check for its phenotype.

4 Notes

1. Excessive heat may alter the compounds. To avoid contamination risks associated with a nonsterile thermometer, one simply learns to check the temperature manually. The temperature of the medium is considered not to be too high, if the hot flask could be held comfortably for about 10 s. Although inherently subjective, this manual temperature assessment is in our experience quite reliable for all antibiotics and DNA-damaging compounds used thus far.
2. For methodological reasons, earlier studies had to minimize the number of mutated genes per candidate. This cautious approach involved using relatively modest doses of UV light, with survival rates typically exceeding 10%. Consequently, to identify the mutations of interest, researchers needed to test a large number of candidates. Additionally, the procedure also involved the tedious and time-consuming process of backcrossing mutants with wild-type strains. This was done to purify the genetic background, i.e., to remove any unintended mutations introduced during the mutagenesis. However, with advancements in gene cloning and whole-genome sequencing, random mutagenesis can be performed using higher doses of UV light, allowing investigators to obtain a preliminary pool of desired mutants in a more efficient and less labor-intensive manner. Also, it is not mandatory nowadays to perform the lengthy process of backcrossing. Once the responsible gene is identified, the gene can be directly and singularly mutated in the wild-type background using gene replacement techniques or CRISPR-Cas technology. Mutants

generated in this way are then used for further analyses. Thus, with the aforementioned technological advancements, the somewhat dated practice of forward genetic screen should be more readily employed as a valuable option in the search for novel genes. Indeed, based on the experimental experience accumulated in the lab, renewed adoption of this approach is strongly encouraged.

3. Because freshly prepared agar plates perform much worse than dry plates, plates should be dried by keeping them at room temperature for roughly 24 h. Plates can be dried much faster in the laminar flow hood with the lid slightly off for about 30 min. Once dried, plates can be used immediately or refrigerated in closed bags. When needed, refrigerated plates should be warmed to room temperature prior to use.
4. UV lamp should be warmed up for around 15 min prior to use. Additionally, petri plate lids prevent UV light from passing through, so make sure to take them off before irradiation. CAUTION: Since UV light is mutagenic, wear safety glasses and gloves to protect eyes and skin from UV light.
5. Screening of mutants can also be achieved by manually picking individual colonies with sterile toothpicks and streaking them onto media containing genotoxic agents, as well as onto control YEPS Petri plates. However, this method has several disadvantages, the most significant being that it is quite time-consuming. Only about 50 candidates can be conveniently and reliably tested per 15 × 100 mm plate. Although a template is usually used to guide the markings, just creating a 50-square grid, with consistent spacing and numbering, on the bottoms of numerous Petri dishes, requires a considerable amount of time. However, there are advantages to using this method, particularly because it generates fewer false-positive candidates and reduces the possibility of cross contamination between the mutants. Also, it is especially suitable for screening when there is prior indication that the mutants will exhibit altered colony morphology—such as being small, shriveled, wrinkled, or brownish—which is often the case with DNA repair mutants. Additionally, the time required for the process can be significantly reduced by utilizing repositionable, self-adhesive 50-square Petri dish grid stickers, which are commercially available and can be quickly and neatly attached to the outer bottom of the dish.
6. Since the main risk of the replica plating technique is cross-contamination between colonies on the original plate, it is essential before further testing that the potential candidates are purified by streaking for single colony isolation.

7. Semiquantitative spot survival assays are a relatively quick and simple way to assess sensitivity of mutants to various DNA-damaging agents. They are easy to perform, require ordinary laboratory equipment, and can be used to test a relatively large number of samples. Although they are semiquantitative, they are still sensitive enough to detect relatively small differences in cell survival. It is absolutely mandatory to have a control agar plate (without the compound) and the parental wild-type strain serving as a positive control. A DNA-repair mutant should also be included, serving as a reference and a negative control. The growth of each strain is determined visually at 24, 48, and 72 h and compared to the same strain grown on the control plate. The relative DNA-damage sensitivity of each mutant has to be appreciated in comparison to the parental wild-type control. If a mutant is extremely sensitive to DNA-damage difference in survival can be observed as soon as 24 h, but the difference is generally clearer at 48 h and better to be confirmed at 72 h. Also, it is standard practice and highly recommended to test 2 or 3 separate/distinct colonies for each mutant. This provides more reliable evaluation of the mutant's sensitivity and allows an assessment of the reproducibility of the phenotype. Two or three different concentrations of the compound may be used to observe a dose-dependent effect. Based on our experience and according to previously published data, the following concentrations of the DNA-damaging compounds should be used for preliminary analyses: MMS (0.005%–0.02%), DEB (0.005%–0.01%), and HU (2–4 mM).
8. Given the genome size of 20 Mb and an average genomic fragment size of 6–8 kb, it is crucial to generate more than 3000 transformants to ensure comprehensive coverage of the entire genome. DNA-repair mutants often struggle with protoplast formation and transformation efficiency, necessitating multiple transformations to ensure an adequate number of transformants and that the entire genome is represented. The antibiotic used for selecting the transformants will depend on the selectable marker present in the vector. It is important to note that the concentration of the antibiotic must be higher in agar medium containing sorbitol than in regular medium, due to the protective effects that sorbitol provides. The following concentrations are used for the respective antibiotics: Hygromycin 150 μg/mL, Nourseothricin (NAT) 60 μg/mL, Gentamicin 200 μg/mL, and Carboxin 4 μg/mL.
9. For quick testing prepare a starting solution for each clone, by picking an average-sized colony and suspend it in 1 mL of

sterile water. Vortex the suspension thoroughly. Starting with these suspensions, make two-step tenfold dilutions and spot 10 μL of all these suspensions onto YEPS agar that contains DNA-damaging compounds.

10. This step is necessary for enriching the plasmid, as *U. maydis* cells typically contain only a few plasmid copies. This is the main reason why *U. maydis-E. coli* shuttle vectors are constructed. Standard strains such as DH5α, XL1-Blue, and DH10B can be used for this purpose. It is usually helpful to isolate total DNA from several resistant clones because of the possibility that they contain plasmids carrying different genomic inserts with overlapping regions. Restriction analysis and sequencing of these plasmids can suggest which gene is required for complementation.
11. Sometimes the arrangement of restriction sites is such that individual genes can be easily deleted. If the restriction map is not suitable for easy manipulations, individual genes may need to be PCR-amplified and cloned into overexpression vectors. Regardless of how plasmids carrying individual genes are created, they are individually introduced back into the original mutant and tested for complementation.

Acknowledgments

This work was supported by the Fund of the Ministry of Science, Technological Development and Innovation of the Republic of Serbia, registration number 451–03–136/2025–03/200042 agreement on the implementation and financing of research in 2025.

References

1. Holliday R (2004) Early studies on recombination and DNA repair in *Ustilago maydis*. DNA Repair (Amst) 3:671–682. https://doi.org/10.1016/j.dnarep.2004.02.002
2. Holloman WK, Schirawski J, Holliday R (2008) The homologous recombination system of *Ustilago maydis*. Fungal Genet Biol 45:S31–S39. https://doi.org/10.1016/j.fgb.2008.04.006
3. Azanjac N, Milisavljevic M, Stanovcic S et al (2024) Suppressors of Blm-deficiency identify three novel proteins that facilitate DNA repair in *Ustilago maydis*. DNA Repair (Amst) 140:103709. https://doi.org/10.1016/j.dnarep.2024.103709
4. Kojic M, Kostrub CF, Buchman AR et al (2002) BRCA2 homolog required for proficiency in DNA repair, recombination, and genome stability in *Ustilago maydis*. Mol Cell 10:683–691. https://doi.org/10.1016/S1097-2765(02)00632-9
5. Kojic M, Yang H, Kostrub CF et al (2003) The BRCA2-interacting protein DSS1 Is vital for DNA repair, recombination, and genome stability in *Ustilago maydis*. Mol Cell 12:1043–1049. https://doi.org/10.1016/S1097-2765(03)00367-8
6. Holliday R (1965) Radiation sensitive mutants of *Ustilago maydis*. Mutat Res 2:557–559. https://doi.org/10.1016/0027-5107(65)90022-9
7. Holloman WK, Schirawski J, Holliday R (2007) Towards understanding the extreme radiation resistance of *Ustilago maydis*. Trends Microbiol 15:525–529. https://doi.org/10.1016/j.tim.2007.10.007

8. Kämper J, Kahmann R, Bölker M et al (2006) Insights from the genome of the biotrophic fungal plant pathogen *Ustilago maydis*. Nature 444:97–101. https://doi.org/10.1038/nature05248
9. Milisavljevic M, Petkovic J, Samardzic J et al (2018) Bioavailability of nutritional resources from cells killed by oxidation supports expansion of survivors in Ustilago maydis populations. Front Microbiol 9:990. https://doi.org/10.3389/fmicb.2018.00990
10. Ikehata H, Ono T (2011) The mechanisms of UV mutagenesis. J Radiat Res 52:115–125. https://doi.org/10.1269/jrr.10175
11. Lederberg J, Lederberg EM (1952) Replica plating and indirect selection of bacterial mutants. J Bacteriol 63:399–406. https://doi.org/10.1128/jb.63.3.399-406.1952

Check for updates

Chapter 9

Synchronous Induction and Microscopic Analysis of Hyphal Growth Using Laboratory Strain AB33

Johannes Postma, Pascal Künzel, Nick Wierckx, Kerstin Schipper and Michael Feldbrügge

Abstract

In nature, hyphal growth of *Ustilago maydis* is strictly coupled to plant infection and initiated after mating of two genetically compatible yeast cells on the plant surface. Compatibility is defined by mating type loci: the biallelic *a* locus mediates recognition and fusion through a pheromone/receptor system, whereas the multiallelic *b* locus, encoding transcription factor subunits bEast (bE) and bWest (bW), establishes and maintains the dikaryotic hypha controlled via the assembly of compatible subunits into active heterodimers after plasmogamy. The laboratory strain AB33 enables a controlled transition from yeast-like cells to monokaryotic hyphae in liquid culture without mating. This relies on a genetic trick where the expression of the compatible subunits bW2 and bE1 is under the control of a nitrate-inducible promoter, allowing synchronous hyphae induction by switching from ammonium- to nitrate-containing medium. This system allows the study of processes associated with hyphal growth independently of plant infection and permits microscopic and microfluidic observations, providing a powerful tool for addressing fundamental cell biological questions and beyond. Here, a protocol for synchronous induction of hyphal growth in liquid cultures is described, together with practical guidelines for obtaining high-quality microscopic images using formaldehyde fixation, agarose pad preparation, and mitochondrial staining with TMRE.

Keywords Hyphal growth, Mating type, *b*-genes, Microscopy, Mitochondrial imaging, Formaldehyde fixation, Agarose pad preparation

1 Introduction

Microscopic analyses of fungal hyphae are of huge importance with respect to cell biological questions. These cells provide the advantage of being eukaryotic with dimensions that allow for detailed microscopic studies. Due to its role as one of the best studied basidiomycete fungal model microorganisms and its genetic tractability, *Ustilago maydis* hyphae are of particular research interest. However, in nature, hyphal growth in *U. maydis* is strictly coupled to sexual reproduction. Mating of compatible yeast cells on the surface of maize plants results in the formation of infectious dikaryotic hyphae and is thus

An N. T. Phan et al. (eds.), *Ustilago maydis*, Methods in Molecular Biology, vol. 3050,
https://doi.org/10.1007/978-1-0716-5364-7_9,

prerequisite for the switch to hyphal growth and consequently also for plant infection. Self–nonself discrimination is, hence, crucial to identify mating partners and to guarantee genetic outbreeding in nature, but this makes detailed laboratory study of hyphal growth prohibitively complex.

The determinants of sexual reproduction are very well understood in *U. maydis*: Two genetically unlinked mating-type loci called *a* and *b* are required for mating. The biallelic *a* locus mediates recognition of compatible haploid sporidia with opposite *a* alleles, namely *a1* and *a2*, via a pheromone/receptor system. Perception of compatible pheromone induces the formation of conjugation tubes that grow along the pheromone gradient towards cells secreting compatible peptide pheromone encoded by the opposite *a* allele. Upon physical contact, compatible cells fuse resulting in plasmogamy [1]. Only in this stage, the second mating type locus, the *b* locus, becomes relevant (Fig. 1). This locus is multiallelic and codes for a dimeric homeodomain transcription factor termed bEast/bWest (bE/bW). Only heterodimeric bE/bW dimers produced from different alleles are functional and activate the transcription of a cascade of hundreds of genes for further transcriptional regulators or direct targets [2, 3]. Eighteen *b* alleles have been discovered to date, and only the combination of subunits from different *b* alleles results in a functional transcription factor (i.e., *b1*/*b2* or *b2*/*b20*), while homodimeric transcription factors are inactive (i.e., *b2*/*b2*). Each subunit is composed of a variable N-terminal and a conserved C-terminal region, with the conserved region contain the homeodomain motifs, while the variable regions are characterized by the natural amino acid variations that define the alleles (Fig. 1, top) [2]. Strikingly, a single amino acid variation in bE is sufficient to result in compatibility with the bW from the identical allele.

Detailed knowledge on the mating system has led to the invention of laboratory strains, in which compatible mating type alleles were artificially combined [4, 5]. Among these, laboratory strains AB31 and AB33 are particularly useful for cell biological studies and to dissect hyphal growth uncoupled from plant infections [4]. Both strains originate from the wild-type isolate FB2 (*a2*/*b2*) [6] that

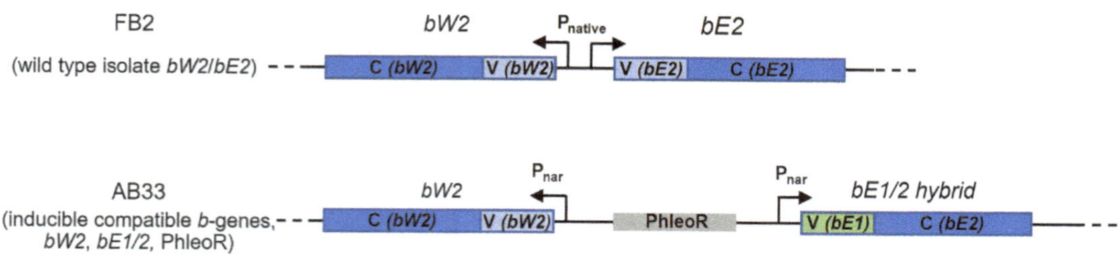

Fig. 1 Genetic architecture of the original and modified mating type locus of AB33. Adapted from Brachmann et al. [4]

carries the allele for bE2 and bW2. In AB31 and AB33 the variable region of $bW2^V$ was replaced by $bW1^V$ and the native bidirectional promoter of the encoding genes was replaced by inducible promoters (Fig. 1, bottom). Expression of the compatible *b* genes in AB31 is controlled by P_{crg}, which is repressed by glucose and induced by arabinose, while AB33 harbors P_{nar}, which is repressed by ammonium and activated by nitrate [4]. In both cases, a simple medium switch induces hyphal growth of previously yeast-like growing haploid sporidia. The resulting monokaryotic fungal hyphae show unipolar tip growth. This growth is characterized by polarized expansion at the cell apex up to a maximal length of about 80 μm, accompanied by vacuolization and septum formation that generates so-called empty sections devoid of cytoplasm, while the nucleus remains centrally positioned (Fig. 2). In essence, these laboratory strains allow for synchronous and controlled hyphal induction and are excellent tools to study the genetic cascade activated by bE/bW, as well as to address cell biological questions like membrane trafficking and mRNA transport in fungal hyphae [7–9]. Here, a detailed protocol for hyphal induction in AB33 is provided (Sect. 3.1).

Hyphal induction in liquid culture offers a major advantage over mating-induced hyphae, as they develop synchronously and without the complexity of cell–cell fusion and the plant context. To fully exploit their potential for live-cell imaging and the study of dynamic cellular processes, immobilization is recommended. This allows for high-resolution observation of individual cells over timeframes of several minutes. Different stainings can easily be applied to dissect

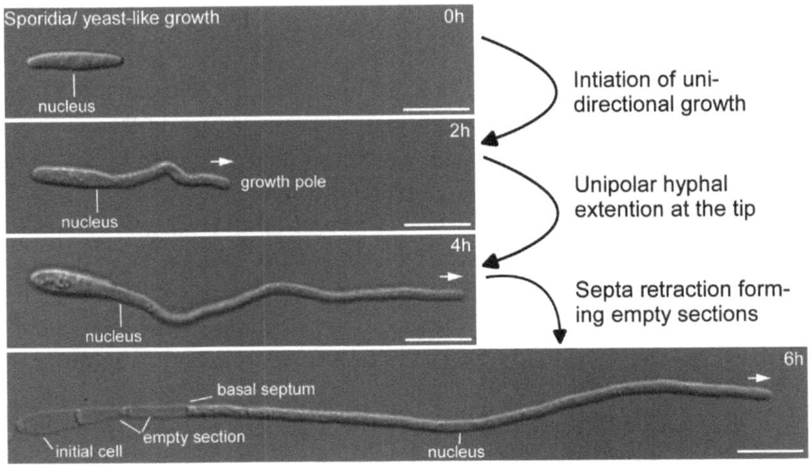

Fig. 2 Time-course analysis of nitrate-induced hyphae formation in strain AB33. The strain was grown in CM-glc until exponential phase in a batch culture. After washing, cells were resuspended in NM-glc, resulting in nitrate-mediated induction of the *nar1* promoter. Micrographs taken at 4 and 6 h postinduction show the progression of hyphal growth. A key morphological feature is the formation of septa, which results in empty hyphal sections that are devoid of cytoplasm. The white arrows indicate the growth direction. Scale bar: 10 μm

the cells in more detail. Moreover, they are suitable for microfluidics applications.

Imaging of AB33 hyphae can be performed with either living or fixed cells. For the latter, fixative agents such as formaldehyde can be used to preserve the native state of the cells without compromising cellular structure. Small cross-linking aldehydes can penetrate the plasma membrane and form intermolecular bridges with free amino groups of soluble proteins. In contrast, precipitating fixatives, such as methanol or acetone destroy membrane structures and may affect the localization and structure of soluble proteins, making them unsuitable for membrane staining and the visualization of genetically encoded fluorescent proteins. Fixation can be useful for halting cell growth, for instance, to capture and preserve distinct stages of hyphal development, and for minimizing secondary effects caused by stress (e.g., oxygen limitation) when there is a long delay between stopping culture growth and imaging. Here, a protocol to stop hyphal growth and preserve cell morphology at high-quality overnight is described (Sect. 3.2). This protocol was adapted from an immunostaining protocol for sporidia and hyphae, as well as for imaging Gfp-tagged proteins in intact hyphae, described by Steinberg et al. [10] and Wedlich-Söldner et al. [11], respectively. Reducing the formaldehyde concentration to 0.5% proved sufficient to completely stop growth and preserve cell morphology overnight while minimizing the loss of genetically encoded fluorophore signal.

Hyphae formation can be examined using a light microscope. This can be done either during or directly after the hyphal induction protocol (Sect. 3.1), or after long-term storage with formaldehyde (Sect. 3.2). In principle, for just a quick check, it is sufficient to directly add 2 μL of cell suspension onto the microscope slide. However, to obtain high-quality images, especially in combination with staining protocols (e.g., Sect. 3.3), it is recommended to image the cells on an even agarose bed (Fig. 3). This ensures that all cells are well distributed and lie in the same optical plane.

Different staining protocols can be applied to visualize cellular compartments in *U. maydis* hyphae. Here, a protocol for the visualization and quantification of mitochondria using the lipohilic cationic dye tetramethylrhodamine ethyl ester (TMRE) is described (Sect. 3.4). TMRE specifically stains active (polarized) mitochondria in a membrane-potential-dependent manner and colocalizes with Gfp translationally fused to a mitochondrial targeting sequence (MTS) (see Fig. 4).

Thus, in addition to the guideline for synchronous hyphal induction (Sect. 3.1), protocols for formaldehyde fixation (Sect. 3.2), agarose pad preparation (Sect. 3.3), and mitochondrial staining with TMRE (Sect. 3.4) are provided, enabling high-quality microscopic analyses of *U. maydis* hyphae.

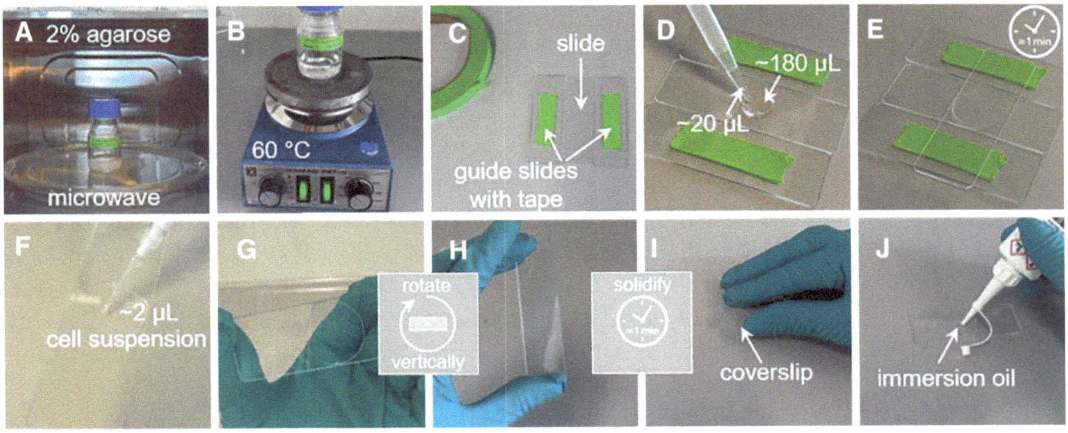

Fig. 3 Preparation of an agarose pad for microscopy. Schematic workflow illustrating the complete procedure, starting with the preparation of 2% (w/v) agarose (**A, B**), followed by the formation of an even agarose pad using guide slides topped with three layers of tape (**C–E**), and concluding with the preparation of a ready-to-use microscope slide with evenly distributed cells (**F–J**). Detailed descriptions of each step are provided in Sect. 3.3

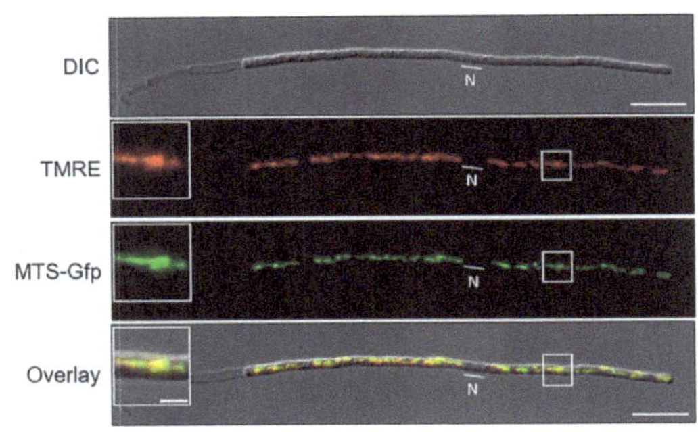

Fig. 4 Microscopic colocalization of genetically encoded mitochondrial marker and TMRE stain. Constitutively expression of Gfp translationally fused to a mitochondrial targeting sequence (MTS) in green overlaps with TMRE staining in red. "N" indicates position of the nucleus. Scale bars: 10 μm (overview), and 2 μm (inset)

2 Materials

2.1 Media

All solutions are prepared using ultrapure water.

Complete Medium (CM): 0.25% (w/v) casaminoacids (Difco), 0.1% (w/v) yeast extract, 1% (w/v) Holliday vitamins, 6.25% (w/v) Holliday salt solution, 0.05% (w/v) Herring sperm DNA (Sigma, D-3159), and 0.15% (w/v) NH_4NO_3. Note that Holliday vitamins and Holliday salt solution are complex mixtures that need to be prepared beforehand. Sterilize by autoclaving at 121°C for 5 min. Adjust the pH to 7.0 with 1 M KOH. Store at room temperature (RT). Caution: The medium needs to be supplemented with the appropriate carbon source prior to use.

Holliday vitamins: 0.1% (w/v) thiamine (Hydrochloride, Serva 36020), 0.05% (w/v) riboflavin (Sigma R-4500), 0.05‰ (w/v) pyridoxine (Monohydr.chl., Sigma p-9755), 0.2‰ (w/v) Ca-pantothenat (Hemi-Ca.salt, Sigma P-2250), 0.05‰ (w/v) aminobenzoic acid (Free Acid, Sigma A 9878), 0.2‰ (w/v) nicotinic acid (free acid, Sigma N-4126), 0.2‰ (w/v) cholinchloride (Sigma C-1879), and 1‰ (w/v) myo-inositol (Sigma I-5125). Store aliquots at −20°C.

Holliday salt solution: 16% (w/v) KH_2PO_4, 4% (w/v) Na_2SO_4, 4% (w/v) KCI, 8% (w/v) $CaCl_2*2H_2O$, 1.32% (w/v) Holliday trace elements, and 2% (w/v) $MgSO_4$. Note that Holliday trace elements is a separate solution. Sterilize by filtering using a 0.22 μm filter. Store at RT.

Holliday trace elements: 0.6% (w/v) H_3BO_3, 0.14% (w/v) $MnCl*4H_2O$, 0.4% (w/v) $ZnCl_2$, 0.4% (w/v) $Na_2MoO_4*2H_2O$, 0.1% (w/v) $FeCl_3*6H_2O$, and 0.04% (w/v) $CuSO_4*5H_2O$. Store at RT.

Nitrate Minimal medium (NM): 0.3% (w/v) KNO_3 and 6.25% (v/v) Holliday salt solution. Sterilize by autoclaving at 121°C for 5 min. Store at RT. Adjust the pH to 7.0 with 1 M KOH. Caution: The medium needs to be supplemented with the appropriate carbon source prior to use.

Ammonium Minimal medium (AM): 0.3% (w/v) $(NH_4)_2SO_4$ and 6.25% (v/v) Holliday salt solution. Sterilize by autoclaving at 121°C for 5 min. Store at RT. Caution: The medium needs to be supplemented with the appropriate carbon source prior to use.

Glucose solution (medium supplement): 1% (w/v) glucose. The solution needs to be filter sterilized (0.22 μm) or autoclaved at 121°C for 5 min.

Arabinose solution (medium supplement): 1% (w/v) arabinose. The solution needs to be filter sterilized (0.22 μm) or autoclaved at 121°C for 5 min.

1× phosphate-buffered saline (PBS): 8 g L^{-1} NaCl, 0.2 g L^{-1} KCl, 1.44 g L^{-1} Na_2HPO_4, 0.245 g L^{-1} KH_2PO_4. Adjust the pH to 7.2 with 1 M HCl.

2.2 Equipment

1. Microscopes used in our lab for differential interference contrast (DIC) or epifluorescence microscopy (*see* **Note 1**).
   (1) Zeiss Axio Observer.Z1 equipped with a CCD camera (Photometrics CoolSNAP HQ2) and objective lenses Plan Neofluar (40×, NA 1.3), Plan Apochromat (63× and 100×, NA 1.4), and α-Plan Apochromat (100×, NA 1.46).
   (2) Zeiss Axio Imager.M1 equipped with a Spot Pursuit CCD camera (Diagnostic Instruments, Sterling Heights, MI) and objective lenses Plan Neofluar (40× and 100×, NA 1.3; 63×, NA 1.25). (https://doi.org/10.1242/jcs.101212)
   (3) OpenFrame Microscopy System from Cairn GmbH equipped with sCMOS sensor KINTEIX-22MM-M-C

(Teledyne Photometrics) and objective lenses Plan Apo (20×, NA 0.8; 60×, NA 1.42).

2. Incubator at 28°C shaking at 200 rpm.

2.3 Strains

Haploid laboratory strains of *U. maydis* that contain inducible *b* alleles like AB31 or AB33 (4).

3 Methods

3.1 Synchronous Induction of Hyphal Growth

Hyphal induction is exemplified in the following using the example of laboratory strain AB33 that carries the P_{nar} promoter to control the *b* genes (*see* **Note 2 and 3**). All steps are conducted under sterile conditions using a laminar flow.

1. Prepare a fresh CM-glc agar plate from glycerol stock and incubate for 2 days at 28°C (*see* **Note 4**).
2. From this plate, inoculate a 3 mL preculture in complete medium supplemented with 1% glucose (CM-glc) using glass tubes and incubate in a rotating wheel at 28°C overnight (*see* **Note 5**).
3. Grow an overnight culture in 10–20 mL CM-glc in a baffled flask filled to maximum 10%. Inoculate the amount to reach an OD_{600} of 0.5–0.8 the next morning (doubling time ~ 2 h). Incubate at 28°C with 200 rpm.
4. Check the OD_{600}, calculate the required volume of the culture to get the final OD_{600} = 0.5 and take it in a 50 mL conical (Falcon) tube (*see* **Notes 6 and 7**).
5. Spin with 1800 × g, 3 min, RT.
6. Discard the supernatant by decanting and remove the residual supernatant completely, e.g., using a 1 mL glass pipette.
7. Add 20 mL Nitrate Minimal medium supplemented with 1% glucose (NM-glc) and resuspend the cell pellet gently by inverting the tubes.
8. Spin with 1800 × g, 3 min, RT (for washing the cells in NM-glc).
9. Discard the supernatant by decanting and remove the residual NM-glc media completely using a 1 mL glass pipette. This is critical for successful hyphal induction, as even trace amounts of residual CM can delay the induction process.
10. Add the respective volume of NM-glc media to obtain a final OD_{600} of 0.5 and resuspend the cell pellet gently by inverting the tube.
11. Transfer the culture into a fresh baffled flask filled to maximum 10%.

12. Incubate the flasks at 28°C with 200 rpm (*see* **Note 8**).
13. Follow hyphal growth microscopically (see example images in Fig. 2 and **Note 9**).

3.2 Formaldehyde Fixation

Due to the toxicity of formaldehyde, all steps are performed inside a fume hood while wearing appropriate personal protective equipment. For detailed handling instructions and disposal requirements, refer to the safety data sheet of the formaldehyde solution.

1. Add 6.76 μL of 37% (w/v) commercial formaldehyde solution directly to 500 μL of cell culture (final concentration: 0.5%) obtained from hyphal growth induction starting at an OD_{600} of 0.5 (see Sect. 3.1; *see* **Note 11**).
2. Incubate for 30 min at 28°C, shaking at 300 rpm or on a rotating wheel.
3. Centrifuge the cells at 2200 × g for 5 min and carefully remove the supernatant.
4. Wash twice with 1 mL phosphate-buffered saline (1× PBS, pH 7.2), to remove excess formaldehyde and to stop the fixation reaction (*see* **Note 12**).
5. Resuspend the pellet in 1 mL 1× PBS.
6. Fixed cells can be stored at room temperature overnight (*see* **Note 18**).

3.3 Agarose Pad Preparation for Microscopy

1. Prepare 2% (w/v) agarose in ultrapure H_2O and heat it in a microwave until fully dissolved (Fig. 3A) (*see* **Note 15**).
2. Keep molten agarose at 60°C on a heating plate while stirring (Fig. 3B).
3. Place the microscope slide between two guide slides, each topped with three layers of tape to define the pad height of about 0.45 mm (Fig. 3C).
4. Pipette 200 μL molten agarose onto the center of the microscope slide, retaining ~10% in the pipette tip to avoid bubble formation (Fig. 3D).
5. Immediately place another microscope slide on top of the molten agarose and allow it to solidify for about 1 min (Fig. 3E).
6. Add 2 μL of cell suspension, avoiding contact between the pipette tip and the agarose pad (Fig. 3F) (*see* **Notes 16 and 17**).
7. Slowly rotate the microscope slide, held vertically, 360° to distribute the droplet and let it stand for ~1 min until partially dry (Fig. 3G, H) (*see* **Note 14**).
8. Place a coverslip at the edge and let it fall into place (Fig. 3I).

9. For oil-immersion objectives, add one drop of immersion oil onto the coverslip (Fig. 3J).

3.4 Mitochondrial Staining with TMRE

Because of the light sensitivity of TMRE, all samples should be protected from light during preparation and incubation, by wrapping tubes in aluminium foil or using black or brown 1.5 mL reaction tubes.

1. Dilute a 10 mM TMRE stock to a 15 μM working solution phosphate-buffered saline (1× PBS, pH 7.2).
2. Add 1.5 μL of the working solution to 1.5 mL of cell suspension in a 1.5 mL reaction tube (final concentration: 15 nM; *see* **Note 19**).
3. Incubate the reaction tube for 20 min at 28°C, shaking at 300 rpm or on a rotating wheel.
4. Harvest cells by centrifugation at 5700 × g for 5 min at room temperature (*see* **Note 10**).
5. Discard the supernatant and wash the cells twice with 1 mL 1× PBS.
6. Resuspend the cells in 1 mL 1× PBS.
7. Perform microscopic observation immediately (*see* **Notes 13 and 20**).

4 Notes

1. Similar microscopic setups can be used to analyze hyphal growth.
2. Strains can be maintained on plates, but only for short-term use. Prolonged growth on solid medium increases the risk of mutations and population heterogeneity, which can compromise subsequent cell biological analyses. For long-term storage, strains should be kept as glycerol stocks at −70 to −80°C. Fresh CM-glc agar plates should be prepared from these stocks and used for no longer than 2 weeks. Check identity of the strains regularly using diagnostic PCR and antibiotic resistances.
3. In case you aim to use AB31 (*b2/b1/2* controlled by P_{crg} promoter), use CM medium for initial growth and induction. Initially, use 1% (w/v) glucose as a carbon source. After washing, replace with 1% (w/v) arabinose. Caution: Shifting the carbon source to arabinose increases the doubling time. Our solution is to duplicate the inoculation volume calculated for 2 h doubling time).

4. Oxygen limitation must be strictly avoided during both preculture and hyphal induction. Therefore, cultures should not be left standing without shaking on rotating wheel or at 200 rpm, and baffled flasks should be used. For main culture and induction in NM-glc medium, flasks should not be filled beyond 10% of their total volume to ensure sufficient aeration.
5. CM-glc medium is preferred for hyphal induction. However, ammonium minimal medium supplemented with 1% (w/v) glucose (AM-glc) can also be used but comes with the disadvantage of very long duplication times.
6. Do not use the culture if the OD_{600} is above 0.8.
7. Culture dilution calculations can be performed with the following equation: $C1V1 = C2V2$, where C1 is the initial concentration, V1 is the initial volume, C2 is the final concentration, and V2 is the final volume.
8. Induction of hyphal growth increases cell adhesion to glass surfaces, which is often visible as a dense layer in the upper part of the baffled flask. This serves as an early indication of successful induction prior to microscopic analysis.
9. For extended live-cell imaging beyond several minutes, microfluidic devices made from PDMS offer a powerful alternative. Continuous media flow enables controlled induction of hyphal growth and allows for precise temporal application of inhibitors or stains. This setup facilitates long-term observation of individual cells and their dynamic responses over several hours.
10. For harvesting *U. maydis* sporidia, centrifugation at $1000 \times g$ is sufficient.
11. For fixation commercially available formaldehyde solutions often contain methanol (ca. 10%) as a stabilizer to prevent polymerization into paraformaldehyde. However, methanol can permeabilize membranes, potentially affecting the staining of membrane-bound proteins. Hence, for applications requiring optimal membrane preservation, using freshly prepared methanol-free formaldehyde solution from paraformaldehyde powder is recommended.
12. Different buffers can be used, but the salt concentration should be isotonic and the pH ideally close to physiological. Avoid amine-containing buffers, such as Tris, as they will react with formaldehyde.
13. For high-quality images and quantitative analysis, immobilization of cells on agarose pads is recommended (Sect. 3.3).
14. Drying time of cells on agarose pads depends on the volume of the cell suspension. If the slide dries out completely, the

coverslip will not stick; if it is not dry enough, the cells will drift during imaging. For the latter, simply wait until the cells have settled; there is no need to prepare a new slide.

15. Agarose can be reused up to five times, but note that repeated boiling can change the concentration due to water loss, which may affect the reproducibility of certain assays. Higher water purity (e.g., Milli-Q) can reduce background noise in sensitive staining. For sensitive cells, a buffer (e.g., minimal salts) can be used instead of water to reduce osmotic stress.
16. Several stains suitable for live-cell imaging can be applied directly on agarose pads (Sect. 3.3) or within PDMS-based microfluidic devices. Effective options include cell wall dyes (e.g., Calcofluor White, Wheat Germ Agglutinin, Congo Red) and membrane-permeable lipid or membrane dyes (e.g., FM4-64, TMRE, CMAC). To ensure high-resolution imaging, unbound stain should be removed by gently washing the sample three times with 1× PBS.
17. For stains that do not penetrate the membrane, such as DAPI, the cells must be chemically fixed before staining, as described in Sect. 3.2.
18. For antibody labeling, the cells need to be permeabilized (e.g., with Triton X-100) (10).
19. To keep consistent quality for each staining experiment, a fresh 10 mM stock of TMRE (aliquots in DMSO stored at −20°C) is recommended.
20. TMRE accumulation depends on the mitochondrial membrane potential, and the resulting fluorescence intensity can be used as a proxy for mitochondrial fitness. For normalization, dyes independent of membrane potential, such as Mitotracker Green FM, can be employed. However, the use of TMRE as a reporter for mitochondrial membrane potential should be evaluated using the uncoupler carbonyl cyanide 3-chlorophenylhydrazone (CCCP, a protonophore) as a control.

Acknowledgments

We are grateful to Simone Krüger and Ute Gengenbacher for excellent technical assistance and to Dr. Andreas Brachmann for conceptualizing and generating the strains with inducible compatible *b* alleles. Thanks to Felix Herrmann for taking photos for Fig. 3. This project was funded in part by the Deutsche Forschungsgemeinschaft (DFG, German Research Foundation)—SFB1535—Project ID 458090666 to NW, KS, and MF (projects A05, B03, and A03, respectively).

References

1. Spellig T, Regenfelder E, Reichmann M et al (1994) Control of mating and development in *Ustilago maydis*. A Van Leeuwenhoek 65:191–197
2. Kämper J, Friedrich MW, Kahmann R (2020) Creating novel specificities in a fungal nonself recognition system by single step homologous recombination events. New Phytol 228:1001–1010
3. Schlesinger R, Kahmann R, Kämper J (1997) The homeodomains of the heterodimeric bE and bW proteins of *Ustilago maydis* are both critical for function. Mol Gen Genet 254:514–519
4. Brachmann A, Weinzierl G, Kämper J et al (2001) Identification of genes in the bW/bE regulatory cascade in *Ustilago maydis*. Mol Microbiol 42:1047–1063
5. Kämper J, Kahmann R, Bölker M et al (2006) Insights from the genome of the biotrophic fungal plant pathogen *Ustilago maydis*. Nature 444:97–101
6. Banuett F, Herskowitz I (1989) Different a alleles of *Ustilago maydis* are necessary for maintenance of filamentous growth but not for meiosis. Proc Natl Acad Sci U S A 86:5878–5882
7. Heimel K, Scherer M, Vranes M et al (2010) The transcription factor Rbf1 is the master regulator for b-mating type controlled pathogenic development in *Ustilago maydis*. PLOS Pathog 6:e1001035
8. Devan SK, Shanmugasundaram S, Müntjes K et al (2024) Deciphering the RNA-binding protein network during endosomal mRNA transport. Proc Natl Acad Sci U S A 121: e2404091121
9. Stoffel NK, Sankaranarayanan S, Müntjes K et al (2025) Microbial iCLIP2: enhanced mapping of RNA-protein interaction by promoting protein and RNA stability. RNA 31:258–272
10. Steinberg G, Wedlich-Söldner R, Brill M et al (2001) Microtubules in the fungal pathogen *Ustilago maydis* are highly dynamic and determine cell polarity. J Cell Sci 114:609–622
11. Wedlich-Söldner R, Bölker M, Kahmann R et al (2000) A putative endosomal t-SNARE links exo- and endocytosis in the phytopathogenic fungus *Ustilago maydis*. EMBO J 19:1974–1986

Chapter 10

U. maydis Protoplast Generation and Transformation

Hariharan V Chinnasamy, Gouri S. Nair and Saravanan Matheshwaran

Abstract

Ustilago maydis is a dimorphic basidiomycete that exists as a yeast-like haploid form and a filamentous dikaryotic form. The yeast-like morphology is exploited for protoplast-based, PEG (polyethylene glycol)-mediated transformation, which is a simple and reliable means for introducing DNA into protoplasts. In *U. maydis*, this approach has been widely used for gene deletion/insertion, complementation, and transgene expression. In short, *U. maydis* yeast-like cells are grown to mid-log phase, and their cell walls are enzymatically digested in the presence of an osmotic stabilizer to generate stable protoplasts. Exogenous naked DNA, usually a linear integration construct with antibiotic resistance cassette or any other selection marker, is incubated with protoplasts, followed by the addition of high-molecular-weight PEG to enable DNA uptake. In this chapter, we point out the theoretical basis and outline of the procedure.

Keywords *Ustilago maydis*, Protoplast, Transformation, Polyethylene glycol (PEG), STC buffer, SCS buffer, Vinoflow, Fungal genetics, Osmotic stabilization

1 Introduction

1.1 Background

The availability of sequenced genomes for many fungi has facilitated the identification of genes whose functions are unknown and require further investigation. Gene deletion is a key strategy for examining the function of specific genes. Practical and straightforward methods are crucial for transforming genetic material to facilitate gene deletion [1]. Additionally, these transformations aid the introduction of mutations in a targeted gene, mutant complementation, insertion of extra gene copies, gene overexpression, conditional gene expression, heterologous gene integration, and creation of marked variant strains. The first stable transformation of a eukaryotic organism was achieved in *Saccharomyces cerevisiae* in 1978 [2]. Shortly thereafter, *Neurospora crassa* was transformed using recombinant plasmid DNA in 1979 [3]. The transformation of the basidiomycete *Ustilago maydis* was first described in 1988. Beyond its role as the pathogen causing corn smut, *U. maydis* has become a model organism for the host–pathogen system for studying plant–fungal interactions [4, 5].

An N. T. Phan et al. (eds.), *Ustilago maydis*, Methods in Molecular Biology, vol. 3050,
https://doi.org/10.1007/978-1-0716-5364-7_10,

Hence, research on *U. maydis* is crucial in understanding plant–fungal interactions and developing disease management techniques, particularly in maize. Additionally, compared with other traditional yeast models, *U. maydis* is evolutionarily closer to humans, due to the presence of many similar proteins and cellular processes (such as long-distance transport, mitosis, and microtubule organization), which makes it particularly useful in exploring aspects of mammalian cell biology that are not well represented in other fungi.

1.2 Transformation

Genetic transformation, or the uptake and expression of an exogenous DNA construct, is a foundational tool in molecular genetics, enabling researchers to introduce foreign DNA for functional analysis, gene editing, and strain improvement. In bacteria and simple yeast, the conventional basic chemical (Ca^{2+}, PEG, or Lithium acetate) or electroporation methods are used to introduce the desired construct. Filamentous fungi typically possess a rigid chitinous cell wall, and to facilitate efficient DNA uptake, this wall must be removed to generate protoplasts; these wall-less cells can then uptake the DNA in the presence of PEG and divalent calcium ion [6]. The first fungal transformations (e.g., *S. cerevisiae* in 1978, *N. crassa* in 1979) used such methods.

Transformation plays a vital role in the research and manipulation of filamentous fungi, and there are multiple techniques and practices available to achieve this. Among the various approaches employed, PEG-mediated protoplast transformation is widely adopted given its relative simplicity and efficiency in facilitating the uptake of DNA through fungal cell walls following enzymatic digestion [6]. Apart from this, another commonly used method is *Agrobacterium*-mediated transformation, which utilizes *Agrobacterium tumefaciens* as a vector to transfer engineered attenuated T-DNA into fungal cells. Although this technique can yield stable and reproducible genetic integration products, it often tends to be time-consuming, requiring multiple intricate steps and careful optimization, and in some cases, may lead to chromosomal rearrangements that interfere with the genetic analysis [7]. Electroporation is also employed as a transformation strategy, where a high-voltage pulse temporarily disrupts the fungal cell membrane to allow DNA entry. This demands precise control over electrical parameters and cell preparation, limiting its applicability across different fungal species [8]. Furthermore, restriction enzyme-mediated integration, which relies on site-specific cleavage of DNA by restriction enzymes followed by incorporation into the genome, has been used but requires careful selection of enzyme type, concentration, and reaction conditions, and is prone to random or off-target insertion events that can compromise the predictability of genetic outcomes [9–11].

In comparison to other methods, PEG-mediated transformation offers significant advantages because of its technical simplicity, use of standard laboratory equipment, straightforward protocols, and results in both high transformation efficiency and stable transformants [12]. It remains the most widely employed method for genetic transformation in filamentous fungi, where transformations are artificially induced by using exogenous DNA and protoplasts stabilized in an osmotic solution. Cooling the protoplasts with divalent cations, like Ca^{2+} (from $CaCl_2$), increases their permeability to DNA, rendering them "competent" for DNA uptake.

U. maydis transformation is an important method that is essential for several experimental approaches, including targeted gene disruptions, protein fusion, and functional complementation. These are used to identify and characterize the effectors and virulent factors, the real-time localization of the intracellular structure over the course of the cell cycle, along with its pathogenesis on the host plant, and complementation to confirm the gene–phenotype relationship.

1.3 Principle

PEG-mediated protoplast transformation involves three main steps—(a) **Protoplast formation**: Enzymatic digestion removes cell walls to expose the plasma membrane; (b) **DNA uptake**: Incubating protoplasts with DNA and PEG in a Ca^{2+}-containing osmotic buffer to enable entry of exogenous DNA; (c) **Regeneration and selection**: Cells regenerate and recover on osmotically favorable transformation media with desirable selectable markers [13].

The primary step is the protoplast formation, which entails removing the cell wall while keeping the plasma membrane intact. Generally, the fungal sample (haploid cells, spores, mycelial cells) is resuspended in a mixture containing the following components:

(i) an enzyme responsible for breaking down the tough cell wall,

(ii) an osmotic balancer (a salt at a concentration complementing the cytoplasm),

(iii) a buffering agent.

Though the composition of these three components may differ depending on the model organism, the enzyme is often considered the most critical element. Usually, the cell wall degradation is achieved using enzymes such as pectinase, zymolase, and lyticase. Critical factors during protoplast preparation include the type and concentration of the enzyme, temperature, incubation time, fungal age, and the enzymatic digestion process. The conditions necessary for preparing protoplasts vary among different fungal species. Furthermore, it has been observed that actively dividing cells (log phase of growth) are the ideal and efficiently undergo lytic digestion, and for *U. maydis*, this typically lies around the OD600 range of 0.8–1.0.

Since the protoplasts lack a cell wall, an osmotic stabilizer is needed to maintain their shape. Therefore, the concentration of the osmotic stabilizer is crucial, as higher concentrations can cause the protoplasts to shrink, while lower concentrations can cause them to burst [14]. A viable protoplast can regenerate on an osmotically favorable medium. Regeneration refers to the ability of a protoplast to newly synthesize the wall material and reform the cell wall. This allows the cells to return to their normal yeast or hyphal form and undergo cell division in the medium.

Initially, PEG was thought to promote the aggregation and fusion of the protoplasts, which would then trap DNA within the fused fungal cells [15]. However, subsequent research has indicated that PEG seldom causes any interaction between DNA and the cell surface and that DNA uptake is not a direct result of protoplast fusion. Rather, PEG disrupts the cell membrane's phospholipid bilayer, thereby facilitating the entry of foreign DNA into protoplasts [12]. Factors such as the concentration of the PEG used, the amount of the DNA, the density of the protoplast, and the duration of transformation can all affect the efficiency of PEG-mediated transformation.

1.4 Selection Strategy

The percentage of cells that successfully uptake foreign DNA during transformation is generally quite low; therefore, it is important to use screening methods that can identify and select for the transformed cells. Primarily, there are two strategies commonly used to screen transformation efficiency and distinguish transformants from non-transformed cells:

1. The first approach involves the use of a *dominant selectable marker*, which allows an otherwise drug-sensitive strain to acquire resistance to a specific antibiotic or chemical agent through the introduced cassette, such that only the transformed cells with the appropriate resistance gene can survive under selective conditions.
2. The second approach is through nutritional complementation, where the introduced DNA cassette restores a missing function in an auxotrophic strain, converting it into a prototroph capable of growing in a nutrient dropout medium.

2 Materials

2.1 Protoplast Preparation

1. D-sorbitol.
2. NaCl.
3. Tris buffer.
4. Polyethylene glycol PEG 4000.
5. Vacuum filter apparatus.
6. 0.22 μm membrane filter.
7. Heparin sodium salt.
8. Centrifuge.

9. Screw-capped centrifuge tubes (50 mL).
10. 500 mL conical flasks.
11. Optical microscope.
12. Pipettes and pipette tips.
13. Petri dishes.

2.2 Transformant Selection

1. Hygromycin B.
2. Carboxin.
3. Geneticin.

2.3 Solutions

NOTE: All buffers/solutions should be sterilized (autoclave) and stored at 4°C unless specified otherwise.

1. **SCS_A buffer**: 1 M Sorbitol, 20 mM Sodium citrate, **pH 5.8**. To make 500 mL of SCS_A, dissolve trisodium citrate (2.95 g) and D-sorbitol (91.1 g) in 500 mL of deionized double-distilled water.
   SCS_B buffer: 1 M sorbitol, 20 mM citric acid. To make 500 mL of SCS_B, add citric acid (1.925 g) and D-sorbitol (91.1 g) in deionized double-distilled water and bring the final volume to 500 mL. Adjust the pH of SCS_A buffer to 5.8 using SCS_B buffer.
2. **STC_A buffer**: 1 M sorbitol, 100 mM $CaCl_2$, 10 mM Tris-HCl, **pH 7.5**. To prepare STC_A buffer (500 mL), dissolve D-sorbitol (91.1 g) in 300 mL of deionized double-distilled water. Once dissolved, adjust the total volume to 445 mL and then autoclave. After the autoclaving process, allow it to cool, then add 50 mL of 1 M $CaCl_2$ solution and 5 mL of 1 M Tris–HCl (pH 7.5) (both preautoclaved).
3. **Buffer_E** (for enzyme mixture): 1 M sorbitol, 25 mM $CaCl_2$, 25 mM Tris-HCl, **pH 7.5**.
   For 100 mL of Buffer E, add 18.2 g of D-sorbitol and dissolve in about 80 mL of deionized double-distilled water. Autoclave after adjusting the volume to 95 mL. After cooling, add 2.5 mL of 1 M Tris–HCl (pH 7.5) and 2.5 mL of 1 M $CaCl_2$.
4. **Vinoflow®** FCE (Novo) for cell wall digestion. Prepare Vinoflow solution at 384 mg/mL in Buffer E.
   To prepare 10 mL, weigh 3.84 g of Vinoflow and dissolve by gentle pipetting in Buffer E, sterilize by passing it through a 0.22 μm filter. Aliquot and store at −20°C. Thawing before use must strictly be done on ice (*see* **Note 1**).
5. **40% Polyethylene glycol in STC:** To make 20 mL, combine 8 g of PEG 4000 with 3.644 g D-sorbitol and 10 mL distilled water and autoclave (*see* **Note 2**). Add 2 mL of 1 M $CaCl_2$, 0.2 mL of 1 M Tris–HCl, pH 7.5 (both preautoclaved), and

sterile deionized water to a volume of 20 mL (*see* **Note 3**). Keep on ice during use.

6. **YPST medium for transformation plates:** 1% yeast extract, 2% peptone, 2% sucrose, 1 M sorbitol, 1.5% agar for solid media. To make 500 mL of YPST, add peptone (10 g), yeast extract (5 g), Sucrose (10 g), and D-sorbitol (91.1 g) and dissolve them. Add 7.5 g of agar and bring the volume up to 500 mL with deionized double-distilled water.
7. **YPSLA medium:** 0.4% peptone, 0.4% yeast extract, 2% sucrose, and 2% agar for plates. To make 500 mL, dissolve yeast extract (2 g), peptone (2 g), sucrose (10 g), and agar (10 g) in approximately 400 mL of deionized double-distilled water, and bring the volume to 500 mL.
8. **YPSL medium:** 0.4% peptone, 0.4% yeast extract, and 2% sucrose. To make 500 mL, add yeast extract (2 g), peptone (2 g), and sucrose (10 g) in about 400 mL of deionized double-distilled water, and make up to 500 mL.

3 Methods

3.1 Protoplast Preparation

1. Inoculate 10 mL of YPSL broth with an isolated colony of *U. maydis* from a freshly streaked YPSL agar plate and add appropriate antibiotics. Incubate for ≥24 h at 28°C and 220 rpm (OD_{600} can exceed 1).
2. Check the culture for contamination by pipetting a 5–10 μL drop of the culture on a slide and covering it using a cover slip. When observed under a microscope (40×), *U. maydis* cells will typically have a cigar shape, while any other contaminants will appear to have a different morphology or have high motility.
3. Inoculate 100 mL of YPSL medium with 1 mL of the overgrown culture ($OD_{600} \geq 1.0$ in 500 mL conical flasks. Incubate for 24 h ($OD_{600} = 0.6$–0.8) at 28°C and 220 rpm.
4. Split the grown culture into 50 mL centrifuge tubes and harvest the cells by centrifuging at 1500–1800 g for 5 min at 18°C.
5. Decant the supernatant and carefully resuspend the pellet in 5 mL of SCS_A buffer (gentle pipetting using cut tips). Pool the cells into a single tube and make up to 30 mL with SCS_A buffer.
6. Centrifuge for 5 min at 1500–1800 g at 18°C to pellet the cells and decant the supernatant.
7. Add 2 mL of SCS_A buffer containing 200 μL of Vinoflow (384 mg/mL in Buffer E) and gently pipette mix for uniform

resuspension; incubate at room temperature with occasional mixing. Monitor the formation of protoplasts microscopically (40×) every 5 min. When 30%–40% of cells exhibit a spherical or swollen ("pinhead-like") morphology, keep the cells on ice (Fig. 1). Over-digestion should be avoided as it compromises protoplast viability.

8. Resuspend gently in 10 mL of ice-cold SCS_A buffer and centrifuge at 1000 g, 4°C for a duration of 5 min. Decant the supernatant (*see* **Note 4**).
9. Resuspend in 5–10 mL of cold SCS_A buffer and gently invert the tubes 5–7 times to remove any traces of the enzyme. Centrifuging at 1000–1500 g, 4°C for 5 min. Discard supernatant and repeat twice.
10. Rinse one time with 10 mL of ice-cold STC_A buffer, centrifuge at 1000–1500 g, 4°C for a duration of 5 min.
11. Resuspend the harvested protoplast in STC_A buffer (1–2 mL) and dispense as 50 μL aliquots in precooled 1.5 mL microcentrifuge tubes. These protoplasts can be used directly for transformation. While fresh protoplasts generally yield higher transformation efficiency, aliquots may be stored at −80°C, after adding DMSO (filter sterilized) with a final concentration of 7%, for later use.

3.2 Protoplast Transformation

1. Prepare the two-layered YPST (YPS-Sorbitol) plates for plating the transformation reaction. The bottom layer contains twice the concentration of appropriate antibiotics, and the top layer (regeneration medium) is without antibiotics. [(Hygromycin B (400 μg/mL) or Carboxin (4 μg/mL) or Geneticin (1 mg/mL)].

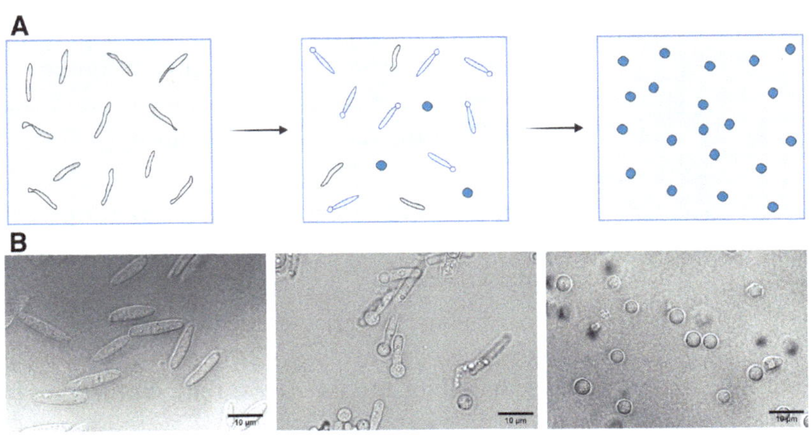

Fig. 1 Stages of protoplast formation: (**a**) Schematic representation of the progression of protoplast formation. (**b**) Microscopic observation of protoplast formation at various time intervals [Magnification—100× and Scale Bar: 10 μm]

2. First, prepare the bottom layer, allow it to solidify, and proceed to the transformation.
3. To the protoplast suspension, add the transformation mixture containing linear DNA (3–5 μg) and 1 μL of heparin (1 mg/mL), without exceeding 5 μL. Incubate on ice for 15 min (*see* **Note 5**).
4. Add 500 μL of 40% PEG/STC and mix by gentle pipetting or by inverting the tube several times. Incubate on ice for 20 min. Meanwhile, pour the regeneration media without antibiotics and allow it to solidify.
5. Plate the transformation reaction on two YPST (YPS-Sorbitol) plates, with 250 μL each, and the plated inoculum must be completely dried inside the laminar hood.
6. Incubate the plates at 28°C. Monitor the plates every 24 h; transformants will typically appear after 3–5 days. The transformants should be patched freshly onto YEPS agar plates containing the respective antibiotic and proceed for screening.

4 Notes

1. In previous reports, Vinoflow was shown to be highly efficient in producing protoplasts compared to Lallzyme MMX (Lallemand Inc., Canada), although needing more prolonged incubation, it yielded protoplasts with high transformation rates. (Lallzyme MMX solution: 0.5 g/mL in Buffer E. To prepare 10 mL, dissolve 5 g of Lallzyme MMX in Buffer E. Gently mix by pipetting. This can be used instantly or aliquoted and stored at −80°C. Thaw at room temperature before use.)
2. This solution is highly concentrated, and proper dissolution might be tricky. To avoid any clumping, combine the PEG 4000, D-sorbitol, and deionized water, then autoclave directly. **Do not make up the volume to the final amount here**.
3. Post-autoclaving, a clear and viscous solution forms, and then the remaining components are added, and then made up to the final volume.
4. Ensure you use a variable speed centrifuge, as centrifuging at high speed at this stage can cause the protoplasts to burst. Gentle handling and use of low temperatures must be ensured in all further steps.
5. We start with 5–10 μg of plasmid DNA to guarantee enough DNA for the transformation after precipitation. Typically, each transformation uses 3–5 μg of linear DNA.

6. Few common error and their possible solutions are:
   i. ***Low protoplast yield***: If the protoplast yield is not good, check the enzyme activity (the enzyme might have lost its activity), culture age (the Culture may be too old or reached late stationary phase, try to harvest cells during log phase), and ensure the buffer osmolarity.
   ii. ***High lysis rate***: The brittle protoplasts can burst if they are centrifuged at high speed or subjected to low centrifugation speed; hence, make sure the stabilizer solution is prepared carefully.
   iii. ***Poor transformation efficiency***: The transformation efficiency varies with the quality of the DNA, PEG concentrations, and incubation time. Try not to keep the cells in the transformation mixture for a longer time, always use high-quality DNA, and use appropriate PEG concentrations.

References

1. García-Pedrajas MD, Nadal M, Denny T, Baeza-Montañez L, Paz Z, Gold SE (2010) DelsGate: A Robust and Rapid Method for Gene Deletion. In: Sharon A (Ed.) Molecular and Cell Biology Methods for Fungi. Methods in Molecular Biology. Humana Press, vol 638 55–76 https://doi.org/10.1007/978-1-60761-611-5_5
2. Hinnen A, Hicks JB, Fink GR (1978) Transformation of yeast. Proc Natl Acad Sci USA 75:1929–1933
3. Case ME, Schweizer M, Kushner SR et al (1979) Efficient transformation of *Neurospora crassa* by utilizing hybrid plasmid DNA. Proc Natl Acad Sci USA 76:5259–5263
4. Steinberg G, Perez-Martin J (2008) *Ustilago maydis*, a new fungal model system for cell biology. Trends Cell Biol 18(2):61–67
5. Wang J, Holden DW, Leong SA (1988) Gene transfer system for the phytopathogenic fungus *Ustilago maydis*. Proc Natl Acad Sci USA 85(865):869
6. Bösch K, Frantzeskakis L, Vraneš M et al (2016) Genetic manipulation of the plant pathogen *Ustilago maydis* to study fungal biology and plant microbe interactions. J Vis Exp 2016(115):54522
7. Bundock P, den Dulk-Ras A, Beijersbergen A et al (1995) Trans-kingdom T-DNA transfer from *Agrobacterium tumefaciens* to *Saccharomyces cerevisiae*. EMBO J 14:3206–3214
8. Weaver JC (1995) Electroporation theory. Concepts and mechanisms. Methods Mol Biol 47:1–26
9. Schiestl RH, Petes TD (1991) Integration of DNA fragments by illegitimate recombination in *Saccharomyces cerevisiae*. Proc Natl Acad Sci USA 88:7585–7589
10. Lu SW, Lyngholm L, Yang G et al (1994) Tagged mutations at the Tox1 locus of *Cochliobolus heterostrophus* using restriction enzyme-mediated integration. Proc Natl Acad Sci USA 91:12649–12653
11. Sweigard J (1996) A REMI primer for filamentous fungi. IS-MPMI Rep Spring 1996:3–5
12. Liu Z, Friesen TL (2012) Polyethylene glycol (PEG)-mediated transformation in filamentous fungal pathogens. Plant Fungal Pathog: Methods Protoc 365–375
13. Schaefer D, Zryd JP, Knight CD et al (1991) Stable transformation of the moss *Physcomitrella patens*. Molec Gen Genet 226:418–424
14. Ning Y, Hu B, Yu H et al (2022) Optimization of protoplast preparation and establishment of genetic transformation system of an arctic-derived fungus *Eutypella* sp. Front Microbiol 13:769008
15. Fincham J (1989) Transformation in fungi. Microbiol Rev 53(1):148–170

Check for updates

Chapter 11

Isolation of Genomic DNA from *Ustilago* Species

Felicia Zlati, Andreas Müsgens, Lars M. Blank, Nick Wierckx and Kerstin Schipper

Abstract

Reliable genomic DNA isolation is essential for investigating the genetic basis of any organism. In fungi such as *Ustilago maydis*, this task is particularly challenging due to the rigid cell wall and the presence of polysaccharides, pigments, and other compounds that can compromise DNA quality as contaminants. This chapter provides practical guidance on overcoming these challenges by detailing robust cell lysis strategies, including enzymatic digestion and mechanical disruption, which form the critical first step in gDNA extraction. Additionally, a set of protocols is presented, tailored to different downstream applications, from PCR amplification to high-throughput short- and long-read sequencing for complex genome assemblies.

Keywords *Ustilago*, Genomic DNA, Sequencing, PCR, Cell lysis, Enzymatic digestion, Mechanical disruption, HMW gDNA

1 Introduction

The complete genome sequence of *Ustilago maydis*, published by Kämper et al. in 2006, was a key milestone in advancing molecular research on this plant pathogenic fungus [1]. This achievement enabled in-depth genomic analyses of previously unknown fungal mechanisms and facilitated rational genetic engineering approaches. Apart from addressing fundamental biology questions, this was particularly valuable for studying and overproducing the diverse range of biotechnological products related to *U. maydis* [1, 2]. Since then, molecular tools for this organism have been extensively optimized, and techniques such as PCR amplification, gene cloning, sequencing, and transformation have become standard practice [3–5]. High-quality genomic DNA is the starting point for all of these molecular biology applications.

Successful extraction of genomic DNA from fungi presents several specific challenges, including the rigid cell wall composed of chitin, glucans, glycoproteins, lipids, and pigments, as well as the

An N. T. Phan et al. (eds.), *Ustilago maydis*, Methods in Molecular Biology, vol. 3050,
https://doi.org/10.1007/978-1-0716-5364-7_11,

presence of nucleases and polysaccharides that can interfere with DNA quality and purity [6]. This chapter describes cell lysis methods such as enzymatic digestion and mechanical disruption, which represent a crucial initial step for gDNA isolation. Cell wall disruption through lyophilization or freezing in liquid nitrogen followed by grinding is also included as an efficient approach for samples that are otherwise difficult to lyse, particularly melanin-producing fungi, including many *Ustilago* spp. [6].

1.1 Genomic DNA for Standard Applications

Isolating genomic DNA is a key first step for a variety of PCR-based applications, including the generation of cloning constructs, colony PCR, and real-time qPCR [4, 7]. Furthermore, genomic sequence information is essential for designing CRISPR–Cas9 target sites, while genomic DNA is required for constructing donor DNA templates used in homology-directed repair (HDR), when applicable [5]. Another major downstream application of genomic DNA isolation is sequencing. Short-read sequencing platforms such as Illumina and Ion Torrent offer very high accuracy rates (>99.5%) and are widely used for applications including whole-genome sequencing, epigenomic analysis, and comparative genomics [8]. For these applications, this chapter describes two commonly used gDNA extraction protocols (Fig. 1). The phenol/chloroform method is cost-effective and does not require a commercial kit. However, it involves handling hazardous chemicals. A second, kit-based protocol offers a safer and more standardized workflow. Both methods yield genomic DNA that is suitable for PCR-based analyses and short-read sequencing.

1.2 HMW Genomic DNA for Long-Read Sequencing

Short-read sequencing technologies are often insufficient to resolve complex genomic features such as long repetitive sequences and structural variants [8]. Long-read sequencing platforms (e.g., PacBio,

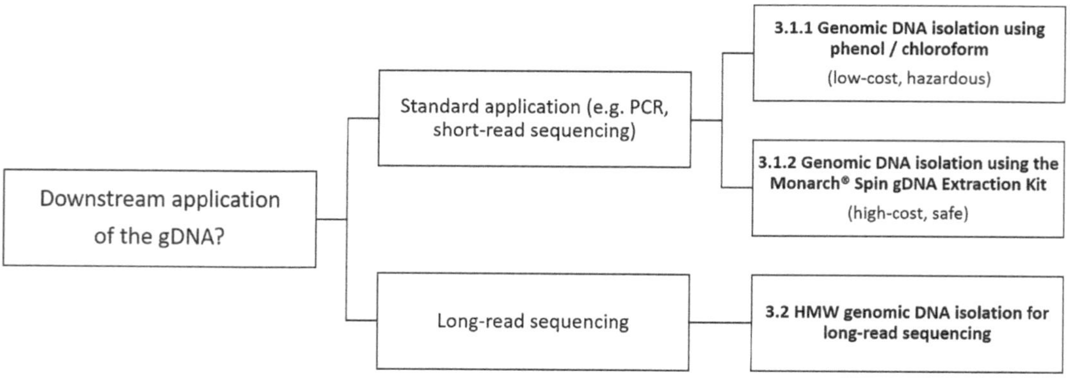

Fig. 1 Flowchart summarizing the genomic DNA isolation strategies described in this chapter. For standard downstream applications such as PCR and short-read sequencing, either the phenol/chloroform method (Sect. 3.1.1) or the Monarch® Spin gDNA Extraction Kit (Sect. 3.1.2) can be used. For high molecular weight DNA required for long-read sequencing, the NucleoBond® HMW DNA Kit protocol is recommended (Sect. 3.2)

Oxford Nanopore) overcome these limitations by producing reads that span several kilobases, enabling accurate assembly of such regions [8]. These advanced technologies are applied in *Ustilago* spp. research for population genomics, phylogenomics, resolving virulence clusters, and performing *de novo* genome assemblies [9, 10]. Long-read sequencing is particularly valuable for complex fungal genomes, as it enables the combination of structural resolution with the high base accuracy of short-read data in hybrid approaches [10]. Consequently, this chapter also includes a protocol specifically designed for isolating high-quality, high-molecular-weight (HMW) genomic DNA suitable for long-read sequencing applications (Fig. 1).

2 Materials

Solutions, media, and buffers should be prepared with high-purity water, obtained from respective purification devices. All chemicals should be purchased in the *per analysis* (p.a.) grade.

2.1 Genomic DNA Isolation for Standard Applications

2.1.1 Solutions for Phenol/Chloroform Genomic DNA Isolation

1) **Phenol/chloroform/isoamyl alcohol**
   Caution: This is a dangerous substance. Strictly apply careful safety measures, including safety goggles, gloves, and a laboratory coat. Use in a fume hood only. Do not inhale or touch without protection. The solution should be stored at 4°C.
2) ***Ustilago*lysis buffer** 10 mM NaCl, 100 mM Tris-HCl, 1% (w/v) sodium dodecyl sulfate (SDS), 2% (w/v) Triton X-100, 1 mM EDTA. Mix the obtained solution 1:1 with Tris-EDTA pH 8.0.
3) **Ethanol***p.a.*
4) **TE/RNase:** 1.31 mM Tris-Base, 8.69 mM Tris-HCl, 10 mM $Na_2EDTA{\bullet}2H_20$. Autoclave, then add 10 mg/mL RNase A. Store at room temperature.

2.1.2 Cell Wall-Degrading Enzyme

A suitable enzyme mix is required to degrade the fungal cell wall when following the **kit-based gDNA isolation** protocol. Commonly used options include:

1) **Zymolyase.**
2) **VinoTaste® Pro** (Novozymes, Lyngby, Denmark).
3) **Lysing Enzymes from *Trichoderma harzianum*** (formerly available from Sigma-Aldrich, Darmstadt, Germany). *Unfortunately, this product has been discontinued. According to*

its product specifications, the key enzymatic activities were cellulase, chitinase, and protease.

Enzyme preparations with comparable activities can be employed as alternatives.

2.1.3 Mechanical Lysis Buffer Solution

Mechanical cell lysis for the **kit-based gDNA isolation** protocol for standard applications requires one of the following buffer solutions:

1) **Phosphate Buffered Saline (PBS):** 8.0 g of sodium chloride, 0.2 g of potassium chloride, 1.44 g of sodium phosphate dibasic, and 0.245 g of potassium phosphate monobasic. Adjust solution to desired pH (typically pH ≈ 7.4). Bring the final volume to 1 L with distilled water.
2) **10 mM Tris–HCl pH 8.0:** 1.21 g of Tris base, adjust the pH to 8.0 with HCl. Bring the final volume to 1 L with distilled water.

2.1.4 Mechanical Lysis Beads

For mechanical lysis via bead beating, one of the following materials is required:

1) **Bead-mill tubes** prefilled with the appropriate type of beads (e.g., MN Bead Tubes Type C, 1–3 mm corundum, suitable for yeast, from MACHEREY-NAGEL, Düren, Germany).
2) **Glass beads** (0.4–0.6 mm diameter) suitable for standard microcentrifuge tubes.

2.1.5 Monarch® Spin gDNA Extraction Kit

The NEB Monarch kit for gDNA extraction (New England Biolabs, Frankfurt am Main, Germany) is recommended following cell lysis. This method proved effective for downstream applications such as PCR and short-read sequencing.

2.1.6 Equipment

The protocols can be carried out using standard laboratory equipment, including:

1) **Centrifuge** compatible with microcentrifuge tubes and equipped with temperature control.
2) **Fume hood.**
3) **Vortex mixer** compatible with microcentrifuge tubes.
    - Alternatively, a **cell lysis** device (e.g., FastPrep®-24, MP Biomedicals, Irvine, USA).
4) **Thermomixer** (e.g., ThermoMixer®, Eppendorf, Hamburg, Germany).

2.2 HMW Genomic DNA Isolation for Long-Read Sequencing

For this protocol, one of the following cell lysis methods should be employed:

- **Enzymatic lysis** using enzymes listed in Sect. 2.1.2.
- **Mechanical lysis** using beads listed in Sect. 2.1.4.

- Lysis by **liquid nitrogen** freezing or **lyophilization**.

2.2.1 NucleoBond® HMW DNA Kit

For long-read sequencing applications, the NucleoBond® HMW DNA Kit (MACHEREY-NAGEL, Düren, Germany) is recommended. This kit yielded high-quality gDNA suitable for long-read sequencing and enabled de novo genome assembly of multiple *Ustilago* spp. [10].

2.2.2 Equipment

1) **Centrifuge** compatible with microcentrifuge tubes.
2) **Vortex mixer** or a **cell lysis** device (see Sect. 2.1.6) for mechanical cell lysis.
3) **Thermomixer.**
4) **Mortar and pestle** for lysis by liquid nitrogen freezing or lyophilization.
5) **Wide bore pipette tips.**

3 Methods

3.1 Genomic DNA Isolation for Standard Applications (e.g., PCR, Short-Read Sequencing)

3.1.1 Genomic DNA Isolation Using Phenol/Chloroform

This protocol has proven proficient for downstream genome sequencing using the Nextera XT kit (Illumina) and a MiSeq sequencer (Illumina, v3 chemistry). The main advantage of this protocol is its cost-effectiveness, as it does not require a commercial kit.

Caution: The protocol involves working with phenol/chloroform, a toxic and dangerous liquid that should only be handled by people with solid laboratory experience. Starting at step 5 the work must be conducted in a fume hood, in combination with careful personal protection measures.

Cultivation in complete medium supplemented with glucose (CM-glc), as described by Holliday (1974) [11], is recommended prior to phenol/chloroform genomic DNA isolation.

1) Inoculate strains from a fresh CM-glc plate in 3 mL CM-glc using glass tubes and incubate at 28°C for 2 days on a rotary wheel.
2) On day 3, use 2 mL of the dense culture, pipette in a 2 mL reaction tube, and add about 200 μL glass beads (*see* **Note 1**).
3) Centrifuge at 16,000 x *g* for 2 min at room temperature.

4) Carefully remove the supernatant with a micropipette and proceed with the pellet.
5) *From now on, work in a fume hood applying careful protective measures*: Add 500 μL phenol/chloroform/isoamyl alcohol (*see* **Note 2**).
6) Add 500 μL *Ustilago* Lysis Buffer, close tightly.
7) Shake the tube for 10 min using a vortex mixer at 1,400 rpm.
8) Centrifuge at 16,000 x *g* for 15 min at room temperature.
9) Carefully transfer 400 μL of the upper, watery phase (without white interphase!) into a new 1.5 mL reaction tube (*see* **Note 3**).
10) Add 1 mL of 100 % ethanol and invert the tube (*see* **Note 4**).
11) Centrifuge for 5 min at 16,000 x *g* at room temperature.
12) Remove the supernatant with a micropipette, shortly spin the tube again (15 s at 16,000 x *g*), and soak off the residual liquid carefully (*see* **Note 5**).
13) Add 50 μL TE/RNase and incubate at 50°C in a thermomixer for about 10 min at 400 rpm agitation.
14) Eventually store the gDNA at −20°C till further use (*see* **Note 6**).

3.1.2 Genomic DNA Isolation Using the Monarch® Spin gDNA Extraction Kit

The following protocol was adapted from the manufacturer's instructions as provided in the Monarch® Spin gDNA Extraction Kit user manual [12]. The adapted protocol was successfully implemented for downstream PCR applications, as well as for short-read genome sequencing on Illumina platforms (*see* **Note 6**).

For this gDNA extraction protocol, *Ustilago* is first cultivated in either complex media (e.g., YEPS [13]) or minimal media (e.g., Modified Tabuchi Medium as described by Geiser et al. [14]). Cell lysis, carried out enzymatically or mechanically, is then required prior to gDNA isolation (*see* **Notes 7 and 9**).

Enzymatic Lysis

1) Harvest approximately 5 • 10^7 yeast cells by centrifugation for 1 min at > 12,000 x *g* (*see* **Note 10**). Discard the supernatant.
2) Resuspend the cell pellet in 100 μL of lytic enzyme solution prepared in digestion buffer (according to the manufacturer's protocol) by vortexing or pipetting up and down.
3) If the lytic enzyme preparation does not already contain RNase A, add 3 μL of RNase A (provided with the kit).
4) Incubate for 30 min at 37°C, or until spheroplast/protoplast formation is complete (can be confirmed by a quick microscopy check).
5) Add 10 μL of Proteinase K (provided with the kit) and vortex briefly.

6) Add 100 µL of Tissue Lysis Buffer and mix thoroughly by vortexing.
7) Incubate at 56°C for a minimum of 30 min in a thermomixer at maximum agitation (~ 1,400 rpm).
8) Proceed to Genomic DNA Binding and Elution according to the kit instructions.

Mechanical Lysis

1) Harvest approximately 5 • 10^7 yeast cells by centrifugation for 1 min at > 12,000 x *g* (*see* **Note 10**). Discard the supernatant.
2) Resuspend the cell pellet in 150 µL of cold PBS or 10 mM Tris–HCl, pH 8.0.
3) Add 150 µL of Tissue Lysis Buffer and mix by vortexing or pipetting up and down.
4) Transfer the suspension to a 1.5 mL microcentrifuge tube containing ~ 100–200 µL glass beads (*see* **Note 1**). Vortex for 15–30 min at maximum speed (~ 2000 rpm), then incubate for 7 min at 95°C in a thermomixer (without agitation).
   - Alternatively, transfer the suspension to a bead-mill tube containing the appropriate amount of beads. Disrupt cells at the appropriate settings (e.g., 30 s at position 6.0 in a Fast-Prep®-24) (*see* **Note 8**).
5) Transfer 200 µL of the homogenized cell lysate to a new tube, taking care not to carry over foam that may have formed.
6) Add 10 µL of Proteinase K (provided with the kit), vortex briefly, and incubate at 56°C for 10 min in a thermomixer at maximum agitation (~ 1,400 rpm).
7) Add 3 µL of RNase A (provided with the kit), vortex briefly, and incubate for a minimum of 5 min at 56°C with vigorous agitation (~ 1,400 rpm).
8) Proceed to Genomic DNA Binding and Elution according to the kit instructions.

Genomic DNA Binding and Elution

After cell lysis, proceed with gDNA isolation and elution by following the instructions from the Monarch® Spin gDNA Extraction Kit. This procedure has yielded gDNA concentrations of up to 50 ng/µL in a 50 µL elution volume (up to 2.5 µg of gDNA) for *Ustilago* spp.

3.2 HMW Genomic DNA Isolation for Long-Read Sequencing

The following protocol was adapted from the manufacturer's instructions for the NucleoBond® HMW DNA Kit, including the cell lysis methods carried out with the kit's buffer solutions [15]. The enzymatic and mechanical lysis approaches are comparable to those described above for gDNA isolation for standard applications.

For optimal efficiency of HMW gDNA isolation, fungal strains should be cultivated in minimal medium (typically 50 mL), such as

Modified Tabuchi Medium with 50 g/L glucose [10, 14], prior to proceeding with this protocol (*see* **Note 11**).

3.2.1 Enzymatic Lysis

1) Weigh 100 mg of zymolyase (1 mg enzyme should correspond to 20–100 U) into a DNase-free microcentrifuge tube. Add 1 mL of Resuspension Buffer HE and mix until the zymolyase is resuspended completely to prepare a stock solution (*see* **Note 12**).
2) Pellet the yeast culture by centrifugation, then transfer up to 50 mg (wet weight) of the pellet into a pre-weighed tube using a spatula (*see* **Note 13**).
3) Add up to 1 mL of Lysis Buffer H1 to the tube. Adjust buffer volume proportionally to pellet size (e.g., resuspend a 25 mg pellet in 0.5 mL of Lysis Buffer H1).
4) Add 100 μL of zymolyase stock solution per 1 mL of Lysis Buffer H1. Adjust proportionally (e.g., add 50 μL stock to 0.5 mL buffer with 25 mg yeast).
5) Vortex well for 5 s.
6) Incubate at 37°C for 2 h, preferably with constant agitation (e.g., 1,400 rpm in a thermomixer).
7) Add fresh Lysis Buffer H1 to reach a final volume of 5 mL.
8) Add 200 μL of liquid Proteinase K (provided with the kit) and mix well.
9) Proceed to Step 3 of the NucleoBond® HMW DNA Kit protocol.

3.2.2 Mechanical Lysis

1) Transfer up to 300 mg (wet weight) yeast sample to a 1.5 mL microcentrifuge tube containing ~ 100–200 μL glass beads (*see* **Note 1**).
   - Alternatively, transfer the sample into a Bead Tube Type C.
2) Add 900 μL of Lysis Buffer H1.
3) Add 200 μL of liquid Proteinase K (provided with the kit) and vortex briefly.
4) Vortex the microcentrifuge tube with glass beads for 15–30 min at maximum speed (~ 2,000 rpm).
   - Alternatively, attach the Bead Tube with a compatible shaker and lyse under one of the following conditions: 5 min at 30 Hz; 30 s at 5 m/s; 5 min at full speed horizontally on a flat bed vortexer (*see* **Note 8**).
5) Centrifuge the Bead Tube/microcentrifuge tube for 2 min at 11,000 x *g*. Transfer the supernatant to a clean 50 mL centrifuge tube using wide-bore pipette tips.
6) Add 4 mL of Lysis Buffer H1 to the sample.
7) Vortex tube for 5 s.

8) Proceed to Step 3 of the NucleoBond® HMW DNA Kit protocol.

3.2.3 Lysis by Liquid Nitrogen Freezing or Lyophilization

1) Pellet the yeast sample and grind the pellet under liquid nitrogen using a mortar and pestle. Alternatively, freeze-dry the yeast pellet using a lyophilizer and subsequently grind it.
2) Transfer up to 30 mg of the ground material into a clean 50 mL centrifuge tube.
3) Add 5 mL of Lysis Buffer H1.
4) Add 200 μL of liquid Proteinase K (provided with the kit).
5) Vortex tube for 5 s.
6) Proceed to Step 3 of the NucleoBond® HMW DNA Kit protocol.

3.2.4 Genomic DNA Extraction and Elution

After completing one of the three lysis methods described above, proceed to Step 3 of the NucleoBond® HMW DNA Kit protocol for gDNA extraction and elution (*see* **Notes 14 and 15**). This approach has yielded gDNA concentrations of up to 250 ng/μL in 150 μL of resuspension buffer (up to 37.5 μg of gDNA) for *Ustilago* spp. The extracted gDNA exhibited high purity, with A260/A280 and A260/A230 ratios of approximately 1.9 and 2.2, respectively. For pure DNA, the expected A260/A280 ratio is 1.8–2.0; lower values indicate protein or phenol contamination [16]. The A260/A230 ratio should be 2.0–2.2, with lower values suggesting residual chemicals such as EDTA or guanidine salts [16]. Ratios above these ranges typically reflect RNA contamination [16]. As shown in Fig. 2, the genomic DNA also displayed a high average fragment length (> 20 kb), making it suitable for long-read sequencing. These

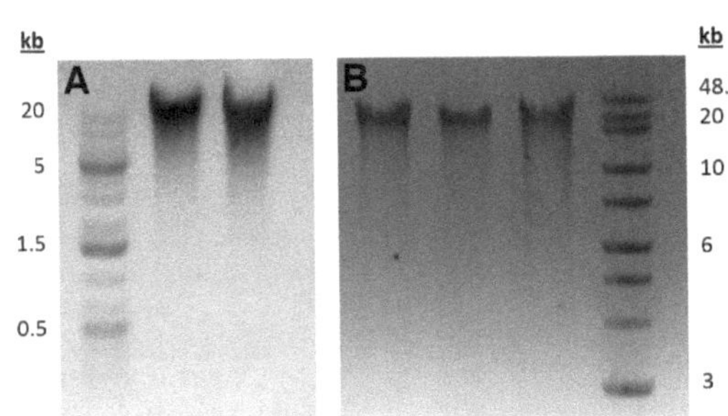

Fig. 2 Genomic DNA samples isolated using the NucleoBond® HMW DNA Kit (mechanical lysis) from MACHEREY-NAGEL (Düren, Germany) analyzed on 1% agarose gels. (**A**) gDNA from *Ustilago triodiae* BRIP 26907a, alongside the GeneRuler 1 kb Plus DNA Ladder (up to 20 kb, Thermo Scientific™, Schwerte, Germany). (**B**) gDNA from *Ustilago maydis* No. 485, *Ustilago maydis* No. 482, and *Sporisorium exsertum* RK 033 (left to right) [10], alongside the Quick-Load® 1 kb Extend DNA Ladder (up to 48.5 kb, New England Biolabs, Frankfurt am Main, Germany)

long-read data, when combined with short-read sequencing, enabled the successful assembly of high-quality *Ustilago* spp. genomes, with N50 values exceeding 1 Mb [10] (*see* **Note 6**).

4 Notes

1. Self-made spoons are very convenient for aliquoting **glass beads** (Fig. 3). For this purpose, use a 1.5 mL reaction tube filled with 200 μL of water. Mark the upper edge of the liquid using a waterproof pen. Remove the water and carefully cut the tube at the marker line, e.g., using a scalpel. Use a 1,000 μL pipette tip to generate a short handle, using a Bunsen burner to carefully melt the two parts together. Eventually, add a second pipette tip for a longer handle. Spilling of glass beads should be avoided due to the resulting slippery surface.
2. When using **phenol/chloroform/isoamyl alcohol** (ROTI®), pipette only the lower, yellowish organic phase, taking care not to disturb the upper aqueous phase.
3. The **phenol/chloroform extraction** step can be repeated to further reduce contaminants and increase DNA purity, if required.
4. A final **salt precipitation** step may be included at the end of the phenol/chloroform protocol to improve DNA purity. For instance, DNA can be precipitated using a solution of ammonium acetate in ethanol to help remove residual impurities.
5. Do not allow the **gDNA pellet** to dry completely during preparation.
6. Carefully check the **quality of the gDNA** before sequencing to ensure reliable results.

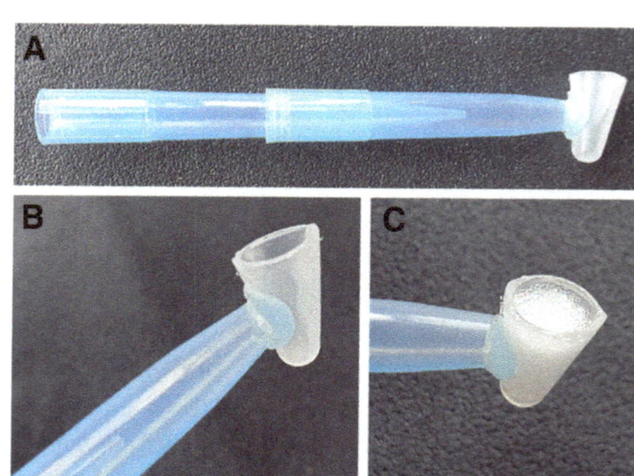

Fig. 3 Self-made device for simple glass bead aliquoting. (**A**) Full device. (**B**) Close-up of an empty device. (**C**) Close-up of the device filled with glass beads (about 200 μL)

7. **Bead beating** generally enhances fungal cell lysis compared to **enzymatic lysis**, resulting in higher DNA yield but shorter average fragment length.
8. For optimal results, the **bead beating** speed and duration should be determined individually for each sample type.
9. If enzymatic and mechanical methods do not provide sufficient cell lysis, **lyophilization (freeze-drying)** or **liquid nitrogen freezing** followed by **grinding** can also be employed for gDNA isolation for standard applications. This approach is particularly effective for otherwise difficult-to-lyse samples and can substantially increase gDNA yield in *U. maydis*.
10. For *Ustilago maydis*, an OD_{600} of 1.0 corresponds to about **1–2 • 10^7 cells/mL** [4].
11. The MN NucleoBond® HMW DNA Kit yielded a significantly more efficient HMW gDNA isolation when fungal samples were obtained from **minimal medium cultures**. Avoid cultivating strains in complex medium prior to using this kit.
12. When using the MN NucleoBond® HMW DNA Kit, a 1 mL stock solution of **zymolyase** is sufficient for 10 reactions, each with ~ 50 mg yeast wet weight. Store the stock at −20°C for up to 6 months. Aliquot to prevent repeated freeze–thaw cycles [15].
13. When performing **enzymatic lysis** with the MN NucleoBond® HMW DNA Kit, the cell material can be pelleted directly in a 50 mL centrifuge tube. Alternatively, the first steps may be carried out in a microcentrifuge tube, with transfer to a larger tube after the 2 h incubation at 37°C.
14. For the MN NucleoBond® HMW DNA Kit, incubation in **Lysis Buffer H1** with Proteinase K at 50°C for ~ 30 min is usually sufficient, but the time can be safely extended up to overnight if needed without compromising DNA integrity or molecular weight [15].
15. Handle isolated HMW DNA with **wide-bore pipette tips** and minimize shearing by avoiding vigorous vortexing and repeated pipetting whenever possible.

Acknowledgments

We thank Simon Wegman for constructive comments and the phenol/chloroform genomic DNA isolation protocol. We further acknowledge Pascal Künzel for his valuable comments and advice.

References

1. Kämper J, Kahmann R, Bölker M et al (2006) Insights from the genome of the biotrophic fungal plant pathogen Ustilago maydis. Nature 444(7115):97–101. https://doi.org/10.1038/nature05248
2. Wierckx N, Miebach K, Ihling N et al (2021) Perspectives for the application of *Ustilaginaceae* as biotech cell factories. Essays Biochem 65(2):365–379. https://doi.org/10.1042/EBC20200141
3. Brachmann A, König J, Julius C et al (2004) A reverse genetic approach for generating gene replacement mutants in *Ustilago maydis*. Mol Genet Genomics 272(2):216–226. https://doi.org/10.1007/s00438-004-1047-z
4. Bösch K, Frantzeskakis L, Vranes M et al (2016) Genetic manipulation of the plant pathogen *Ustilago maydis* to study fungal biology and plant microbe interactions. J Vis Exp 115:54522. https://doi.org/10.3791/54522
5. Schuster M, Schweizer G, Reissmann S et al (2016) Genome editing in *Ustilago maydis* using the CRISPR-cas system. Fungal Genet Biol 89:3–9. https://doi.org/10.1016/j.fgb.2015.09.001
6. Fazio AT, Avanzato MV, Cinto IE et al (2023) A simple method for optimal DNA extraction from different filamentous fungi species growing on earthen walls of "vale historico paulista", Sao Paulo, Brazil. Stud Conserv 68(3):380–387. https://doi.org/10.1080/00393630.2022.2025705
7. Kämper J (2004) A PCR-based system for highly efficient generation of gene replacement mutants in *Ustilago maydis*. Mol Genet Genomics 271(1):103–110. https://doi.org/10.1007/s00438-003-0962-8
8. Goodwin S, McPherson JD, McCombie WR (2016) Coming of age: ten years of next-generation sequencing technologies. Nat Rev Genet 17(6):333–351. https://doi.org/10.1038/nrg.2016.49
9. Schweizer G, Haider MB, Barroso GV et al (2021) Population genomics of the maize pathogen *Ustilago maydis*: demographic history and role of virulence clusters in adaptation. Genome Biol Evol 13(5):evab073. https://doi.org/10.1093/gbe/evab073
10. Ullmann L, Wibberg D, Busche T et al (2022) Seventeen ustilaginaceae high-quality genome sequences allow phylogenomic analysis and provide insights into secondary metabolite synthesis. J Fungi 8(3):269. https://doi.org/10.3390/jof8030269
11. Holliday R (1974) *Ustilago maydis*. In: King RC (ed) Bacteria, bacteriophages, and fungi. Springer, Boston, MA. https://doi.org/10.1007/978-1-4899-1710-2_31
12. New England Biolabs (2024) INSTRUCTION MANUAL Monarch® Spin gDNA extraction kit. https://www.neb.com/en/-/media/nebus/files/manuals/manualt3010.pdf?rev=624b140ff2834dcb9eecffa430909083 Accessed 29 Aug 2025
13. Becker J, Hosseinpour Tehrani H, Gauert M et al. (2020) An *Ustilago maydis* chassis for itaconic acid production without by-products. Microb Biotechnol 13(2):350–362. https://doi.org/10.1111/1751-7915.13525
14. Geiser E, Wiebach V, Wierckx N et al (2014) Prospecting the biodiversity of the fungal family *Ustilaginaceae* for the production of value-added chemicals. Fungal Biol Biotechnol 1: 2. https://doi.org/10.1186/s40694-014-0002-y
15. MACHEREY-NAGEL (2022). User manual nucleo bond® HMW DNA. https://www.mn-net.com/media/pdf/1b/f4/18/Instruction-NucleoBond-DNA-HMW.pdf. Accessed 29 Aug 2025
16. Jaudou S, Tran ML, Vorimore F et al (2022) Evaluation of high molecular weight DNA extraction methods for long-read sequencing of Shiga toxin-producing Escherichia coli. PLOS One 17(7):e0270751. https://doi.org/10.1371/journal.pone.0270751

Check for updates

Chapter 12

An Efficient Colony PCR Method for Construct Verification in the Basidiomycete *Ustilago maydis*

Pascal Künzel, Andreas Müsgens, Lars M. Blank and Nick Wierckx

Abstract

Rapid advances in AI-driven annotation and automation are increasing the demand for fast and reliable screening protocols for reverse genetic approaches. Traditional methods for genomic DNA extraction from the basidiomycete *Ustilago maydis* are often time-consuming and require expensive reagents and hazardous chemicals, making them unsuitable for large-scale mutant screening. Existing colony PCR protocols using crude lysates without DNA precipitation are fast but often fail in terms of reproducibility because cellular components can interfere with polymerase activity or DNA stability. These limitations can be addressed by adapting the yeast GC-prep protocol of Blount et al., based on glass-bead lysis and boiling in the presence of the metal-chelating resin Chelex-100, and combining it with the impurity-tolerant Phire Hot Start II DNA Polymerase. Here, we present a straightforward and reliable protocol for DNA isolation and storage in the screening of *U. maydis* mutants, enabling the reproducible amplification of 0.5–5 kb fragments directly from fresh colonies grown on agar plates.

Keywords Colony PCR, *Ustilago maydis*, Yeast, Fungal genetics, DNA isolation, PCR, Mutant screening, Glass beads, Chelex-100

1 Introduction

The corn smut fungus *Ustilago maydis* is a well-established model organism with a fully sequenced and excellently annotated genome of about 6,900 genes, many of which have been functionally characterized [1]. Its efficient homologous recombination machinery enables targeted genome modifications with >50% efficiency when using 1 kb of flanking sequences upstream and downstream of the target locus [2]. *U. maydis* provides a powerful genetic toolbox for straightforward reverse genetics including promoters, markers, reporters, and landing pads [2, 3]. Functional gene redundancy often requires multiple gene editing events, which can be realized by iterative FLP/FRT marker recycling or multiplex CRISPR-Cas9 targeting [4, 5]. In addition, AI-driven genome annotation and protein structure-function prediction rapidly expand the catalogue of candidate genes and pathways,

An N. T. Phan et al. (eds.), *Ustilago maydis*, Methods in Molecular Biology, vol. 3050,
https://doi.org/10.1007/978-1-0716-5364-7_12,

the demand for integrating high-throughput gene replacement with efficient mutant screening protocols is accelerating.

Screening for correct transformants carrying the desired genetic modification is commonly performed by polymerase chain reaction (PCR). While established protocols for genomic DNA isolation from *U. maydis* yield high-quality DNA, they are often too time-consuming and costly for large-scale applications. As an alternative, colony PCR can be carried out directly from colonies on agar plates by directly using crude cell extract without subsequent DNA purification as template for PCR. Since PCR tolerates cellular impurities such as proteins or cell debris, high-purity DNA is usually not necessary unless downstream applications such as cloning or sequencing are intended. However, published colony PCR protocols for *U. maydis* often suffer from poor reproducibility and are generally not recommended for amplicons larger than 600 bp. Consequently, verification commonly relies on a two-step diagnostic PCR, which consists of an initial colony PCR using crude cell lysates, the first diagnostic PCR, followed by a second PCR using purified genomic DNA, the second diagnostic PCR. The first diagnostic PCR serves as an initial prescreen, using primers within the targeted locus (designated P1 and P2 primers) to yield a product only for the progenitor strain, thereby preventing the erroneous exclusion of correct transformants due to false-negative results. The second diagnostic PCR uses a primer pair located outside the homologous flank and inside the integration

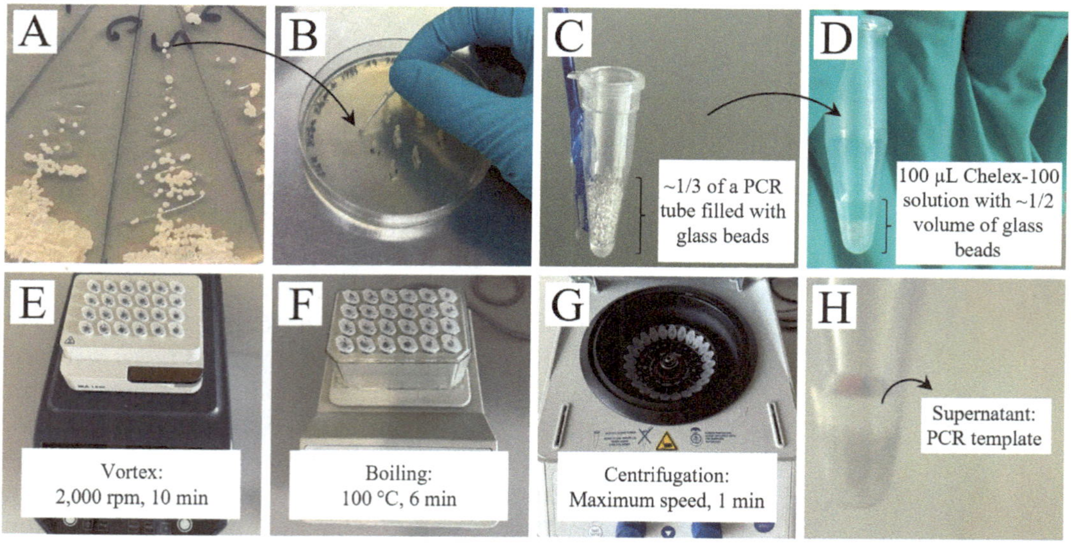

Fig. 1 Experimental workflow of the colony PCR protocol for *Ustilago maydis*. (**A**) A single colony is picked with a sterile toothpick (*see* **Note 1**) and (**B**) transferred to a master plate (*see* **Notes 4 and 5**). (**C**) A PCR tube is filled to one-third with glass beads (*see* **Note 2**), and (**D**) combined with 100 µL of 5% (w/v) Chelex-100 in a reaction tube (*see* **Note 3**). Cell material from the master plate (**B**) is transferred into the prepared reaction tube (**D**) using a 10 µL pipette tip. (**E**) Samples are vortexed at 2,000 rpm for 10 min and (**F**) boiled at 100 °C for 6 min. (**G**) After centrifugation at maximum speed for 1 min, (**H**) the clear supernatant is used directly as a template for the Phire Plant Direct PCR Master Mix (*see* **Notes 6–9**)

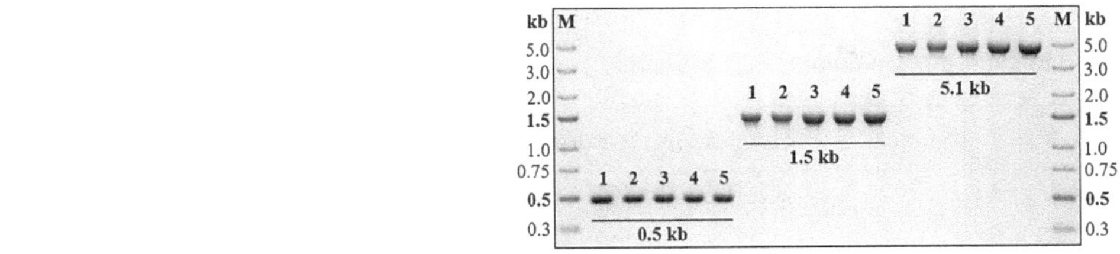

Fig. 2 Evaluation of colony PCR protocol in terms of reproducibility and amplicon size range for five replicates. Three different primer pairs targeting the *upp3* locus (UMAG_11908), all sharing the same forward primer (oPK104) combined with different reverse primers (oPK105–107), yielding amplicons of 0.5 kb, 1.5 kb, and 5.1 kb (see Chap. 2.1). Template DNA obtained with the colony PCR protocol described in this study was directly used with the Phire Plant Direct PCR Master Mix. **M**: O'GeneRuler Express DNA Ladder

construct (e.g., antibiotic resistance gene), thus an amplification product is obtained only for correct transformants [3, 6].

The protocol presented here provides robust and consistent results with amplicons up to 5 kb (see Fig. 2), enabling the direct amplification with primer pairs from the second diagnostic PCR based on colony material, and thereby the identification of correct transformants in a single step. Here, the GC-prep protocol originally developed for yeast by Blount et al., which relies on mechanical disruption with glass beads (see Fig. 1E) and boiling (see Fig. 1F) in the presence of the metal-ion chelator Chelex-100, was adjusted for *U. maydis* [7]. This chelating resin binds divalent cations with high affinity and removes essential cofactors from nucleases, preserving DNA integrity and reducing interference with the PCR. Essential for obtaining robust and reproducible results is the choice of polymerase. Phire Hot Start II DNA Polymerase was selected for its high tolerance to impurities. In addition, the enzyme offers rapid amplification at 20 s per kb and is comparably priced to One*Taq*® DNA Polymerase. Our protocol reliably produces consistent amplicons up to 5 kb (see Fig. 2) and shows low sensitivity to variations in cell number, making it particularly suitable for colony PCR where colony size and manually picked cell material often vary. The method can also be adapted to a 96-well plate format as demonstrated for yeast [7], although this has not yet been systematically tested for *U. maydis*. The protocol is inexpensive, robust, and compatible with high-throughput mutant screening workflows. It consists of only three steps: glass bead lysis, boiling, and centrifugation, after which the supernatant can be directly used as a template in the Phire Plant Direct PCR Master Mix (see Fig. 1).

2 Materials

Chelex-100 solution: Weigh 10 g of Chelex® 100 Sodium form (Merck, 95577-100G-F) into a 50 mL conical tube and fill up with deionized water to a final volume of 50 mL to obtain a final concentration of 5% (w/v).

Table 1
Oligonucleotides used to evaluate the colony PCR protocol

Name	Oligonucleotide sequence 5′ → 3′
oPK104_um11908.fw	GCGTCATTGGTTCTGGAAG
oPK105_um11908.rv-1	GAACAACTCACGACTTTCCAC
oPK106_um11908.rv-2	TTGATGCCAATACCGACTCG
oPK107_um11908.rv-3	GCAAAAAGTCGCAGAAGCAG

Glass beads suitable for 1.5 mL reaction tube: BioSpec Products Mini-BeadBeater Glass Mill Beads, 0.5 mm; 1 lb (Cole-Parmer, 36270-51)
Polymerase: Phire Plant Direct PCR Master Mix (Thermo Scientific™, F160L)

2.1 Oligonucleotides

All oligonucleotides used in this study were obtained from Eurofins Genomics Germany GmbH (Ebersberg, Germany) in salt-free form, and their sequences are listed in Table 1. They were designed to bind to the *upp3* locus (UMAG_11908), which encodes an endopeptidase and has been described as a suitable genomic landing pad for the integration of homologous constructs with stable and homogeneous expression [8, 9]. To evaluate the performance of our colony PCR protocol, amplicons of different lengths were generated from a fresh colony of the *Ustilago maydis* strain MB215 (MiCat #216) by combining the forward primer oPK104 with reverse primers oPK105 (500 bp), oPK106 (1489 bp), or oPK107 (5107 bp) (see Fig. 2).

2.2 Equipment

The protocol described in this study can be carried out using standard laboratory equipment, including:

1. **Microcentrifuge**
2. **Vortex mixer** with microcentrifuge tube adapter
3. **Heating block**

3 Methods

1. Transfer single colonies from an agar plate to a master plate with a numbered grid system (see Fig. 1A, B, *see* **Note 1**). Grow the master plate until the cell material is visible (typically 1–2 days at 30 °C).
2. Pipette 100 μL of a 5% (w/v) Chelex-100 solution into a 1.5 mL reaction tube and add 0.5 mm glass beads until about half of the liquid volume is filled with beads (see Fig. 1C, D;

see **Notes 2 and 3**). Genomic DNA isolation for each clone should be performed in duplicates. Prepare two reaction tubes for each clone. The resulting DNA isolates will later be combined for a single PCR.

3. Transfer cell material from a fresh *U. maydis* master plate into each of the two reaction tubes using a 10 μL pipette tip (see Fig. 1B; *see* **Notes 4 and 5**). Briefly rub the tip between the glass beads and leave the tip in place for at least 1 min.
4. Remove the pipette tip from the tubes and vortex the samples at 2,000 rpm for 10 min in a vortex mixer (see Fig. 1E).
5. Incubate the samples at 100 °C for 6 min in a heating block (see Fig. 1F).
6. Centrifuge the samples at maximum speed (typically 16,000 × *g*) for 1 min in a microcentrifuge (see Fig. 1G).
7. Pool 10 μL of the clear supernatant from each duplicate into one reaction tube (*see* **Note 6**). Use 3 μL as PCR template (see Fig. 1H). The remaining supernatant can be stored at 4 °C as backup.
8. Perform PCR using the Phire Plant Direct PCR Master Mix, following the manufacturer's instructions (*see* **Notes 7–9**). For 12 PCRs, a 100 μL master mix was prepared, containing 71 μL 2× Phire Plant Direct PCR Master Mix, 3 μL of each 10 μM oligonucleotide, and 23 μL nuclease-free water. For each PCR, 7 μL of this master mix were combined with 3 μL of template, yielding a final oligonucleotide concentration of approximately 200 nM and a 1× master mix.
9. Examine PCR products by agarose gel electrophoresis.
10. Positive candidates should be verified by sequencing.

4 Notes

1. Use a sterile toothpick to transfer a small amount of cell material and streak a short line onto the assigned grid position on the master plate. The master plate provides fresh biomass for colony PCR and serves as short-term storage for subsequent inoculations.
2. The required amount of glass beads is approximately one-third of the volume of a standard PCR tube. A practical tip is to create a small scoop by gluing an inoculation loop onto a PCR tube (see Fig. 1C), which facilitates handling and dosing of the beads. Plastic storage bottles with a narrow opening can also be used for precise dosing.
3. Always use safe-lock reaction tubes. Standard tubes will burst open during heating, causing complete sample loss.

4. Alternative tools can be used for colony picking; however, wooden toothpicks may absorb liquid from the reaction mix. A small visible amount of cells at the tip of a pipette tip is sufficient, although in our hands larger amounts of cell material tend to yield more consistent results (personal communication with Felicia Zlati).
5. Fresh colonies harvested on the first day they become visible are recommended; however, the protocol usually works with older plates as well (personal communication with Felicia Zlati).
6. Only use the clear supernatant after centrifugation, as carry-over of Chelex-100 resin may inhibit PCR.
7. The annealing temperatures of Phire Hot Start II DNA Polymerase differ from those of many common DNA polymerases and can be calculated using the Thermo Fisher *Tm Calculator for Primers.*
8. Phire Hot Start II DNA Polymerase yielded distinct amplification products with the tested templates (see Fig. 2), whereas One*Taq*® DNA Polymerase did not (data not shown). The Q5® High-Fidelity DNA Polymerase, in contrast, can be used as an alternative, but requires doubling the elongation time (personal communication with Florian Altegoer).
9. There are several strategies for primer design, depending on the type of genetic manipulation to be performed. For homologous recombination-based knock-outs or knock-ins, the best practice is to design primers that bind outside the homologous flanks in the genomic region, and to verify the resulting amplicon by sequencing. Always use the progenitor strain as a control to ensure that the bands differ in size and to test for heterogeneity. For the latter, an additional control PCR can be performed with a primer that binds inside the native locus in combination with a primer outside the integration site, which should yield a product in the progenitor strain only. If multiple copy insertions are expected, or if the construct is too large, one primer can be placed outside the flank and the other within the integration cassette. Testing for multiple copy insertion requires quantitative PCR or Southern blot analysis.

Acknowledgments

The authors thank Felicia Zlati, Lorenzo Garcia Tejada, Florian Altegoer, and Kerstin Schipper for sharing protocols and providing valuable advice, and Elena von Helden for helpful comments on the manuscript. This work was supported by the Deutsche Forschungsgemeinschaft (DFG, German Research Foundation)—SFB 1535—Project ID 458090666.

References

1. Kämper J, Kahmann R, Bölker M et al (2006) Insights from the genome of the biotrophic fungal plant pathogen *Ustilago maydis*. Nature 444:97–101. https://doi.org/10.1038/nature05248
2. Terfrüchte M, Joehnk B, Fajardo-Somera R et al (2014) Establishing a versatile Golden Gate cloning system for genetic engineering in fungi. Fungal Genet Biol 62:1–10. https://doi.org/10.1016/j.fgb.2013.10.012
3. Brachmann A, König J, Julius C et al (2004) A reverse genetic approach for generating gene replacement mutants in *Ustilago maydis*. Mol Genet Genomics 272(2):216–226. https://doi.org/10.1007/s00438-004-1047-z
4. Khrunyk Y, Münch K, Schipper K et al (2010) The use of FLP-mediated recombination for the functional analysis of an effector gene family in the biotrophic smut fungus *Ustilago maydis*. New Phytol 187(4):957–968. https://doi.org/10.1111/j.1469-8137.2010.03413.x
5. Schuster M, Schweizer G, Reissmann S et al (2016) Genome editing in *Ustilago maydis* using the CRISPR-Cas system. Fungal Genet Biol 89:3–9. https://doi.org/10.1016/j.fgb.2015.09.001
6. Bösch K, Frantzeskakis L, Vraneš M et al (2016) Genetic manipulation of the plant pathogen *Ustilago maydis* to study fungal biology and plant microbe interactions. J Vis Exp 115:e54522. https://doi.org/10.3791/54522
7. Blount BA, Driessen MRM, Ellis T (2016) GC preps: fast and easy extraction of stable yeast genomic DNA. Sci Rep 6:26863. https://doi.org/10.1038/srep26863
8. Müntjes K (2020) Towards visualizing translation by single-molecule imaging of nascent peptides within hyphae of *Ustilago maydis*. Dissertation, Heinrich Heine University Düsseldorf
9. Müntjes K, Magnus P, Hüsemann L et al (2020) Establishing polycistronic expression in the model microorganism *Ustilago maydis*. Front Microbiol 11:1384. https://doi.org/10.3389/fmicb.2020.01384

Southern Blotting: A Valuable Molecular Technique to Verify Gene Replacement in *Ustilago maydis*

Shivam Chaudhary, Lukas Dorian Dittiger, and Jan Schirawski

Abstract

One of the most often applied procedures in the quest for gene function analysis in *Ustilago maydis* is to replace the target gene with an antibiotic resistance cassette by double homologous recombination with a linear gene replacement construct. Unfortunately, many unwanted events can also happen, which would go unnoticed if relying on PCR to verify the resulting strains. Therefore, it is recommended to verify the proper gene replacement using the easily doable and highly informative Southern blotting technique. This chapter provides a detailed and robust protocol for the verification of proper gene replacement in *U. maydis* mutants by non-radioactive Southern blotting.

Keywords Gene replacement, Southern blotting, *Ustilago maydis*, Genomic DNA, Capillary blot, Digoxigenin, Non-radioactive, Agarose gel, Nylon membrane

1 Introduction

To investigate the function of a gene, it is common to replace the gene of interest with an antibiotic resistance cassette [1–3]. To confirm the correct integration of the marker at the intended locus, it is tempting to use PCR since it is a fast and easily conductible experiment. However, even a well-designed PCR approach is not sufficient to unambiguously prove the intended modification. For example, it often fails to detect ectopic integrations, multiple integrations, or partial integration of the gene replacement construct (Fig. 1). Proper verification of the genomic modification is essential to draw conclusions on gene function, as additional ectopic construct integration might lead to phenotypes caused by off-target gene disruption.

To circumvent the limitations of PCR-based strain verification, Southern blotting can be a valuable, easily doable, as well as informative procedure [4–6]. It relies on the detection of size differences of restriction fragments generated by a total digest of genomic DNA. The procedure can be subdivided into several steps, namely,

An N. T. Phan et al. (eds.), *Ustilago maydis*, Methods in Molecular Biology, vol. 3050,
https://doi.org/10.1007/978-1-0716-5364-7_13,

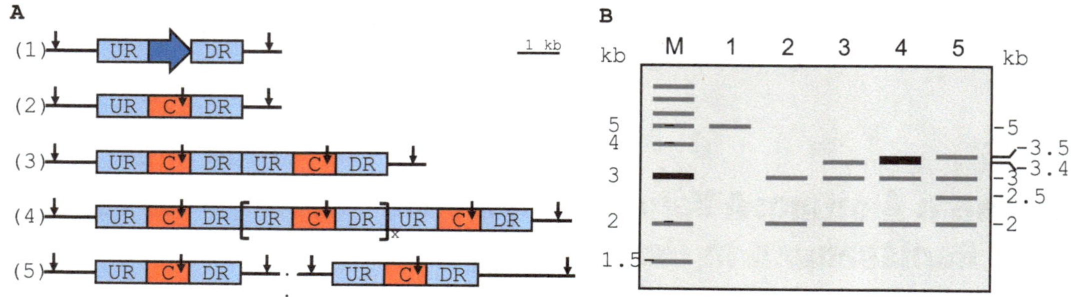

Fig. 1 Identification of unintended events occurring during gene replacement in *Ustilago maydis* that are easily detectable by Southern blotting. (**A**) Schematic representation of gene replacement as well as exemplary additional integration events of the gene replacement construct into the targeted or an ectopic locus. (1) Native locus containing the upstream region (UR), the gene of interest (blue filled arrow), and the downstream region (DR). (2) Successful replacement of the native gene with a gene replacement construct containing UR, an antibiotic resistance cassette (C), and DR. (3) Single integration of an additional gene replacement construct downstream of the correctly positioned gene replacement construct. (4) Integration of multiple additional gene replacement constructs downstream of the correctly positioned gene replacement construct. (5) Additional ectopic integration of the gene replacement construct in an unrelated locus of the genome. Strains carrying the gene replacement construct would all be identified as clean gene replacement strains using a PCR approach designed to identify correct recombination in UR and DR. Figure drawn to size. The vertical arrows indicate cutting sites of the chosen restriction enzyme. (**B**) Schematic result of a developed Southern blot of strains carrying the loci described in (**A**). When using the gene replacement construct as a probe and an enzyme cutting once in the construct but not in the gene, the expected banding patterns of the example depicted in (**A**) are as follows. A single band at 5 kb for the progenitor strain (1), two bands (at 3 kb and 2 kb) for proper gene replacement strains (2). Appearance of an additional band at the size of the full gene-replacement construct (3.4 kb in this example) indicates a single additional integration (3), while appearance of a very prominent band at the size of the full gene-replacement construct indicates multiple additional integration events. Additional ectopic construct integration leads to the appearance of two additional bands (in this example, at 3.5 kb and 2.5 kb). M, size marker. Band sizes of the marker are indicated on the left, and expected band sizes of the investigated strains are indicated on the right

restriction digest of the genomic DNA, size separation of the restriction fragments by agarose gel electrophoresis, transfer of the DNA fragments to a positively charged nylon membrane, hybridization to a labeled DNA probe, followed by probe detection [5].

In this chapter, it is described how to design the probe and choose a reasonable restriction enzyme, as well as which control to include, and give a detailed protocol to perform Southern blotting using non-radioactive probe labeling to verify gene replacement strains of *Ustilago maydis*.

2 Materials

All solutions are prepared in deionized water and are stored at room temperature unless specified otherwise.

2.1 Genomic DNA Isolation

1. Potato Dextrose (PD) agar: 2.4% (w/v) PD broth, 10 mM Tris–HCl at pH 8, 2% (w/v) agar. Autoclave for 20 min at 121°C.
2. YEPS-light medium: 1% (w/v) yeast extract, 0.4% (w/v) Tryptone, 0.4% (w/v) Saccharose. Autoclave for 20 min at 121°C.
3. Sterile glass beads (0.25–0.5 mm) (*see* **Note 1**).
4. Ustilago lysis buffer: 10 mM Tris HCl at pH 8, 100 mM NaCl, 1 mM EDTA at pH 8, 35 mM SDS, 34 mM Triton X–100.
5. 1x TE buffer at pH 8: 10 mM Tris base, 1 mM Na_2-EDTA.2H_2O. Adjust the pH to 8.0 with HCl. Autoclave at 121°C for 20 min.
6. TE-Phenol-Chloroform: 40% Phenol, 40% Chloroform, 20% 1x TE buffer at pH 8. Prepare freshly one day before use and store overnight at 4°C to separate phases.
7. IKA VXR basic Vibrax® (IKA-Werke GmbH & Co. KG, Cat. #: 0002819000).
8. Pure ethanol: 99.8% (v/v) ethanol.
9. 70% ethanol: 70% (v/v) ethanol.
10. TE-RNaseA: 50 μg/ml RNaseA in 1x TE buffer at pH 8.

2.2 Probe Generation

1. Sodium chloride (NaCl).
2. DIG-High Prime (Roche, Cat. # 11585606910).
3. 0.5 M EDTA at pH 8.0.

2.3 DNA Digestion and Precipitation

1. Restriction enzyme.
2. Restriction enzyme buffer (Manufacturer's choice).
3. 3 M Sodium acetate at pH 5.2.
4. Pure ethanol: 99.8% (v/v) ethanol.
5. 70% ethanol: 70% (v/v) ethanol.
6. 1x TE buffer at pH 8: 10 mM Tris base, 1 mM Na_2-EDTA.2H_2O. Adjust the pH to 8.0 with HCl. Autoclave at 121°C for 20 min.
7. 6x DNA loading dye: 1.46 M Sucrose, 1.87 mM Bromophenol blue.
8. 1x TAE buffer: 40 mM Tris base, 20 mM Acetic acid, 1 mM EDTA at pH 8.0. Autoclave at 121°C for 20 min.
9. Agarose.
10. DNA stain.
11. 1 kb DNA ladder.

2.4 Capillary Blot Setup

1. 0.25 M HCl.
2. 0.4 M NaOH.
3. Plastic tray.
4. Blotting sheets (e.g., Whatman 3MM Blotting paper).
5. Microcentrifuge tube storage box.
6. 2X SSC buffer: 300 mM NaCl, 30 mM trisodium citrate dihydrate.
7. Positively charged nylon membrane.
8. Paper towels.
9. Parafilm.

2.5 Probe Hybridization

1. Sodium phosphate buffer 1 M at pH 7.0: Add 1 M NaH_2PO_4 to 1 M Na_2HPO_4 until the pH reaches 7.0.
2. Southern hybridization buffer: 50% (v/v) 1 M Sodium phosphate buffer at pH 7, 35% (v/v) 0.694 M SDS. Keep at 37°C to avoid salt precipitation.

2.6 Signal Detection

1. Southern wash buffer: 10% (v/v) 1 M Sodium phosphate buffer at pH 7, 5% (v/v) 0.694 M SDS.
2. DIG I buffer: 0.1 M Maleic acid, 0.15 M NaCl. Adjust to pH 7.5 with appx. 200 mL of 5 M NaOH.
3. DIG wash buffer: 0.3% (v/v) Tween20 in DIG I buffer.
4. Blocking solution: 10% (w/v) blocking reagent in DIG I buffer. Resuspend while heating. Autoclave at 121°C for 20 min.
5. DIG II buffer: 10% (v/v) blocking solution in DIG I buffer. Prepare fresh before use.
6. Antibody solution: 0.01% Anti-Digoxigenin-AP, Fab fragments in DIG II buffer.
7. DIG III buffer: 0.1 M NaCl, 0.0234 M $MgCl_2.6H_2O$. Adjust to pH 9.5 with appx. 500 mL of 1 M Tris–HCl at pH 9.5.
8. CDP-Star solution: 10% (v/v) CDP-Star in DIG III buffer. Store at −20°C.

2.7 Laboratory Equipment

1. Thermal cycler.
2. Tabletop centrifuge.
3. Agarose gel electrophoresis system.
4. Agarose gel imaging system.
5. Hybridization oven and UV crosslinker.
6. Hybridization tubes.
7. Forceps.
8. Chemiluminescence imaging system.

3 Methods

3.1 Experimental Design

The method relies on fragment size differences between the progenitor strain and the mutant. These differences should be reliably detectable, and fragment sizes should be in a range that can be separated by agarose gel electrophoresis. Therefore, the choice of the restriction enzyme depends on the targeted locus that is specified by the probe and requires some thought.

It is recommended to use the gene replacement construct as a probe (*see* **Note 2**). The gene replacement construct usually consists of an upstream homologous region, a resistance cassette, and a downstream homologous region [1, 2]. The gene replacement construct can either be directly labeled using random hexamer-directed DIG High Prime labeling, or can serve as a template for PCR labeling using the outside primers used for gene replacement construct generation in the presence of DIG-dUTP in the nucleotide mix.

To detect difference in the restriction pattern detectable with this probe, look for enzymes that result in fragments ranging in size between 1 and 10 kb and lead to a different restriction pattern between your progenitor strain and the resulting strain. A good idea is to select enzymes that cut once or several times in your gene of interest and not at all in the gene replacement construct, or to choose enzymes that cut once or twice only in the resistance cassette but not in the homologous arms of the replacement construct and not in the gene of interest (Fig. 1A). Such enzymes will lead to a pattern with a difference in the number of expected fragments, which may be easier to detect than having to rely on small restriction size differences (Fig. 1B; *see* **Note 3**).

The design can be easily adapted to other genome modifications like gene insertion in a defined locus, or any modification leading to the appearance or disappearance of restriction sites.

3.2 Isolation of Genomic DNA

1. Freshly streak your putative mutant and progenitor strains from their stock at −80°C onto PD-agar plates and incubate for two days at 28°C.
2. Use a single colony to inoculate 5 ml of YEPS-light medium and incubate until you have a turbid culture.
3. Pellet cells at 16,000 × *g* for 2 min, add ~100 μl of glassbeads (*see* **Note 4**, Fig. 2) and freeze at −80°C for at least 1 h.
4. Add 600 μl of TE-Phenol-Chloroform and 500 μl of Ustilago lysis buffer. Vibrax for 10 min at maximum speed at room temperature.
5. Centrifuge the mixture at 16,000 g for 25 min at 4°C.
6. Carefully transfer the upper phase to a labeled 1.5 ml microcentrifuge tube. Do not disturb the interphase (*see* **Note 5**).

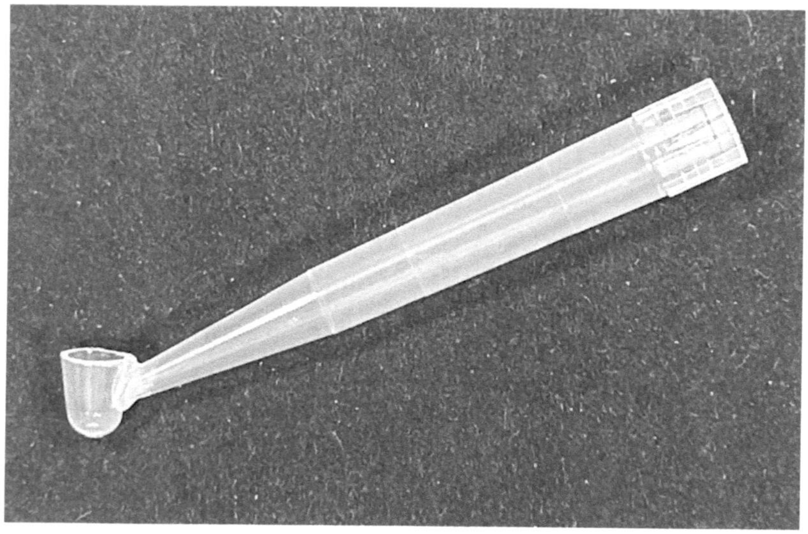

Fig. 2 Picture of a self-made spatula useful for aliquoting 100 µl of glass beads. The spatula is made by cutting the bottom of a 1.5 ml plastic microcentrifuge tube at the 100-µl mark to generate a small cup, and gluing the cup to a 1 ml plastic pipette tip that has been melted at the slim end prior to cup attachment

7. Add 900 µl of pure ice-cold ethanol, mix by inverting several times, and centrifuge at 16,000 g for 15 min at 4°C.
8. Discard the supernatant and resuspend the pellet in 500 µl 70% ice-cold ethanol.
9. Centrifuge at 16,000 × *g* for 5 min at 4°C. Discard the supernatant and let the pellet dry for 15 min at room temperature.
10. Resuspend the pellet in 50 µl TE-RNaseA and incubate at 50°C for 10 min gently shaking.
11. Check the integrity and concentration of the isolated genomic DNA by agarose gel electrophoresis (*see* **Note 6**). Store the isolated gDNA at 4°C for up to one month, or at −20°C for long-term storage, or directly proceed with the next step.

3.3 DNA Digestion and Precipitation

1. Set up a restriction digest reaction with approximately 5 µg of genomic DNA in a total volume of 100 µl using 1 µl of your chosen restriction enzyme and incubate overnight. Choose the buffer as well as the incubation temperature as recommended by the manufacturer of the enzyme.
2. Digested DNA samples need to be precipitated and concentrated so that they can fit into the pockets of an agarose gel. For this, add 300 µl of the pure ice-cold ethanol and 10 µl of ice-cold 3 M sodium acetate at pH 5.2 to the digested samples. Mix well and centrifuge at 16,000 × *g* for 30 min at 4°C.

3. Resuspend the pellet in 500 μl of 70% ethanol and centrifuge at 16,000 g for 5 min at room temperature. Carefully remove the supernatant so as not to disturb or lose the DNA pellet.
4. Dissolve the pellet in 15 μl of 1x TE buffer at pH 8.0 and mix with 3 μl of 6x DNA loading dye.
5. Transfer the gDNA samples into the wells of a 0.7% TAE agarose gel prepared in 1x TAE buffer and run at 6 V/cm for about 90 min until the loading dye has crossed 2/3 of the total gel length. Image the gel using a suitable imaging system before proceeding with the capillary blot to confirm proper digest of the genomic DNA (*see* **Note 7**).

3.4 Probe Generation

1. To obtain the template for probe generation, first amplify your gene replacement construct using PCR and gel-purify the amplified fragments with an appropriate gel purification kit.
2. Use a total of 500 ng of the probe template and dilute with deionized water to a final volume of 16 μl.
3. Denature the DNA by incubating at 95°C for 5 min followed by quick-cooling on ice/NaCl for 1 min.
4. Add 4 μl of DIG-High Prime and incubate at 37°C overnight.
5. To stop the reaction, add 1 μl of 0.5 M EDTA at pH 8.0 and incubate for 10 min at 65°C. Store the probe at −20°C.

3.5 Transfer of DNA by Capillary Blot

To transfer the DNA from the agarose gel to the positively charged nylon membrane, it is recommended to use a capillary blot, because it is an easy-to-use method that does not depend on expensive equipment.

1. To facilitate transfer of larger DNA fragments, the DNA is first fragmented in the gel by soaking the gel in 0.25 M HCl for approximately 15–20 min until the bromophenol blue turns yellow. Stir carefully during incubation to speed up buffer exchange in the gel. If using a belly dancer, be sure to use it on very low speed only so as not to damage the soft agarose gel.
2. Remove the HCl and add 0.4 M NaOH. Incubate again for 15–20 min with slight stirring until the bromophenol blue turns blue again. Treatment with NaOH leads to the denaturation of the double-stranded DNA into single-stranded DNA, an important step that allows single-stranded probe hybridization.
3. To set up the capillary blot, build a support at least equivalent to the size of your gel, for example, by stacking two empty plastic microcentrifuge tube storage boxes on top of each other in a plastic tray. Place a bridge of blotting paper over the plastic boxes so that it touches the bottom of the tray below on both sides (Fig. 3). Prewet the blotting paper with 0.4 M NaOH.

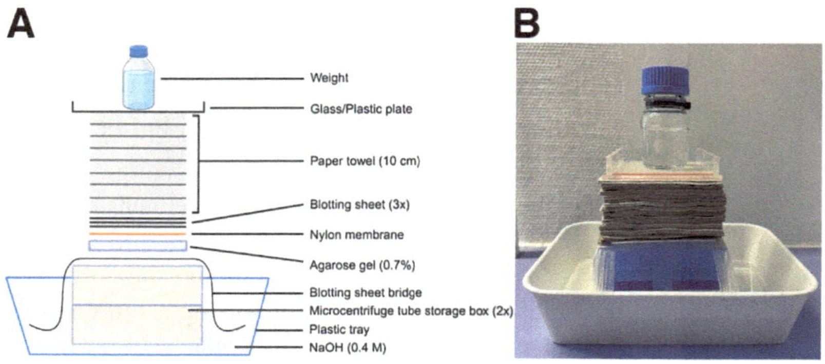

Fig. 3 Set-up of a capillary blot for Southern transfer. (**A**) Schematic representation of the assembled capillary blot used for the transfer of DNA from agarose gel to a positively charged nylon membrane. The gel is placed between the blotting sheet bridge and the membrane. On top of the membrane, three wet blotting sheets, a stack of dry paper towels, a plate, and a lightweight are positioned. The blotting sheet bridge paper reaches in a solution of 0.4 M NaOH. The image was created using BioRender.com. (**B**) Picture of a typical capillary blot setup

4. Place the agarose gelagarose gel on top of the blotting paper.
5. Cut the membrane to the same size as the gel. Pre-wet the nylon membrane in 2X SSC buffer and carefully place it on top of the gel using forceps. When handling the membrane, use gloves and forceps at all times and try to touch the membrane as little as possible (*see* **Notes 8** and **9**).
6. Cut three blotting sheets to the same size as the membrane, soak them briefly in 2× SSC buffer, and place on top of the membrane (*see* **Note 9**).
7. Cover all areas of the blotting paper next to the gel with a plastic covering (e.g., parafilm strips) to avoid direct contact of the later-to-position dry paper towels to the blotting paper bridge.
8. Fill the plastic tray to two-thirds with 0.4 M NaOH.
9. To initiate the capillary force, add a 10-cm stack of paper towels on top of the wet blotting sheets. Put a glass plate on the paper towel to add pressure and stabilize the whole setup, and place a light weight (e.g., a half-filled 100 ml glass bottle) on the glass plate. Let stand for 4 h up to overnight (*see* **Note 10**).
10. Disassemble the capillary blot by removing the pieces and layers from top to bottom (*see* **Note 11**). Carefully place the membrane on a clean surface (e.g., a small autoclave bag) with the DNA side facing up.
11. Expose the membrane to UV at 120 mJ/cm^2 for 20 s (*see* **Note 12**).

3.6 Probe Hybridization

1. Preheat the hybridization buffer to 65°C.
2. Place the membrane on a clean surface, use forceps to carefully roll up the membrane, and place it with the DNA side facing inwards inside the hybridization tube so that it tightly clings to the inner wall of the tube and does not overlap with itself (*see* **Note 13**).
3. Add 20 ml of hybridization buffer in the tube to prehybridise the membrane, incubate in the hybridization oven for 30 min at 65°C while rotating.
4. After 20 min of prehybridization, denature the probe at 95°C for 10 min and add the denatured probe to the prehybridization solution avoiding direct contact of the concentrated probe to the membrane.
5. Incubate at 65°C with constant rotating for at least 6–8 h.

3.7 Signal Detection

1. Pour the probe solution into a 50-ml tube and store at −20°C (*see* **Note 14**). The following wash and incubation steps (steps 2 – 8) are all performed by rotating in the hybridization oven at the indicated temperatures (*see* **Note 13**).
2. Wash the membrane twice with 30 ml Southern wash buffer at 65°C for 15 min each.
3. Discard the Southern wash buffer and replace it with 20 ml DIG wash buffer and incubate for 10 min at room temperature.
4. Replace the DIG wash buffer with 30 ml DIG II buffer and incubate at room temperature for 30 min.
5. Discard DIG II buffer and pour 20 ml of antibody solution into the hybridization tube. Incubate for 30 min at room temperature.
6. Collect the antibody solution (*see* **Note 14**). Wash the membrane twice with 20 ml DIG wash buffer at room temperature for 15 min each.
7. Remove the DIG wash buffer, add 30 ml DIG III buffer, and incubate for 5 min at room temperature.
8. Discard DIG III solution, pour 30 ml CDP-Star solution into the hybridization tube (*see* **Note 13**) and incubate at 37°C for 5 min. Afterward, collect the CDP-Star solution (*see* **Note 14**).
9. Carefully take out the membrane from the hybridization tube and seal it immediately in a clean, see-through autoclave bag and remove air bubbles that occur on the membrane.
10. Image your membrane with a chemiluminescence imaging system.

4 Notes

1. To get rid of DNA and DNase from glass beads, soak unsterile glass beads in 25% HCl for 60 min with occasional shaking. Afterward, wash the glass beads several times with sterile water until the pH of the washing fluid is neutral. Before using, sterilize the glass beads in an oven at 180°C for at least 6 h.
2. In case you are generating a complementation or overexpression strain, use your complementation or overexpression construct as a probe.
3. Keep in mind that the procedure relies on restriction fragment size differences, which means you need to apply the genomic DNA of the progenitor strain as a control on your Southern gel.
4. You can easily generate a plastic spatula for transferring 100-μl glass bead aliquots. For this, take a 1.5-ml plastic microcentrifuge tube and cut the bottom at 100 μl. To attach a handle to your small cup, take a 1-ml pipette tip, melt the slim end of the tip with a fire lighter and gently press the liquid plastic to your cup. Let it solidify prior to use.
5. Most of the genomic DNA is located very close to the interphase. Therefore, your yield of genomic DNA depends on how well you manage to collect the upper phase. In case you transferred parts of the interphase, continue the protocol at step 4.
6. The genomic DNA should be represented by a clear band between the largest marker fragment and the gel pocket. A smear of DNA instead of a clear band indicates degradation of your DNA during isolation and marks this sample as unsuitable for Southern analysis.
7. Ideally, the digested DNA should show a smear with some distinguishable higher intensity bands representing mitochondrial DNA.
8. To place the membrane, hold it with forceps at the sides, such that it bends down in the middle, position the hanging down part first on the gel and carefully lower your forceps to place the membrane evenly on the gel. Do not move the membrane once it has touched the gel, as this can lead to the appearance of smear or overlaying banding patterns on the developed Southern blot.
9. To remove air bubbles between the gel and membrane, and between the membrane and the blotting sheets, use a rod-shaped device (e.g., a 5-ml glass pipette or a 10-ml serological plastic pipette) to gently roll over the membrane or the

blotting sheets in one direction from one end of the gel to the other.

10. To speed up the process of transfer, regularly replace the wet paper towels by dry ones. In this case, incubation may be shortened to one hour only.
11. It is convenient to mark the gel pockets on the membrane for easier interpretation of banding patterns. For this, do not remove the membrane from the gel, but remove the membrane together with the gel, turn the stack upside down on a clean surface (e.g., a small autoclave bag), take a soft pencil and mark the pockets with the pencil through the gel on the membrane.
12. The membrane can get dry at this stage and can be stored at 4°C for several days before proceeding with probe hybridization.
13. Instead of using a hybridization oven and hybridization tubes, the prehybridization, hybridization and signal detection steps can also be done by sealing the membrane together with the desired volume of solutions in an autoclave bag and incubating on a rotary device. Make sure to exclude air bubbles inside the sealed bag. Exchange buffers through an opening by cutting a corner of the bag and resealing.
14. The used probe, the used antibody, and the used CDP-Star solution can be reused four to five times if stored at –20°C. Before reusing these solutions, ensure that they are thawed completely. Freshly denature the probe solution by placing it in boiling hot water for 10 min prior to use.

References

1. Kämper J (2004) A PCR-based system for highly efficient generation of gene replacement mutants in *Ustilago maydis*. Mol Genet Genomics 271(1):103–10. https://doi.org/10.1007/s00438-003-0962-8 PMID: 14673645
2. Brachmann A, König J, Julius C et al (2004) A reverse genetic approach for generating gene replacement mutants in *Ustilago maydis*. Mol Genet Genomics 272(2):216–26. https://doi.org/10.1007/s00438-004-1047-z Erratum in: Mol Genet Genomics. 2004 Nov;272(4):488. PMID: 15316769
3. Bösch K, Frantzeskakis L, Vraneš M et al (2016) Genetic manipulation of the plant pathogen *Ustilago maydis* to study fungal biology and plant microbe interactions. J Vis Exp 30(115):54522. https://doi.org/10.3791/54522 PMID: 27768088; PMCID: PMC5092128
4. Codner GF, Erbs V, Loeffler J et al (2021) Universal Southern blot protocol with cold or radioactive probes for the validation of alleles obtained by homologous recombination. Methods 191:59–67. https://doi.org/10.1016/j.ymeth.2020.06.011 PMID: 32599056; PMCID: PMC10790599
5. Southern EM (1975) Detection of specific sequences among DNA fragments separated by gel electrophoresis. J Mol Biol 98(3):503–17. https://doi.org/10.1016/s0022-2836(75)80083-0 PMID: 1195397
6. Ghareeb H, Drechsler F, Löfke C et al (2015) SUPPRESSOR OF APICAL DOMINANCE1 of *Sporisorium reilianum* modulates inflorescence branching architecture in maize and Arabidopsis. Plant Physiol 169(4):2789–804. https://doi.org/10.1104/pp.15.01347 PMID: 26511912; PMCID: PMC4677912

Chapter 14

Golden Gate Assembly for Genome Editing in *Ustilago maydis*

Andreas Müsgens, An N. T. Phan, Kerstin Schipper and Lars M. Blank

Abstract

Golden Gate cloning has become a key strategy for genetic manipulation of *Ustilago maydis*. Originally established for this organism by Terfrüchte et al. (2014), the approach uses type IIS restriction enzyme BsaI to assemble homology arms and integration module in a single reaction, yielding a plasmid from which a linear DNA fragment for transformation can be excised. In this chapter, the Golden Gate method for *U. maydis* is presented, with a focus on primer design and the DNA assembly workflow, taking practical considerations into account. By structuring the method into a step-by-step guide, it is made accessible for researchers new to *U. maydis* aiming to generate genome engineered mutants in this intriguing fungus.

Keywords *Ustilago maydis*, Gene deletion, Genomic modifications, Golden Gate cloning, Type IIS restriction enzyme, BsaI

1 Introduction

Golden Gate cloning utilizes type IIS restriction enzymes to enable the directed assembly of multiple DNA fragments in a single reaction. In contrast to traditional cloning approaches, restriction and ligation are combined in a one-pot reaction, which allows the rapid construction of even complex plasmids. Type IIS enzymes, such as BsaI, cleave DNA outside of their recognition sequence, enabling the design of unique, predefined sticky ends for directed assembly and preventing the introduction of unwanted sequences at junction sites.

The principle for Golden Gate cloning was first described in 2008 by Engler et al. [1] and has since become widely used in molecular biology. It was expanded into modular cloning toolboxes, such as the MoClo system [2] and GoldenBraid [3], which provide standardized libraries of functional DNA elements. These toolkits have been applied to a variety of organisms and contexts, including metabolic engineering in bacteria [4], gene stacking in plants [5, 6], and fungal genetics [7].

An N. T. Phan et al. (eds.), *Ustilago maydis*, Methods in Molecular Biology, vol. 3050,
https://doi.org/10.1007/978-1-0716-5364-7_14,

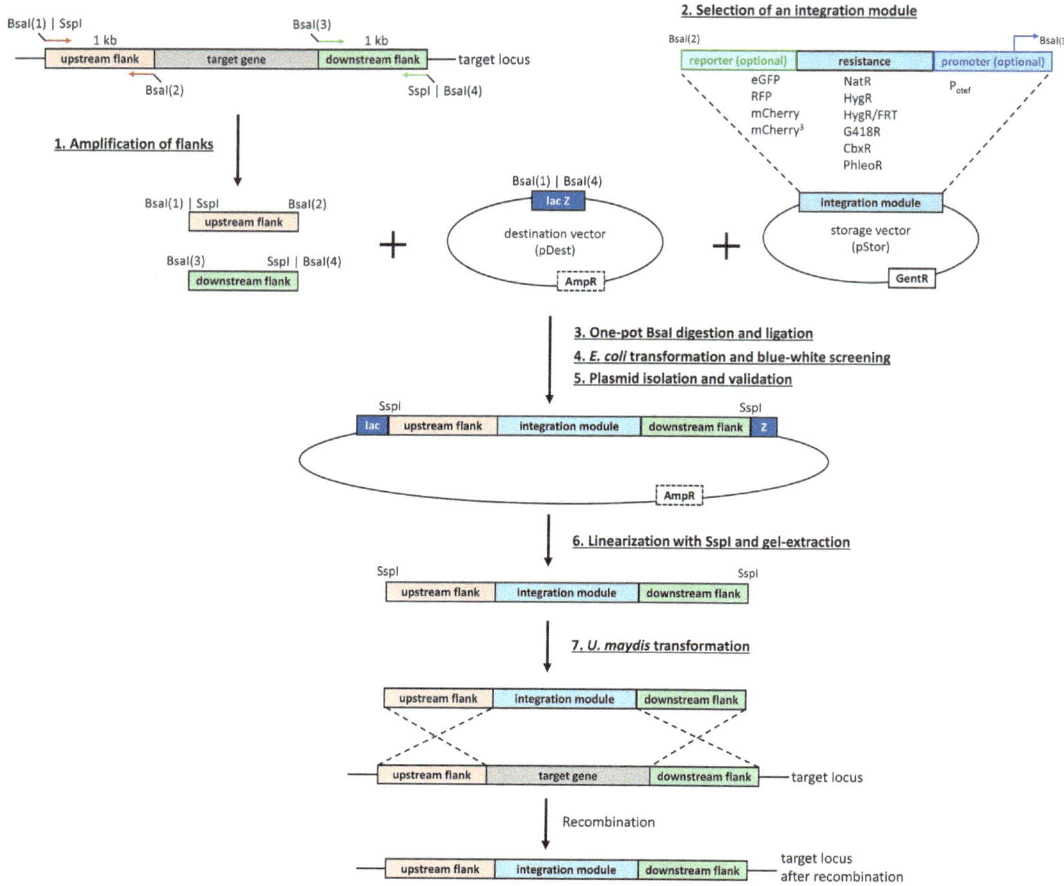

Fig. 1 Schematic workflow of a Golden Gate-based gene deletion strategy in *U. maydis*. The procedure starts with PCR amplification of ~1 kb upstream and downstream flanks of the genomic target locus. These flanks define the site of integration and ensure specific homologous recombination. In parallel, a suitable integration module is selected from a storage vector. For gene deletions, this module typically contains a resistance marker for selection in *U. maydis* (NatR, HygR, G418R, CbxR, PhleoR) and may optionally include a reporter gene (e.g., GFP, RFP, mCherry). The destination vector carries a *lacZ* site, which allows blue–white screening after *E. coli* transformation. In a one-pot Golden Gate assembly, upstream and downstream flanks, the chosen integration module, and the destination vector are reassembled. BsaI digestion generates defined overhangs, and T4 DNA ligase joins the fragments in the correct orientation and order. After *E. coli* transformation, correct assemblies are identified by blue–white screening, and candidate plasmids are validated by plasmid isolation and sequencing. For recombination in *U. maydis*, the verified plasmid is linearized with SspI, thereby releasing the complete integration construct, which consists of the upstream flank, integration module, and downstream flank. Since half of the SspI site was present in the genomic DNA, the flanks are 100% homologous and do not contain any unwanted sequence. After gel purification, the linear DNA is introduced into *U. maydis* cells. Homologous recombination between the genomic locus and the cloned flanking regions results in deletion of the target gene and insertion of the integration module

For *Ustilago maydis*, Golden Gate assembly was adapted by Terfrüchte et al. [7]. Building on an existing library of resistance cassette modules [8, 9], the introduction of BsaI sites streamlined the generation of integration constructs, enabling a direct route to linear DNA fragments for homology-directed integration into the *U. maydis* genome. The workflow established by Terfrüchte and colleagues is the basis and primary reference for this chapter.

In practice, three main components are required for a one-pot Golden Gate reaction (Fig. 1): (i) PCR-amplified upstream and downstream homology arms flanking the genomic target locus, (ii) a storage vector (pStor) carrying the selected integration module, and (iii) a destination vector (pDest) that provides the cloning backbone and a *lacZ* site for blue/white screening. During the reaction, the homology arms, the integration module, and the destination vector are assembled in the correct order by BsaI digestion and ligation. The resulting plasmid carries the complete integration construct. In a final step, the integration construct is excised with SspI (or SwaI), yielding a linear fragment with perfect homologous ends that can be used directly for protoplast transformation in *U. maydis*.

Today, the system is an established standard for reverse genetics in *U. maydis*. Beyond simple gene deletions, its modular design also enables promoter replacements, transcriptional fusions, and gene integrations [7]. The following protocol outlines the generation of integration constructs in *U. maydis* using Golden Gate cloning.

2 Materials

Prepare all solutions, buffers, media, and plates with ultrapure water. Growth media are autoclaved for sterilization, unless otherwise specified.

2.1 Equipment

1. PCR cycler
2. Agarose gel chamber with power supply
3. Gel imager
4. Table-top centrifuge
5. Thermoblock or water bath
6. Rotary shaker

2.2 DNA Preparation Kits

1. Plasmid Miniprep Kit
2. PCR Cleanup Kit
3. DNA Gel Extraction Kit

or other established routes for DNA purification

2.3 PCR

1. *Ustilago maydis* gDNA
2. CloneAmp HiFi PCR Premix (Takara)

2.4 Agarose Gel

1. TAE-buffer: 40 mM Tris-HCl, 2 mM EDTA, pH 8
2. 1% (w/v) agarose in TAE buffer

3. DNA gel stain dye
4. DNA ladder and loading dye

2.5 Enzymes

1. BsaI-HF (New England Biolabs)
2. T4 DNA Ligase (Roche) (*see* **Note 1**)
3. SspI or SwaI

2.6 Media and Buffers

1. T4 DNA Ligase reaction buffer (Roche) (*see* **Note 1**)
2. CutSmart Buffer (New England Biolabs)
3. LB (lysogeny broth): 1% (w/v) peptone, 0.5% (w/v) yeast extract, 1% sodium chloride (w/v), add 2 % (w/v) agar for solid plates

2.7 Antibiotics for E. coli Selection

1. Ampicillin
2. Gentamicin

2.8 Other Chemicals

2% (w/v) X-Gal (5-Brom-4-Chlor-3-Indolyl-Beta-D-Galactopyranosid) in DMSO (protect from light, store at 4°C)

2.9 Plasmids and Strains

1. pUMa1467 (destination vector)
2. Storage vector of choice (listed here are only plasmids for simple gene deletions):
   pUMAa1506 (NatR), pUMa1507 (HygR), pUMa1522 (HygR/FRT), pUMa1545 (G418R), pUMa1777 (CbxR), pUMa1788 (PhleoR).
   Plasmids are accessible through the Institute for Microbiology at Heinrich-Heine University Düsseldorf, Germany (http://www.mikrobiologie.hhu.de/ustilago-community.html) [7].
3. Chemo-competent *E. coli*

3 Methods

3.1 Amplification of Upstream and Downstream Flanks

3.1.1 Primer Design (see Note 2)

Two sets of primers are required to amplify the upstream flank (UF) and the downstream flank (DF); each should be about 1 kb in size. Ensure that both flanking regions are free of BsaI recognition sites (5′ GGTCTC 3′) (*see* **Note 3**). Design the forward UF primer to contain an overhang with BsaI(1) and the first half of the SspI recognition site (5′ AAT↓ATT 3′). The SspI site is split between the primer and the genomic region. Within the potential flank sequence, the

Table 1

5′ primer overhangs for amplification of upstream and downstream flanks. Gray: BsaI recognition site. Underlined: resulting overhang after BsaI digestion. Red italics: one half of the SspI recognition site.

	Primer overhangs(5′ – 3′ orientation)
Upstream forward	GGTCTCGCCTGAAT
Upstream reverse	GGTCTCCAGGC
Downstream forward	GGTCTCCGGCC
Downstream reverse	GGTCTCGCTGCAAT

start of the primer's binding region must, therefore, begin with the second half of the SspI recognition site, search for the trinucleotide "ATT" on the upper strand (*see* **Note 4**). Design the reverse UF primer to carry a BsaI(2) overhang, and the forward DF primer to include a BsaI(2) overhang. The reverse DF primer has overhangs for one-half of the SspI recognition site and BsaI(4). Here, search for "AAT" (upper strand) as a start point for the primer's binding region. Table 1 summarizes the 5′ overhang sequences for all primers, and Fig. 2 illustrates the design principle using the UF forward primer as an example. Alternatively to using SspI, other blunt cutters such as SwaI (5′ ATTT↓AAAT 3′) can be selected following the same logic.

3.1.2 PCR Amplification of Flanks

Obtain *U. maydis* gDNA as template for the amplification of upstream and downstream flanks (*see* **Note 5**). CloneAmp HiFi PCR Premix is recommended for high fidelity and fast amplification. Use gel-extraction to remove by-products.

3.2 Preparation of Destination and Storage Vector

3.2.1 Selection of an Integration Module

The original catalogue of Terfrüchte et al. [7] lists a large variety of different integration modules (Fig. 1). The integration modules are accessible via a corresponding storage vector (pStor). For standard gene deletions, the resistance markers NatR, HygR, G418R, CbxR, and PhleoR (*see* **Note 6**) are available. Optionally, modules for genetic modification can also include reporter genes or promoters, for transcriptional fusions or promoter exchange studies. Grow *E. coli* strains carrying the pStor plasmid on LB with gentamicin and generate clean plasmid DNA by kit purification.

3.2.2 Destination Vector

The destination vector pUMa1467 (pDest) will harbor the final assembly product integrated into the *lacZ* site. pUMa1467 has an ampicillin resistance for selection in *E. coli*. Generate clean plasmid DNA by kit purification.

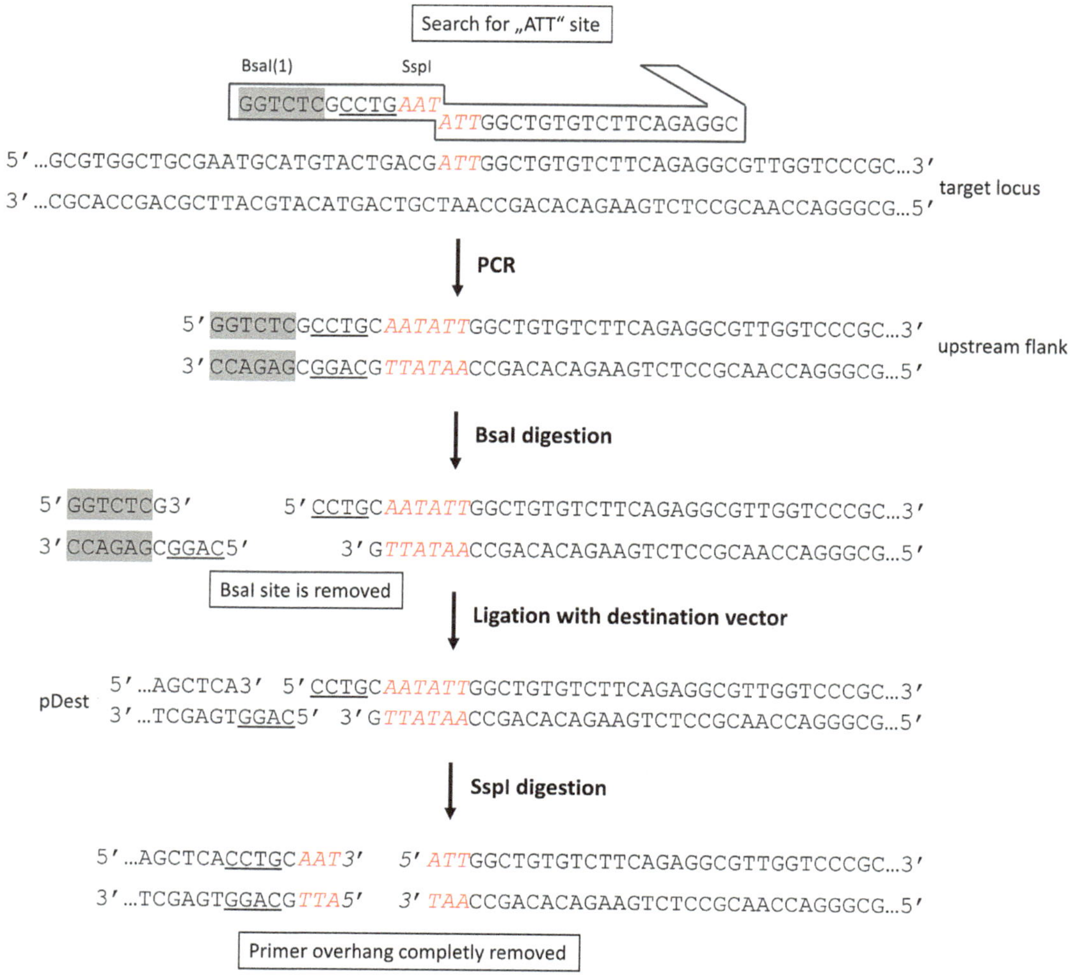

Fig. 2 Detailed look at primer design and the Golden Gate principle. The example genomic DNA sequence illustrates a potential start site for either the upstream or downstream forward primer. In this case, the upstream forward primer is shown, carrying the first of four unique BsaI overhang pairs (BsaI(1)–BsaI(4)) that define the orientation of the assembled fragments. The genomic sequence contains the triplet "ATT" (highlighted in red and italics), making it a suitable starting point for primer design. By adding "AAT" to the primer overhang, a complete SspI recognition site is created. After PCR amplification, the BsaI site is cleaved off, leaving a 5′ overhang on the upstream flank. The corresponding complementary overhang is generated by digestion of the destination vector. Before transformation, SspI digestion completely removes the initial primer-overhang, ensuring that no unwanted sequences remain and that the homology arm corresponds to the genomic sequence with 100% identity

3.3 One-Pot BsaI Digestion and Ligation

This step combines BsaI digestion and ligation. Setup a total reaction volume of 15 μL containing 75 ng storage vector (*see* **Note 7**), 75 ng destination vector, 40 ng upstream flank, 40 ng downstream flank, 1 × T4 DNA ligase buffer, 0.5 μL BsaI-HF (10 U), and 0.75 μL T4 DNA ligase (0.75 U). Incubate the mixture in a PCR cycler using the following temperature settings: (37°C 2 min, 16°C 5 min) × 30, 37°C

5 min, 50°C 5 min, 80°C 5 min, 16°C terminal hold. Immediately before transformation, incubate the reaction mixture for 5 min at 55°C. As a negative control, exclude one PCR flank only.

3.4 *E. coli* Transformation, Validation, and Linearization of Integration Construct

1. For transformation of *E. coli*, add 2 μL of the one-pot reaction mix into chemically competent cells (*see* **Note 8**). Perform the required steps for transformation. Plate onto LB-agar plates containing 60 μL X-Gal (2% (w/v) in DMSO) and 100 μg/mL ampicillin (*see* **Note 9**).
2. Prepare an overnight culture of white colonies and isolate the plasmid. Correct plasmids show the integration module flanked by upstream and downstream sequence in the pDest backbone. Confirm the assembly by restriction analysis and/or plasmid sequencing of three white colonies (*see* **Note 10**).
3. Cut out the integration construct using SspI (or SwaI). Purify the resulting linear DNA fragment via gel extraction before *U. maydis* transformation.

3.5 *U. maydis* Transformation and Verification of Gene Deletion

Conduct transformation of *U. maydis* protoplasts according to standard protocols [10]. Select for growth on the integrated resistance and verify the gene deletion using Colony PCR and sequencing, or Southern blot analysis (*see* **Note 11**). Compare two positive clones to make sure that no undetected side effects of genetic modification have occurred.

4 Notes

1. Alternatively, the NEBridge® Ligase Master Mix (New England Biolabs) is specifically designed for Golden Gate assembly and can be recommended.
2. It is strongly recommended to conduct careful virtual cloning of all steps.
3. When a BsaI site is present in the flank, amplify the flank in two parts and use fusion PCR to introduce a single-nucleotide mismatch that removes the site.
4. When no suitable blunt-end cutter is available, generate the required linear fragment for transformation directly by PCR or by using a sticky-end generating enzyme.
5. Strain UM521 is the reference strain for *U. maydis*, with a very well-annotated genome [11]. For many studies, it may be a preferred strain background for amplifying genetic elements.
6. PhleoR should not be routinely used as it carries the risk of generating side-mutations. Mutants generated with PhleoR should be carefully verified.

7. Depending on the outcome of the reaction, the respective DNA amounts need to be adapted. For example, if the resulting colonies are largely blue, the amount of destination vector might need to be reduced. If flanks vary strongly in size, one flank can be set to 40 ng, and the other is added in equimolar amounts.
8. Commercially available competent cells can often be aliquoted, allowing more transformations than recommended by the manufacturer.
9. Prepare an additional X-Gal LB agar plate and plate a higher cell density after brief centrifugation. This is especially useful with self-made competent *E. coli*.
10. Whole-plasmid or amplicon sequencing can speed up the verification of the integration construct.
11. For basic research questions, it is recommended to stick to Southern blot analysis.

Acknowledgments

This work was supported by the Deutsche Forschungsgemeinschaft (DFG, German Research Foundation) - SFB1535 - Project ID 458090666. The laboratory of LMB is partially funded by the Deutsche Forschungsgemeinschaft (DFG, German Research Foundation) under Germany's Excellence Strategy Cluster of Excellence 2186 "The Fuel Science Center" - ID: 390 919 832 and by the Werner Siemens Foundation in the frame of the WSS Research Centre "catalaix."

References

1. Engler C, Kandzia R, Marillonnet S (2008) A one-pot, one-step, precision cloning method with high throughput capability. PLOS One 3:e3647. https://doi.org/10.1371/journal.pone.0003647
2. Weber E, Engler C, Gruetzner R et al (2011) A modular cloning system for standardized assembly of multigene constructs. PLOS One 6:e16765. https://doi.org/10.1371/journal.pone.0016765
3. Sarrion-Perdigones A, Falconi EE, Zandalinas SI et al (2011) Golden Braid: an iterative cloning system for standardized assembly of reusable genetic modules. PLOS One 6:e21622. https://doi.org/10.1371/journal.pone.0021622
4. Wingler LM, Corn JE (2011) Application of Golden Gate cloning for bacterial metabolic engineering. In: Methods in molecular biology, vol 765. Humana Press, Totowa, pp 245–254. https://doi.org/10.1007/978-1-62703-764-8_9
5. Patron NJ, Orzaez D, Marillonnet S et al (2015) Standards for plant synthetic biology: a common syntax for exchange of DNA parts. New Phytol 208:13–19. https://doi.org/10.1111/nph.13532
6. Engler C, Youles M, Gruetzner R et al (2014) A Golden Gate modular cloning toolbox for plants. ACS Synth Biol 3:839–843. https://doi.org/10.1021/sb4001504
7. Terfrüchte M, Joehnk B, Fajardo-Somera R et al (2014) Establishing a versatile Golden Gate cloning system for genetic engineering in fungi. Fungal Genet Biol 62:1–10. https://doi.org/10.1016/j.fgb.2013.10.012

8. Brachmann A, König J, Julius C et al (2004) A reverse genetic approach for generating gene replacement mutants in *Ustilago maydis*. Mol Genet Genomics 272:216–226. https://doi.org/10.1007/s00438-004-1047-z
9. Kämper J (2004) A PCR-based system for highly efficient generation of gene replacement mutants in *Ustilago maydis*. Mol Genet Genomics 271:103–110. https://doi.org/10.1007/s00438-003-0962-8
10. Bösch K, Frantzeskakis L, Vraneš M et al (2016) Genetic manipulation of the plant pathogen *Ustilago maydis* to study fungal biology and plant–microbe interactions. J Vis Exp 115:e54522. https://doi.org/10.3791/54522
11. Kämper J, Kahmann R, Bölker M et al (2006) Insights from the genome of the biotrophic fungal plant pathogen *Ustilago maydis*. Nature 444:97–101. https://doi.org/10.1038/nature05248

Check for updates

Chapter 15

Generation of Gene Knock-Out Mutants in *Ustilago maydis* Using Cas9hf Nuclease

Jessica Tiefenbacher and Björn Sandrock

Abstract

Gene disruption of nonessential genes became more convenient with the adaptation of the CRISPR-Cas9 system for *Ustilago maydis* by the group of Regine Kahmann (MPI Marburg, Germany) in 2016. In our group, we have developed the system further to create defined marker-free gene deletions. Therefore, we used the CRISPR-Cas9 system together with oligonucleotides composed of 40 nucleotides of upstream and downstream flanking regions. Here, we describe the entire way from the decision, which gene of interest should be deleted, to a marker-free *U. maydis* mutant strain.

Keywords *Ustilago maydis*, Gene deletion, CRISPR-Cas, pSM2, Marker-free strain, Nonessential gene

1 Introduction

Since the establishment of *U. maydis* as model organism for plant–pathogen interactions, in the early 1980s, the need for gene deletion strategies was obvious. Fungal antibiotics together with resistance genes were used to create knock-out (ko) mutants for individual genes. Therefore, approximately 1000 base pairs (bp) of the left and the right borders of the gene of interest were cloned upstream and downstream of the resistance cassette, respectively (Fig. 1A). Several groups established a PCR-based method, a reverse genetic approach and Golden Gate cloning systems [1–4] to facilitate cloning and amplification of the ko cassettes. The protocols also allowed the use of up to five resistance genes for multiple rounds of gene deletions. In addition, the recovery of resistance markers was adapted using a FLP-mediated recombination system [5].

The Clustered regularly interspaced short palindromic repeats (CRISPR) together with the CRISPR-associated proteins (Cas)

An N. T. Phan et al. (eds.), *Ustilago maydis*, Methods in Molecular Biology, vol. 3050,
https://doi.org/10.1007/978-1-0716-5364-7_15,

were identified as an immune defense system to react on viral infection in bacteria and archeobacteria by first memorizing the foreign DNA by incorporation of DNA fragments in the CRISPR array [6–8]. During a second infection, the recognition of the foreign DNA leads to cleaving of or binding to this DNA at specific protospacer adjacent motifs (PAM) [9, 10]. Meanwhile, many types and subtypes of CRISPR-Cas systems have been identified allowing gene silencing, gene disruption, large chromosomal deletions, or gene editing [11–14].

Using the constitutive expression of the codon-optimized Cas9 nuclease together with a single guide RNA (sgRNA) expressed from an RNA polymerase III promoter, the system was adapted to induce double strand brakes (DSB) at suitable PAM sites and to create gene disruption by nonhomologous end-joining (NHEJ) in *U. maydis* [15]. Further development of the targeting vector allowed the multiple gene disruption [16].

Here, we describe the defined gene deletion in *U. maydis* using the modified high-fidelity Cas9 (Cas9hf) nuclease, the sgRNA towards the gene of interest and the donor DNA composed of 40 bp upstream and downstream of the ko region (Fig. 1B) [17].

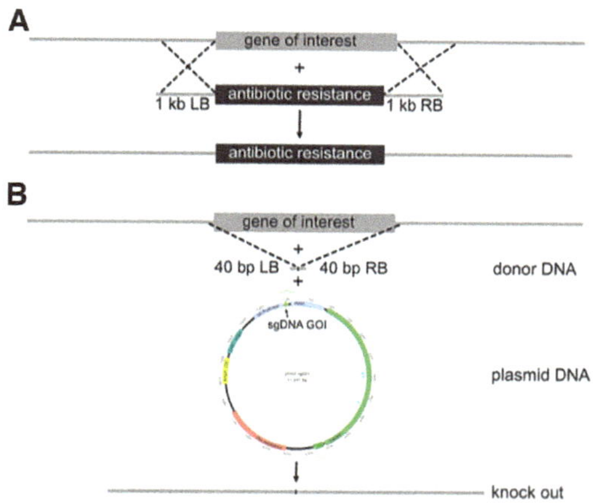

Fig. 1 Knock-out strategies in *U. maydis*. (**A**) Homologous recombination using antibiotic resistance marker genes. Cells are transformed with the knock-out cassette consisting of approximately 1000 bp left (LB) and right border (RB) fused to an antibiotic resistance gene allowing selection on the corresponding antibiotic substance. (**B**) Homologous recombination using the Cas9 nuclease. Cells are transformed with a plasmid containing the Cas9 nuclease and a module expressing the small guide RNA for the gene of interest (GOI) together with the donor DNA composed of 40 bp left (LB) and right border (RB)

2 Materials

All buffers, media, plates, and solutions should be prepared with deionized water unless indicated otherwise.

2.1 Equipment

1. Minifuge centrifuge for 50 mL falcon-tubes, bench centrifuge for 1, 5 mL, and 2 mL Eppendorf-tubes.
2. Vibrax shaker.
3. Vortex mixer.
4. Ice bath.
5. Thermo mix or water bath.
6. PCR machine.
7. Photometer for measuring optical density (OD_{600}).
8. Nanodrop for measuring DNA concentration.
9. Chambers for culture growth at 28°C and 37°C.
10. Autoclave for sterile media preparation and waste inactivation.
11. Agarose gel facility together with Gel-Doc system.
12. Hood for working with organic solvents.
13. Organic and agarose gel waste opportunities.
14. Microscope (40×).
15. Microwave.

2.2 Strains

For cloning experiments, the *E. coli* strain Top10 (F^- mcrA Δ(mrr-hsdRMS-mcrBC) ϕ80lacZΔM15 ΔlacX74 recA1 araD139 Δ(ara leu) 7697 galU galK rpsL (StrR) endA1 nupG) was used, but DH5α (F^- ϕ80lacZΔM15 Δ(lacZYA-argF) U169 recA1 endA1 hsdR17 (rK^-, mK^+) phoA supE44 λ^- thi-1 gyrA96 relA1) is also suitable.

For *U. maydis* strains, which are not based on the sequenced strain 521, make sure that the sequence of the gene of interest (GOI) has no SNPs (single nucleotide polymorphisms) in the PAM region.

2.3 Agarose Gel

50× TAE-buffer: Dissolve 242 g Tris-Base, 57.1 mL glacial acetic acid and 100 mL 0.5 M EDTA-NaOH, pH 8.0 in a total volume of 1 L with H_2O.

1% Agarose Gel: Dilute the 50× TAE-buffer to 1× TAE. Dissolve 1 g agarose in 100 mL 1× TAE by heating in a microwave for approximately 2 min. After cooling to 60°C (10–15 min), add 3 μL Midori green Advance (or any other DNA staining solution) and pour the gel in an appropriate horizontal chamber (*see* **Note 1**).

6× Loading dye: 50 % (w/v) glucose, 1 mM EDTA-NaOH, pH 8.0, Orange G (few granules)

Running a gel: The agarose gel is placed in the gel chamber and filled with 1× TAE-buffer. After loading, the probes and a size marker the gel should run at 120 mA for 1 h.

2.4 Media and Buffers for Working with E. coli Cells

LB medium: Dissolve 10 g/L Tryptone, 5 g/L Yeast extract, and 10 g/L NaCl. Adjust the pH to pH 7.0. Autoclave for 20 min at 121°C.

SOB medium: Dissolve 5 g/L Yeast extract, 20 g/L Trypton, 10 mM NaCl (from 1 M NaCl), and 2.5 mM KCl (from 1 M KCl) in H_2O and autoclave.

CCMB80-buffer: Mix 10 mM Potassium Acetate (KAc) (from 1 M KAc pH 7.0), 80 mM $CaCl_2$, (from 1 M $CaCl_2$), 20 mM $MnCl_2$ (from 1 M $MnCl_2$), 10 mM $MgCl_2$ (from 1 M $MgCl_2$), and 10% Glycerin in H_2O. If it is necessary, use 0.1 N HCl for pH 6.4 (*see* **Note 2**). Sterilize through a 0.22 μm filter.

LB_{amp} plates: Dissolve 10 g/L Tryptone, 5 g/L Yeast extract, and 10 g/L NaCl. Adjust the pH to pH 7.0. Add 15 g/L Agar–Agar, a magnet stirrer and autoclave for 20 min at 121°C. After cooling to 60°C, add 1 mL Ampicillin (amp; 100 mg/mL in H_2O) to get a final concentration of 100 μg/mL and mix thoroughly. Pour the medium in petri dishes. 1 L is sufficient for approximately 35 plates.

LB_{amp} medium: Add 100 μL Ampicillin (100 mg/mL) to 100 mL LB medium and store at 4°C.

Destroyer lysis buffer: Mix 10 mM Tris-HCl, pH 8.0 (from 1 M Tris-HCl, pH 8.0), 1 mM EDTA-NaOH, pH 8.0 (from 0.5 M EDTA-NaOH, pH 8.0), 15% (w/v) Sucrose, 2 mg/mL Lysozyme (from 100 mg/mL), 200 μg/mL RNAse A (from 10 mg/mL RNAse A), and 100 μg/mL BSA (from 20 mg/mL BSA) in H_2O. Prepare aliquots of 10 mL and store at −20°C.

LB-glycerol: Dissolve 400 mL Glycerol [87% (v/v)], 5 g Tryptone, 2.5 g Yeast Extract, and 5 g NaCl in 500 mL with H_2O. Adjust the pH to 7.0 and autoclave in aliquots of 50 mL.

2.5 Media and Buffers for Working with U. maydis Cells

$YEPS_{light}$ medium: Dissolve 10 g/L **Y**east **E**xtract (Difco), 4 g/L Bacto **P**eptone (Difco), and 4 g/L **S**ucrose in H_2O and adjust to 1 L. Prepare aliquots of 200 mL and sterilize by autoclaving.

PD plates: Dissolve 24 g/L **P**otato **D**extrose (Difco) in demineralized water (!, no bidest) (*see* **Note 3**) and add 20 g/L Bacto agar (Difco). After autoclaving and cooling to 60°C, pour the plates.

PD_{cbx} plates: Carboxin (cbx) is prepared as 2 mg/mL stock solution in Methanol. Working concentration is 2 μg/mL Carboxin.

SCS-buffer: 20 mM **S**odium **c**itrate pH 5.8 and 1 M **S**orbitol. Prepare SCS-1 (20 mM Sodium citrate and 1 M Sorbitol) and SCS-2 (20 mM Citric acid and 1 M Sorbitol). Mix 1 volume SCS-1 and approximately 4 volumes SCS-2 in order to adjust the pH of the solution to 5.8. Sterilize by autoclaving.

SCS-buffer with 4 mg/mL Novozyme or 20 mg/mL Lysing enzyme: Mix immediately before use and sterilize through a 0.22 μ m filter.

STC-buffer: Mix 10 mM Tris-HCl, pH 7.5 (from 1 M Tris-HCl, pH 7.5), 100 mM $CaCl_2$ (from 1 M $CaCl_2$), and 1 M sorbitol (from 2 M sorbitol) in H_2O. Sterilize by autoclaving.

Regeneration agar (reg agar): Dissolve 10 g/L yeast extract (Difco), 20 g/L Bacto Peptone (Difco), 20 g/L sucrose, and 182.2 g/L sorbitol in H_2O. Add 15 g/L agar (Difco) and autoclave in 300 mL batches.

Bottom agar: Use 10 mL reg agar per plate and add the antibiotic at twofold concentration (Carboxin: 4 μg/mL) after cooling to 60°C. Make sure the agar is distributed well and without air bubbles. Top agar: Use 10 mL reg agar per plate and pipette as mentioned above.

STC-PEG buffer: Give 10 g PEG4000 (40% w/V) in a 50 mL plastic tube and add STC-buffer to approximately 25 mL. Dissolve the PEG (takes some time) by continuously shaking or rotating and adjust the volume to exactly 25 mL with STC. Sterilize through a 0.45 μm filter or autoclave (*see* **Note 4**).

NSY-Glycerol-medium: Dissolve 400 mL Glycerol [87% (v/v)], 4 g Nutrient Broth, 0.5 g Yeast Extract, and 2.5 g Sucrose in 500 mL with H_2O. Adjust the pH to 7.0 and autoclave in aliquots of 50 mL.

Acidified glass beads (425–600 μm)

Usti-lysis buffer: Mix 100 mM NaCl (from 1 M NaCl), 10 mM Tris-HCl, pH 8.0, 1 mM EDTA-NaOH, pH 8.0, 2% (v/v) Triton-X100, and 1% SDS [from 10% (w/v) SDS] in H_2O.

TE: Mix 10 mM Tris-HCl, pH 8.0 (from 1 M Tris-HCl, pH 8.0), and 1 mM EDTA-NaOH pH 8.0 (from 0.5 M EDTA-NaOH, pH 8.0) in H_2O.

PCI: Phenol:Chloroform:Isoamylalcohol (25:24:1) equilibrated with TE.

TER: Mix 10 mM Tris-HCl, pH 8.0, 1 mM EDTA-NaOH pH 8.0, 10 μg/mL RNaseA (from 10 mg/mL) in H_2O.

3 Methods

3.1 Cloning

3.1.1 Gene of Interest (GOI) Analysis via E-CRISP (as Proof of Principle the Gene UMAG_01436 is Used)

1. To identify suitable PAM sites in the gene of interest the E-CRISP tool is used (http://www.e-crisp.org).
2. The DNA sequence of the GOI is added in the FASTA format field. The start application is set to strict (only NGG PAM, only G as 5' base, off-target tolerates many mismatches and ignores nonseed region, introns, purpose is knockout) (*see* **Note 5**).
3. For UMAG_01436, the 20 base pair (bp) sequence with the highest E-score was 5'-GTACT ACGTC TACTT GCCCG NGG-3' and is almost in the middle of the gene sequence [18].

3.1.2 Primer Design

The primers for the knock-out project include six oligonucleotides, for which the design is explained (Fig. 2, Table 1).

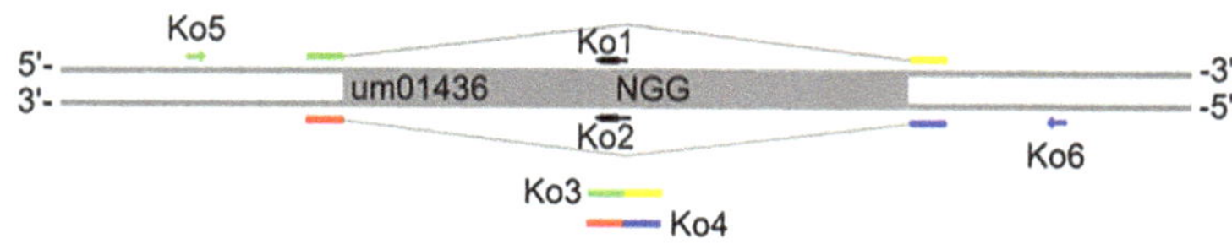

Fig. 2 Schematic drawing of the oligonucleotides ko1–6 used for the gene depletion of um01436. sgDNA: ko1 and ko2; 80 bp donor DNA: ko3 and ko4; Test-PCR and sequencing: ko5 and ko6. NGG marks the PAM-site

Table 1
List of oligonucleotides used for the gene depletion of um01436 (ko1–6) and for sequencing of pSM2-derived plasmids (MJ422). The color code is the same as in Fig. 2

Primer	Length	DNA Sequence (5′ -> 3′)
Ko1	32 nt	**AAC** GTACT ACGTC TACTT GCCCG **GTTTTAGAG**
Ko2	36 nt	**CTAGCTCTAAAAC** CGGGC AAGTA GACGT AGTAC **GTT**
Ko3	80 nt	CTACTCCATCTACATCAACTTGAAAGACGGCATCATCAAC ATAGGTTGGTGCACGTCTTTTGGATAAAGTACCGAACTTT
Ko4	80 nt	AAAGTTCGGTACTTTATCCAAAAGACGTGCACCAACCTAT GTTGATGATGCCGTCTTTCAAGTTGATGTAGATGGAGTAG
Ko5	22 nt	GCTCTTAACAGCGCTTGATTCC
Ko6	22 nt	CGTGAATCCAAATTCAGACTCG
MJ422	24 nt	ATTCGTGATTTACACCAAACACGC

1. Ko1 and Ko2 are needed for the sgDNA hybridization for cloning in pSM2. Specific flanking sequences should be added to Ko1 and Ko2 to prepare the hybridized sgDNA to fit into the SnaBI-XbaI-sites of pSM2. Both recognition sequences are disrupted after insertion of the sgDNA.
2. Ko3, Ko4 are two oligonucleotides for the donor DNA sequence composed of 40 nucleotides upstream and downstream of the GOI.
3. Ko5, Ko6 are two test primers 250 bp upstream and downstream of the GOI. These primers are also usable as sequencing primer to verify the knock-out.

3.1.3 Vector Preparation

1. For the cloning process, the pSM2 vector should be prepared (*see* **Note 6**). Therefore, 500 ng pSM2 should be digested with ten units (1 μL) of the restriction enzymes SnaBI and XbaI in a total volume of 50 μL for at least 4 h at 37°C.
2. The linearized vector should be purified via a 0.8%–1.0% agarose-gel to remove the stacking insert of 19/23 bp (*see* **Note 7**).

3.1.4 Competent *E. coli* Top10 Cells (According to a Protocol of Hanahan et al. [19])

1. For the preparation of competent, *E. coli* cells streak them on LB plates and incubate overnight (o/n) at 37°C.
2. Transfer some cells to 5 mL LB medium and incubate under shaking at 37 °C o/n.
3. On the next morning, inoculate 100 mL SOB-Medium 1:80 with this *E. coli*-culture.
4. Incubate under vigorous shaking at 22°C for approximately 5–6 h.
5. Meanwhile prepare cups and prefreeze at −80°C.
6. At an OD_{550} of ~0.3–0.4, transfer the culture in two sterile ice-cold 50 mL Falcon-tubes and centrifuge (10 min, 3000 g, 4°C).
7. Discard the medium, resuspend each cell pellet carefully in 15 mL ice-cold CCMB80-buffer and incubate for 20 min on ice.
8. After centrifugation (10 min, 3000 g, 4°C), remove the supernatant and resuspend the cell pellets carefully in 4 mL ice-cold CCMB-buffer without air-bubbles.
9. Prepare aliquots à 100 µL (around 80 aliquots) and freeze at −80°C or use directly for transformation. The quality of the competent cells will be and should be $>1 \times 10^7$ colony forming units (cfu)/µg pUC18 (*see* **Note 8**).

3.1.5 Hybridization of Oligonucleotides

1. All oligonucleotides are dissolved in H_2O to gain a concentration of 100 pmol/µL.
2. Aliquots of the primers used for PCR and sequencing (Ko5, Ko6) are further diluted with H_2O to 10 pmol/µL.
3. To hybridize the complementary oligonucleotides Ko1/Ko2 500 pmol of each oligo should be prepared in a total volume of 100 µL containing 50 mM NaCl.
4. The complementary oligos Ko3/Ko4 should be prepared in the same way.
5. Boil these solutions for 10 min at 99°C and cool slowly to room temperature.
6. DNA hybrids should be stored at −20°C.

3.1.6 Ligation and Transformation in Competent *E. coli* Cells

1. 40 ng of the digested pSM2-vector and 5 µL of the Ko1/Ko2 hybridized oligo solution are ligated in a total volume of 30 µ L using T4 DNA ligase at 18°C o/n. As control, the empty vector is used without hybridized oligos.
2. For transformation thaw, one aliquot of competent *E. coli* cells per ligation reaction. 10 µL of each ligation reaction are pipetted to the cells and incubated on ice for 20 min.

3. After a heat-shock (42°C, 90 s), the cells can be directly plated on LB_{amp} plates (*see* **Note 9**). Colonies appear after 16 h at 37°C.

3.1.7 Analysis of Plasmid-DNA

Depending on the number of colonies that appear in the control transformation, two different strategies were used:

1a. Less than 20% colonies in the control transformation:

1. Transfer four single colonies to 3 mL LB_{amp} medium containing tubes and incubate them o/n at 37°C on a rotary shaker.
2. For purification of the plasmid-DNA, a Plasmid-DNA-Kit (suitable for large plasmids!) can be used following the manufacturer's protocol.

1b. More than 20% colonies in the control transformation:

1. Transfer ten single colonies to 3 mL LB_{amp} medium containing tubes and incubate them o/n at 37°C on a rotary shaker.
2. For rapid purification of the plasmid-DNA, the Destroyer-Lysis method was used. Transfer 1 mL of each culture in an Eppendorf-tube and centrifuge [1 min, 16,000 g, room temperature (RT)].
3. Discard the supernatant, resuspend the cells in 50 μL Destroyer lysis buffer and incubate for 5 min at RT.
4. Boil the cells for 1 min at 99 °C and cool on ice immediately.
5. After 1 min on ice centrifuge (6 min, 16,000 g, RT).

2. Digest 500 ng (1a) or 7 μL (1b) Plasmid DNA with AatII and XbaI. Because the XbaI-site is disrupted in the novel plasmids, the pSM2-vector should be digested as control, too.
3. The digested DNA is separated on a 0.8%–1.0% agarose-gel. Two bands (approximately 9200 bp and 2000 bp) point to plasmids containing the sgDNA. For pSM2, three bands should appear (9200 bp, 1580 bp, and 420 bp) (Fig. 3) (*see* **Note 10**).
4. Plasmid DNA of positive clones identified in 1b should be purified using a Plasmid-DNA-Kit following the manufacturer's protocol.
5. Plasmids must be sequenced using a suitable primer (e.g., MJ422).
6. Backup plates should be spotted for positive clones on LB_{amp} plates and grown o/n at 37°C.

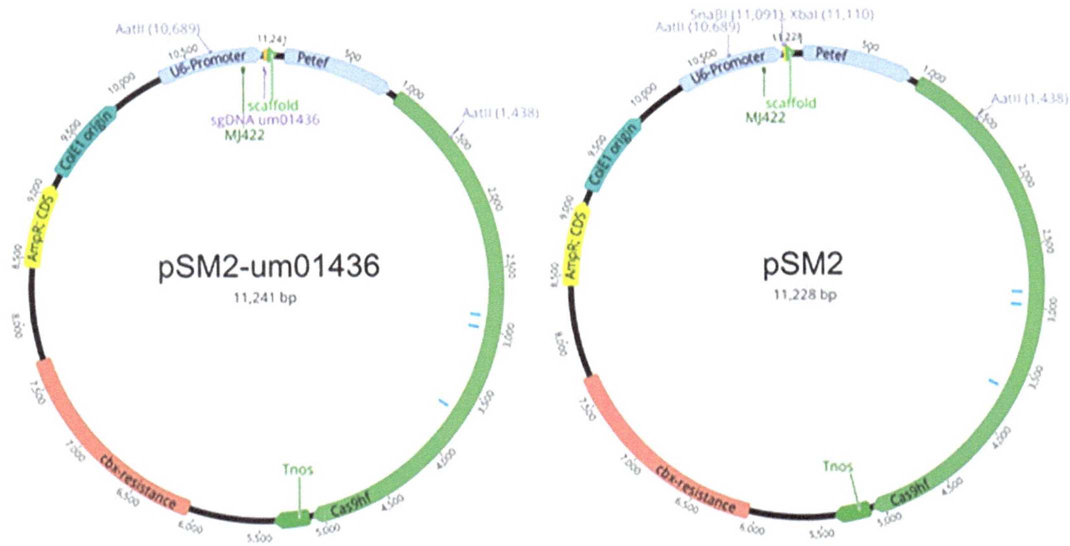

Fig. 3 Plasmids pSM2-um01436 and pSM2 together with the AatII and XbaI restriction sites

7. Positive clones with the correct sequence of the sgDNA should be inoculated in 3 mL LB_{amp} medium and grown o/n at 37°C.
8. On the next day, 1 mL culture is mixed with 1 mL LB-Glycerol-medium and stored at −80°C.

3.2 Transformation of *U. maydis*

3.2.1 Protoplast Preparation of *U. maydis* (According to a Protocol of Schulz et al. [20])

1. To gain protoplasts of *U. maydis* inoculate 3 mL $YEPS_{light}$ medium in the morning with a single colony and incubate on a rotary shaker at 28°C.
2. After 6 h, check the OD_{600} of the culture. When the OD is approximately 0.5 (5×10^6 cfu/mL), dilute the culture 1:1000 and 1:3000 in 50 mL $YEPS_{light}$ medium and incubate at 28°C o/n under vigorous shaking (*see* **Note 11**).
3. In the next morning, transfer the culture with an OD_{600} of 0.4–0.8 to a 50 mL Falcon tube and centrifuge (5 min, 2300 g, 20°C).
4. Discard the supernatant, resuspend the pellet in 25 mL SCS buffer and centrifuge again (5 min, 2300 g, 20°C).
5. Meanwhile prepare the Lysing enzyme/Novozyme solution in SCS buffer.
6. Discard the supernatant and resuspend the cell pellet in 2 mL Lysing enzyme/Novozyme solution.

7. Incubate at room temperature for 10–15 min, until 70% of the cells round up and begin to form protoplasts (*see* **Note 12**).
8. Add 10 mL SCS buffer to stop the protoplastation and centrifuge (10 min, 1200 g, 16°C).
9. Discard the supernatant and resuspend the cell pellet carefully without air bubbles in 10 mL SCS (10 min, 1200 g, 16°C).
10. Repeat this washing step.
11. Discard the supernatant and resuspend the cell pellet carefully in 10 mL STC buffer.
12. After the final centrifugation step (10 min, 1200 g, 16°C), discard the supernatant and resuspend the cells slowly in 500 μL (ice cold) STC buffer.
13. Immediately prepare aliquots of 60 μL and freeze them at –80°C. Protoplasts can also be used directly (stored on ice for up to 5 h). The quality of the protoplasts should be around 1000 cfu/μg pSM2 (*see* **Note 13**).

3.2.2 Transformation

1. Approximately 60 min in advance, boil the regeneration agar (Reg Agar) in a microwave and let it cool to approximately 60°C.
2. Take 12 mL Reg Agar and add the Carboxin in twofold concentration (4 μg/mL) and pour the bottom agar (10 mL) (*see* **Note 14**).
3. After thawing, the protoplasts on ice add 100 ng pSM2-sgDNA together with 5 μL of the complementary oligos Ko3/Ko4 (25 pmol) and incubate on ice for 10 min.
4. Pipette slowly 500 μL STC/PEG onto the protoplasts, mix carefully by pipetting up and down without air bubbles and incubate on ice for further 15 min.
5. During this incubation step, add another 10 mL of Reg Agar without carboxin on the bottom agar.
6. Plate the whole sample on the Reg Agar plates and distribute the solution by rotating the petri dish.
7. Incubate for 1 day at 28°C before the plate can be turned (*see* **Note 15**). After two more days, colonies should appear.

3.3 Identification of Deletion Mutants

After the double strand brake (DSB) of the genomic DNA at the PAM site by Cas9 nuclease two different processes can happen to repair the DSB: NHEJ or homologous recombination. Therefore, 12 colonies were selected and screened for deletion of the gene of interest.

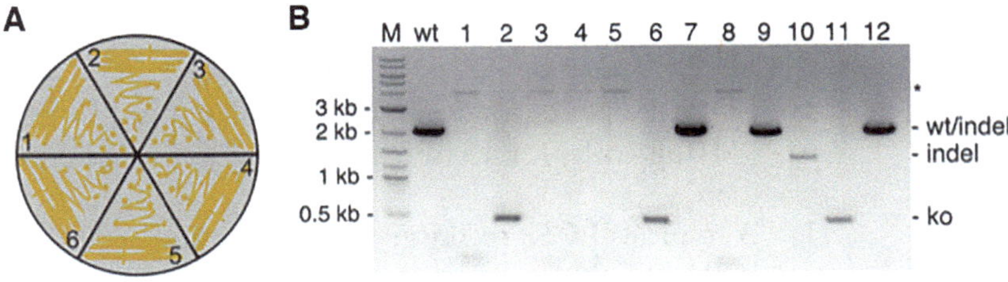

Fig. 4 (**A**) Scheme for singling out six transformants on one PD_{cbx} plate. (**B**) Test PCR using ko5 and ko6. Ko: successful knock-out; indel: NHEJ Δ542 bp; wt/indel: wt or NHEJ with few bp; *: chromosomal loss of telomers by DSB of Cas9hf (um01436 is located at the end of chr. 2 of *U. maydis*)

3.3.1 Colony Selection and gDNA Preparation

1. Warm up two PD_{cbx} plates to RT and mark six fields on each plate (Fig. 4A).
2. Twelve individual colonies from the transformation plate should be streaked out using a 0.1 mL glass pipette to gain single colonies after three days at 28°C (Fig. 4A).
3. Prepare 13 tubes with 3 mL $YEPS_{light}$.
4. Inoculate one single colony from fields 1–12 and the wt strain for preparation of genomic DNA. Grow them o/n at 28°C.
5. On the next morning, the phenotype of the mutants can be studied using a standard microscope with 40× objective.
6. Before starting with the gDNA extraction spot 3 μL of each culture on PD and PD_{cbx} plates and let them grow for 2 days at 28°C.
7. Transfer 0.2 mL glass beads in a 2 mL Eppendorf tube (safe-lock) and add 1.8 mL culture.
8. After centrifugation (2 min, 14,000 g), discard the supernatant (pellets can be frozen at −20°C).
9. Under the hood (Steps 9–14) add 500 μL Usti-lysis buffer and 500 μL PCI (*see* **Note 16**). Place the mixture on a Vibrax shaker (10 min, 1500 rpm).
10. Make sure that all cell pellets are dissolved and centrifuge (10 min, 14,000 g).
11. In the meantime, transfer 0.95 mL EtOH to a 1.5 mL Eppendorf tube.
12. Add 400 μL of the upper aqueous phase without any interphase and mix the solution by inverting the tube several times. gDNA should be visible.
13. Incubate for 5 min and centrifuge to precipitate the gDNA (10 min, 14,000 g, 20°C).
14. Remove the supernatant, wash the DNA pellet with 500 μL 70% EtOH and centrifuge again (5 min, 14,000 g).

15. Remove the supernatant, collect residual EtOH by a short centrifugation step and pipette this out.
16. Dry the pellet for 5 min and dissolve it in 100 µL TE+RNase (10 µg/mL) for 15 min at 55°C; spin the tube briefly.

3.3.2 PCR Analysis to Verify Gene Deletion

1. A typical PCR reaction using Phusion™-DNA polymerase looks like this:
   1 µL gDNA, 1 µL Primer ko5 (10 pMol/µL), 1 µL Primer ko6 (10 pMol/µL), 30 µL H_2O, 10 µL GC-buffer, 5 µL DMSO, 1 µL dNTPs, and 1 µL Phusion polymerase.
   Such PCR protocol normally works:
   Step 1: 98°C 2 min; Step 2: 98°C 10 s; Step 3: 55°C 20 s; Step 4: 72°C 100 s (depending on the size of the ORF to be deleted) (35× Steps 2–4); Step 5: 72°C 5 min.
2. Stop each reaction with 10 µL DNA loading buffer (6×) and load 15 µL on a 1% TAE-agarose gel and run the gel for 45 min. Gene deletion products should be 500 bp, NHEJ products can differ in size (Fig. 4B).
3. Positive candidate PCR products should be purified and sequenced using ko5 or ko6 as sequencing primers (*see* **Note 17**).

3.4 Marker-Free Strain Isolation for Glycerol Stock Conservation

1. The pSM2 plasmid is an ARS plasmid and might get lost during growth in media without carboxin. To get completely rid of the plasmid, the selected ko clones should be inoculated from the PD plate (without Carboxin) in 3 mL $YEPS_{light}$, grown o/n at 28°C and streaked out again on PD and PD_{cbx} plates.
2. This should be repeated until no growth could be observed on the PD_{cbx} plates (*see* **Note 18**).
3. A carboxin sensitive colony should be inoculated from the corresponding PD plate in 3 mL $YEPS_{light}$ and grown o/n at 28°C.
4. On the next day, 1 mL culture is mixed with 1 mL NSY-Glycerin-Medium and stored at −80°C.

4 Notes

1. Wear protective eyewear in case of bumping and a hot hand to transport the Erlenmeyer flask containing the agarose solution. Do not use a magnetic stirrer. Midori green Advance ™ binds to DNA, and is a nontoxic and nonhazardous alternative to Ethidium bromide.
2. In our lab, the pH is always below pH = 6.4 (in the range of 5.5–6.2). Therefore, there is no need to adjust the pH with

1 N HCl. It is important to know that above pH 6.4 MnO_2 starts to precipitate.

3. Using double distilled water, the PD agar is often very soft and not suitable for singling out cells.
4. Sterilizing using 0.45 μm filters is more exhausting than autoclaving but should be preferred.
5. The result page orders suitable sequences according to their E-scores (E = efficiency). If there are no results, it is possible to set the start application to medium although in our hands all genes, which received no results in the "strict"-setting could not be deleted using "medium" PAM sequences.
6. pSM2 contains an ARS, an expression cassette for the Cas9hf nuclease, a Carboxin resistance marker gene and an expression cassette for the sgRNA under the control of the U6 snRNA promoter [17].
7. The digested vector is larger than 10 kb. Common gel extraction kits have a special section for DNA fragments >1000 bp.
8. 0.1 ng pUC18 (approximately 3 kB) should be used in a control transformation. Other small plasmid vectors with similar size could be used, too.
9. After the heat shock, no regeneration step is necessary and can be omitted. Ampicillin blocks the development of an intact cell wall, but gene expression is not inhibited like it is for other antibiotics such as kanamycin or chloramphenicol.
10. It is important to mention that sometimes the largest band of 9400 bp is reduced in size due to recombination events in the ETEF-promotor.
11. The doubling time for *U. maydis* wt cells is approximately 120 min.
12. If you see clumps of cell under the microscope when checking for rounding up, the cells should be resuspended further.
13. For testing, the transformation efficiency 100 ng of supercoiled pSM2 should be used resulting in approximately 100 cfu. Alternatively, 100 ng pNEBUC can be used as control [2].
14. To avoid air bubbles, take up 11 mL in a glass pipette and pour 10 mL in the petri dish. This avoids the last drops with air bubbles. Be fast. The agar hardens quickly.
15. You need to check that the shelves in the incubator are really horizontal. Otherwise, the solution is collected on one side of the agar plate.
16. The Phenol/chloroform/isoamylalkohol-solution should contain an upper aqueous phase and can be stained yellowish with some granules of 8-hydroxychinolin.

17. One primer (e.g., ko5) is sufficient for sequencing of a 500 bp product in case of homologous recombination. NHEJ products should be sequenced from both sites. Due to the location of um01436 near to the end of chr. 2, the DSB induced by Cas9hf lead to loss of the telomeric region. In all cases labeled with *, the neighboring gene um01438 was also lost.
18. Normally, the complete plasmid loss occurs after three rounds of inoculation and spotting on PD and PD_{cbx} plates.

Acknowledgments

This work was supported by the grant SA 1018/5-1 from the Deutsche Forschungsgemeinschaft (DFG) to B.S.

References

1. Kämper J (2004) A PCR-based system for highly efficient generation of gene replacement mutants in *Ustilago maydis*. Mol Genet Genomics 271(1):103–110. https://doi.org/10.1007/s00438-003-0962-8 , Feb, Epub 2003 Dec 12, PMID: 14673645
2. Brachmann A, König J, Julius C et al (2004) A reverse genetic approach for generating gene replacement mutants in *Ustilago maydis*. Mol Genet Genomics 272(2):216–226. https://doi.org/10.1007/s00438-004-1047-z , Sep, Epub 2004 Aug 17. Erratum in: Mol Genet Genomics 272(4):488. Nov PMID: 15316769
3. García-Pedrajas MD, Nadal M, Kapa LB et al (2008) DelsGate, a robust and rapid gene deletion construction method. Fungal Genet Biol 45(4):379–388. https://doi.org/10.1016/j.fgb.2007.11.001 , Apr, Epub 2007 Nov 19, PMID: 18248826
4. Terfrüchte M, Joehnk B, Fajardo-Somera R et al (2014) Establishing a versatile golden gate cloning system for genetic engineering in fungi. Fungal Genet Biol 62:1–10. https://doi.org/10.1016/j.fgb.2013.10.012 , Jan, Epub 2013 Nov 8, PMID: 24211735
5. Khrunyk Y, Münch K, Schipper K et al (2010) The use of FLP-mediated recombination for the functional analysis of an effector gene family in the biotrophic smut fungus *Ustilago maydis*. New Phytol 187(4):957–968. https://doi.org/10.1111/j.1469-8137.2010.03413.x , Sep, Epub 2010 Jun 18, PMID: 20673282
6. Mojica FJ, Díez-Villaseñor C, Soria E et al (2000) Biological significance of a family of regularly spaced repeats in the genomes of archaea, bacteria and mitochondria. Mol Microbiol 36(1):244–246. https://doi.org/10.1046/j.1365-2958.2000.01838.x , Apr, PMID: 10760181
7. Bhaya D, Davison M, Barrangou R (2011) CRISPR-cas systems in bacteria and archaea: versatile small RNAs for adaptive defense and regulation. Annu Rev Genet 45:273–297. https://doi.org/10.1146/annurev-genet-110410-132430 , PMID: 22060043
8. Makarova KS, Wolf YI, Iranzo J et al (2020) Evolutionary classification of CRISPR-cas systems: a burst of class 2 and derived variants. Nat Rev Microbiol 18(2):67–83. https://doi.org/10.1038/s41579-019-0299-x , Feb, Epub 2019 Dec 19, PMID: 31857715; PMCID: PMC8905525
9. Mojica FJM, Díez-Villaseñor C, García-Martínez J et al (2009) Short motif sequences determine the targets of the prokaryotic CRISPR defence system. Microbiology (Reading) 155(Pt 3):733–740. https://doi.org/10.1099/mic.0.023960-0 , Mar, PMID: 19246744
10. Shah SA, Erdmann S, Mojica FJ et al (2013) Protospacer recognition motifs: mixed identities and functional diversity. RNA Biol 10(5):891–899. https://doi.org/10.4161/rna.23764 , May, Epub 2013 Feb 12, PMID: 23403393; PMCID: PMC3737346
11. Wiedenheft B, Sternberg SH, Doudna JA (2012) RNA-guided genetic silencing systems in bacteria and archaea. Nature 482(7385):331–338. https://doi.org/10.1038/nature10886 , Feb 15, PMID: 22337052
12. Cong L, Ran FA, Cox D et al (2013) Multiplex genome engineering using CRISPR/Cas

systems. Science 339(6121):819–823. https://doi.org/10.1126/science.1231143 , Feb 15, Epub 2013 Jan 3, PMID: 23287718; PMCID: PMC3795411

13. Sander JD, Joung JK (2014) CRISPR-Cas systems for editing, regulating and targeting genomes. Nat Biotechnol 32(4):347–355. https://doi.org/10.1038/nbt.2842 , Apr, Epub 2014 Mar 2, PMID: 24584096; PMCID: PMC4022601
14. Lu M, Yu C, Zhang Y et al (2024) Structure and genome editing of type I-B CRISPR-Cas. Nat Commun 15(1):4126. https://doi.org/10.1038/s41467-024-48598-2 , May 15, PMID: 38750051; PMCID: PMC11096372
15. Schuster M, Schweizer G, Reissmann S et al (2016) Genome editing in *Ustilago maydis* using the CRISPR-Cas system. Fungal Genet Biol 89:3–9. https://doi.org/10.1016/j.fgb.2015.09.001 , Apr, Epub 2015 Sep 11, PMID: 26365384
16. Schuster M, Trippel C, Happel P et al (2018) Single and multiplexed gene editing in *Ustilago maydis* using CRISPR-Cas9. Bio Protoc 8(14):e2928. https://doi.org/10.21769/BioProtoc.2928 , Jul 20, PMID: 34395750; PMCID: PMC8328660
17. Wege SM, Gejer K, Becker F et al (2021) Versatile CRISPR/Cas9 systems for genome editing in *Ustilago maydis*. J Fungi (Basel) 7(2):149. https://doi.org/10.3390/jof7020149 , Feb 18, PMID: 33670568; PMCID: PMC7922307
18. Heigwer F, Kerr G, Boutros M (2014) E-CRISP: fast CRISPR target site identification. Nat Methods 11(2):122–123. https://doi.org/10.1038/nmeth.2812 , Feb, PMID: 24481216
19. Hanahan D, Jessee J, Bloom FR (1991) Plasmid transformation of *Escherichia coli* and other bacteria. Methods Enzymol 204:63–113
20. Schulz B, Banuett F, Dahl M et al (1990) The *b* alleles of *U. maydis*, whose combinations program pathogenic development, code for polypeptides containing a homeodomain-related motif. Cell 60:295–306

Chapter 16

FLP/FRT-Mediated Recombination for Resistance Marker Recycling

Lorenzo García Tejada, An N. T. Phan, Kerstin Schipper and Lars M. Blank

Abstract

The availability of selection markers for metabolic engineering of *Ustilago* species is limited. The inducible flippase (FLP) recombinase system allows for the recycling of a selection marker after a successful recombination event. Following a traditional marker-based homologous recombination of DNA for deletion or insertion of a gene using special constructs with two flanking Flippase-Recognition Targets (FRT) at both sides of the selection marker, FLP-mediated recombination allows for marker removal after transformation. This enables its recycling for a further transformation approach. In *Ustilago maydis*, the FLP protein has been codon optimized and configured under the control of an arabinose-inducible promoter in a self-replicating plasmid for episomal expression of the recombinase. This FLP/FRT system facilitates advanced genetic modifications in *Ustilago* spp. by enabling repeated genetic manipulation with one marker, thus streamlining molecular genetic methodologies.

Keywords FLP/FRT, Recombination, Flippase, Antibiotic marker recycling, Gene deletion

1 Introduction

The FLP/FRT system originates from the two-micron plasmid of *Saccharomyces cerevisiae*, where it serves as an evolutionary mechanism to ensure correct copy number of the plasmid and its adequate segregation between mother and daughter cells [1, 2]. It consists of two elements: a recombinase ("Flippase") enzyme encoded by the FLP gene and its target recognition sequences, FRT (Flippase-Recognition Target) [3]. The FRT sequences consist of a conserved 8 bp asymmetric core region flanked by 13 bp inverted, imperfect palindromic sequences (Fig. 1A). The standard sequence reads: 5′–GAAGT TCCTA TTCtc tagaa aGAAT AGGAA CTTC–3′. The palindromic repeats serve as the anchor point for FLP dimers to bind, whereas the core region directs the directionality of the sequence, and therefore, the expected outcome of recombination events [2, 4]. Strategies for the application of the FLP/FRT system for directed mutagenesis include insertion, excision, inversion, and translocation

An N. T. Phan et al. (eds.), *Ustilago maydis*, Methods in Molecular Biology, vol. 3050,
https://doi.org/10.1007/978-1-0716-5364-7_16,

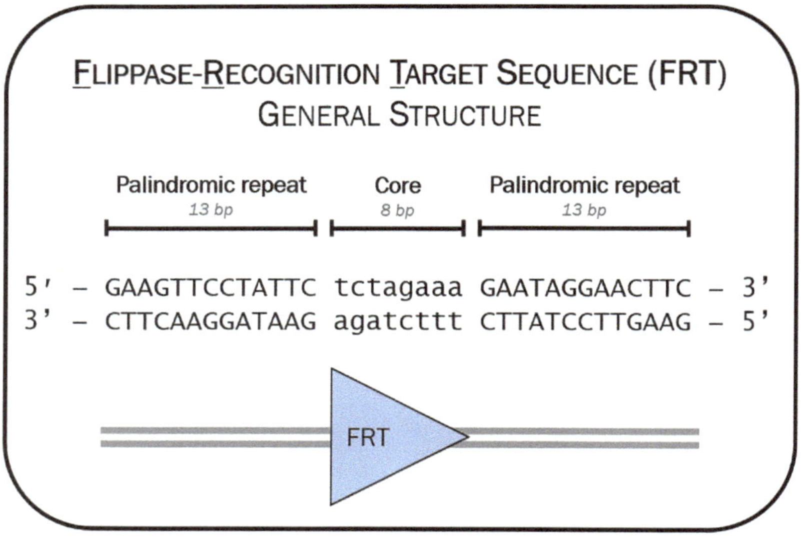

Fig. 1 General structure of the standard FRT sequence. The 13 bp inverted palindromic repeats flank an 8 bp core region that gives the directionality of the sequence (represented by a triangle diagram)

of gene sequences [5]. It has been adapted for use in bacteria [6, 7], yeast [8], fungi [9], plant [10–12], or animal cells [13–15].

In *Ustilago* spp., it has been utilized as a mean of marker-based gene deletion that allows for the resistance marker to be excised from the genome in a second step (Fig. 2) [16]. This process relies on the homologous recombination of a hygromycin B resistance cassette (*HygR*) between two homology arms (upstream/downstream flank) that flank the gene to delete, therefore, allowing for cultivation of deletion mutants through selective pressure cultivation with hygromycin B. The *HygR* cassette is further flanked by two FRT sites between each homology arm, which enable its recycling out of the genome through the action of an arabinose-induced FLP recombinase encoded in a free-replicating expression plasmid with a carboxin resistance (*CbxR*) cassette. Genotypically confirmed mutants—either by diagnostic PCR or sequencing—are serially cultured in nonselective media to achieve plasmid loss, thereby resulting in a deletion strain with no antibiotic marker.

It should be noted that one copy of the FRT sequence remains in the genome after deletion, so simultaneous or successive engineering attempts might lead to undesired chromosomal rearrangements. Khrunyk et al. [16] showed that it was possible for two regionally spaced, standard FRT sequences in the *U. maydis* genome to produce the deletion of a higher length of chromosomal DNA (0.43 Mbp) with the expression of an active FLP enzyme. To address this issue, the authors explored a series of mutations in the highly conserved core region of the FRT region, as it was shown in other organisms that these led to more stable integration products with lower

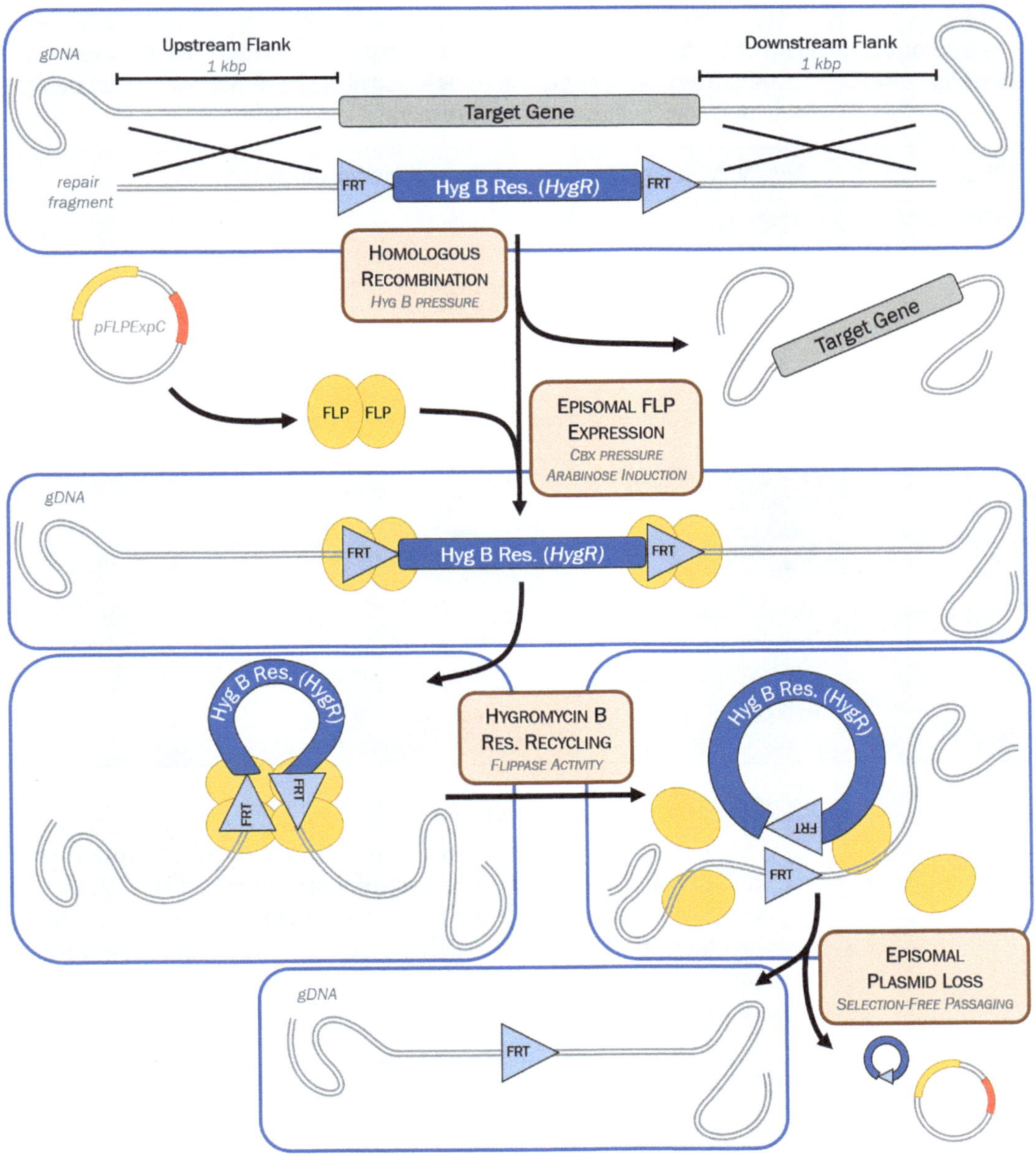

Fig. 2 Graphical representation of a gene deletion event through homologous recombination of a repair template harboring a hygromycin B resistance marker (*HygR*) flanked by FRT sequences, and its recycling through episomal expression of a heterologous FLP recombinase. The process yields a genomic deletion with a single FRT site scar in lieu of the original target gene

recombination rates between identical FRT sites (Table 1) [17, 18]. These point mutations were shown to have a lower recombination efficiency than the standard FRT sequence through loss of the *HygR* cassette, while not experiencing the same observed chromosomal rearrangement.

Table 1

Collection of the standard FRT 8 bp core sequence (wt) and mutations thereof with their respective recombination efficiency, where available, as evidenced by Khrunyk et al. [16, Supplementary Data], or Terfrüchte et al. [25]. Point mutations in the sequences are marked in bold letter

FRT site	Core point mutation	Recombination efficiency
FRT (wt)	tctagaaa agatcttt	68% (*n*=259) [16]
FRTm1	tc***g***agaaa ag***c***tcttt	40% (*n*=267) [16]
FRTm2	tc***a***agaaa ag***t***tcttt	21% (*n*=205) [16]
FRTm3	tc***c***agaaa ag***g***tcttt	42% (*n*=264) [16]
FRTm4	tct***g***gaaa aga***c***cttt	14% (*n*=272) [16]
FRTm6	tct***t***gaaa aga***a***cttt	Successful [25] Efficiency unreported
FRTm10	tctag***c***aa agatc***g***tt	Successful [25] Efficiency unreported
FRTm11	tctag***g***aa agatc***c***tt	Successful [25] Efficiency unreported

This system has been utilized to understand the function of redundant genes in the genome of *Ustilago* spp. [16, 19, 20], as well as to engineer different *Ustilago* species for the biotechnological production of molecules, such as itaconic acid, sesquiterpenoids, or single chain antibodies [21–24].

2 Material

2.1 Strains and Plasmids

E. coli DH5α is recommended as the expression host for standard molecular biology steps.

For the cloning of the deletion repair fragment in *E. coli*, any high copy number plasmid with an *E. coli* ORI and a corresponding selectable marker are suitable. The pJET1.2/blunt vector is an example of an already linearized vector with ampicillin resistance for insertion of DNA.

A target *Ustilago* spp. strain, with respective protoplast suspensions (*see* **Note 1**).

For the arabinose-inducible expression of the flippase enzyme, the chapter describes the use of the pFLPexpC plasmid containing *CbxR* [16].

2.2 Media and Solutions

1. YEPS—1% (w/v) yeast extract, 1% (w/v) peptone, 2% (w/v) sucrose + 2% (w/v) agar–agar for solid media.
   - Other complex media that allow for unrestricted growth of the respective *Ustilago* spp. are also acceptable, such as complete medium (CM) or potato-dextrose (PD).
2. YEPAra—1% (w/v) yeast extract, 1% (w/v) peptone, 1% (w/v) arabinose
3. Saline solution - 0.9% (w/v) NaCl.
4. Carboxin stock solution (recommended 2 mg/mL in methanol; store at 4°C).
5. Hygromycin B stock solution (recommended 50 mg/mL in water or 200 mg/mL in methanol; store at 4°C, protect from light).

2.3 Equipment

1. PCR cycler.
2. Electrophoresis chamber and agarose-gel imaging device.
3. High-fidelity polymerase such as Q5® or CloneAmp™.
4. Plasmid extraction, DNA purification, and DNA assembly kits.
5. Microvolume spectrophotometer for determination of DNA concentration.
6. Wardrobe incubator (30°C).
7. Petri Dishes.
8. Shake flasks (100 mL).
9. Orbital shaking incubator (30°C, 200 rpm, 25 mmø, 70% humidity).

3 Methods

3.1 Deletion Construct Design

1. The method described aims for DNA assembly via homologous overlaps to generate the required deletion vectors. Alternative strategies exist (*see* **Note 2**).
2. Identify the target region to be excised from the genome of the chosen *Ustilago* species. Identify the 1 kb flanking regions directly upstream and downstream from it. Design appropriate primers to amplify both regions.
3. Modify the forward primer of the upstream homology arm and the reverse primer of the downstream homology arm to

include 20–40 overlap regions of the MCS in the expression plasmid of choice in *E. coli*.

4. Modify the reverse primer of the upstream homology arm to include the FRT sequence (or its complementary if primer anneals to the reverse strand) (*see* **Note 3**).
5. Repeat Step 2 for the forward primer of the downstream homology arm (or its complementary if primer anneals to the reverse strand) (*see* **Note 4**).
6. Design primers to amplify a hygromycin resistance cassette (*HygR*), and introduce the FRT sequences on both primers, keeping in mind they should both have the same orientation.

3.2 Expression Plasmid Construction

7. Perform genomic DNA extraction of the target *Ustilago* species.
8. Amplify the homology arms and the hygromycin cassette with the respective designed primers through PCR using a high-fidelity polymerase (*see* **Note 5**).
9. Purify the products and confirm the sequences by size through electrophoresis in an agarose gel or sequencing. if necessary (*see* **Note 6**).
10. Construct the expression plasmid through a DNA assembly method, e.g., Gibson assembly®, Aqua cloning, and transform competent *E. coli* strains with the corresponding antibiotic pressure. The resulting structure of the deletion construct would, therefore, be: Flank1-FRT-*HygR*-FRT-Flank2.
11. Perform colony PCR to confirm correct construct assembly (*see* **Note 7**).
12. Isolate positive clones and propagate them. Perform plasmid extraction and sequence if necessary (*see* **Note 8**).
13. On the purified expression plasmid, perform a PCR amplification of the entire deletion product with primers on each side of the homology arms. Confirm size through electrophoresis in an agarose gel and purify the PCR products. Determine concentration of the resulting DNA solution through spectrophotometry (*see* **Note 9**).

3.3 Transformation of Ustilago spp. Protoplasts for Target Gene Deletion

14. Use 500–2000 ng of the purified deletion construct to transform protoplasts of the target strain (*see* **Note 10**).
15. Spread the transformation cell solution on REG-Agar plates with hygromycin B as selection agent (*see* **Note 11**).
16. Incubate plates at 30°C for 3–7 days, or until single colonies are observable (*see* **Note 12**).
17. Propagate single colonies by restreaking or dropping onto a YEPS Agar master plate with hygromycin B and incubate at 30°C overnight.

18. Perform colony PCR of positive clones by amplification of the entire deletion construct, the *HygR* cassette, or a combination of both. Phenotypic assays are also recommended if possible for the deleted gene (*see* **Note 13**).
19. Make cryosolutions of positive deletion mutants by mixing 1:1 exponentially-grown cultures and sterile 80% (v/v) glycerol.

3.4 Recycling of the Resistance Marker

20. Recycling of the resistance marker is achieved through transformation of the strains with an episomal expression plasmid containing the *flp* gene, e.g., pFLPexpC [16]. The plasmid includes an ORI, a reporter LacZ and an ampicillin resistance gene (Amp^R) for *E. coli*, an autonomous replicating signal for *Ustilago* (UARS), a carboxin resistance cassette ($CbxR/ip^R$) and the codon optimized*flp* gene for *U. maydis* under the control of an arabinose-inducible promoter (P_{crg1}) and the T_{nos} terminator (*see* **Note 14**).
21. Extract the plasmid from *E. coli*, confirm size and identity through electrophoresis in an agarose gel and sequence, if necessary. Determine DNA concentrations of the plasmid through spectrophotometry.
22. Generate protoplasts of the *Ustilago* spp. strain harboring the deletion construct.
23. Use 500–2000 ng of the purified FLP-expressing plasmid for transformation of the *Ustilago* spp. protoplasts.
24. Spread the transformation cell solution on REG Agar plates with carboxin as selective agent and incubate for 3–7 days at 30°C (*see* **Note 15**).
25. Propagate single colonies by restreaking or dropping onto a YEPS agar master plate with carboxin and incubate at 30°C overnight.
26. Select positive growth clones and inoculate a 100 mL shake flask with 10 mL of liquid YEPS with 2 µg/mL carboxin. Grow overnight at 30°C and 200 rpm (ø 25 mm) shaking (*see* **Note 16**).
27. Wash the cells with 0.9% (w/v) NaCl and use the resuspended solution to inoculate a 100 mL shake flask with 10 mL of liquid YEPAra with 2 µg/mL carboxin to an OD_{600} of 0.7 in order to induce expression of the FLP enzyme.
28. Incubate cells at 30°C and 200 rpm for 16–24 h.
29. Plate serial dilutions of the cultivation broth on YEPS plates and incubate at 30°C overnight to obtain isolated colonies.
30. Propagate single colonies onto a YEPS master plate and incubate overnight. Alternatively, propagate colonies simultaneously on YEPS plates with hygromycin B to test for sensitivity to the antibiotic.

31. Perform colony PCR of positive clones by amplification of both homology arms to test for loss of the hygromycin resistance. Sizes expected exclude the size of the *HygR* cassette sequence.
32. Make cryo-solutions of positive mutants lacking the *HygR* sequence and/or unable to grow on YEPS plates with hygromycin B.

3.5 Loss of FLP Expression Plasmid

33. Inoculate the positive mutants with tested recycling of the *HygR* cassette in a 100 mL shake flask with 10 mL of liquid YEPS. Incubate overnight at 30°C and 200 rpm.
34. Plate the cell culture in YEPS and incubate overnight at 30°C.
35. Test resistance of the obtained single colonies to carboxin by replica plating or restreaking on YEPS-Cbx plates and incubating overnight.
36. Repeat Steps 32–34 if no colony is found with complete loss of the carboxin resistance.
37. Make cryo-solutions of positive mutants sensitive to carboxin, signaling plasmid loss.
38. A final check of the strain can be performed by testing its sensitivity to hygromycin and carboxin on YEPS agar plates, both of which should be able to limit growth of the cells.

4 Notes

1. An *Ustilago* spp. strain for which there is genomic data available, either a reference genome or actual strain sequencing data, is recommended for this transformation, since the gene targeting requires its previous identification for amplification. In the event no genomic data is available, ortholog regions can be used from closely related strains, but extra attention should be paid to resulting phenotypes.
2. Alternatively, a Golden Gate system for the assembly of the construct can also be used, which has already been described for fungal species, including *Ustilago* spp. [25].
3. If multiple deletion events are planned for the same strain with this method, mutant FRT sites are advised for enhanced transformation stability, though lower recombination efficiency is to be expected (Table 1).
4. Important: For deletions, both FRT sites should have the same orientation in the resulting DNA for correct recombination through FLP activity, which is given by the 8 bp core region (Fig. 1).

5. Annealing temperature should only be considered for the primer sections that anneal to the target DNA—FRT sites and backbone overhangs should be excluded from these calculations.
6. Depending on the original vector from which the hygromycin cassette was amplified, template digestion with appropriate restriction enzymes is advised to ensure correct selective pressure in the assembly stage, e.g., DpN-I digestion if the vector was amplified from a plasmid propagated in *E. coli* with the same selection agent than the expression plasmid used for cloning of the deletion construct. Otherwise, if the same selection marker is used for both plasmids, *E. coli* cells might propagate the storage plasmid used for template as opposed to the assembled construct, returning a false positive colony in the transformation plate.
7. Total construct size ranges from 4 to 5 kbp depending on the specific sequence of the *HygR* cassette. Amplification of the whole construct is recommended, though if lower cycle times are desired, overlapping PCR products between homology arms and the *HygR* cassette would also point at potentially correct assembly results.
8. The correct orientation in the FRT sites is critical for the recycling of the hygromycin resistance gene. Therefore, whole plasmid sequencing is highly advised.
9. PCR amplification might carry the risk of introducing point mutation on the DNA construct depending on the fidelity of the polymerase used, and strain verification should be performed with extra care to rule out these potential mutations. To avoid this problem, engineering restriction sites to the flanking sites of the construct would allow for excision through use of the respective restriction enzymes and successive purification, though larger volume restrictions might need to be performed to reach the target transformation DNA quantities.
10. Different *Ustilago* species have different efficiencies of transformation depending on the amount of DNA used, e.g., *U. maydis* might have sufficient transformant colonies with 500–1000 ng, whereas *U. cynodontis* might experience increased efficiency with 1500–2000 ng of DNA.
11. A concentration of 200 μg/mL is recommended, but the concentrations should be adjusted to the resistance specifications of each particular strain, for which a prior drop test on different antibiotic concentrations is recommended.
12. Depending on the volume of the PEG-protoplast-DNA solution that is used for transformation, incubation of the places with the agar at the bottom might be beneficial to prevent

dripping on to the lid if incubated upside down. Lower volumes of solution (<1 mL) might not drip if spread evenly and allowed to settle and could be incubated upside down, but higher volume would require incubations facing up to ensure proper settlement of the transformed cells onto the agar.

13. Amplification of the construct or a part thereof carries the potential risk of false positives on DNA that has not integrated in the genome or has integrated in a wrong locus. Propagation of positive growth colonies onto a master plate with the adequate selection pressure would minimize this risk, through to eliminate it altogether, using primer constructs that anneal to the genomic DNA immediately outside of the flanking regions would allow to amplify the entire target fragment to compare total size differences between the native gene and the construct. Similarly, a combination of primers that anneal on the genome and inside the *HygR* would attest to positive integration of the construct in the target locus.
14. Introduction of the inducible FLP cassette into the repair fragment alongside the *HygR* cassette is also possible and has been demonstrated in literature [16, Supplementary Information, S5]. This shortens the time for the entire process, since it removes the need for a second transformation of the inducible plasmid and its successive loss after marker recycling, though the size of the repair fragment significantly increases.
15. Concentration of carboxin for mutant selection ranges between different *Ustilago* species, e.g., *U. cynodontis* requires 2–4 μg/mL carboxin for transformation, whereas *U. maydis* or *U. trichophora* benefit from higher concentrations up to 10 μg/mL for the reduction of background growth of false positive colonies.
16. Whereas transformation efficiency is increased depending on the particular sensitivity of each *Ustilago* spp. to the antibiotic in solid media, as mentioned previously, for liquid culture, lower concentrations of the antibiotic allow for sufficient selection pressure while enabling observable growth rates. Nevertheless, a longer lag phase and/or lower growth rates are typically expected.

5 Funding

Funded by the European Union's Horizon 2020 research and innovation programme under the Marie Skłodowska-Curie grant agreement No. 101034323 (Talent4BBI), cofunded by BiOrbic Research Ireland Centre for Bioeconomy; and by the Deutsche

Forschungsgemeinschaft (DFG, German Research Foundation)—SFB1535—Project ID 458090666.

References

1. Reynolds AE, Murray AW, Szostak JW (1987) Roles of the 2 microns gene products in stable maintenance of the 2 microns plasmid of *Saccharomyces cerevisiae*. Mol Cell Biol 7(10):3566. https://doi.org/10.1128/MCB.7.10.3566
2. Umlauf SW, Cox MM (1988) The functional significance of DNA sequence structure in a site-specific genetic recombination reaction. EMBO J 7(6):1845. https://doi.org/10.1002/J.1460-2075.1988.TB03017.X
3. Sadowski PD (1995) The flp recombinase of th 2-μm plasmid of *Saccharomyces cerevisiae*. Prog Nucleic Acid Res Mol Biol 51(C):53–91. https://doi.org/10.1016/S0079-6603(08)60876-4
4. Senecoff JF, Rossmeissl PJ, Cox MM (1988) DNA recognition by the FLP recombinase of the yeast 2 μ plasmid: a mutational analysis of the FLP binding site. J Mol Biol 201(2):405–421. https://doi.org/10.1016/0022-2836(88)90147-7
5. Voziyanov Y, Grainge I, Jayaram M (2005) Applications of fungal site-specific recombination as a tool in biotechnology and basic biology. Appl Mycology Biotechnol 5(C):189–210. https://doi.org/10.1016/S1874-5334(05)80010-3
6. Schweizer HP (2003) Applications of the *Saccharomyces cerevisiae* Flp-FRT system in bacterial genetics. J Mol Microbiol Biotechnol 5(2):67–77. https://doi.org/10.1159/000069976
7. Ji X, Wang Y, Li J et al (2017) Application of FLP-FRT system to construct unmarked deletion in *Helicobacter pylori* and functional study of gene hp0788 in pathogenesis. Front Microbiol 8(Nov):263607. https://doi.org/10.3389/FMICB.2017.02357/TEXT
8. Park YN, Masison D, Eisenberg E et al (2011) Application of the FLP/FRT system for conditional gene deletion in yeast *Saccharomyces cerevisiae*. Yeast (Chichester, England) 28(9):673. https://doi.org/10.1002/YEA.1895
9. Kopke K, Hoff B, Kuck U (2010) Application of the *Saccharomyces cerevisiae* FLP/FRT recombination system in filamentous fungi for marker recycling and construction of knockout strains devoid of heterologous genes. Appl Environ Microbiol 76(14):4664. https://doi.org/10.1128/AEM.00670-10
10. Lyznik LA, Rao KV, Hodges TK (1996) FLP-mediated recombination of FRT sites in the maize genome. Nucleic Acids Res 24(19):3784–3789. https://doi.org/10.1093/NAR/24.19.3784
11. Hu Q, Kononowicz-Hodges H, Nelson-Vasilchik K et al (2008) FLP recombinase-mediated site-specific recombination in rice. Plant Biotechnol J 6(2):176–188. https://doi.org/10.1111/J.1467-7652.2007.00310.X;WGROUP:STRING:PUBLICATION
12. Rao MR, Moon HS, Schenk TMH et al (2010) FLP/FRT recombination from yeast: application of a two gene cassette scheme as an inducible system in plants. Sensors 10(9):8526–8535. https://doi.org/10.3390/S100908526
13. Long DP, Zhao AC, Chen XJ et al (2012) FLP recombinase-mediated site-specific recombination in silkworm, *Bombyx mori*. PLOS One 7(6):e40150. https://doi.org/10.1371/JOURNAL.PONE.0040150
14. Phan QV, Contzen J, Seemann P et al (2017) Site-specific chromosomal gene insertion: Flp recombinase versus Cas9 nuclease. Sci Rep 7(1):1–12. https://doi.org/10.1038/s41598-017-17651-0
15. Shah R, Li F, Voziyanova E et al (2015) Target-specific variants of Flp recombinase mediate genome engineering reactions in mammalian cells. FEBS J 282(17):3323. https://doi.org/10.1111/FEBS.13345
16. Khrunyk Y, Münch K, Schipper K et al (2010) The use of FLP-mediated recombination for the functional analysis of an effector gene family in the biotrophic smut fungus *Ustilago maydis*. New Phytol 187(4):957–968. https://doi.org/10.1111/J.1469-8137.2010.03413.X;-JOURNAL:JOURNAL:14698137;WGROUP:STRING:PUBLICATION
17. Barrett AR, Kang Y, Inamasu KS et al (2008) Genetic tools for allelic replacement in *Burkholderia* species. Appl Environ Microbiol 74(14):4498–4508. https://doi.org/10.1128/AEM.00531-08/SUPPL_FILE/AEM00531_08_SUPPLEMENTARY_FIG.ZIP
18. Storici F, Coglievina M, Bruschi CV (1999) A 2-μm DNA-based marker recycling system for multiple gene disruption in the yeast *Saccharomyces cerevisiae*. Wiley Online Libr 15:271–283. https://doi.org/10.1002/(SICI)1097-0061(19990315)15:4
19. Schmitz L, Kronstad JW, Heimel K (2019) Conditional gene expression reveals stage-specific functions of the unfolded protein response in the *Ustilago maydis*–maize pathosystem.

Mol Plant Pathol 21(2):258. https://doi.org/10.1111/MPP.12893
20. Hosseinpour Tehrani H, Tharmasothirajan A, Track E et al (2019) Engineering the morphology and metabolism of pH tolerant *Ustilago cynodontis* for efficient itaconic acid production. Metab Eng 54:293–300. https://doi.org/10.1016/J.YMBEN.2019.05.004
21. Becker J, Hosseinpour Tehrani H, Gauert M et al (2020) An *Ustilago maydis* chassis for itaconic acid production without by-products. Microb Biotechnol 13(2):350–362. https://doi.org/10.1111/1751-7915.13525
22. Sarkari P, Reindl M, Stock J et al (2014) Improved expression of single-chain antibodies in *Ustilago maydis.* J Biotechnol 191:165–175. https://doi.org/10.1016/J.JBIOTEC.2014.06.028
23. Hosseinpour Tehrani H, Becker J, Bator I et al (2019) Integrated strain- and process design enable production of 220 g L^{-1} itaconic acid with *Ustilago maydis.* Biotechnol Biofuels 12(1):263. https://doi.org/10.1186/S13068-019-1605-6
24. Lee J, Hilgers F, Loeschke A et al (2020) *Ustilago maydis* serves as a novel production host for the synthesis of plant and fungal sesquiterpenoids. Front Microbiol 11:558071. https://doi.org/10.3389/FMICB.2020.01655/XML
25. Terfrüchte M, Joehnk B, Fajardo-Somera R et al (2014) Establishing a versatile golden gate cloning system for genetic engineering in fungi. Fungal Genet Biol 62:1–10. https://doi.org/10.1016/j.fgb.2013.10.012

Chapter 17

Conditional Gene Deletion in *Ustilago maydis* Via CRE Recombinase In Planta

Antonio de la Torre and José Pérez-Martín

Abstract

Site-specific recombinases have been extensively utilized in higher eukaryotes, particularly in animals, for various applications such as chromosomal translocations, large deletions, site-specific integration, and, notably, tissue-specific conditional knockouts. This technique has also been applied in simpler eukaryotes, including fungi and protozoa. However, in fungal research, especially involving phytopathogenic fungi, its use has often been limited to the recycling of marker genes during transformation experiments. In this study, detailed instructions are provided for generating conditional gene deletions in the corn phytopathogenic fungus *Ustilago maydis,* while infecting corn plants. This is accomplished through the transcriptional control of the recombinase by fungal promoters that are specifically activated during the biotrophic phase of fungal growth. This method enables gene deletions at defined developmental stages within plant tissues. These tools will prove valuable for investigating the roles of virulence-specific functions within the *U. maydis*-maize pathosystem under in planta conditions.

Keywords Cre recombinase, *Ustilago maydis*, Corn smut disease, Fungal development, Conditional mutants

1 Introduction

Site-specific recombination (SSR) has become the preferred method for gene modification or deletion in many model systems [1]. The two most widely used recombinases are Cre and FLP. Each recognizes short sequences, known as *lox* and *FRT*, respectively, which are 34 base pairs long. Depending on the relative orientation of these recognition motifs, SSR can mediate either the excision or inversion of intervening sequences. When the gene of interest is flanked by recombinase recognition sites in a direct orientation, the presence of the corresponding recombinase induces the deletion of the modified gene [2]. Additionally, the controlled expression of recombinase genes using tissue- or effector-specific promoters allows gene deletion to occur only in specific tissues or under certain conditions [3].

An N. T. Phan et al. (eds.), *Ustilago maydis*, Methods in Molecular Biology, vol. 3050,
https://doi.org/10.1007/978-1-0716-5364-7_17,

A codon-optimized version of Cre recombinase for *Ustilago maydis* has been developed, which is capable of excising *lox*-flanked sequences from a genomic locus with nearly 100% efficiency. Cre activity can be controlled during the biotrophic stages of *U. maydis* by placing the recombinase gene under the control of a promoter that is active only when the fungus is growing inside the plant. Using this tool, the conditional deletion of several key regulators in *U. maydis* during the biotrophic phase of growth is achieved, which enabled us the description of plant-specific roles for these regulators [4].

To obtain targeted conditional deletion of a specific gene or chromosomal region, two distinct cassettes carrying one *lox* site and antibiotic resistance markers, are inserted into the corresponding 5′ and 3′ flanking regions of the gene of interest (upstream and downstream cassettes, respectively). The upstream cassette is composed of the promoter for the *hxt1* gene (hexose transporter 1), enabling high and constitutive expression levels during the plant infection process [5]. This promoter is positioned upstream of a *lox* site, followed by a G418/Geneticin-resistance marker. The downstream cassette includes a *lox* site positioned upstream of a promoter-less, NLS-GFP-encoding open reading frame, along with a Nourseothricin-resistance marker. Once these cassettes are integrated at their respective sites—5′ upstream and 3′ downstream of the gene of interest—the excision of the intervening region (composed of the gene of interest and flanked by two *lox* sites in direct conformation) facilitated by Cre recombinase activity, results in the fusion of the *hxt1* promoter to the NLS-GFP coding sequence. Consequently, cells in which the targeted DNA region has been deleted can be identified using fluorescence microscopy by the presence of green fluorescence localized in the nucleus.

To provide recombinase activity, the upstream and downstream *lox*-cassettes are inserted into the genome of the SG200CRE strain. In this solopathogenic strain, the native *mig2_1* open reading frame was replaced with a sequence encoding Cre through homologous recombination [6]. The *mig2_1* gene is not essential for plant infection or during the biotrophic stage; however, it is activated early in pathogenic development after the infective hyphae penetrate the plant surface [7].

2 Materials

Prepare and store reagents at room temperature (RT) unless otherwise specified. Also, prepare all solutions using ultrapure water and analytical grade reagents.

2.1 Construction of lox-Containing Cassettes

1. Plasmids pPhxt1LOXG418 and pLOXGFPNat [4].
2. PCR reagents: High-fidelity DNA polymerases, 10 mM dNTPs solution and primers (10 μM each).
3. PCR fragment cloning kit.

2.2 Transformation of Ustilago maydis cells

1. *Ustilago maydis* SG200CRE [4].
2. YEPS: 2% Bactopeptone, 1% Yeast extract, 1% Sucrose.
3. SCS: 20 mM sodium Citrate, 1 M sorbitol pH: 5.8.
4. Protoplasting solution: 6 mg/mL Novozym resuspended in SCS. Filter sterilize through a 20 μm filter.
5. STC: 10 mM Tris HCl, 1 M Sorbitol, 100 mM $CaCl_2$, pH 7.5.
6. STC-PEG: 40% Polyethylene glycol (MW 3350) in STC buffer.
7. Regeneration agar: 2% Bactopeptone, 1% Yeast extract, 2% Sucrose, 1 M Sorbitol, 15% Agar.

2.3 Infection of Corn Plants

1. Potato Dextrose (PD) medium: Mix 24 g PD medium with 1 L distilled water.

3 Methods

The procedure for obtaining conditional mutants of *U. maydis* in planta involves three main steps. First, two independent plasmids are constructed, each containing cassettes carrying one *lox* site and antibiotic resistance markers. These cassettes will be inserted into the corresponding 5′ (upstream cassette) and 3′ (downstream cassette) flanking regions of the gene of interest. For that, the cassettes will be integrated by homologous recombination upon transformation of protoplasts of a solopathogenic *U. maydis* strain that has been engineered to express Cre recombinase specifically when the fungus is growing inside a plant. Finally, the modified strain is used to infect corn plants, and any resulting defects associated with the gene mutation of interest are assessed.

3.1 Construction of lox-Containing Cassettes

To generate the constructions that insert the cassettes carrying the *lox* sites flanking the gene of interest, a pair of DNA fragments (approximately 1 Kb) surrounding the region where the cassette will be inserted are amplified using PCR (homology fragments). These fragments will be the recombination templates to insert the *lox*-containing cassette. PCR primers included *Sfi*I sites with different five-nucleotide core sequences (internal borders) and *Pme*I sites (outer borders). The advantage of utilizing *Sfi*I is that the five core nucleotides between its two identical recognition sequences can be selected freely (5′-GGCCNNNN↓NGGCC-3′). This allows for the use of distinct core sequences in each cassette, enabling directional cloning with a single enzyme [8, 9]. The 5′ and 3′ homology fragments are digested with *Sfi*I and subsequently ligated to the desired cassette. The specific cassette (either the upstream or downstream module) is isolated from the corresponding plasmid (pPhxt1LOXG418 and pLOXGFPNat, respectively) upon *Sfi*I restriction enzyme digestion. Both cassettes are flanked by recognition sites for this enzyme and internal core sequences compatible with the corresponding end of the respective

5′ and 3′ homology fragments. The ligation product containing the three fragments (two homology fragments and the *lox*-containing cassette) is then isolated and directly ligated to pJET1.2 (or any other vector suitable for blunt-end cloning). The resulting plasmid, which carries the complete integration cassette, is digested with *Pme*I to release the integration cassette, which will be used for transforming the protoplasts of *U. maydis*.

The detailed steps are as follow:

3.1.1 Upstream lox Cassette (pPhxt1LOXG418)

1. To obtain the flanking 5′ and 3′ homology fragments, design appropriate primers to amplify approximately 1 Kb fragments flanking the desired integration site. For the left flanking region, the 5′ primer should be 5′ AGCGTTTAAAC$(N)_{27}$ 3′, which includes the *Pme*I recognition site. The 3′ primer should be 5′ CATGGCC**TAGAT**GGCC$(N)_{27}$ 3′, containing the *Sfi*I recognition site followed by 27 nucleotides complementary to the sequence upstream of the insertion site. For the right flanking region, use a 5′ primer with the sequence 5′ CATGGCC**GGAGC**GGCC$(N)_{27}$ 3′, and the 3′ primer should be 5′ AGCGTTTAAAC$(N)_{27}$ 3′.
2. Amplify the flanking fragments using PCR (*see* **Note 1**). Purify the fragments from agarose gels using an appropriate kit. Prepare a solution with a concentration of 20–50 ng/μL for each fragment.
3. Digest the pPhxt1LOXG418 plasmid with *Sfi*I. Isolate the 4.67 Kb fragment from agarose gels using an appropriate kit and prepare a solution of the isolated fragment at around 20–50 ng/μL.
4. Mix equal amounts of the three fragments (200 ng each) in a final volume of 44 μL. Add 5 μL of 10× T4 ligase buffer and 1 μL of T4 ligase. Incubate at room temperature for at least 6 h.
5. Isolate the band corresponding to the ligation of the three fragments using agarose gels and then ligate it to the pJET1.2 vector according to the instructions provided in the kit.
6. Transform *E. coli* cells with the ligation mixture and screen for recombinant bacteria.
7. Purify the corresponding recombinant plasmid.

3.1.2 Downstream lox Cassette (pLOXGFPNat)

Carry out the same above steps with the following differences:

1. Left flanking site: 5′ primer: 5′ AGCGTTTAAAC$(N)_{27}$ 3′/3′ primer: 5′ CATGGCC**AACGC**GGCC$(N)_{27}$ 3′.
2. Right flanking site: 5′ primer: 5′ CATGGCC**TGA GT**GGCC$(N)_{27}$ 3′/3′ primer: 5′ AGCGTTTAAAC$(N)_{27}$ 3′.
3. Digest with *Sfi*I the pLOXGFPNat. Isolate from agarose gels the 2.52 Kb fragment using appropriated kit. Prepare a solution of isolated fragment around 20–50 ng/μL.

3.2 Transformation of Ustilago maydis cells

3.2.1 Preparation of Protoplasts

1. Growth a preculture in 2 mL YEPS from a fresh plate of the SG200CRE strain. Incubate in agitation for 24 h at 28°C.
2. Dilute the culture in 50 mL YEPS in a baffled flask. Incubate on an orbital shaker at 28°C until the culture reach an OD_{600} around 0, 8 (*see* **Note 2**).
3. Recover the cells of the complete 50 mL culture by centrifugation (10 min, 4000 × *g*) and discard the supernatant. Resuspend the pellet in 50 mL sodium citrate, sorbitol solution (SCS), centrifuge 5 min, 4000 × *g*, and discard supernatant.
4. Prepare 3 mL of a protoplasting solution composed of 6 mg/mL *Trichoderma* lysing enzymes (Novozym) in SCS (*see* **Note 3**).
5. Resuspend the pellet in 2 mL protoplasting solution. Incubate for 5–10 min at RT and check under the microscope until more than 70% of the cells are round.
6. Add 10 mL cold (4°C) SCS, mix by inversion carefully and centrifuge at 2000 × *g* for 5 min. Discard supernatant.
7. Repeat above washing step three times more.
8. After last washing step, resuspend the protoplast pellet in 10 mL cold sorbitol, TrisHCl, $CaCl_2$ solution (STC); centrifuge at centrifuge at 2000 × *g* for 2 min, and discard supernatant.
9. Resuspend the pellet in 0.5 mL cold STC, make 50 μL aliquots in precooled 1, 5 mL tubes. Freeze the aliquots at −80°C until further use.

3.2.2 Transformation of Protoplasts

1. Digest around 2 μg of plasmid carrying the upstream *lox* cassette with *Pme*I. Once digested, precipitate the DNA solution and resuspend in 9 μL of TE10:1 and add 1 μL of heparin (15 mg/mL).
2. Thaw on ice protoplasts (1 tube/transformation) and add 10 μL of the above DNA solution.
3. Incubate for 30 min on ice.
4. Prepare the first layer of selection plates (two plates per transforming strain). For that, mix liquified (around 50°C) regeneration agar (1.5% YEPS agar plus 1 M sorbitol) with the selection antibiotic (1 mg/mL G418, final concentration). Dispose 12 mL of this mix into a Petri dish and let cold down at RT.
5. Add 500 μL STC/40%PEG solution to the transformation tube and mix by inverting the tube. Incubate 30 min on ice.

6. Add the second layer to selection plates. Pore 12 mL of liquified (around 50°C) regeneration agar (1.5% YEPS agar plus 1 M sorbitol) without the selection antibiotic on top of first layer plates and let cold down at RT.
7. Spread the transformation mix in selection plates (100 μL in one plate and the rest in the other plate). Incubate for 3–4 days at 28°C.
8. Select and check by diagnostic PCR the strains carrying the integrated transgene.
9. Once obtained the first integrant, repeat the procedure with the downstream *lox* cassette. Follow the same steps but using 300 μg/mL (final concentration in the bottom layer) nourseothricin as selection antibiotic.

3.3 Infection of Corn Plants

1. Grow maize seedlings for 1 week at a light cycle of 16 h, 200 μE, 28°C/22°C (day/night). They should be at the three-leaf stage and 10–15 cm tall at the day of infection.
2. The day before inoculation incubated in a culture flask 50 mL of PD medium with the *U. maydis* strain to be tested. Add G418 at a final concentration of 500 μg/mL (*see* **Note 4**). Incubate overnight in agitation at 28°C.
3. Centrifuge the cells of 50 mL culture at 4000 × *g* for 5 min and discard the supernatant. Wash 3× with sterile H_2O. Resuspend the cells in sterile H_2O to an OD600 of 3.0.
4. Load the concentrated culture in a 1 mL syringe. Inject 500 μL of the cell suspensions into the stem of each 7 day old maize seedlings.
5. Score the phenotype of each plant after 14 days of incubation according to healthy, chlorosis, anthocyanin formation, small tumors, large tumors, dead plants.

4 Notes

1. Carrying out amplification using high-fidelity DNA polymerases such as Q5 (NE Biolabs) or Velocity (Bioline) is recommended. Both DNA polymerases generated blunt-ended amplicons.
2. Doubling time of wildtype *U. maydis* cells in YEPS medium is about 2 h for wild-type strains. It is recommendable to allow three duplications minimum, meaning that starting OD_{600} should be not higher than 0.1.
3. The solution must be fresh for optimal enzymatic activity.

4. When cultured axenically without antibiotic selection pressure, we observed that a small fraction of the cells (approximately 4% of the population) exhibited a nuclear-localized GFP signal. This suggests that the basal activity of the *mig2_1* promoter under axenic conditions is sufficient to induce recombination at a low rate. To ensure that the plants were inoculated with cells carrying the nonrecombined *lox*-flanked allele, the cells were grown in a medium containing G418. The G418-resistance was utilized to integrate the upstream *lox* cassette, which would be lost following Cre-mediated recombination of the gene of interest.

Acknowledgments

This work was supported by Grant PID2020-120242GB-I00, supported by MCIN/ AEI/10.13039/501100011033. The funders had no role in study design, data collection and analysis, decision to publish, or preparation of the manuscript.

References

1. Gorman C, Bullock C (2000) Site-specific gene targeting for gene expression in eukaryotes. Curr Opin Biotechnol 11:455–46
2. Branda CS, Dymecki SM (2004) Talking about a revolution: the impact of site-specific recombinases on genetic analyses in mice. Dev Cell 6:7–28
3. Orban PC, Chui D, Marth JD (1992) Tissue- and site-specific DNA recombination in transgenic mice. Proc Natl Acad Sci U S A 89:6861–6865
4. de la Torre A, Jurca M, Hofmann K et al (2022) Robust Cre recombinase activity in the biotrophic smut fungus *Ustilago maydis* enables efficient conditional null mutants in planta. Genetics 220:iyab152
5. Schuler D, Wahl R, Wippel K et al (2015) Hxt1, a monosaccharide transporter and sensor required for virulence of the maize pathogen *Ustilago maydis*. New Phytol 206:1086–1100
6. Schmitz L, Kronstad JW, Heimel K (2020) Conditional gene expression reveals stage-specific functions of the unfolded protein response in the *Ustilago maydis*-maize pathosystem. Mol Plant Pathol 21:258–271
7. Basse CW, Kolb S, Kahmann R (2002) A maize-specifically expressed gene cluster in *Ustilago maydis*. Mol Microbiol 43:75–93
8. Brachmann A, Konig J, Julius C et al (2004) A reverse genetic approach for generating gene replacement mutants in *Ustilago maydis*. Mol Genet Genomics 272:216–226
9. Kämper J (2004) A PCR-based system for highly efficient generation of gene replacement mutants in *Ustilago maydis*. Mol Genet Genomics 271:103–110

Chapter 18

Efficient Introduction of Genetic Elements at Specific Insertion Loci

Sainath Shanmugasundaram, Lasse van Wijlick, Matias D. Zurbriggen, Michael Feldbrügge and Kerstin Schipper

Abstract

While some genetic modifications are conducted at the native loci of the manipulated genes, "neutral" loci for insertion of (heterologous) genetic elements are equally important for a versatile model system. This, e.g., applies for insertion of genetic complementation constructs or for transplantation of heterologous metabolic pathways. Here, the most frequently used insertion loci of *Ustilago maydis* is presented, namely the so-called *ip* locus and three loci encoding secreted proteases like the *upp3* locus. The strategy of counter selection provides additional benefits in strain generation as it significantly speeds up the identification of correct mutants. Considerations and strategies for construct design and strain generation based on published standard protocols are provided, followed by a detailed example for *upp3* locus insertion combined with counter selection. In essence, these versatile strategies for genetic manipulation have the power to simplify and speed up genetic engineering significantly.

Keywords Insertion locus, Transcriptional unit, Protease, Iron-sulfur protein, Counter-selection

1 Introduction

Reliable genomic insertion loci are of high value for any microbial model system. Insertion of genetic material at loci that support reliable expression while lacking impacts on cellular fitness is crucial for fundamental research, as well as for biotechnological and synthetic biology applications. For example, these sites can be used to insert genetic constructs like expression units for the implementation of heterologous genes or pathways or to complement deletion mutants. Genes for metabolic enzymes of amino acid biosynthesis are often targeted (*URA3*, *LEU2*, *HIS3*, *TRP1*), e.g., in *Saccharomyces cerevisiae*, yielding auxotrophic mutants that rely on amino acid supplementations. For multicopy insertion of genes in *S. cerevisiae* or *Pichia pastoris* (now *Komagataella phaffii*), highly repetitive rDNA sequences are frequently exploited [1]. Neutral positions like intergenic regions

An N. T. Phan et al. (eds.), *Ustilago maydis*, Methods in Molecular Biology, vol. 3050,
https://doi.org/10.1007/978-1-0716-5364-7_18,

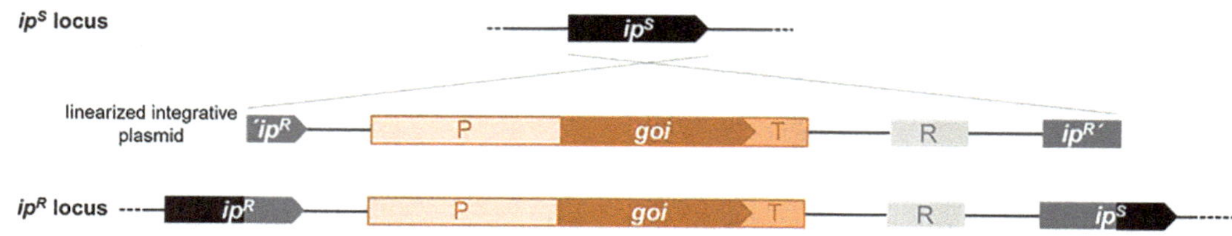

Fig. 1 Schematic representation of *ip* locus insertion (Genetic elements not to scale). ip^S gene does not mediate carboxin resistance, ip^R gene mediates carboxin resistance. *goi* gene of interest, *P* promoter, *T* terminator, *R* resistance cassette for selection in *E. coli*

or nonessential loci often discovered in reverse genetic approaches have been described for multiple fungal microorganisms like in *Candida* [2], *Fusarium* [3, 4], or *Trichoderma* species [5], mainly in the context foreign gene insertion.

In *Ustilago maydis*, in the past, mainly the *ip* locus has been used to integrate plasmids carrying genetic elements to be inserted (Fig. 1). This strategy targets the *ip* gene that is encoded in the genome and encodes an iron–sulfur protein of the mitochondrial respiration chain. *ip* insertion is based on the observation, that a single-point mutation in the protein (H253L) renders the cells resistant to the antibiotic carboxin [6–8]. Specialized integrative plasmids based on the p123 backbone [9] carry the mutated version of the *ip* locus and are exploited to insert expression cassettes. For genomic integration, the integrative plasmids are linearized within the *ip* gene yielding homologous regions at the ends of the linear plasmid. The linearized plasmid is then transformed. Homologous recombination results in the insertion of the complete linearized plasmid and reconstitution of two *ip* copies, one without the mutation (ip^S) and one with the mutation (ip^R). Resulting strains are thus carboxin resistant (Fig. 1). Importantly, there is also the chance that the homologous recombination event is repeated, yielding multiple insertions (Fig. 1). This is a very useful tool, for instance, to increase the protein yield by multiple insertion of gene expression units. However, the downside of this locus is the instability of the integration; environmental pressures, such as UV mutagenesis, or the expression of toxic proteins can lead to the excision of the integrated cassette (unpublished observations).

More recently, alternative insertion loci have been identified in the framework of studies that aimed at minimizing the extracellular proteolytic activity [10, 11]. Up to eight different proteases have been deleted at a time, resulting in yeast-like growing cells that maintained normal morphology and growth behavior in complete and rich medium [11]. Of the targeted proteases, mainly *pep4*, *upp1*, and *upp3* were studied for insertion of genetic constructs (Table 1) [12–15]. This comes with the advantage, that upon insertion, genes for secreted proteases are deleted, which reduces the proteolytic potential of the strains. It must be noted, that insertion in protease

Table 1

Frequently used insertion loci. Insertion of genetic constructs in these loci lead to deletion of the respective gene, and hence, reduces the proteolytic activity in the culture supernatant

Protease-encoding gene	Gene(umag number)	Remarks
pep4	*umag_04926*	Population heterogeneity in AB33 hyphae observed, preferentially use in yeast-like cells
upp1	*umag_02178*	
upp3	*umag_11908*	*upp3* deletion reduces virulence in SG200, preferentially use in yeast-like cells or AB33 hyphae

loci has only been confirmed for axenic cultures involving yeast-like cells and synthetic AB33 hyphae. Note that a detailed study for the applicability of the insertion loci (Table 1) in other strain backgrounds or for plant infection studies is currently missing. SG200upp3Δ deletion strains, e.g., are reduced in virulence. Hence, this point needs carefully evaluation or alternatively, an alternative locus should be used.

A very convenient tool that can be easily applied in combination with the protease loci is counter selection. This strategy is very powerful for efficient strain generation based on the fact that cells that retain a distinct genetic element (selection marker, gene, plasmid) are killed or inhibited while cells that lose it in the process of DNA insertion survive. In *U. maydis* counter selection based on two different selection markers is mainly used (Fig. 2). In the present example, a strain lacking the *upp3* gene was generated using a construct with flanking regions in the 5′ and 3′ regions of the *upp3* gene and a nourseothricin resistance cassette (NatR), resulting in replacement of *upp3* by NatR. For counter selection, identical flanks are fused to the genetic elements to be inserted and to a second resistance cassette mediating, e.g., hygromycin resistance. Transformants can efficiently be selected for nourseothricin sensitivity and hygromycin resistance. This strategy significantly speeds up genetic manipulation, as the frequency of positive candidates in Southern blot analysis is strongly increased as compared with strains generated without counter selection. The procedure has been used successfully multiple times, confirming its applicability for axenic culture experiments.

2 Materials

Solutions, media, and buffers should be prepared with high purity water (H_20_{bidest}), obtained from corresponding purification devices. All chemicals should be purchased in the *per analysis* (p.a.) grade.

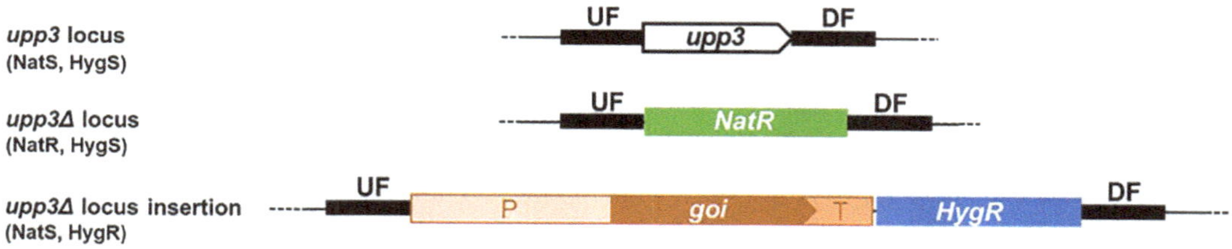

Fig. 2 Schematic representation of a counter selection strategy at the *upp3* locus (Genetic elements not to scale). In brief, a suitable counter selection strain is first generated by replacing the native *upp3* gene by a nourseothricin resistance cassette (NatR). In the next step, new genetic information can efficiently be introduced by using identical flanks for homologous recombination but a different resistance cassette, in the present example mediating hygromycin resistance (HygR). *goi* gene of interest, *P* promoter, *T* terminator, *NatR/HygR* resistance cassettes composed of constitutive promoters, the resistance gene and a terminator mediating nourseothricin/hygromycin resistance. UF *upp3* upstream flank, *DF upp3* downstream flank. Flanks should have a length of approximately 1 kb

2.1 Cultivation and Patch Plates

1. Complete medium supplemented with glucose (CM-glc) for cultivation.

   Prepare complete medium (CM) as mentioned below. For solid CM-glc plates, add 2% (w/v) agar–agar. Autoclave the solution for 20 min at 121°C. After autoclaving, add 1% (v/v) glucose (see below) and appropriate antibiotics (see below) to the medium at room temperature (RT). Use the following working concentrations: carboxin, 2 μg/mL; hygromycin, 200 μg/mL; nourseothricin, 150 μg/mL; geneticin (G418), 500 μg/mL.

   Complete Medium (CM): 0.25% (w/v) casaminoacids (Difco),0.1% (w/v) yeast extract, 1% (w/v) Holliday vitamins, 6.25% (w/v) Holliday salt solution, 0.05% (w/v) Herring sperm DNA (Sigma, D-3159), 0.15% (w/v) NH_4NO_3. Note that Holliday vitamins and Holliday salt solution are complex mixtures that need to be prepared beforehand. Sterilize by autoclaving at 121°C for 5 min. Adjust the pH to 7.0 with 1 M KOH. Store at room temperature (RT). Caution: The medium needs to be supplemented with the appropriate carbon source prior to use.

   Holliday vitamins: 0.1% (w/v) thiamine (Hydrochloride, Serva 36020), 0.05% (w/v) riboflavin (Sigma R-4500), 0.05‰ (w/v) pyridoxine (Monohydr.chl., Sigma p-9755), 0.2‰ (w/v) Ca-pantothenat (Hemi-Ca.salt, Sigma P-2250), 0.05‰ (w/v) aminobenzoic acid (Free Acid, Sigma A 9878), 0.2‰ (w/v) nicotinic acid (free acid, Sigma N-4126), 0.2‰ (w/v) cholinchloride (Sigma C-1879), 1‰ (w/v) myo-inositol (Sigma I-5125). Store aliquots at −20°C.

   Holliday salt solution: 16% (w/v) KH_2PO_4, 4% (w/v) Na_2SO_4, 4% (w/v) KCl, 8% (w/v) $CaCl_2 {*} 2H_2O$, 1.32% (w/v) Holliday trace elements, 2% (w/v) $MgSO_4$. Note that Holliday trace elements is a separate solution. Sterilize by filtering using a 0.22 μm filter. Store at RT.

Holliday trace elements: 0.6% (w/v) H_3BO_3, 0.14% (w/v) $MnCl*4H_2O$, 0.4% (w/v) $ZnCl_2$, 0.4% (w/v) $Na_2MoO_4*2H_2O$, 0.1% (w/v) $FeCl_3*6H_2O$, 0.04% (w/v) $CuSO_4*5H_2O$. Store at RT.

Glucose solution (medium supplement): 1% (w/v) glucose. The solution needs to be filter sterilized (0.22 μm) or autoclaved at 121°C for 5 min.

2. Stock solutions of the most frequently used antibiotics in *U. maydis*. Caution: Antibiotics used in fungi are also active in humans. Apply safety measures.

   Carboxin stock solution: 5 mg/mL (in methanol). Filter sterilize, store at 4°C.

   Hygromycin stock solution: 50 mg/mL (in phosphate-buffered saline, PBS). Filter sterilize, store at 4°C, protect from light.

   Nourseothricin (ClonNat): 200 mg/mL (in H_20_{bidest}). Filter sterilize, store in aliquots at −20°C.

   Geneticin (G418): 50 mg/mL. Filter sterilize, store in aliquots at −20°C.

2.2 Strains

1. For *ip* insertion, a carboxin sensitive (CbxS) strain is needed. Since in nature the *ip*^{S} gene is prevalent, wild-type strains are sensitive (*see* **Note 1**).
2. For insertion in protease *loci* via counter selection, a respective deletion strain must be generated using established protocols [16–18] Standard molecular biology techniques such as restriction/ligation, Golden Gate, or Gibson cloning can be applied for construct assembly. It is important to select a resistance cassette for deletion, that differs from the resistance cassette that is used for insertion of the genetic construct.

2.3 Equipment

1. Clean bench.
2. Cultivation equipment: glass tubes, 28°C cupboards or rooms, rotary wheel, and culture shaker or incubator.

3 Methods

3.1 Strain Generation

Use virtual cloning to carefully plan and verify your genetic construct carrying the heterologous gene. There is freeware like Benchling that provides all required tools (https://www.benchling.com/).

1. (Optional) If not available, generate a deletion strain, in which you replace the locus intended to be used for insertion of a

genetic construct by a resistance cassette following published strategies (*see* **Note 2**) [16, 17, 19].

2. Generate a plasmid carrying the expression unit for your gene of interest (*goi*) (promoter, *goi*, terminator) and a resistance cassette (HygR, G418R, NatR, or CbxR) that differs from the resistance cassette used in the deletion strain. In our example, the *upp3* gene was replaced by a G418 resistance cassette. The expression unit needs to be encased by about 1 kb flanking regions for genome targeting by homologous recombination, in our example the 5′ and 3′ regions of *upp3* were amplified by PCR. The construct consisting of a HygR cassette, the expression unit for the *goi* encased by 1 kb flanking regions for homologous recombination at the *upp3* locus and parallel *upp3* deletion are assembled using standard molecular cloning techniques such as restriction-ligation, Golden Gate, Gibson, or AQUA cloning [16, 17, 19, 20] (Fig. 2).
3. Linearize the plasmid for transformation using an appropriate restriction enzyme to obtain a single fragment encompassing the genetic construct (*see* **Note 3**).
4. Transform the *U. maydis* strain suitable for counter selection using published protocols [18] (*see* **Note 4**) and incubate on 28°C until colonies appear on the RegLight plates.
5. The colonies from the Reg-light plates grown on the new resistance are patched on CM-glc plates supplemented with the antibiotic for the newly introduced resistance, in the present example hygromycin. Simultaneously, the colony is also patched on a second plate supplemented with the antibiotic compatible with the progenitor strain, in the present example, G418 (*see* **Note 5**).
6. Patch two previously generated and confirmed control strains harboring the respective resistances used in the experiment. In the present example, a G418R strain and a HygR strain were included.
7. Incubate the plates on 28°C for about 2 days until colony growth can be judged (Fig. 3).
8. Positive clones, i.e., transformants that integrated the genetic construct while replacing the previous resistance cassette, grow on hygromycin, and fail to grow on G418 (Fig. 3).
9. Use the preselected transformants for verification using Southern blot analysis as described in published protocols [18] (*see* **Note 6**).

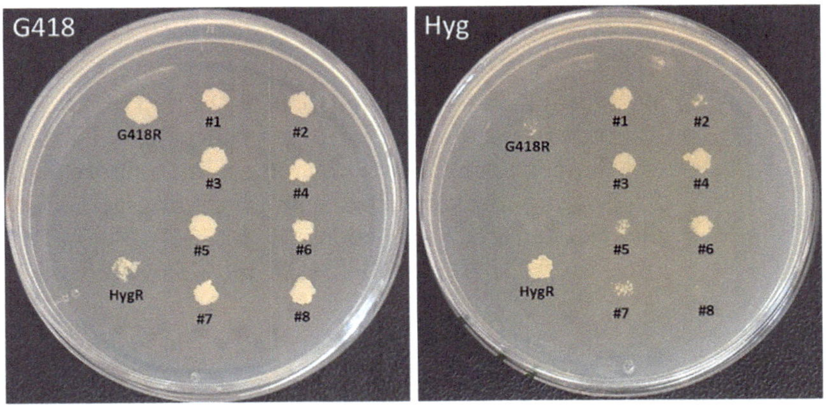

Fig. 3 Counter selection using two different antibiotics. The pictures show CM-glc plates supplemented with G418 (left) or Hyg (right). Colony patches have been applied to the showing colony patches. Correct strains are resistant to hygromycin (Hyg) and sensitive to G418. Strains #2, #5, #7, and #8 show the expected pattern. Pre-existing Hyg resistant (HygR) and G418 resistant (G418R) strains were used as controls

4 Notes

1. The most frequently used genetically engineered laboratory strains such as AB33, AB33P5Δ, AB33P8Δ, SG200, and CL13 are also CbxS. Remember that studies on the impact of protease deletions on infection are lacking.
2. A large plasmid collection for the assembly of gene deletion constructs is available from the Institute of Microbiology at HHU Düsseldorf.
3. It is essential that the genetic construct for transformation is linear and complete. For homologous recombination via 1 kb flanks, 100% homologous flank ends are advantageous and strategies for their generation exist [17].
4. A negative control use sterile H_20_{bidest} in the transformation process to simplify identification of true transformants. If available, a plasmid lacking a resistance cassette or any homology region would be even more appropriate.
5. The colonies should first be patched on the plate supplemented with the antibiotic of the progenitor resistance using, e.g., sterile pipet tips. The identical tip is then used to also patch the colony on the plate with the antibiotic for the newly acquired resistance.
6. Southern blot analysis is the gold standard. You may also use diagnostic PCR for verification.

Acknowledgments

We greatly acknowledge the work of Dr. Kira Müntjes on the establishment of the protease loci. We further thank Ute Gengenbacher and Simone Krüger for excellent technical support. The project was funded in part by the Deutsche Forschungsgemeinschaft (DFG, German Research Foundation)—SFB1535—Project ID 458090666 (projects A03, B03 to MF/MDZ and KS, respectively).

References

1. Zheng H, Wang K, Xu X et al. (2022) Highly efficient rDNA-mediated multicopy integration based on the dynamic balance of rDNA in *Saccharomyces cerevisiae*. Microb Biotechnol 15:1511–1524
2. Nemeth T, Papp C, Vagvolgyi C et al. (2020) Identification and characterization of a neutral locus for knock-in purposes in *C. parapsilosis*. Front Microbiol 11:1194
3. Tong S, An K, Chen W et al. (2023) Identification of neutral genome integration sites with high expression and high integration efficiency in *Fusarium venenatum* TB01. Synth Syst Biotechnol 8:141–147
4. Darino M, Urban M, Kaur N et al. (2024) Identification and functional characterisation of a locus for target site integration in *Fusarium graminearum*. Fungal Biol Biotechnol 11: 2
5. Qin L, Jiang X, Dong Z et al. (2018) Identification of two integration sites in favor of transgene expression in *Trichoderma reesei*. Biotechnol Biofuels 11: 142
6. Stock J, Sarkari P, Kreibich S et al. (2012) Applying unconventional secretion of the endochitinase Cts1 to export heterologous proteins in *Ustilago maydis*. J Biotechnol 161:80–91
7. Keon JP, White GA, Hargreaves JA (1991) Isolation, characterization and sequence of a gene conferring resistance to the systemic fungicide carboxin from the maize smut pathogen, *Ustilago maydis*. Curr Genet 19:475–481
8. Broomfield PL, Hargreaves JA (1992) A single amino-acid change in the iron-sulphur protein subunit of succinate dehydrogenase confers resistance to carboxin in *Ustilago maydis*. Curr Genet 22:117–121
9. Aichinger C, Hansson K, Eichhorn H et al. (2003) Identification of plant-regulated genes in *Ustilago maydis* by enhancer-trapping mutagenesis. Mol Genet Genomics 270:303–314
10. Sarkari P, Reindl M, Stock J et al. (2014) Improved expression of single-chain antibodies in *Ustilago maydis*. J Biotechnol 191:165–175
11. Terfrüchte M, Wewetzer S, Sarkari P et al. (2018) Tackling destructive proteolysis of unconventionally secreted heterologous proteins in *Ustilago maydis*. J Biotechnol 284:37–51
12. Lee J, Hilgers F, Loeschke A et al. (2020) *Ustilago maydis* serves as a novel production host for the synthesis of plant and fungal sesquiterpenoids. Front Microbiol 11:1655
13. Heucken N, Tang K, Husemann L et al. (2023) Engineering and implementation of synthetic molecular tools in the basidiomycete fungus *Ustilago maydis*. J Fungi (Basel) 9:480
14. Müntjes K, Philipp M, Hüsemann L et al. (2020) Establishing polycistronic expression in the model microorganism *Ustilago maydis*. Front Microbiol 11:1384
15. Reindl M, Stock J, Hussnaetter KP et al. (2020) A novel factor essential for unconventional secretion of chitinase Cts1. Front Microbiol 11:1529
16. Brachmann A, König J, Julius C et al. (2004) A reverse genetic approach for generating gene replacement mutants in *Ustilago maydis*. Mol Genet Genomics 272:216–226
17. Terfrüchte M, Joehnk B, Fajardo-Somera R et al. (2014) Establishing a versatile Golden Gate cloning system for genetic engineering in fungi. Fungal Genet Biol 62:1–10
18. Bösch K, Frantzeskakis L, Vranes M et al. (2016) Genetic manipulation of the plant pathogen *Ustilago maydis* to study fungal biology and plant microbe interactions. J Vis Exp 115:54522
19. Kämper J (2004) A PCR-based system for highly efficient generation of gene replacement mutants in *Ustilago maydis*. Mol Genet Genomics 271:103–110
20. Beyer HM, Gonschorek P, Samodelov SL et al. (2015) AQUA cloning: a versatile and simple enzyme-free cloning approach. PLOS One 10:e0137652

Chapter 19

Viability of *Ustilago maydis* by Flow Cytometry

Juan Pablo Pardo, Eda P. Tenorio and Lucero Romero-Aguilar

Abstract

Amine-reactive dyes are useful for determining cell viability and are widely used in mammalian cells. These chromophores easily cross the damaged cell membranes, e.g., during apoptosis or autophagy processes, or due to physicochemical treatments, and thus can react irreversibly with the free amines in the cytosol. In live cells, chromophores react with exposed amines in the cell membrane but are excluded from the cytosol. Therefore, unfixed dye is washed after staining incubation, and the bound dye to dead cells remains associated. Since amine-reactive dyes are excited by lasers, dead cells can be clearly distinguished from live cells in a sample by flow cytometry. This protocol describes how to determine the percentage of dead and live cells in a population of *Ustilago maydis* by measuring the fluorescence emitted by the amine-reactive viability dye Ghost Dye™ Violet 450 using flow cytometry.

Keywords Cell viability, *Ustilago maydis*, Amine-reactive dye, Flow cytometry

1 Introduction

Flow cytometry is a powerful biophysical technique used for the comprehensive analysis of individual cells or particles within a heterogeneous population as they traverse a focused laser beam. This methodology facilitates the rapid assessment of cells in suspension, yielding valuable insights into their physical and chemical characteristics, such as cell size, internal complexity as revealed by granularity, viability status, and the quantitative assessment of both surface and intracellular marker expression [1–6]. Flow cytometry involves the hydrodynamic focusing of cells into a single-file stream, ensuring their sequential passage through the interrogation point [1, 7]. As each cell intercepts the laser beam, it scatters light and emits fluorescence, which are subsequently detected by an array of detectors strategically positioned around the flow cell [7]. The scattered light provides information about the cell size (forward scatter, FSC) and granularity or internal complexity (side scatter, SSC), whereas fluorescence signals reveal the presence and abundance of specific molecules within or on the cell surface [1]. The data acquired from

An N. T. Phan et al. (eds.), *Ustilago maydis*, Methods in Molecular Biology, vol. 3050,
https://doi.org/10.1007/978-1-0716-5364-7_19,

flow cytometry experiments are typically represented as histograms, dot plots, or contour plots, which visually depict the distribution of cells based on their light scattering and fluorescence properties [1].

Assessing cellular viability is crucial for evaluating the impact of stressful factors. These factors can be substrates, such as acetate in *U. maydis* [8] or toxic agents like paraquat [9], H_2O_2, or menadione, all of which increase the production of reactive oxygen species.

Although cell death may be the outcome of the toxicity of these chemical or physical factors, there are other possibilities. For instance, the agent may compromise the ability of cells to divide without causing cell death in the short term. This is the case with the effect of paraquat on *U. maydis* [9]. After 4 h of incubation in the presence of paraquat, *U. maydis* cells maintained their respiratory capacity and the integrity of the plasma membrane but were unable to undergo cell division on a rich YPD-agar solid medium [9]. This scenario can be described as "cell vitality," where some of the physiological/metabolic capabilities of the cells are preserved, but they are unable to divide. Cell viability and cell vitality represent distinct aspects of cellular functions, and both are essential for evaluating the physiological state of a cell following its exposure to stressful factors, whether chemical or physical.

There are numerous techniques to investigate the viability of cells. One prominent approach involves analyzing the number of cell colonies on Petri dishes (CFU). By quantifying the total number of cells using a Neubauer chamber, the percentage of viable cells within the population can be estimated by CFU. Additionally, alternative methods for assessing cell viability rely on the ability of dyes to stain cellular components when the cell membrane is compromised. An example of such a technique is the use of fixable amine reactive dyes, which covalently react with amine groups and allow for the fixation of cells without losing the dye. This enables an effective and reliable viability analysis through flow cytometry. Other dyes used in flow cytometry to determine cell viability are reported in Table 1, each of them has different binding mechanisms and specific excitation and emission wavelengths, allowing convenient incorporation in multiparametric analysis. In this work, we describe a simple protocol to measure cell viability using the amine reactive dye Ghost Dye Violet 450 and flow cytometry.

2 Materials

All reagents are of analytical grade and are described below, together with the equipment required for the assay.

2.1 Equipment

1. Flow cytometry instrument: Required for cell viability analysis.
2. Microcentrifuge tubes: For preparing the samples.

Table 1
Membrane permeability indicators

Dyes	Mechanism	Common applications	Notes
Propidium Iodide (PI)	Binds to DNA and RNA through an interbase intercalating [10].	Live/dead discrimination.	PI is considered a potential carcinogen.
		Cell cycle analysis.	Cell cycle and ploidy determination require RNAse incubation prior to staining.
		Ploidy determination.	Incompatible with intracellular staining protocols where prior fixation is required.
7-Aminoactinomycin D (7-AAD)	Binds to DNA and RNA through an interbase intercalating mode [10].	Live/dead discrimination.	Cell cycle and ploidy determination require RNAse incubation prior staining.
		Cell cycle analysis.	Incompatible with intracellular staining protocols where prior fixation is required.
SYTOX™ Green/Blue/Red	Binds DNA and RNA through dual binding mode [10].	Live/dead discrimination.	Cell cycle and ploidy determination require RNAse incubation prior staining.
		Cell cycle analysis.	Incompatible with intracellular staining protocols where prior fixation is required.
[a]Viability Fixable dyes	Reacts covalently to free amine groups [11].	Live/dead discrimination.	Compatible with intracellular staining protocols where prior fixation is required.
			Available from multiple brands.
			Multiple spectral profiles.
			Easy incorporation in multiparametric flow cytometry analyses.

[a]Ghost dye, Zombie aqua™ dye, Fixable Live/dead eFlour Fixable viability dye

3. Pipettes and pipette tips: For accurate addition of samples and reagents.
4. Cuvettes or Neubauer chamber: For spectrophotometric measures or counting the cells.

5. Spectrophotometer: For measurement of absorbance at 600 nm.
6. Vortex mixer: For mixing reagents and samples.
7. Incubator: For culturing the cells and the incubation of samples with the dye.
8. Ultrasonic system.

2.2 Solutions

1. **Phosphate-Buffer Saline (PBS), pH 7.4:** 8 g NaCl, 0.2 g KCl, 1.44 g Na_2HPO_4, and 0.245 g KH_2PO_4. Add distilled water to a volume of 800 mL. Adjust the pH to 7.4. Brought the volume to 1 L. Filter the PBS with a 0.45 μm PVDF (Polyvinylidene Fluoride) membrane.
2. **Yeast Peptone Dextrose (YPD) culture medium**: 5 g dextrose, 5 g peptone, and 5 g yeast extract. Dissolve them in 1 L of distilled water. Steam sterilize the solution.
3. **Preparation of the GD450 solution:** GD450 (Ghost Dye™ Violet 450 Viability Dye) is obtained from Tonbo Biosciences in anhydrous DMSO (1 μL/test). The supplier's recommendations are: store the vial at −20°C. Before opening, equilibrate to room temperature and protect from light and moisture. GD450 is excited at 405 nm, and the emission is detected at 450 nm. Thaw the dye at room temperature. Give the reagent a quick spin in a tabletop centrifuge before opening. Mix 6 μL of GD450 stock reagent with 1 mL of PBS. Per sample, use 250 μL of this solution.
4. **Formaldehyde solution:** 1 mL of 38% formaldehyde in 9 mL of distilled water.

3 Methods

3.1 Preparation of Cells

Culture the cells for a fixed period, typically 8–24 h in either YPD media or the media of your interest.

1. Collect an aliquot containing 2 million cells/mL (2×10^6 cells/mL) and centrifuge at 2400 × *g* for 5 min at 4°C without sterility (*see* **Note 1**).
2. Wash the cells twice with 1 mL of cold-filtered PBS pH 7.4.
3. Spin down the cells at 2400 × *g* for 5 min at 4°C. Discard the supernatants.
4. Resuspend the cells with 250 μL of the GD450 solution (*see* **Note 2**).
5. Vortex the cell suspension and then incubate for 1 h in the dark at 28°C with shaking at 200 rpm (*see* **Note 3**).

6. After incubation, spin down the cells at 2400×*g* for 5 min at 4°C. Wash twice with cold-filtered PBS and discard the supernatants (*see* **Note 4**).
7. Resuspend the cells with 1 mL of cold-filtered PBS.
8. Preserve the cells in the dark at 4°C or proceed to fix the cells.

3.2 Fixation of the Cells in Suspension

1. Fix the stained cells in suspension by adding formaldehyde at a final concentration of 3.8%.
2. Incubate for 15 min at room temperature in the dark to preserve the dye.
3. At the end of the incubation, spin down the cells at 2400×*g* for 5 min at 4°C and discard the supernatant.
4. Resuspend the cells in 1 mL of cold-filtered PBS and wash the cells twice using the same volume of PBS. Repeat the centrifugation step.
5. Finally, resuspend the cells in 1 mL of cold-filtered PBS.
6. Keep the cells in the dark until data acquisition or cryopreserve for later use.

3.3 Preparation of Positive Control

Positive control cells correspond to dead cells (highly fluorescent cells).

1. Collect an aliquot containing approximately 2 million cells/mL and centrifuge at 2400×*g* for 5 min at 4°C without sterility (*see* **Note 1**).
2. Wash the cells twice with 1 mL of cold-filtered PBS. Resuspend the cells in 1 mL PBS. (*see* **Note 2**).
3. Incubate the cell suspension for 10 min in a boiling water bath.
4. Centrifuge the boiled cells at 2400×*g* for 5 min at 4°C. Discard the supernatant.
5. Add 250 μL of GD450 staining solution and follow the steps described in Sect. 3.1.

3.4 Preparation of Negative Control

Negative control corresponds to unstained cells. However, the cells have autofluorescence, even when they are not incubated with the staining solution.

1. Harvest an aliquot containing approximately 1.5 million cells/mL.
2. Wash the cells twice with cold-filtered PBS and resuspend the pellet with 1 mL filtered PBS.
3. Incubate the cell suspension at 28°C for 1 h, similar to the other samples.
4. After incubation, follow the steps described in Sect. 3.2.

3.5 Data Acquisition

1. Before the acquisition of events, mix the cell suspension using a vortex.
2. Sonicate briefly the cell suspension to avoid aggregates. Here, we use the compact high-performance ultrasonic system (Fisher Scientific FS3).
3. Flow cytometry For flow cytometry experiments, the MACSQUANT analyzer (Miltenyi Biotec Bergisch Gladbach, Germany) is used. Set the parameters to acquire 20,000 events per condition (Fig. 1). Acquire the events at low speed to minimize multiple cells detected as one. Establish the limits of detection using the negative and the positive controls (Fig. 2) (*see* **Note 5**).

3.6 Data Analysis

1. Analyze the samples of interest using FlowJo Software V.10 (BD).
2. Define the population(s) of interest first by their FSC (Forward Scatter Side) and SSC (Side Scatter Side) characteristics using a dot plot (Fig. 3A).
3. Perform a double-parameter analysis: to this end, create a dot plot to analyze GD450 fluorescence intensity and FSC. This will show whether there is a relationship between cell size and viability (Fig. 3B).

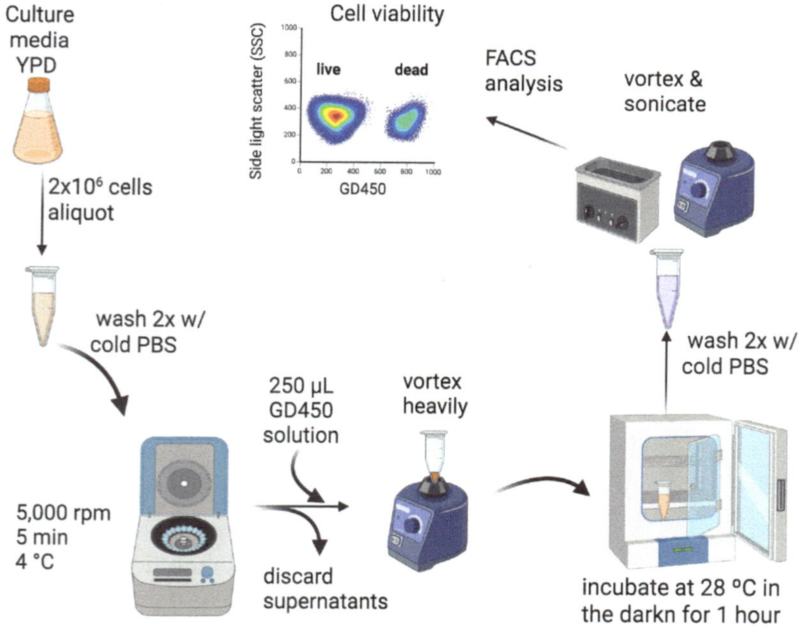

Fig. 1 Scheme of sample preparation. An aliquot of 2 million cells/mL is taken, and then the sample is washed twice with 1 mL of filtered PBS and incubated with the GD450 solution for 1 h. In the end, the sample is washed twice. Finally, it is resuspended in 1 mL and analyzed with the flow cytometer

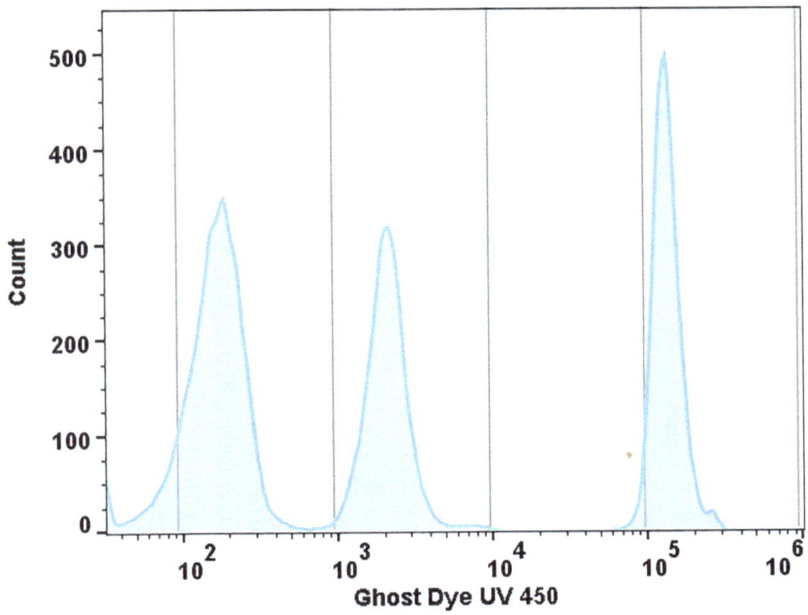

Fig. 2 Detection limits using unstained, negative, and positive control cell populations. Representative flow cytometry histogram displaying fluorescence intensity distribution of a mixed sample, analyzing three distinct populations: Unstained cells (left peak) exhibit low autofluorescence ($1–1 \times 10^3$). The negative control (viable cells) forms a central peak ($1 \times 10^3–1 \times 10^4$). Positive control (non-viable cells) shows high fluorescence intensity, ($>2 \times 10^4$, right peak)

4. In both cases, it is possible to analyze a single sample per graph or overlay the samples of interest for an easier comparison, e.g., samples exposed to different conditions (Fig. 3C) or unstained, negative, and positive controls (Fig. 2) (*see* **Note 6**).

4 Notes

1. Optical density provides an instant approximation of cell concentration (cells per mL). Count the cells in a Neubauer counting chamber and correlate this value with the optical density. For *U. maydis*, an optical density of 1 at 600 nm typically corresponds to 20 million cells/mL.
2. PBS is used for washing and staining incubation to minimize culture media autofluorescence and ensure optimal signal detection.
3. Incubation periods, washing steps, GD450, and sample storage are critical parameters in determining *U. maydis* viability. A high-quality specimen is desired to avoid pitfalls of nonspecific staining.

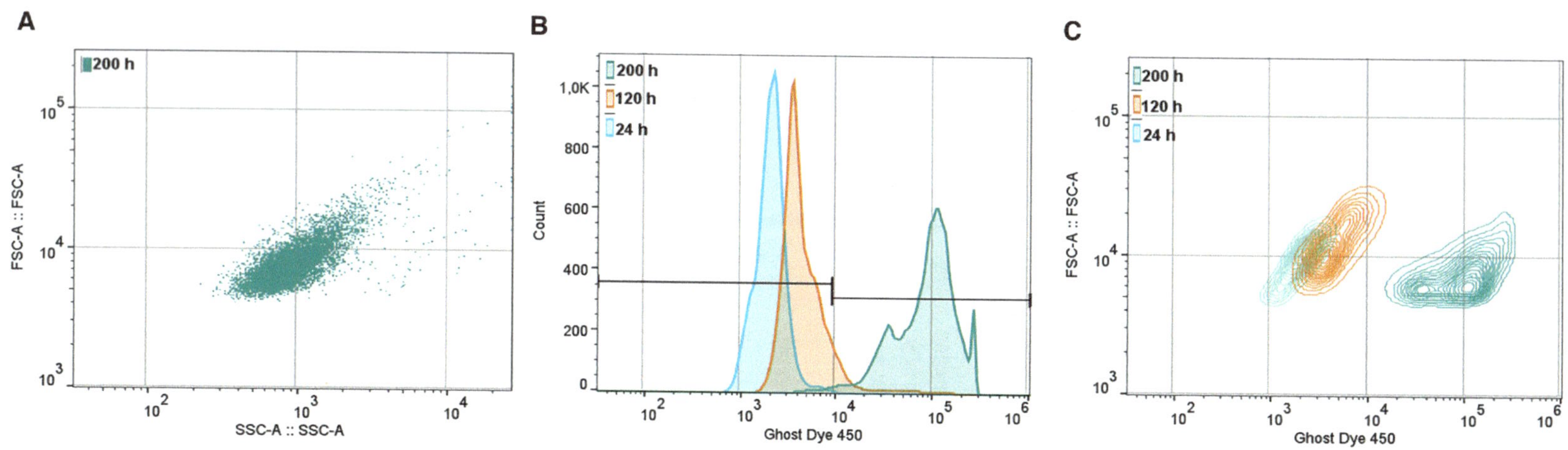

Fig. 3 Representative viability analysis. *U. maydis* cells were cultured and collected at different timepoints and stained with GD450 to determine cell viability. Cells were fixed, acquired in the flow cytometer, and analyzed. (**A**) The population of interest was gated according to its FSC and SSC characteristics. Cells within the gate were further analyzed using a single-parameter histogram (**B**) or a dot plot showing the relation between fluorescence and FSC (**C**). In both cases, data collected at different timepoints were overlayed: 24 h (light blue), 120 h (brown), and 200 h (positive control, green)

4. Avoid aggregates by centrifuging at low speed, and dissolve the aggregates by sonicating the cell suspension.
5. Each instrument's photodetectors have different sensitivity characteristics, and scale capture parameters are instrument-specific. The *U. maydis* cells have autofluorescence that increases after fixation. Autofluorescence must be considered when setting the limits of negative and positive controls.
6. Cells cultivated for 200 h in the presence of 0.5% glucose have a damaged plasma membrane; thus, the chromophore can diffuse into the cytosol and react with the amino groups of the proteins, and these cells can be used as the positive control.

Acknowledgments

The authors wish to acknowledge PAPIIT IN202525, Proyectos de Investigación e Innovación Tecnológica from Universidad Nacional Autónoma de México [PAPIIT-UNAM].

References

1. Adan A, Alizada G, Kiraz Y et al (2017) Flow cytometry: basic principles and applications. Crit Rev Biotechnol 37:163–176
2. Bouix M, Ghorbal S, Picque D et al (2022) A rapid method for the assessment of the vitality of microorganisms using flow cytometry. Cytom Part J Int Soc Anal Cytol 101:577–587
3. Lloyd D (2001) Flow cytometry of yeasts. Curr Protoc Cytom 9:11.10.1
4. Martínez-Pastor F, Mata-Campuzano M, Alvarez-Rodríguez M et al (2010) Probes and techniques for sperm evaluation by flow cytometry. Reprod Domest Anim Zuchthyg 45(Suppl 2):67–78
5. McKinnon KM (2018) Flow cytometry: an overview. Curr Protoc Immunol 120:5.1.1–5.1.11
6. Volejníková A, Hlousková J, Sigler K et al (2013) Vital mitochondrial functions show profound changes during yeast culture ageing. FEMS Yeast Res 13:7–15
7. Dean PN, Hoffman RA (2007) Overview of flow cytometry instrumentation. Curr Protoc Cytom 39:1.1.1–1.1.8
8. Romero-Aguilar L, Hernández-Morfín KD, Guerra-Sánchez G et al (2023) Metabolic changes and antioxidant response in Ustilago maydis grown in acetate. J Fungi Basel Switz 9:749
9. Romero-Aguilar L, Vázquez-Meza H, Guerra-Sánchez G et al (2022) The mitochondrial alternative oxidase in Ustilago maydis is not involved in response to oxidative stress induced by paraquat. J Fungi Basel Switz 8:1221
10. Poot M (2004) Nucleic acid probes. Curr Protoc Cytom 4:Unit 4.3
11. Perfetto SP, Chattopadhyay PK, Lamoreaux L et al (2006) Amine reactive dyes: an effective tool to discriminate live and dead cells in polychromatic flow cytometry. J Immunol Methods 313:199–208

Chapter 20

Evaluating Gene Expression in *Ustilago maydis* Using RT-qPCR

Amanda M. Seto and Barry J. Saville

Abstract

Gene expression analysis detects changes in the transcript level of a gene in response to a specific condition. Analysis of these changes can reveal links to changes in the growth, development or physiology of an organism. In *Ustilago maydis*, reverse transcriptase quantitative PCR (RT-qPCR) has primarily been used to examine the changes in messenger RNA (mRNA) levels. This chapter outlines a method for assessing gene transcript level changes using RT-qPCR and the comparative C_T ($2^{-\Delta\Delta CT}$) method. Successful RT-qPCR experiments rely on several factors including primer design and primer efficiency. In addition, the comparative C_T method reports relative gene expression levels, which relies on the selection of a reference gene with stable mRNA levels in the cell types being examined. The presented protocol includes considerations for primer design, validation of reference genes, and assessing primer efficiency. Guidance is provided on synthesizing cDNA, setting up RT-qPCR reactions, and analyzing data using the comparative C_T method.

Keywords Gene expression, Transcript level determination, Primer design, Primer efficiency, Comparative CT method, RT-qPCR, *Ustilago maydis*

1 Introduction

Understanding changes in transcript levels enables researchers to gain valuable insight into gene expression and function. In *Ustilago maydis*, gene deletion studies or altering the expression of a gene can reveal defects in growth, pathogenesis, teliospore development, and/or expression of the gene's downstream targets. These investigations can provide insight into the roles genes have in pathways and processes [1–4]. One method for quantifying messenger RNA (mRNA) levels is by reverse transcriptase quantitative polymerase chain reaction (RT-qPCR). This is a widely used method because of its speed, specificity, sensitivity, and practicality [5, 6]. Successful RT-qPCR reactions require one to consider, among other things, primer design, efficiency of the reverse transcriptase, and quality of the RNA [7].

An N. T. Phan et al. (eds.), *Ustilago maydis*, Methods in Molecular Biology, vol. 3050,
https://doi.org/10.1007/978-1-0716-5364-7_20,

RT-qPCR involves the conversion of RNA to cDNA using reverse transcriptase, the amplification of cDNA, and the detection and quantification of the amplicon through fluorescence detection [8]. This protocol provides guidelines for the following methods: (1) designing primers for performing RT-qPCR, in which detection occurs through the intercalation of a fluorescent dye, such as SYBR Green, (2) the use of probes that are fluorescently labeled, (3) how to validate a reference gene suitable for RT-qPCR, (4) how to determine the primer efficiency of the designed primers or primer/probe set, (5) converting DNase I treated RNA to cDNA, (6) how to setup an RT-qPCR experiment, and (7) how to calculate the relative quantity of a gene of interest using the comparative C_T ($2^{-\Delta\Delta CT}$) method.

1.1 RT-qPCR with Fluorescent Dyes

RT-qPCR performed with an intercalating dye, such as SYBR Green, is a less specific method for quantifying gene transcript levels. The dye is fluorogenic and binds to double-stranded DNA. During the extension phase of PCR, the fluorogenic dye binds to the PCR product and emits a fluorescent signal. During the denaturation and annealing stages of PCR, the dye is unbound and the fluorescence is undetectable [9]. The advantages of using this detection method are its simple and easy setup, and it is an economical choice for RT-qPCR experiments. However, since fluorescent dyes are nonspecific, multiplexing experiments are not possible and can be a disadvantage to some researchers. Another disadvantage of the lack of specificity is that the fluorogenic dye will bind to both specific and nonspecific PCR products, such as primer dimers. This can lead to overestimating the quantity of the target gene [9, 10]. To mitigate these effects, researchers need to optimize primer designs, RT-qPCR reaction conditions, and perform a melt curve analysis. The protocol within this chapter outlines methods for designing primers, RT-qPCR reaction conditions, and PCR cycling conditions for a melt curve analysis.

1.2 RT-qPCR with Fluorescently Labeled Probes

Performing RT-qPCR with a primer and probe set where the probe is fluorescently labeled is another method for quantifying gene transcript levels. The designed primers and probes are sequence-specific, and the chemistry of the probe relies on fluorescence resonance energy transfer (FRET). There are several fluorophore technologies to choose from, and the most common are TaqMan hydrolysis probes, Molecular Beacons, Scorpions, and dual hybridization probes. The advantages of these technologies are that the signal produced is specific to the amplified product and the ability to multiplex reactions. In comparison to fluorogenic dyes, using fluorescently labeled probes is more expensive as each gene target requires its own primer and probe set [9, 10].

The presented protocol provides guidance on designing primer and probe sets for hydrolysis probes and RT-qPCR cycling conditions. Hydrolysis probes utilize the 5′-3′ exonuclease activity of the DNA polymerase during PCR. These probes are labeled with a fluorescent reporter and quencher, and these labels are in close proximity to each other when the probe is intact. The quencher absorbs the fluorescence of the reporter, resulting in no signal being produced. During PCR amplification, the probe hybridizes to the target sequence and during the extension phase of PCR, the probe is cleaved, releasing the fluorescent reporter and allowing for a fluorescent signal to be detected [9, 10].

1.3 Reference Gene Validation

Comparing gene transcript levels between biological samples using RT-qPCR relies on reliable and accurate results. The presented protocol utilizes the comparative C_T ($2^{-\Delta\Delta CT}$) method for quantifying gene transcript levels. This method relies on normalizing RT-qPCR data to a reference gene to assess technical variations [11]. Reference genes, also known as housekeeping genes, are used as internal controls for calculating relative gene expression. The transcript level of the reference gene should not vary between experimental samples [12]. It is highly recommended that reference genes be validated to ensure little to no variability across experimental cell types. For example, in *U. maydis*, *gapdh* is commonly used as a reference gene in RT-qPCR experiments. However, transcriptome analysis indicated that the *gapdh* transcript level is variable between the haploid and teliospore cell types [13] and is, therefore, unsuitable to use as a reference gene in experiments that contain these cell types. Reference gene validation experiments supported this and alternative reference genes were validated in teliospore time course experiments [14]. A method for validating candidate reference genes and validation programs is outlined in this protocol.

1.4 Determining Primer Efficiency

The precision of RT-qPCR experiments is dependent on primer efficiency. The higher the primer efficiency is, the more robust and precise the PCR assay is. Relative quantification of gene transcript levels relies on normalizing C_T values to a reference gene and calculating the differences between experimental samples. The precision of these calculations is dependent on the PCR amplification of the two genes having comparable efficiencies [15]. A method for determining the primer efficiency of designed primers is presented.

1.5 First-Strand cDNA Synthesis

Reverse transcription of total RNA converts it to complementary DNA (cDNA), which is the template used in the qPCR portion of the experiments. To generate cDNA, reverse transcriptase, primers, deoxynucleoside triphosphates (dNTPs), $MgCl_2$, and RNase inhibitors are used to facilitate this reaction [6]. A method for generating cDNA using oligo $d(T)_{16}$ primers is outlined here.

1.6 RT-qPCR

RT-qPCR is used to quantify gene transcript levels to understand gene expression changes under specific experimental conditions.

The method relies on the detection of a fluorescent signal produced during the amplification phase of PCR. This is reported as the PCR cycle where fluorescence increases and surpasses a defined threshold cycle (C_T). In general, low CT values indicate that there is an increased amount of the target sequence in the sample, whereas higher CT values indicate lower amounts of the target sequence [8]. A method for performing RT-qPCR is outlined in this chapter.

1.7 Comparative C_T ($2^{-\Delta\Delta CT}$) Method

The comparative C_T method is a method for quantifying gene transcript levels. It is a relative quantification method that calculates the change in transcript level of a target gene relative to a calibrator group (e.g., untreated sample, time 0 in a time course, or wild-type sample) [11, 16]. This chapter provides an overview of how to calculate relative gene transcript levels from the C_T values downloaded from the qPCR instrument.

2 Materials

2.1 Designing RT-qPCR Primers for Fluorescent Dye Detection

1. Primer design software (e.g., OligoArchitect Online).
2. Sequence of the gene of interest.

2.2 Designing RT-qPCR Primers and Fluorescent Labeled Probes

1. Primer design software (e.g., Primer Express or PrimerQuest Tool).
2. Sequence of the gene of interest.

2.3 Reference Gene Validation

1. qPCR instrument.
2. qPCR primers or primers and probes for candidate reference genes (e.g., *gapdh*, actin, 18s rRNA, and *UMAG_00175*).
3. cDNA from samples that are representative of the experimental conditions.
4. qPCR master mix: the master mix contains all components for qPCR and must be compatible with the type of primers being used (*see* **Note 1**).
5. Nuclease-free water.
6. 96-well PCR reaction plate: ensure that the PCR plate selected is suitable for your qPCR instrument.
7. Optical adhesive plate seals.
8. Reference gene validation program (e.g., geNorm, NormFinder, or BestKeeper).

2.4 Primer Efficiency

1. qPCR instrument.
2. qPCR master mix: the master mix contains all components for qPCR and must be compatible with the type of primers designed (*see* **Note 1**).
3. cDNA of a cell type where the target gene is known to have high expression.
4. qPCR primers or primers and probes to be tested.
5. Nuclease-free water.
6. 96-well PCR reaction plate: ensure that the PCR plate selected is suitable for your qPCR instrument.
7. Optical adhesive plate seals.

2.5 First-Strand cDNA Synthesis

1. Reverse transcription reagents: these reagents may come in a kit that typically include a buffer, dNTP mixture, RNase inhibitor, reverse transcriptase, and primers such as oligo $d(T)_{16}$, or random hexamers. First-strand cDNA synthesis may be performed with gene-specific primers if experimental conditions allow (*see* **Note 2**).
2. Nuclease-free water.
3. Thermocycler.

2.6 Quantitative Polymerase Chain Reaction

1. qPCR instrument.
2. qPCR master mix: a master mix that contains all components suitable for qPCR and is compatible with the type of primers designed (*see* **Note 1**).
3. 96-well PCR reaction plate: ensure that the PCR plate selected is suitable for your qPCR instrument.
4. Optical adhesive plate seals.
5. cDNA from experimental samples.

2.7 RT-qPCR Data Analysis

1. Comparative C_T ($2^{-\Delta\Delta CT}$) method for relative quantification.
2. The C_T values downloaded from the qPCR instrument.

3 Methods

3.1 Primer Design for RT-qPCR with Fluorescence Dyes

1. Gene-specific primer design is a critical step for successful RT-qPCR experiments. Primers are designed as close to the 3′ end of the predicted mRNA as the gene sequence allows based on GC content, level of repeated sequences, and other design program-specific factors. The amplicon produced by primers designed in this manner are less likely to be affected by reverse transcriptase synthesized cDNAs that are not full-length.

Designing primers for RT-qPCR experiments using fluorescence dyes (e.g., SYBR Green), select primers using a free open-source primer design program (e.g., OligoArchitect Online, https://www.oligoarchitect.com/).

2. In the primer parameters, set the annealing temperature (T_m) to 60°C ± 5°C. When selecting sets of primers, select primer sets that have a T_m close to one another. This allows for reference and target genes to be amplified in the same RT-qPCR run.
3. Set the primer length to 18–25 nucleotides and the amplicon size between 90 and 120 bp.

3.2 Designing RT-qPCR Primers and Fluorescent Labeled Probes

1. Designing RT-qPCR primers and fluorescent labeled probes requires the use of primer design software such as Primer Express or the online tool PrimerQuest (https://www.idtdna.com/pages/tools/primerquest). Similar to RT-qPCR primers for experiments that use fluorescence dyes, the primer pair and nested probe should be designed to a region close to the 3′ end of the predicted mRNA sequence.
2. The primers should be between 18 and 25 nucleotides, and the amplicon size between 90 and 120 bp.
3. Probes should be located close to either the forward or reverse primer, but do not overlap the primer. The probe length should be between 13 and 30 bp, and the T_m is typically 5°C–10°C higher than the primers.
4. Consideration should be given when selecting a reporter dye for the probe. The reporter dyes should differ for reference and target genes if a goal is to multiplex RT-qPCR reactions.

3.3 Reference Gene Validation

1. Validation of the reference gene for RT-qPCR is crucial if measuring differential gene expression with the comparative C_T ($2^{-\Delta\Delta CT}$) method. Select several candidate reference genes to test for the experiment (*see* **Note 3**).
2. Perform RT-qPCR using samples that represent the experimental conditions or cell types to be used (e.g., haploid cells and teliospores).
3. Use a reference gene validation program (e.g., geNorm, NormFinder, or BestKeeper) to identify the optimal reference gene among the candidate genes (*see* **Note 4**).

3.4 Primer Efficiency

1. Prepare five dilutions using a tenfold dilution series of the cDNA (*see* **Note 5**) from a cell type where the expression of the target gene is known to be high (*see* **Note 6**).

2. Perform RT-qPCR with the primer set and using the tenfold dilution series. Ensure that each dilution template is aliquoted in triplicate on the RT-qPCR plate.

3. Download the RT-qPCR data from the qPCR instrument to obtain the C_T values.
4. Plot the C_T values versus the dilution factor. Fit the data points to a straight line and determine the slope of the line.
5. Calculate the amplification efficiency (*see* **Note 7**, see results following RT-qPCR Fig. 1) using the formula:

$$E = 10^{-\frac{1}{slope}} - 1 \qquad (1)$$

where

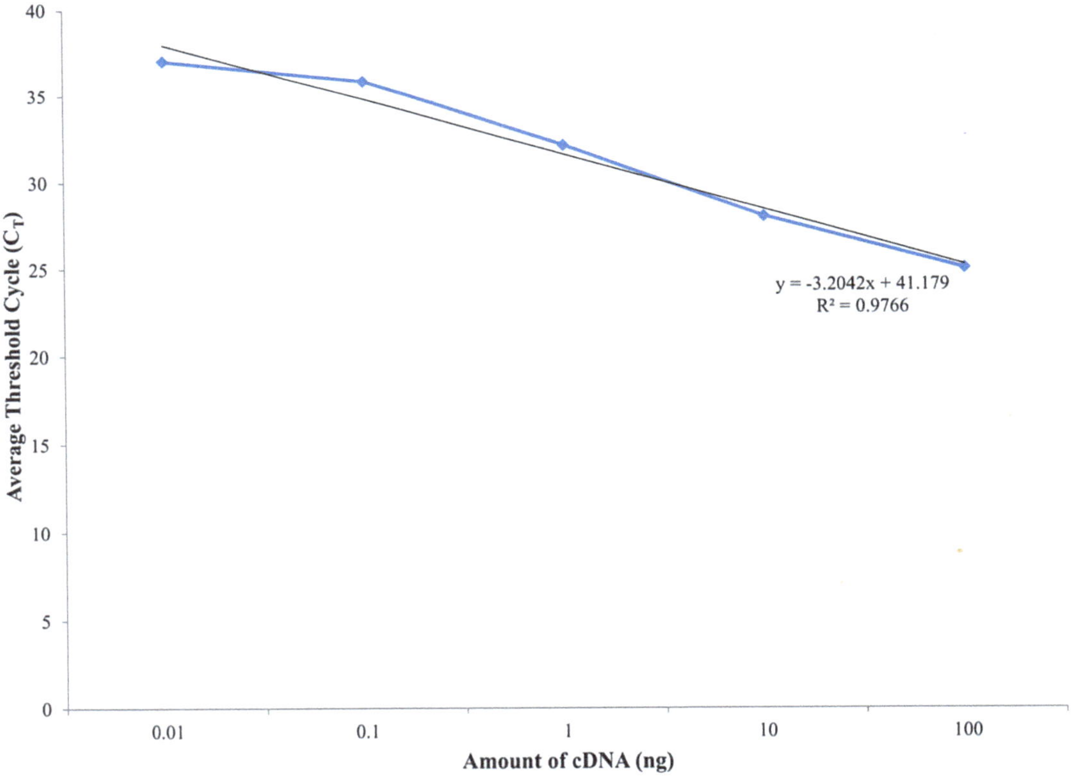

Fig. 1 Primer efficiency test for primer and probe set designed for *UMAG_03234* using cDNA generated from haploid cells. A tenfold serial dilution series was created, and RT-qPCR was performed using the primer and probe set for *UMAG_03234*. The threshold cycles (C_T) were obtained from data downloaded from the qPCR instrument. Each reaction was performed in triplicate, and the average C_T was calculated. The results were plotted with the serial dilution on the x-axis and the average C_T on the y-axis. The line of best fit is generated (black line in figure), and the slope of the line was calculated to be −3.2042. The primer efficiency was calculated to be approximately 105%

$$slope = \frac{(y - b)}{x} \tag{2}$$

3.5 First-Strand cDNA Synthesis

1. DNase I treated RNA should be used to perform first-strand synthesis reactions (*see* **Note 8**).
2. Reverse transcribe 200 ng of total RNA using the reverse transcription reagents: RNA concentrations are normalized to 100 ng/μL and 2 μL of RNA is used to perform reverse transcription in 10 μL reactions. To the RNA, 8 μL of the reverse transcription master mix (*see* **Note 9**) is added. For every 10 μL reaction, a typical reverse transcription master mix contains: 1 μL of 10× reverse transcription buffer (1× final concentration), 2.2 μL of 25 mM $MgCl_2$ (5.5 mM final concentration), 2 μL of 2.5 mM dNTPs (500 μM final concentration), 0.5 μL of 50 μM Oligo $d(T)_{16}$ (2.5 μM final concentration), 0.2 μL of 20 U/μL RNase inhibitor (0.4 U/μL final concentration), 0.25 of 50 U/μL reverse transcriptase (1.25 U/μL final concentration), and the remaining volume (1.85 μL) made up of nuclease-free water. Gently mix reactions and centrifuge briefly to collect the reaction contents.
3. A no reverse transcriptase control (NRT) reaction must be included if the total RNA has not been previously screened for the presence of gDNA contamination (*see* **Note 10**).
4. The first-strand synthesis reactions are placed in a thermocycler using cycle conditions specific to the reverse transcriptase used. For example: 25°C for 10 min, 50°C for 30 min, 95°C for 10 min, and followed by a 4°C hold.
5. Following reverse transcription, 30 μL of nuclease-free water is added to each reaction to dilute the cDNA. Gently mix each dilution and briefly centrifuge to collect the contents.

3.6 Quantitative Polymerase Chain Reaction

1. If all primers (reference and target genes) are designed with a similar T_m, RT-qPCR reactions for all genes can be performed on the same 96-well plate.
2. Three replicates should be performed for each RT-qPCR reaction on the same plate (*see* **Note 11**). A no template control (NTC) and positive control (such as gDNA) are included (*see* **Note 12**).
3. If using a fluorescence dye such as SYBR Green, each 20 μL RT-qPCR reaction contains 2 μL of cDNA and 18 μL of a master mix containing: 10 μL of 2× premix master mix (1× final concentration), 2 μL each of 5 μM forward and reverse primers designed in Sect. 3.1 (0.5 μM final concentration), and 4 μL of nuclease-free water.

4. If using a fluorescent probe, each 20 μL RT-qPCR reaction contains 2 μL of cDNA and 18 μL of a master mix containing: 10 μL of 2× Universal Master Mix (1× final concentration), 1 μ L each of 4 μM forward and reverse primers designed in Sect. 3.2 (200 μM final concentration), 2 μL of 2 μM probe designed in Sect. 3.2 (200 μM final concentration), and 4 μL of nuclease-free water.
5. RT-qPCR plates are sealed with an optical adhesive seal and gently mixed. Plates are briefly centrifuged to collect all contents and remove bubbles (*see* **Note 13**).
6. For RT-qPCR plates containing a fluorescent dye, the qPCR instrument is setup with the following program: 50°C for 2 min, 95°C for 2 min, followed by 40 cycles of 95°C for 15 s, and 60°C for 1 min (*see* **Note 14**). PCR cycling conditions are immediately followed up with the following continuous melt (dissociation) curve program (*see* **Note 15**): 1.6°C/s ramp rate at 95°C for 15 s, 1.6°C/s ramp rate at 60°C for 1 min, and 0.15°C/s at 95°C for 15 s.
7. For RT-qPCR reactions that contain fluorescent probes, the plate is placed in the qPCR instrument with the following program: 50°C for 2 min, 95°C for 10 min, followed by 40 cycles of 95°C for 15 s, and 60°C for 1 min.
8. At the end of the RT-qPCR run, the data are collected and exported from the qPCR instrument.

3.7 RT-qPCR Data Analysis Using the Comparative C_T Method

1. The RT-qPCR data can be exported from the qPCR instrument and will contain threshold cycle (C_T) values for each sample. These data can be viewed in a program such as Microsoft Excel.
2. The ΔC_T is first calculated for the experimental sample and the calibrator sample. The calibrator sample is typically one of the samples in the experiment. For example, the calibrator sample is 0 h in a time course experiment. If comparing wild type versus mutant samples, the calibrator sample is the wild type sample. The ΔC_T is calculated with the following formulas:

$$\Delta CTsample = (CT\ of\ target\ gene - CT\ of\ reference\ gene) \quad (3)$$

$$\Delta CTcalibrator = (CT\ of\ target\ gene - CT\ of\ reference\ gene) \quad (4)$$

3. For the comparative C_T method (*see* **Note 16**), the relative quantity (RQ) is calculated with the formula:

$$RQ = 2^{-\Delta\Delta CT} \quad (5)$$

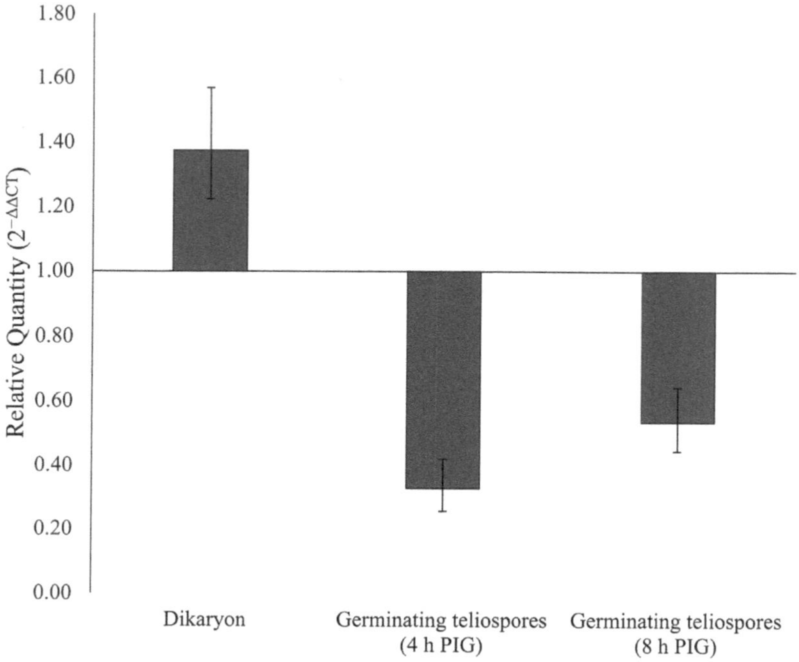

Fig. 2 Transcript levels for *uded1* in the 518 × 521 dikaryon, and germinating teliospores at 4 h and 8 h postinduction of germination. RT-qPCR was performed using primers and probes and relative quantities (RQ) were calculated using the comparative C_T ($2^{-\Delta\Delta CT}$) method. *UMAG_00175* was used as the reference gene and dormant teliospores was set as the calibrator sample. Bars indicate the RQ minimum and RQ maximum values (95% confidence interval, $n = 3$)

where

$$-\Delta\Delta CT = -[\Delta CTsample - \Delta CTcalibrator] \quad (6)$$

4. The relative quantity can be determined using Eq. 5 in Step 3, after which the results may be visualized (see results following analysis of RT-qPCR, Fig. 2). Experimental samples that have an RQ that is greater than 1 indicate that the gene transcript is upregulated. If the RQ value is less than 1, the gene transcript is downregulated in relative to the calibrator sample [11].

4 Notes

1. A suitable premix qPCR master mix that is appropriate for the chosen detection method can be used. The master mix will contain all components necessary for a successful RT-qPCR reaction and typically includes a buffer, dNTPs, $MgCl_2$, and DNA polymerase. Other components that may be included: a

fluorescent dye (e.g., SYBR Green) if using primers designed for fluorescent dye detection and a passive reference dye (e.g., ROX), which corrects for variation in liquid volumes in wells.

2. Choosing to generate cDNA with gene-specific primers is dependent on the experiment. If the RNA is of good quality, quantity is not a limiting factor, and only a single gene is of interest for your experiment, priming with gene-specific primers is a specific and sensitive method for first-strand cDNA synthesis [17].

3. The comparative C_T method measures relative expression and is normalized to a reference gene. The expression of a reference gene should not be variable between different sample types (e.g., haploid cells vs teliospore) or experimental conditions (e.g., untreated vs treated). The use of a reference gene allows for normalization of the data to account for differences in amounts of starting material, variance in the quality of the starting material, and differences in the preparation of the RNA and first-strand synthesis. The reference gene is exposed to the same experimental conditions as the target genes, and therefore, the transcript must remain constant or minimally affected between cell types or by the treatment conditions [18]. For example, in *U. maydis*, *gapdh* is an unsuitable reference gene in gene expression experiments that compare transcript levels of target genes in haploid cells vs teliospore samples, or dormant vs germinating teliospores, as the expression of *gapdh* is variable [13, 14, 19]. Validation of the reference gene is imperative to calculating accurate relative quantities of RNA using the comparative C_T method.

4. There are three reference gene validation programs available that are free to download at the time of writing: geNorm [12], NormFinder [20], and BestKeeper [21]. Each validation program will rank candidate reference genes and indicate which of the candidates is the most stable given the samples tested.

5. When determining primer efficiency, it is best to use cDNA for gene expression RT-qPCR experiments. The cDNA samples are typically prepared in the same manner as your experimental samples and will indicate how efficient your primer sets are.

6. Transcriptome data or prior knowledge from previous experiments can be used to determine the cell type that the target gene is known to be high in. For example, if the gene of interest is *UMAG_04080*, RNA-seq data from Seto et al. [13] showed that the transcript is the highest in the dormant teliospore. Based on these data, cDNA generated from dormant

teliospores would be the most suitable template for carrying out the primer efficiency test.

7. Acceptable primer efficiency is 90%–110%. An efficiency of 100% will have a slope of −3.32, which corresponds to a twofold increase in PCR product every 3.32 cycles. Lower efficiencies will have a steeper slope, and higher efficiencies will have a shallower slope. For example, a slope of −3.58 has a primer efficiency of approximately 90%, whereas a slope of −3.1 has an efficiency of approximately 110%. If primer efficiencies fall outside of the acceptable range, consider adjusting the concentration of the primers and probes and annealing temperature of the qPCR run. Ensure that all pipettes are calibrated and free from contamination. If the primer efficiencies are still not improved, consider redesigning the primers.
8. A DNase I treatment of RNA should be performed before first-strand cDNA synthesis. This treatment degrades contaminating genomic DNA (gDNA) in the RNA samples. PCR reactions will amplify contaminating gDNA in RNA samples and will lead to inaccurate quantification of RNA in gene expression experiments using RT-qPCR. Ensure that RNA samples are devoid of gDNA before performing first-strand cDNA synthesis by performing PCR on the DNase I treated RNA samples. RNA is not a template for DNA polymerase and will not amplify using standard PCR conditions, but gDNA will. If a PCR product is produced in a reaction containing RNA as a template, an additional DNase I treatment and PCR screen are required.
9. A master mix is used to ensure that the concentration of all the components are the same in each tube of samples. When calculating the master mix, ensure that you multiply each component by the number of reactions plus one or two. The extra reactions account for pipetting error that can occur during dispensing of the master mix into the samples.
10. A no reverse transcriptase control is usually included during first-strand cDNA synthesis and will indicate if gDNA is present in the sample during PCR. However, if the RNA has been previously DNase I treated and PCR screened to ensure there is no presence of gDNA, this control can be removed during the setup of first-strand cDNA synthesis.
11. Technical replicates are multiple wells that contain the cDNA sample and PCR reagents. Replicates ensure that the RT-qPCR reaction is precise and will identify potential outliers, which can be removed.
12. No-template and positive PCR controls for each gene of interest and reference gene is included in each RT-qPCR run.

The negative control will not produce a PCR amplicon. The production of an amplicon indicates a reagent in the master mix is contaminated. The positive control contains a template (e.g., gDNA) that will produce a PCR product. Amplification of a PCR product of the correct length confirms that the PCR reaction is performing as expected.

13. After setting up an RT-qPCR plate, bubbles may be present at the bottom of the well. It is important to ensure that wells do not contain bubbles as they can interfere with the fluorescence readings during a qPCR run. This can result in an abnormal reading. Centrifugation of the plate can ensure that bubbles at the bottom of a well are eliminated.

14. The PCR cycling conditions should be followed according to the manufacturer's recommendations. For example, if the T_m of the primers are ≥60°C, the annealing temperature is set to 60°C. If the T_m of the primers are <60°C, set the annealing temperature to the T_m of the primers.

15. When using fluorescent dyes such as SYBR green, a melt (dissociation) curve is performed at the end of an RT-qPCR run. SYBR Green is nonspecific and will bind to any double-stranded DNA in the sample. The melt curve program and analysis are performed as a quality control step to verify that fluorescence is from a single amplicon. The observation of a single peak in the melt curve indicates that a single product was amplified; however, it is recommended that the RT-qPCR products are also visualized by agarose gel electrophoresis to verify the correct length amplicon was produced. Multiple peaks on a melt curve can indicate nonspecific amplification or the presence of primer-dimers. Nonspecific priming could be resolved by redesigning primers and primer–dimer formation can be prevented by either adjusting primer concentrations, annealing temperatures, or redesigning primers.

16. The comparative C_T ($2^{-\Delta\Delta Ct}$) method of determining RNA levels in samples is a relative quantification method. It is a ratio that compares the expression level of the target gene relative to the reference gene. This ratio is then used to compare the gene expression of two different samples (e.g., untreated vs treated samples). When data is exported from the qPCR instrument, the threshold cycle (C_T) value is reported. This is the number of cycles that are required for the fluorescent signal to surpass the defined threshold. A low CT value indicates higher quantities of the target RNA in the sample. In contrast, high C_T values indicate lower target RNA quantities in the sample [22].

Acknowledgments

Funding provided to BJS by the Natural Sciences and Engineering Research Council (NSERC) of Canada and the Ministry of Ontario Research and Innovation, Ontario Research Fund: Research Excellence (ORF-RE) award.

References

1. Cheung HYK, Donaldson ME, Storfie ERM et al (2021) Zfp1, a putative $Zn(II)_2Cys_6$ transcription factor, influences *Ustilago maydis* pathogenesis at multiple stages. Plant Pathol 70(7):1626–1639. https://doi.org/10.1111/ppa.13398
2. Donaldson ME, Saville BJ (2013) *Ustilago maydis* natural antisense transcript expression alters mRNA stability and pathogenesis. Mol Microbiol 89(1):29–51. https://doi.org/10.1111/mmi.12254
3. Doyle CE, Cheung HYK, Spence KL et al (2016) Unh1, an *Ustilago maydis* Ndt80-like protein, controls completion of tumor maturation, teliospore development, and meiosis. Fungal Genet Biol 94:54–68. https://doi.org/10.1016/j.fgb.2016.07.006
4. Seto AM, Saville BJ (2025) Characterization of RNA helicase genes in *Ustilago maydis* reveals links to stress response and teliospore dormancy. Int J Mol Sci 26(6):2432. https://doi.org/10.3390/ijms26062432
5. Harshitha R, Arunraj DR (2021) Real-time quantitative PCR: a tool for absolute and relative quantification. Biochem Mol Biol Educ 49(5):800–812. https://doi.org/10.1002/bmb.21552
6. Bong D, Sohn J, Lee S-JV (2024) Brief guide to RT-qPCR. Mol Cells 47(12):100141. https://doi.org/10.1016/j.mocell.2024.100141
7. Rodríguez A, Rodríguez M, Córdoba JJ et al (2015) Design of primers and probes for quantitative real-time PCR methods. In: Basu C (Ed.)Eds. PCR primer design. Springer, New York, 31–56. 10.1007/978-1-4939-2365-6_3
8. Nolan T, Hands RE, Bustin SA (2006) Quantification of mRNA using real-time RT-PCR. Nat Protoc 1(3):1559–1582. https://doi.org/10.1038/nprot.2006.236
9. Nestorov J, Matić G, Elaković I et al (2013) Gene expression studies: how to obtain accurate and reliable data by quantitative real-time RT PCR. J Med Biochem 32(4):325–338. https://doi.org/10.2478/jomb-2014-0001
10. VanGuilder HD, Vrana KE, Freeman WM (2008) Twenty-five years of quantitative PCR for gene expression analysis. BioTechniques 44(5):619–626. https://doi.org/10.2144/000112776
11. Livak KJ, Schmittgen TD (2001) Analysis of relative gene expression data using real-time quantitative PCR and the $2^{-\Delta\Delta CT}$ method. Methods 25(4):402–408. https://doi.org/10.1006/meth.2001.1262
12. Vandesompele J, De Preter K, Pattyn F et al (2002) Accurate normalization of real-time quantitative RT-PCR data by geometric averaging of multiple internal control genes. Genome Biol 3(7):0031. https://doi.org/10.1186/gb-2002-3-7-research0034 research0034
13. Seto AM, Donaldson ME, Saville BJ (2025) Exploring mechanisms of gene expression control during *Ustilago maydis* teliospore germination. Can J Plant Pathol 47(1):80–97. https://doi.org/10.1080/07060661.2024.2413557
14. Seto AM (2013) Analysis of gene transcripts during *Ustilago maydis* teliospore dormancy and germination. Master's Thesis, Trent University, Peterborough, ON
15. Bustin SA, Benes V, Garson JA et al (2009) The MIQE guidelines: minimum information for publication of quantitative real-time PCR experiments. Clin Chem 55(4):611–622. https://doi.org/10.1373/clinchem.2008.112797
16. Schmittgen TD, Livak KJ (2008) Analyzing real-time PCR data by the comparative CT method. Nat Protoc 3(6):1101–1108. https://doi.org/10.1038/nprot.2008.73
17. Jozefczuk J, Adjaye J (2011) Chapter six—quantitative real-time pcr-based analysis of gene expression. In: Jameson D, Verma M, Westerhoff HV (Eds.) Methods in enzymology. Academic Press, vol 500 Cambridge, MA, 99–109 https://doi.org/10.1016/B978-0-12-385118-5.00006-2
18. Radonić A, Thulke S, Mackay IM et al (2004) Guideline to reference gene selection for quantitative real-time PCR. Biochem Biophys Res Commun 313(4):856–862. https://doi.org/10.1016/j.bbrc.2003.11.177
19. Sacadura NT, Saville BJ (2003) Gene expression and EST analyses of *Ustilago maydis* germinating teliospores. Fungal Genet Biol

40(1):47–64. https://doi.org/10.1016/s1087-1845(03)00078-1
20. Andersen CL, Jensen JL, Ørntoft TF (2004) Normalization of real-time quantitative reverse transcription-PCR data: a model-based variance estimation approach to identify genes suited for normalization, applied to bladder and colon cancer data sets. Cancer Res 64(15):5245–5250. https://doi.org/10.1158/0008-5472.Can-04-0496
21. Pfaffl MW, Tichopad A, Prgomet C et al (2004) Determination of stable housekeeping genes, differentially regulated target genes and sample integrity: BestKeeper—excel-based tool using pair-wise correlations. Biotechnol Lett 26(6):509–515. https://doi.org/10.1023/b:bile.0000019559.84305.47
22. Ruiz-Villalba A, Ruijter JM, van den Hoff MJB (2021) Use and misuse of Cq in qPCR data analysis and reporting. Life 11(6):496

Check for updates

Chapter 21

Isolation of RNA for RNA Sequencing in *Ustilago maydis*

Amanda M. Seto and Barry J. Saville

Abstract

In the last 15 years, RNA sequencing (RNA-seq) has emerged as a popular tool for studying gene expression changes in an organism. In fungal plant pathology, this is a comprehensive method for identifying transcriptome changes to both the fungal pathogen and its host. RNA-seq generates a vast amount of data that can increase our understanding of the interaction between the fungus and its host, or how the fungal species adapts to various environmental conditions. This chapter emphasizes the importance of sample preparation for RNA-seq. It also describes a protocol for isolating high quality RNA from *Ustilago maydis* haploid, dikaryon, and teliospore samples, removing genomic DNA contamination and screening RNA to ensure its removal, and it provides a method for cDNA library preparation. Successful RNA-seq experiments require careful handling of RNA during isolation and its downstream steps. Careful handling enables RNA sequencing to generate vast amounts of information on transcriptome changes, which can provide meaningful insights into the growth and development of *U. maydis*.

Keywords Precautions in RNA isolation, *Ustilago maydis*, Total RNA isolation, RNA-seq, cDNA library creation

1 Introduction

Pathogens, such as pests, weeds, and microbes, account for approximately 20%–40% of global crop losses. The greatest crop loss is due to fungal pathogens, and the diseases that they cause are economically important, causing annual global losses of approximately $220 billion USD [1–3]. Identifying the molecular mechanisms involved in fungal plant pathogenesis will increase our understanding of disease development and help in the development of methods for mitigating disease spread.

The development of high-throughput methods for studying gene expression has produced vast datasets identifying, which genes are expressed in an array of experimental conditions. It has increased our understanding of how cells adapt their transcriptomes to changing environments and provided insight into the degree and direction of transcript level change during disease development and progression

An N. T. Phan et al. (eds.), *Ustilago maydis*, Methods in Molecular Biology, vol. 3050,
https://doi.org/10.1007/978-1-0716-5364-7_21,

[4, 5]. In the last 15 years, RNA sequencing (RNA-seq) has emerged as a popular tool for studying the changes in the transcriptome of an organism [5, 6]. It is a comprehensive method for identifying transcript level changes on a global scale. In fungal pathology, transcriptome analysis of both the fungal pathogen and its host can provide insight into the effects the fungal pathogen has on its host and identify mechanisms, by which fungi adapt to the environment inside the host and its defense mechanisms [5, 7, 8].

Transcriptome analysis of *Ustilago maydis* during fungal growth, pathogenesis, and teliospore germination can provide insights into the molecular mechanisms that are involved. Following appropriate experimental design and completion, the isolation of high-quality RNA is important to accurately identify the transcript level changes induced by the experiment. The isolation of high-quality RNA ensures successful cDNA library creation and accuracy of the data generated from RNA-seq [5]. The focus of any isolation must be to maintain RNA integrity. In this chapter, methods for isolating total RNA, DNase I treatment of RNA, screening the RNA for gDNA contamination, and cDNA library construction are outlined.

1.1 Isolating High-Quality RNA from Ustilago maydis Samples

Isolating high-quality RNA from experimental samples is the first step of preparing samples for RNA-seq. Maintaining the quality of RNA provides more reliable interpretations of gene expression patterns observed in RNA-seq experiments. This is achieved through: (1) effective disruption or lysis of the cell wall, (2) denaturation of nucleoprotein complexes, (3) inactivation of nucleases, and (4) ensuring that the RNA is free from contaminants such as proteins, lipids, genomic DNA (gDNA), or carbohydrates [9]. Total RNA can be isolated with a commercial kit or by organic extraction. Commercial kits are reliable, quick, efficient, and easy methods for isolating total RNA [10]. Organic RNA extractions are conventional methods for isolating high-quality RNA. The presented protocol uses a guanidinium thiocyanate-phenol-chloroform extraction method for isolating RNA from *U. maydis* samples. Although this method is more labor-intensive, protocols for isolating from *U. maydis* cell types are well established. This includes haploid cells from axenic cultures and frozen down cultures [11–13], infected plant tissues [14–16], and teliospore samples [11, 12, 17].

Preventing RNA degradation when handling RNA during RNA isolation and subsequent steps after is important. The presence of ribonuclease (RNase) contamination is one of the major sources of RNA degradation and can have an impact on RNA quality and experimental results. Examples of sources of RNase contamination are careless handling of buffers and pipettors [18]. It is recommended that readers familiarize themselves with potential sources of RNase contamination and how to inhibit RNases [18].

1.2 DNase I Treatment of RNA Samples

Before RNA samples are prepared for RNA-seq, the total RNA must be devoid of any gDNA. The presence of gDNA in RNA samples can cause false positives in downstream applications as it can be amplified. The removal of gDNA is accomplished by using DNase I to digest the contaminating gDNA from the RNA sample [19]. A method for digesting gDNA from total RNA is outlined in this chapter.

1.3 Screening DNase I-Treated RNA for Genomic DNA Contamination

Ensuring that the DNase I-treated RNA is free from gDNA contamination is an important step following treatment. A protocol that uses PCR to screen for gDNA contamination following DNase I treatment is outlined in this chapter.

1.4 cDNA Library Construction

Preparing the RNA for RNA-seq is the next step. In this chapter, some details on preparing cDNA libraries are covered. An important step enabling the determination of mRNA level changes is the removal of ribosomal RNA. This can be done through rRNA depletion (e.g., REF), however, the method outlined here is for poly(A)$^+$ enrichment from total RNA.

2 Materials

2.1 *Ustilago maydis* Total RNA Isolation

1. *Ustilago maydis* samples: This protocol has been optimized for samples such as haploid cells, dikaryons, and dormant and germinating teliospores using a bead mill homogenizer. Other methods, including freezing in liquid nitrogen and grinding with a mortar and pestle can also be used to homogenize these *U. maydis* cell types and infected plant tissue.
2. RNase Away.
3. Nuclease-free water.
4. Screw cap tubes containing Lysing matrix C
5. Bead mill homogenizer
6. TRIzol reagent.
7. Chloroform.
8. 75% Ethanol.
9. 10×BPTE Electrophoresis Buffer: 100 mM PIPES, 300 mM Bis-Tris, and 10 mM EDTA. Prepare 1 L of Diethylpyrocarbonate (DEPC) water by combining 1 mL DEPC, 9 mL of 100% ethanol, and 990 mL of water (0.1% DEPC *v/v* final concentration). Weigh 24.2 g of PIPES, 50.2 g of Bis-Tris and add to 500 mL of DEPC-treated water. Add 16 mL of 0.5 M EDTA (pH 8.0) and bring the volume up to 800 mL. Ensure that all contents are dissolved before transferring the solution to a DEPC-treated 1 L bottle. Let the bottle sit at room temperature overnight before autoclaving.

10. Glyoxal reaction mixture: Add 6 mL DMSO, 2 mL deionized glyoxal, 1.2 mL of 10× BPTE Buffer, 0.6 mL 80% glycerol, and 0.2 mL ethidium bromide (10 mg/mL in water). Make aliquots of the glyoxal reaction mixture and store at −80°C for long-term storage.
11. RNA size marker.

2.2 DNase I Treatment of Total RNA

1. Total RNA (approximately 30 μg).
2. RNase-free DNase I, includes: 10× Buffer and DNase I (2000 U/mL).
3. 250 mM EDTA (RNase-free).
4. Isopropanol.
5. RNA precipitation solution: 0.8 M disodium citrate, 1.2 M NaCl. Prepare 1 L of DEPC water by combining 1 mL DEPC, 9 mL of 100% ethanol, and 990 mL of water (0.1% DEPC *v/v* final concentration). Weigh 42.10 g of disodium citrate and 14.03 g of NaCl. Dissolve in 100 mL DEPC water before bringing the volume up to 200 mL. Allow the solution to sit at room temperature overnight before autoclaving to sterilize.
6. 75% EtOH.
7. Nuclease-free water.

2.3 Test RNA for Genomic DNA Contamination

1. Thermocycler.
2. Hot Start Taq polymerase kit includes: 10× buffer, 25 mM $MgCl_2$, and 5 U/μL Taq polymerase.
3. 10 mM dNTP mix.
4. Primers for *gapdh*: the primers are designed to span the intron in *gapdh* to distinguish between genomic DNA amplification and cDNA amplification.
5. Genomic DNA template.
6. 50× TAE Buffer: Combine 242 g of Tris-base, 57.1 mL of glacial acetic acid, 100 mL of 0.5 M EDTA (pH 8.0) in 500 mL of water before bringing the volume to 1 L.
7. DNA molecular size marker.
8. 6× gel loading buffer: 0.25% bromophenol blue (w/v), 0.25% xylene cyanol FF (w/v), 30% glycerol (v/v) in water.
9. UV transilluminator.

2.4 cDNA Library Construction

1. RNA library preparation kit: this protocol outlines a method for creating cDNA libraries with mRNA enrichment. The RNA samples are prepared with the TruSeq Stranded mRNA kit (Illumina), which uses oligo d(T)-beads to capture

poly(A) tail mRNAs and reagents to create cDNA libraries. Two kits must be purchased: the TruSeq Stranded mRNA Library Prep kit and the Index Adapter Component that is suitable for your experimental samples. The library preparation kit contains: RNA purification beads, the reagents DTE, DTA, DTL, bead binding buffer, bead washing buffer, elution buffer, and fragment prime finish mix. The core library preparation kit contains reagents for adapter ligation and first- and second-strand cDNA synthesis.

2. 1.5 mL RNase/Dnase-free tubes.
3. AMPure XP beads.
4. Ethanol.
5. Microseal "B" adhesive seals.
6. Nuclease-free ultrapure water.
7. Superscript II Reverse Transcriptase.
8. Tris-HCl (10 mM, pH 8.5).
9. Tween 20.
10. 96-well 0.3 mL PCR plates.
11. 96-well Thermocycler.
12. RNase Away.
13. Magnetic stand for 96-well plates.

3 Methods

3.1 *Ustilago maydis* Total RNA Isolation and Quality Assessment

1. Rigorous RNA handling procedures must be followed to avoid exogenous RNase contamination (*see* **Note 1**).
2. *Ustilago maydis* samples are resuspended in TRIzol according to the manufacturer's protocol and transferred to a screw cap tube containing lysing matrix C.
3. Incubate the sample at room temperature for 5 min. Transfer the tube to a bead mill homogenizer to homogenize the sample. *U. maydis* cells are lysed using the settings 6.5 m/s for 45 s (*see* **Note 2**). The sample is transferred to ice for 1 min before repeating the homogenization step.
4. The sample is incubated at room temperature for 5 min and then centrifuged at 12,200 × *g* for 10 min.
5. Alternatively, *U. maydis* samples can be homogenized with liquid nitrogen and an RNase-free mortar and pestle (*see* **Notes 3 and 4**). Once homogenized, the sample can be resuspended in TRIzol.

6. RNA isolation is performed following the manufacturer's TRIzol protocol.
7. RNA quality is assessed by electrophoresis using glyoxylated RNA that is run on an agarose gel. Up to 10 μg of RNA can be used on the gel.
8. A total of 10 μL of the glyoxal reaction mixture is added to 1 μ L RNA and 1 μL RNA ladder. The samples are incubated at 65°C for 10 min and then briefly incubated on ice. Centrifuge the tubes briefly to collect contents.
9. A 1.5% agarose gel is prepared with 1× BPTE. The gel is loaded with the glyoxalated RNA and separated using electrophoresis.
10. The gel is visualized with a UV transilluminator to determine the quality of RNA samples (*see* **Note 5**).
11. Determine the quantity of the RNA with a spectrophotometer.

3.2 DNase I Treatment of Total RNA

1. Approximately 30 μg of total RNA is treated with DNase I according to the manufacturer's instructions.
2. A 100 μL reaction is prepared: 30 μg of total RNA, 10 μL of 2000 U/mL DNase I (20 U), 10 μL of 10× DNase I Reaction Buffer (1× final concentration), and nuclease-free water to bring the reaction volume to 100 μL.
3. Incubate the sample at 37°C for 30 min and inactivate the DNase I reaction by adding 2 μL of 250 mM EDTA (5 mM final concentration), then incubating the sample at 75°C for 10 min. Place the sample on ice for 1 min. Centrifuge the sample briefly to collect the contents.
4. Purify the RNA by precipitating the RNA with 25 μL of RNA precipitation solution and 50 μL isopropanol. Invert the tube several times to mix. Incubate the sample at room temperature for 10 min followed by a 30 min incubation at −20°C for 20 min. Centrifuge sample at 12,200 × *g* at 4°C for 15 min.
5. Remove the supernatant and wash the RNA pellet with 1 mL of 75% EtOH. Centrifuge the sample at 7500 × *g* for 5 min. Allow the pellet to air dry.
6. Dissolve the pellet in 50 μL of nuclease-free water.
7. Determine the RNA concentration with a spectrophotometer.

3.3 Test RNA for Genomic DNA Contamination

1. PCR reactions are performed on the DNase I-treated RNA using a hot start DNA polymerase following the manufacturer's protocol using the *gapdh* primers (*see* **Note 6**).
2. A 23 μL PCR master mix (*see* **Note 7**) is added to 2 μL of DNase I-treated RNA. For every 25 μL PCR reaction, the master mix contains: 2.5 μL of 10× buffer (1× final

concentration), 2.0 μL of 2.5 mM dNTPs (200 μM final concentration), 1.0 μL each of 5 mM *gapdh* primers (0.2 μM final concentration), 0.125 μL of 5 U/μL hot start Taq DNA Polymerase (0.625 U), 13.375 μL of water, and 2 μL of DNase I-treated RNA.

3. Place PCR reaction in a thermocycler using the following conditions: 95°C for 10 min, 40 cycles of 95°C for 30 s, 65°C for 15 s, and 72°C for 1 min, followed by 72°C for 10 min and a 4°C hold.
4. Add 5 μL of 6× gel loading buffer to the PCR reaction and load 10 μL of the sample for electrophoresis at 90 V for 60 min. An ethidium bromide-stained agarose gel is prepared with 1× TAE buffer. A DNA molecular size marker is also loaded onto the gel.
5. A UV transilluminator is used to visualize the gel. If genomic DNA has been successfully removed from the RNA sample, *gapdh* will not amplify with PCR. If a PCR amplicon is produced, then gDNA is still present in the RNA sample and will require that the DNase I treatment be repeated (*see* **Note 8**).

3.4 cDNA Library Construction

1. Sufficient high-quality RNA is required for RNA-sequencing. Ensure that you have 0.1–1 μg of total RNA per sample.
2. The quality of the RNA is further assessed with a Bioanalyzer to obtain an RNA integrity number (RIN) (*see* **Note 9**).
3. For poly(A) tail-enriched RNA samples, the manufacturer's protocol for the TruSeq Stranded mRNA kit is followed.
4. First-strand and second-strand synthesis are performed following the manufacturer's protocol.
5. The 3′ ends are adenylated, and the adapters are ligated to the ends following the manufacturer's protocol.
6. PCR is performed to enrich for DNA fragments and purified to create the cDNA library following the manufacturer's protocol.
7. cDNA libraries can then be normalized and pooled according to the manufacturer's protocol before submission for RNA-sequencing.

4 Notes

1. Avoiding RNase contamination is an important step when handling RNA. The most common sources of RNase contamination come from contaminated reagents and micropipettors

[18]. All reagents should be handled with gloves that have been immediately removed from the box, and it is critical to avoid touching the fingertips of the gloves. All reagents and buffers should be stored in small single-use aliquots. Glassware should be baked at 300°C or DEPC-treated, and plasticware should be DEPC-treated to remove RNases. Autoclaving alone will not inactivate RNases on these wares. It is also advisable to use a surfactant such as RNase Away to wipe down benchtop surfaces, micropipettors, and plasticware (e.g., microtube racks) during RNA isolation [18].

2. These bead mill settings have been successfully used to homogenize *U. maydis* haploid cells that have been previously frozen down with liquid nitrogen, axenic *U. maydis* cultures, and dikaryon samples. For teliospore samples, it is important to include the second homogenization step using the bead mill homogenizer to ensure disruption of the teliospore cell walls.

3. As with removing RNases from glassware and plasticware, mortars and pestles must be baked at 300°C for at least 4 h. Each sample should be homogenized using separate mortars and pestles.

4. *Ustilago maydis* haploid cells, dikaryon, and infected plant tissues have been successfully homogenized with a mortar and pestle. Using a mortar and pestle to homogenize teliospore samples has not been verified with the presented protocol, and it is recommended that a bead mill homogenizer be used for best results.

5. The quality of eukaryotic RNA is assessed on an agarose gel. High-quality RNA will have two clear and sharp bands. There should be no or minimal smearing present. The upper band is the 28S rRNA, and the lower is the 18S rRNA. If the RNA isolated is not excessively degraded, the 28S rRNA band should be approximately twice as bright as the 18S rRNA band.

6. A PCR screen should include a negative no template control and a positive control (*U. maydis* gDNA). If the RNA sample is devoid of gDNA contamination, the PCR reaction will not produce an amplicon as RNA is not a template for the DNA polymerase. If gDNA is still present in the RNA sample, a PCR amplicon will be produced in the same size as the positive gDNA control. This is due to the *gapdh* primers being designed to span the intron.

7. A master mix is created to ensure that the concentration of all reagents is the same for each tube of samples. The master mix is calculated by multiplying each component by the number of PCR reactions plus one or two. The addition of extra

reactions to the calculations accounts for pipetting errors that may occur when adding the master mix to the samples.

8. If gDNA is still present in the RNA sample, the DNase I treatment must be repeated. The treatment can be scaled up or down linearly. Repeat the PCR screen following the second DNase I treatment.
9. The RNA integrity number (RIN) is assessed using a Bioanalyzer and gives a value that ranges from 1 to 10, where one is degraded RNA and ten is the most intact RNA [20]. For RNA-seq, samples with an RIN of at least seven are acceptable.

Acknowledgments

Funding provided to BJS by the Natural Sciences and Engineering Research Council (NSERC) of Canada and the Ministry of Ontario Research and Innovation, Ontario Research Fund: Research Excellence (ORF-RE) award.

References

1. Savary S, Ficke A, Aubertot J-N et al. (2012) Crop losses due to diseases and their implications for global food production losses and food security. Food Secur 4(4):519–537. https://doi.org/10.1007/s12571-012-0200-5
2. Faraji Rad Z (2023) Microneedle technologies for food and crop health: recent advances and future perspectives. Adv Eng Mater 25(4):2201194. https://doi.org/10.1002/adem.202201194
3. Khan MR, Sharma RK (2020) Fusarium-nematode wilt disease complexes, etiology and mechanism of development. Indian Phytopathol 73:615–628. https://doi.org/10.1007/s42360-020-00240-z
4. Gondane A, Itkonen HM (2023) Revealing the history and mystery of RNA-seq. Curr Issues Mol Biol 45(3):1860–1874. https://doi.org/10.3390/cimb45030120
5. Kukurba KR, Montgomery SB (2015) RNA sequencing and analysis. Cold Spring Harbor Protocols 2015(11):pdb.top084970. https://doi.org/10.1101/pdb.top084970
6. Wang Z, Gerstein M, Snyder M (2009) RNA-seq: a revolutionary tool for transcriptomics. Nat Rev Genet 10(1):57–63. https://doi.org/10.1038/nrg2484
7. Mendes-Giannini MJS, Taylor ML, Bouchara JB et al. (2000) Pathogenesis II: fungal responses to host responses: interaction of host cells with fungi. Med Mycol 38(Suppl_1):113–123. https://doi.org/10.1080/mmy.38.s1.113.123
8. Enguita FJ, Costa MC, Fusco-Almeida AM, et al. (2016) Transcriptomic crosstalk between fungal invasive pathogens and their host cells: opportunities and challenges for next-generation sequencing methods. J Fungi (Basel) 2(1):7. https://doi.org/10.3390/jof2010007
9. Tan SC, Yiap BC (2009) DNA, RNA, and protein extraction: the past and the present. BioMed Res Int 2009(1):574398. https://doi.org/10.1155/2009/574398
10. Ali N, Rampazzo RCP, Costa ADT et al. (2017) Current nucleic acid extraction methods and their implications to point-of-care diagnostics. BioMed Res Int 2017:9306564. https://doi.org/10.1155/2017/9306564
11. Donaldson ME, Ostrowski LA, Goulet KM et al. (2017) Transcriptome analysis of smut fungi reveals widespread intergenic transcription and conserved antisense transcript expression. BMC Genom 18(1):340. https://doi.org/10.1186/s12864-017-3720-8
12. Doyle CE, Cheung HYK, Spence KL et al. (2016) Unh1, an *Ustilago maydis* Ndt80-like protein, controls completion of tumor maturation, teliospore development, and meiosis. Fungal Genet Biol 94:54–68. https://doi.org/10.1016/j.fgb.2016.07.006

13. Seto AM, Saville BJ (2025) Characterization of RNA helicase genes in *Ustilago maydis* reveals links to stress response and teliospore dormancy. Int J Mol Sci 26(6):2432. https://doi.org/10.3390/ijms26062432
14. Cheung HYK, Donaldson ME, Storfie ERM et al. (2021) Zfp1, a putative $Zn(II)_2Cys_6$ transcription factor, influences *Ustilago maydis* pathogenesis at multiple stages. Plant Pathol 70(7):1626–1639. https://doi.org/10.1111/ppa.13398
15. Lanver D, Müller AN, Happel P et al. (2018) The biotrophic development of *Ustilago maydis* studied by RNA-seq analysis. Plant Cell 30(2):300–323
16. Schurack S, Depotter JRL, Gupta D et al. (2021) Comparative transcriptome profiling identifies maize line specificity of fungal effectors in the maize–*Ustilago maydis* interaction. Plant J 106(3):733–752. https://doi.org/10.1111/tpj.15195
17. Seto AM, Donaldson ME, Saville BJ (2025) Exploring mechanisms of gene expression control during *Ustilago maydis* teliospore germination. Can J Plant Pathol 47(1):80–97. https://doi.org/10.1080/07060661.2024.2413557
18. Green MR, Sambrook J (2019) How to win the battle with RNase. Cold Spring Harb Protoc 2019(2):pdb.top101857. https://doi.org/10.1101/pdb.top101857
19. Dotti I, Bonin S (2011) DNase treatment of RNA. In: Stanta G (Ed.)Eds. Guidelines for molecular analysis in archive tissues. Springer, Berlin, Heidelberg, 87–90 https://doi.org/10.1007/978-3-642-17890-0_18
20. Schroeder A, Mueller O, Stocker S et al. (2006) The RIN: an RNA integrity number for assigning integrity values to RNA measurements. BMC Mol Biol 7(1):3. https://doi.org/10.1186/1471-2199-7-3

Check for updates

Chapter 22

Determination of Spontaneous Mutation Rate by Fluctuation Analysis Using 5-Fluoroorotic Acid (5-FOA)

Mira Milisavljevic and Milorad Kojic

Abstract

Spontaneous mutations serve as critical indicators of genome stability. Regardless of the mechanism, the rate of mutation in a certain genetic background can often provide insight into the underlying biological processes that impact the accuracy of DNA replication or the efficacy of DNA repair. A common method for estimating mutation rates is the Luria–Delbrück fluctuation assay. Here, a protocol is presented for determining mutation rates in *U. maydis* by fluctuation analysis, using 5-fluoroorotic acid (5-FOA) as a selective agent. The protocol starts with a number of small parallel inoculums of a given strains, which are then grown to high-density cultures. Subsequently, the cultures are screened for mutants by plating onto solid medium containing 5-FOA. The median numbers of 5-FOA-resistant mutants observed on the plates are used to calculate the mutation rates.

Keywords Spontaneous mutation rate, Fluctuation analysis, 5-FOA, Genetic stability

1 Introduction

Inactivation of genes essential for DNA repair and homologous recombination (HR) presents a serious threat to genome stability. Core proteins in these pathways are vital for resistance to genotoxic agents, while auxiliary factors, though sometimes less obvious in their effects, may play a significant role when combined with other components that contribute to stability. When studying genes responsible for safeguarding the genome, the initial evidence of their importance emerges from studying the sensitivity of mutants to various DNA-damaging agents. For those mutants exhibiting sensitivity, the genome stability can be assessed by measuring the rate of spontaneous mutations.

Spontaneous mutations are rare events, and detecting them requires analyzing many cells over multiple generations. It is important to recognize that a single mutation event can yield varying numbers of mutated cells in a series of parallel cultures of the same strain,

An N. T. Phan et al. (eds.), *Ustilago maydis*, Methods in Molecular Biology, vol. 3050,
https://doi.org/10.1007/978-1-0716-5364-7_22,

depending on when during the growth of the cultures the mutation occurs. Consequently, analyzing just the total number of mutants across all cultures would fail to provide an accurate representation of the mutation rate. Therefore, for a precise determination of mutation rates, fluctuation analysis, originally described by Luria and Delbrück [1], is the preferred method. This approach utilizes the Poisson distribution to account for mutation events.

The following assumptions must be met for this estimation procedure to be valid:

1. The mutation probability is constant throughout a cell's lifetime.
2. This probability does not vary between growth phases.
3. The proportion of mutants remains small.
4. The initial cell count is negligible compared to the final population size.
5. The growth rates of mutants and nonmutants are the same.
6. Reverse mutations and cell death are negligible.
7. All mutants are detected.
8. No new mutants arise after selection is imposed [2–4].

The mutation rate is a more accurate measure than mutant frequency (sometimes incorrectly called mutation frequency) as it reflects actual mutation events rather than just their outcomes (proportion of mutants). It does not depend on the timing of mutation occurrence, is less influenced by population dynamics and selective pressures, and can be standardized across experiments, allowing for more reliable comparisons.

In *Ustilago maydis*, the 5-FOA assay is employed to detect mutations in the orotidine-5′-monophosphate decarboxylase gene (UMAG_04214), irrespective of the mutation type [5]. The assay involves fluctuation analysis of multiple parallel cultures for each tested strain, followed by analysis as described in [6, 7].

5-FOA is a fluorinated derivative of the pyrimidine precursor orotic acid, which becomes toxic to cells when converted to 5-fluorouracil by the orotidine-5′-monophosphate decarboxylase. When mutations inactivate the enzyme, mutant cells can grow on 5-FOA containing medium.

2 Materials

2.1 Strains of *U. maydis* and Media

1. Frozen glycerol stocks of *U. maydis* wild-type strain and mutant strains.
2. Growth medium (YEPS): 1% yeast extract, 2% peptone, 2% sucrose. Solid medium contains 1.8% agar.

3. 5-Fluoroorotic acid (5-FOA). 5-FOA is soluble in water upon heating (*see* **Note 1**).

2.2 Equipment

1. Basic equipment for *U. maydis* culture, such as Erlenmeyer flask, bottles, orbital shakers, inoculating loops, plate spreaders, flame for sterilizing the tools, and so forth.
2. Incubators set at 30°C.
3. Centrifuges: Centrifuge (benchtop) and Microcentrifuge.
4. Light microscope equipped with a 40× long-distance working objective.
5. Neubauer hemocytometer.
6. Vortex mixer.
7. Tubes: microcentrifuge, 1.5, 15,and 50 mL screw cap tubes.
8. 15 × 100 mm plastic Petri dishes.

3 Methods

3.1 Experimental Procedure

1. Starting from a frozen stock streak (both wild-type and mutants) onto solid YEPS medium for single colony isolation.
2. Start 7–15 clonal overnight cultures for wild-type and each strain to be tested, by inoculating entire single colony from a freshly streaked plate into 1–2 mL of YEPS medium (*see* **Note 2**).
3. Grow in a rotating shaker (~200 rpm) at 30°C to saturation (stationary phase) at approximately $4–5 \times 10^8$ cells/mL. In cases where certain mutant strains exhibit slower growth than the wild-type, such strains should be inoculated earlier than regularly growing ones, so that all samples may be analyzed simultaneously.
4. Determine cell densities microscopically using a Neubauer hemocytometer. The cell count will be an approximate value since both viable and dead cells are counted. If the cultures appear to have similar densities, count only a few cultures to estimate the number of cells. This will help in determining the culture volume to be plated on 5-FOA medium. Given that the spontaneous mutation rate of the wild-type strain (UCM350) is approximately 3.5×10^{-8} mutations/cell/generation, at least 1 or 2×10^8 cells should be plated on the 5-FOA agar medium. To accurately determine the number of viable cells, make dilutions and plate them on YEPS solid medium (see Step 6).
5. If the cell density is above 1×10^8/mL, plate 300–500 μL (>1×10^8 cell) of the culture onto YEPS agar medium supplemented with 1 mg/mL of 5-FOA. If the cell density is

lower, adjust it to 2–4 × 10^8/mL, and then plate 300–500 µL. Incubate the plates at 30°C for 4–5 days.

6. To obtain an accurate count of viable cells plated, prepare tenfold serial dilutions from the original cultures (~10^8/mL), and plate 100–300 µL of 10^3/mL dilution (the fifth dilution), onto solid growth medium. Incubate the plates at 30°C for 3 days. For the most accurate assessment of viable cells, it is advisable to plate each culture on three petri dishes and calculate the average number of colonies (*see* **Note 3**).

3.2 Calculating Spontaneous Mutation Rate

The calculation of mutation rate is based on the method of the median, which is applied when most of the cultures produce mutants (*see* **Note 4**).

1. Determination of culture cell density: Count the number of colonies on each plate. Calculate the culture cell density using the volume and dilution that was plated. For instance, if you count 470 colonies from 100 µL of the fifth dilution (approximately 10^3/mL) made from a microscopically estimated culture density of around ~10^8/mL, the actual original culture density would be 4.7 × 10^8/mL (Colony count × Fraction of volume plated (1000 µL/100 µL) × Dilution factor ($10^8/10^3$) = 470 × 10 × 10^5). Calculate the average culture density from all plates (*see* **Note 5**).
2. Count number of colonies on 5-FOA plates. Find the median value and then determine the number of colonies per 1 mL (e.g., if 200 µL is plated, and 23 colonies counted, in 1 mL there is 23 × 1000/200 = 115 colonies). This value will represent the experimentally determined median number of mutants (**r0**). Avoid counting very small colonies, as these are likely, the result of mutations that occurred on the solid medium.
3. In the Table 3 from [6], the r0 values are provided alongside their corresponding **r0/m** values. To calculate the mean number of mutations per culture (**m**), use the formula (**m = r0/(r0/m)**). Then, the mutation rate can be calculated using the formula (**rate = m × ln2/average cell density**). Using the above example, since the value of r0 is 115 and the corresponding r0/m from Table 3 is 4.5, then m can be calculated as follows: (m) = 115/4.5 = 25.55. And given an average cell density of 4.7 × 10^8, the mutation rate is 25.55 × 0.693/4.7 × 10^8 = 3.77 × 10^{-8} /cell/generation (mutations per cell per generation)
4. To estimate significant differences in mutation rate between different strains, it is useful to determine confidence interval. A confidence level of 95% is typically applied. For number of cultures used in fluctuation test, determine the mutation rate as described in Step 9, using the individual culture r0 and the

average cell number. Next, refer to Table B11 from [8] to determine the marginal mutation rates. For example, if using 7 cultures, take the second and sixth values from the table to establish the lower and upper limits of the 95% confidence interval. If two strains do not overlap in their confidence intervals, it indicates a significant difference in their mutation rates.

4 Notes

1. 5-FOA should not be heated above 50°C. To prepare solid YEPS medium, first make and sterilize a 2× concentrated YEPS-agar medium and cool it to approximately 50°C. Prepare an equal volume of a 2 mg/mL 5-FOA water solution by stirring it in a water bath at 50°C for 30 min, or until no traces of 5-FOA remain visible. Mix well the 2× YEPS and the 5-FOA water solution, then pour the warm medium into petri dishes. Uracil in the final concentration of 62 mg/L can be added to the medium.
2. For each strain, a minimum of seven colonies are to be analyzed. An odd number of colonies should be used for the calculation of the median value. A more precise calculation is obtained when a greater number of colonies are analyzed.
3. Alternatively, although less precise, semiquantitative assay can be used for determination of cell number, by making six tenfold serial dilutions from initial overnight culture and spotting 10 μL of each on YEPS-agar plate.
4. If majority of cultures produce no mutants, then statistical method of p0 should be used to estimate the mean number of mutation events [1, 6, 7]. The formula is rate = [–ln (fraction of cultures with no recombinants)]/(average total number of cells). For example, if five of nine plates are with 0 mutants, and average cell number is 4.7×10^8 then the rate = $[-\ln(5/9)]/4,7 \times 10^8 = 1.25 \times 10^{-9}$ mutations/cell/generation.
5. Cultures which cell densities differ from average density by more than two standard deviations should be excluded, as well as jackpots.

Acknowledgments

This work was supported by the Fund of the Ministry of Science, Technological Development and Innovation of the Republic of Serbia, registration number 451–03–136/ 2025–03/200042 agreement on the implementation and financing of research in 2025.

References

1. Luria SE, Delbrück M (1943) Mutations of bacteria from virus sensitivity to virus resistance. Genetics 28:491–511. https://doi.org/10.1093/genetics/28.6.491
2. Lang GI, Murray AW (2008) Estimating the per-base-pair mutation rate in the yeast Saccharomyces cerevisiae. Genetics 178:67–82. https://doi.org/10.1534/genetics.107.071506
3. Rosche WA, Foster PL (2000) Determining mutation rates in bacterial populations. Methods 20:4–17. https://doi.org/10.1006/meth.1999.0901
4. Lang GI (2018) Measuring mutation rates using the Luria-Delbrück fluctuation assay. Methods Mol Biol 1672:21–31. https://doi.org/10.1007/978-1-4939-7306-4_3
5. Azanjac N, Milisavljevic M, Stanovcic S et al (2024) Suppressors of Blm-deficiency identify three novel proteins that facilitate DNA repair in Ustilago maydis. DNA Repair (Amst) 140:103709. https://doi.org/10.1016/j.dnarep.2024.103709
6. Lea DE, Coulson CA (1949) The distribution of the numbers of mutants in bacterial populations. J. Genetics 49:264–285. https://doi.org/10.1007/BF02986080
7. Spell RM, Jinks-Robertson S (2004) Determination of mitotic recombination rates by fluctuation analysis in Saccharomyces cerevisiae. Methods Mol Biol 262:3–12. https://doi.org/10.1385/1-59259-761-0:003
8. Altman DG (1999) Practical statistics for medical research. Chapman and Hall/CRC, Boca Raton, FL

Check for updates

Chapter 23

Insertion Pool Sequencing (iPool-Seq) for the High-Throughput Quantification of *Ustilago maydis* Mutant Fitness

Komal Pervaiz, Riaz Tabassum and Armin Djamei

Abstract

Insertion pool sequencing (iPool-Seq) is a robust functional genomic technique that combines insertion mutagenesis with next-generation sequencing to comprehensively analyze the genome-wide gene function in maize infecting fungus *Ustilago maydis*. This high-throughput method tracks the abundance of individual mutants in pooled populations under defined selective conditions. It relies on the construction of saturated mutant libraries, where each mutant carries a unique insertion, followed by tagmentation for efficient DNA fragmentation, tagging with unique molecular identifiers (UMIs), and affinity purification to enrich the pathogen DNA from host tissues. Sequencing of genome insertion junctions then allows determination of mutant representation in *Ustilago maydis*. Together, this strategy provides a powerful framework to uncover the genes essential for fungal development and virulence within the host.

Keywords iPool-Seq, *Ustilago maydis*, Next-generation sequencing, Genome insertion junction, Unique molecular identifiers

1 Introduction

Classical gene characterization in microbes using individual gene-deletion mutants is a slow approach due to one-by-one phenotyping [1–4]. To overcome the limitations of previous functional genomic approaches, Uhse [5] established iPool-Seq method in 2018 and demonstrated its applications in maize pathogenic fungus *Ustilago maydis*. This method enables high-throughput identification of mutants with growth deficiencies under selecting conditions [6]. This involves pooling together a library of mutant strains and their quantification by next-generation sequencing (NGS) before and after selection which, in the context of identifying virulence-deficient mutants, takes the form of plant infection [6]. However, a key challenge in NGS-based virulence factor screens is the underrepresentation of pathogen genetic material within eukaryotic hosts,

An N. T. Phan et al. (eds.), *Ustilago maydis*, Methods in Molecular Biology, vol. 3050,
https://doi.org/10.1007/978-1-0716-5364-7_23,

making detection difficult. To address this, iPool-Seq integrates tagmentation with unique molecular barcodes followed by affinity purification to selectively isolate pathogen insertion-mutant DNA from infected tissues. In a proof-of-concept experiment, iPool-Seq successfully identified 28 virulence factors from an initial pool of 195 candidate effector mutants including 23 that were previously uncharacterized [7]. Due to its sensitivity and quantitative nature, iPool-Seq is adaptable to any insertional mutagenesis library and is particularly suited for genetically complex environments, such as pooled infections in eukaryotic hosts. Furthermore, its broad applicability suggests potential for uncovering novel gene functions beyond plant–pathogen interactions. We outline the key requirements to undertake iPool-Seq using, as an example, a *U. maydis* mutant library.

In general, iPool-Seq allows to sequence flanking regions of an insertion cassette in a quantitative manner allowing negative selection screens on insertional mutant pools. The virulence is determined by the relative abundance of each deletion mutant in the postinfection output pool relative to the preinfection input pool, and then comparing these ratios to virulence of a set of known neutral mutants to detect significant deviations from expected behavior.

In the methods section, commands will be shown in monospaced font, which meant to be entered in a Unix-style terminal, either directly on a Linux system or through an SSH connection to server. Outside of such commands, filenames are printed monospaced and underlined. Placeholders representing experiment or library names that must be provided by the user are indicated in italics.

The iPool-Seq pipeline is developed using "Snakemake," a workflow engine that ensures reproducible and scalable data analysis [8]. It makes an "easy to run" pipeline by just providing the necessary input files according to the instruction in Step 6. All other intermediate steps will be executed automatically up to final output Step 13. It is not strictly necessary, but it is a good practice to proceed step-by-step by checking the results of key intermediate steps (like mapping and KO assignment) for validity before proceeding. It is recommended for this pipeline to execute and validate each command according to the suggested troubleshooting, which allows for flexible execution rather than requiring a strict step-by-step approach. For instance, once all necessary data have been provided in Step 6, it is possible to proceed directly to Step 13, during which all preceding steps will be executed automatically. Nevertheless, as it is considered best practice to verify the outputs of key intermediate stages, such as read mapping and knockout (KO) assignments. We recommend to follow the workflow sequentially and conducting the validations and checks outlined in the Troubleshooting section for each step.

2 Illumina Sequencing Library Preparation Using Genomic DNA from Insertional Mutant Pool

Sect. 2 not called “Materials,” slightly different format than 1. Introduction, 2. Materials, etc.

Genomic DNA (gDNA) is extracted from the insertional pool of mutant library both before infection (input) and after infection (output). The purified DNA is then processed using the iPool-Seq library preparation protocol to generate Illumina-compatible sequencing libraries (Fig. 1A–D). *U. maydis* gDNA extraction from the pooled input mutants grown in axenic culture follows a protocol originally developed for yeast [9]. Following a standard *U. maydis* infection, material is collected at the desired stage of infection and total gDNA extracted using the method described by Uhse [10].

Figure and Figure caption embedded in the text.

2.1 Materials

1. Oligonucleotides: 100 μm Tn5 adaptor P1 and Tn5 adaptor P2, 5 μM PCR1-F, PCR1-R, 5 μm PCR2-F and PCR-R2 (*see* Table 1 and **Note 1**).
2. Reassociation buffer: 10 mM Tris-HCl, 1 mM EDTA, 50 mM NaCl, pH 8.0.

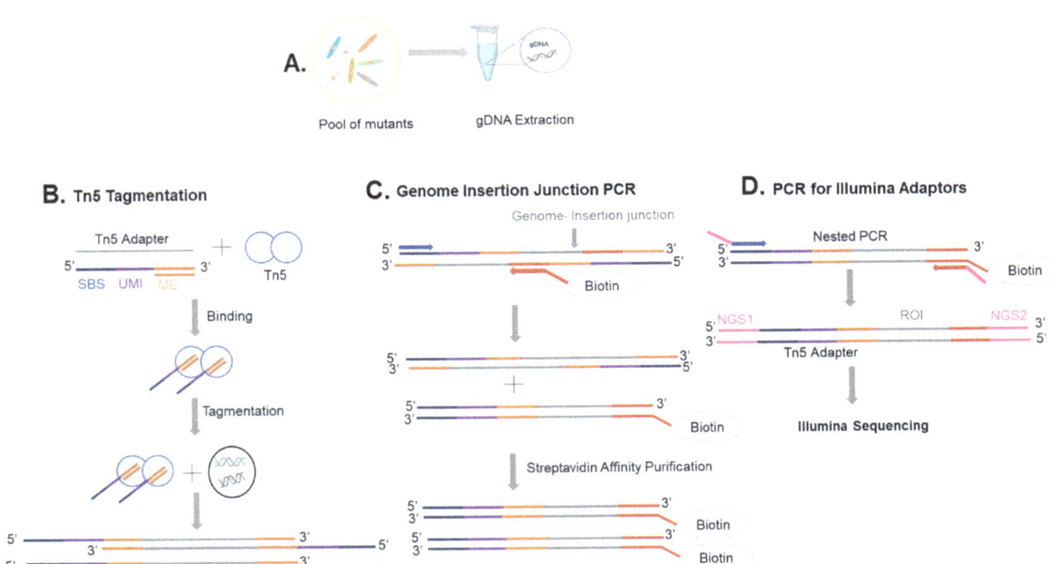

Fig. 1 Workflow for iPool-Seq library preparation based on Tn5 tagmentation (**A**) *U.maydis* mutants are pooled and genomic DNA (gDNA) is extracted. (**B**) Extracted gDNA is fragmented using hyperactive Tn5 transposase preloaded with custom adapters containing an Illumina sequencing primer binding site (SBS, blue), a unique molecular identifier (UMI, purple), and a mosaic end (ME, orange). (**C**) The genome insertion cassette is PCR-amplified using biotinylated primers targeting the insertion sequence (red) and an adapter-specific primer binding to the SBS site. Biotinylated PCR products are then enriched via streptavidin affinity purification. (**D**) In a nested PCR step, P5 (NGS1, Pink) and P7 (NGS2, Pink) adapters are added to generate libraries compatible with Illumina sequencing

Table 1
Oligonucleotides used in this study for tagmentation, specific enrichment PCR, and Illumina sequencing

Oligonucleotide	sequence
Tn5 adaptor P1	5′-CACGACGCTCTTCCGATCTNNNNNNNNNNNNACTTGTGTATAAGAGTCAG-3′
Tn5 adaptor P2	5′-[phos]CTGACTCTTATACACAAGT-3′
PCR1-F	5′-GAGATCTACACTCTTTCCCTACACGACGCTCTTCCGATC-3′
PCR1-R	5′-[BioTEG]CCAGATGTCCTGTGGTATCCTGTG-3′
PCR2-F	5′-AATGATACGGCGACCACCGAGATCTACACTCTTTCCCTACAC-3′
PCR2-R	5′-CAAGCAGAAGACGGCATACGAGATNNNNNNGTGACTGGAGTTCAGACGTGTGCTCTTCCGATCTCCTGTGGTATCCTGTGGCG-3′
Illumina Rd1	5′-ACACTCTTTCCCTACACGACGCTCTTCCGATCT-3′
Illumina Rd2	5′-GTGACTGGAGTTCAGACGTGTGCTCTTCCGATCT-3′

3. Purified Tn5 transposase (1 μg/μL), prepared according to prior protocol [11]; store at −20°C in aliquots for <12 months.
4. 100% glycerol.
5. Input and output gDNA (described above).
6. 5× TAPS buffer: 250 mM TAPS-NaOH, 125 mM $MgCl_2$, 50% (v/v) DMF, pH 8.5 (*see* **Note 2**).
7. Nuclease-free water.
8. SPRI magnetic beads.
9. Phusion High-Fidelity DNA Polymerase.
10. 10 mM dNTPs.
11. Dynabeads M-270 Streptavidin (Invitrogen).
12. B&W buffer: 5 mM Tris-HCl, 1 M NaCl, 0.5 mM EDTA, pH 7.5.
13. 2× B&W buffer: 10 mM Tris-HCl, 2 M NaCl, 1 mM EDTA, pH 7.5.
14. Thermocycler.
15. PCR tubes.
16. Rotator.
17. Magnetic stand.
18. Eppendorf.
19. Gel electrophoresis chamber.

Table embedded in the text.

2.2 Methods

2.2.1 Adaptor Ligation

1. Combine and mix well the following components in a PCR tube to prepare a total reaction of 100 μL.
   a. 25 μL of 100 μM Tn5 Adaptor P1 (see Table 1).
   b. 25 μL of 100 μM Tn5 Adaptor P2 (see Table 1).
   c. 50 μL reassociation buffer.
2. Place the tube in a preheated thermocycler and perform primer annealing starting at 90°C, decreasing the temperature by 1°C/min.

2.2.2 Transposon Assembly

1. Combine and mix well the following components in a PCR tube to prepare a total volume reaction of 100 μL (*see* **Note 1**).
   a. 25 μL Purified Tn5 Transposase (1 μg/μL).
   b. 25 μL of annealed adaptors (refer to Step 1).
   c. 50 μL of 100% glycerol.
2. Incubate the reaction in thermocycler for 30 min at 37°C.

2.2.3 Tagmentation of gDNA

1. Assemble the reaction as follows:
   a. 250–500 ng of input and output gDNA.
   b. Tn5 transposon assembly (refer to Step 3) to a final concentration of 150 ng/μL.
   c. 4 μL of 5× Tagmentation buffer (TAPS) (*see* **Note 2**).
   d. Make 20 μL of total reaction using Nuclease-free water.
2. Incubate the reaction in thermocycler for 10 min at 55°C.
3. Stop the Tn5 reaction by adding 0.02% SDS and incubate the reaction again at 55° for 7 min. This step is necessary to inactivate the Tn5 protein to avoid over digestion of DNA.
4. Clean fragmented PCR product with SPRI magnetic beads (Agencourt AMPure XP beads) according to manufactures instructions (*see* **Note 3**).
5. Ensure successful fragmentation via gel electrophoresis (*see* **Note 4**).
6. Determine the dsDNA concentration for downstream PCRs (*see* **Note 5**).

Stopping point: Store the fragmented DNA at −20°C until ready for specific PCR.

2.2.4 Targeted PCR of Mutant Cassette-Genome Junction

1. Run PCR using Following recipe: 5× Phusion HF Buffer (5 μL), 10 mM dNTPs (1 μL), 5 μM PCR1-F (1 μL), 5 μM PCR1-R (1 μL), Phusion High-Fidelity DNA Polymerase (0.5 μL), Fragmented and clean DNA (from Step 7) (250 ng) and Nuclease-free water (total 25 μL).
2. Place the PCR tubes in preheated thermocycler using the following programme;

Initial Denaturation	98°C for 1 min
Amplification (20 Cycles)	Denaturation: 98°C for 15 s Annealing: 65°C for 15 s Extension: 72°C for 30 s
Final Extension	72°C for 1 min

3. Purify PCR reactions using SPRI magnetic beads following the manufacturer's protocol.
4. Pool clean eluates. Optional but recommended: Measure dsDNA concentration before proceeding to the next streptavidin purification step.
5. **Stopping point:** Freeze the Cleaned PCR product at −20°C until next step.

2.2.5 Streptavidin Purification for Target Enrichment

6. Wash 30 μL of Dynabeads M-270 Streptavidin with 200 μL of B&W buffer in a 1.5-mL Eppendorf tube for each input and output sample while on a magnetic stand.
7. Resuspend the beads in 2× B&W buffer, using the same volume as the clean input and output PCR1 eluates.
8. Combine the eluates with the beads and incubate at room temperature for 15 min with continuous rotation to enrich biotinylated PCR amplicons.
9. Place the tubes on the magnetic stand for 1 min and discard the supernatant.
10. While keeping the tubes on the magnetic stand, wash the beads with 200 μL of B&W buffer.
11. Repeat the washing step three more times. Finally, resuspend the beads in 34 μL of nuclease-free water and proceed with nested PCR.
12. Stopping point: Store the enriched PCR fragments at −20°C until ready for nested PCR.

2.2.6 Nested PCR on Enriched Fragments

13. In PCR tubes, add 5× Phusion HF Buffer (10 μL),10 mM dNTPs (1 μL), 5 μM PCR2-F (2 μL), 5 μM PCR2-R (2 μL), Phusion High-Fidelity DNA Polymerase (1 μL), nuclease-free water containing beads (see Step 20) (34 μL).
14. Run PCR with following conditions:

Initial Denaturation	98°C for 1 min
Amplification (20 Cycles)	Denaturation: 98°C for 15 s Annealing: 65°C for 15 s Extension: 72°C for 30 s
Final Extension	72°C for 1 min

15. Place the tubes on a magnetic stand for 1 min, then transfer the supernatant containing PCR2 amplicons into fresh PCR tube.
16. Purify the PCR reactions using SPRI magnetic beads following the manufacturer's protocol.
17. Measure the dsDNA concentration and proceed with Illumina sequencing (*see* **Note 6**).

3 Bioinformatic Analysis of *Ustilago maydis* Mutants

This protocol describes the analysis of insertion pool data using the pipeline developed by Simon Uhse et al. [6] to identify insertional mutants that exhibit significantly altered virulence.

3.1 Materials

1. Workstation running Linux or Windows 10 or 11 with Windows Subsystem for Linux (WSL), with 64-bit CPU and minimum 8 GB RAM and with free disk space equal to at least five times the size of the raw sequencing data.
2. iPool-Seq analysis pipeline (https://github.com/Cibiv/ipoolseq-pipeline?tab=readme-ov-file).
3. Reference genome of *U. maydis* (or any other organism, for which an insertion mutant library is analyzed) in FASTA format.
4. FASTA files containing sequences at 5′ end (named "5p") and 3′ end (named "3p") of knockout (KO) cassette (*see* **Note 7**).
5. List of deletion mutants of *U. maydis* as GFF2 file listing KO cassette insertion positions (*see* **Note 8**).
6. Input and output sequenced libraries for each sample (*see* **Note 9**).
7. Any Web browser.
8. PDF viewer.

3.2 Methods

3.2.1 iPool-Seq Pipeline Installation

1. Download and install iPool-Seq analysis pipeline on a Linux workstation or Windows 10 with Windows Subsystem for Linux (WSL) by executing:

```
VER=latest-release
URL=http://github.com/Cibiv/ipoolseq-pipeline
curl -L -O $URL/archive/$VER.tar.gz
tar xzf $VER.tar.gz
cd ipoolseq-pipeline-$VER
```

For the manual installation, download the iPool-Seq latest release from https://github.com/Cibiv/ipoolseq-pipeline/tree/master. Then, in terminal window, change the current working directory to the directory containing the downloaded file. Then, use the tar command to unpack the archive:

```
tar xzf $VER.tar.gz
cd ipoolseq-pipeline-$VER
```

2. All the required software packages by iPool-Seq pipeline can be installed and activated by Bioconda [12] environment with the command:

```
/install-environment.sh
source. /environment/bin/activate
```

For manual installation (*see* **Note 10**).

3. Optional: For confirmation of successful pipeline installation, just run one replicate of Uhse et al. [5] (*see* **Note 11**).

3.2.2 Setting Up Reference Genome, KO Cassette file, KO List, and Libraries

4. To organize your analysis, first choose a clear name for your experiment (e.g., *your_design*). This name will be used to label both your data and configuration folders. Then, create the necessary directory structure by running:

```
mkdir -p data/your_design
mkdir -p cfg/your_design
```

Rename your experimental input files (such as for reference genome= reference.fa, KO insertion cassette = cassette.fa and GFF2 file = knockouts.gff).

5. Copy the required input files into your experiment's folder (cfg/your_design/) as: The reference genome (*U. maydis* or any desired organism) file to *cfg/your_design/reference.fa* and the KO cassette ends FASTA file containing the sequences at 5′ end (named "5p") and 3′ end (named "3p") to *cfg/your_design/cassette.fa*, and the list of deletional mutants (GFF2 format) enlisted with KO cassette insertion positions to *cfg/your_design/knockouts.gff*

6. For each single experimental replicate (e.g., your_replicate), you'll have two sequencing libraries (*see* **Note 12**):

   - **Input (e.g., before infection input pool)** → name it your_replicate-in
   - **Output (e.g., post infection pool)** → name it your_replicate-out

3.2.3 Trimming Reads and UMI Extraction (Per Library)

7. For the trimming (*see* **Note 13**) of sequencing adaptors and other nongenomic sequences [UMI and KO cassette overlap (Fig. 2A) from the raw sequencing reads for library *your_lib*], run

```
snakemake data/your_design/your_lib.trim.1.fq.gz
```

To troubleshoot, the reads trimming and other filtering parameters (*see* **Note 14**).

8. Optional (but recommended): For the confirmation of successful trimming of first and second reads, generate FastQC [13] reports by running the command.

```
snakemake data/your_design/your_lib.fastqc.1.html
snakemake data/your_design/your_lib.fastqc.2.html
```

3.2.4 Mapping and Assignment to Insertional KOs (Per Library)

9. To map and assign trimmed reads (*see* **Note 15**) to KOs (Fig. 2B) for the sequencing library *your_lib*, do

```
snakemake data/your_design/your_lib.assign.bam
```

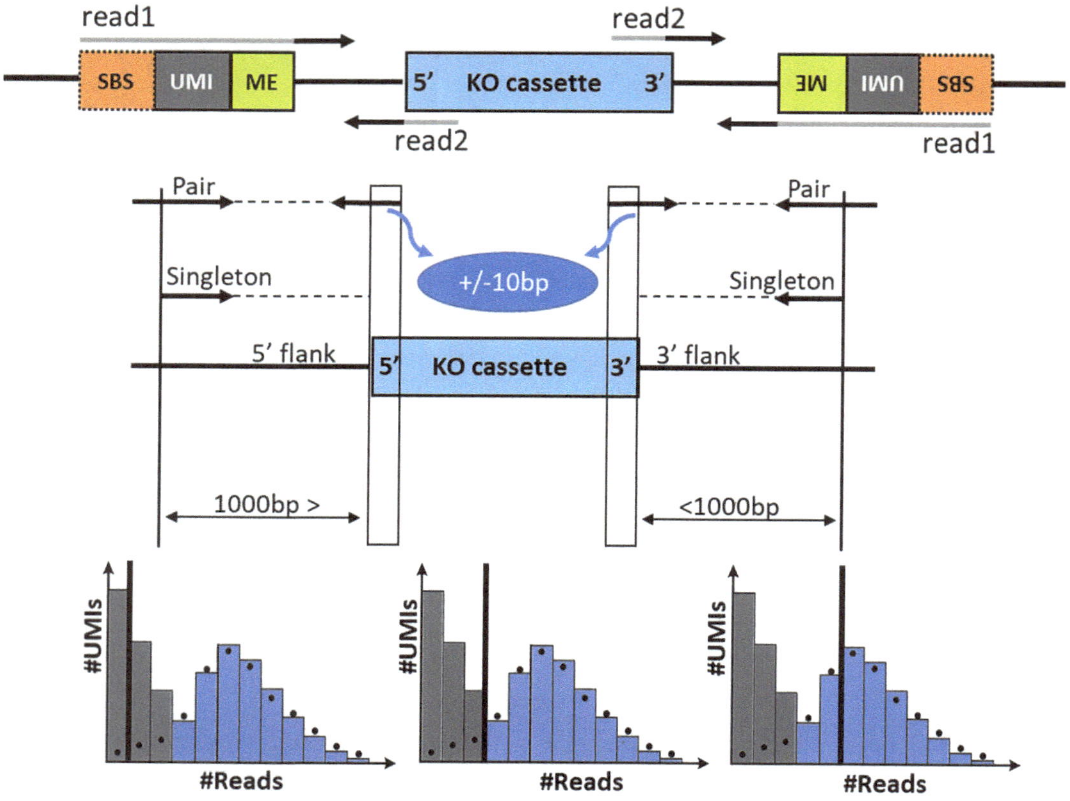

Fig. 2 Schematic overview of iPool-Seq Results (**A**) sequenced fragments layout on both sides of knock-out (KO) cassette insertions. All nongenomic part (SBS, UMI, ME, and KO cassette ends) of reads are trimmed before mapping. The unique molecular identifier (UMI) consists of 12 random bases, ME = 5′-ACTTGTGTATAAGAGTCAG-3′. (**B**) Required mapping locations and directions for read pairs with either both mates or one mate mapped to be assigned to a specific KO. (**C**) Data indicating that the chosen TRUmiCount threshold is too low (left), optimal (middle), or too high (right)

3.2.5 Determining KO Abundances (Per Library)

10. To produce a table of genome abundance estimates (*see* **Note 16**) for the sequencing library *your_lib*, do

```
snakemake data/your_design/your_lib.count.tab
```

For the description of the columns of data/your_design/your_lib.count.tab, see (Table 3). During this step, the TRUmiCount algorithm also produces the accompanying PDF report data/your_design/your_lib.count.pdf [14].

3.2.6 TRUmiCount Threshold Adjustment for Phantom (i.e., Amplification Artifacts) Rejection (Per Library)

11. Adjust the suitable TRUmiCount's threshold for phantom rejection (*see* **Note 17**).

3.2.7 Identification of Differentially Virulent KOs (Per Replicate)

12. To calculate the log-fold changes of KO virulence for your_replicate and *p*-values (*see* **Note 18**) to determine how significantly

these log-fold changes differ from zero (i.e., no change compared to the neutral reference set), run

```
snakemake data/your_design/your_replicate.dv.tab
```

3.2.8 Downstream Analysis for Data Interpretation

13. Interpret the results from iPool-pipeline by combining data from multiple replicates (*see* **Note 19**).

4 Notes

1. The nucleotide sequences may need to be adjusted depending on the insertional mutant library being screened. Adapter P1 consists of 12Ns forming the UMI, which should be random. The 6 Ns in PCR2-R constitute the (single-index) library multiplexing barcode. Enrichment of insertion mutant flanks is specific for the sequence 5′-CCAGATGTCCTGTGGTATCCTGTGGCG-3′, and read2 will start with the underlined part of PCR2-R. Illumina Rd1 and Rd2 are the standard Illumina TruSeq sequencing primers.
2. This step is critical and crucial for proper binding of TN5 to adaptors. Because the reaction contains 50% glycerol and it is highly viscous. To ensure the proper mixing, flick the tube gently two–three times.
3. Dimethylformamide (DMF) in TAPS buffer is highly toxic and volatile. Open the DMF bottle in fume hood and take all necessary safety precautions. TAPS buffer can be stored for 4 weeks at room temperature in the dark in a cabinet for flammable liquids.
4. SPRI bead-based gDNA purification is recommended to minimize DNA loss and ensure diverse sequencing library preparation. The 1.5× SPRI beads to DNA ratio gives optimal results. Column-based purification systems can be used as an alternative to SPRI magnetic beads; however, they typically result in lower DNA recovery and less diverse sequencing libraries.
5. We recommend using a fragment analyzer for precise fragmentation assessment.

   Quantification of dsDNA via fluorescence assay, e.g., Qubit or Pico Green fluorometric assay is recommended instead of Nanodrop. We also recommend to start with minimal concentration of 20 ng/μL.
6. A minimum dsDNA final concentration should be 0.1 ng/μL to enable Illumina sequencing. We highly recommend fragment length analysis and quality control of library preparation via qPCR prior to Illumina sequencing.

7. It is important to consider that sequences from the KO cassette FASTA file should represent the part of read2 that overlap with the cassette. This sequence is the part of 3′ end of PCR2-R (Table 1) primer and complementary to the end of cassette. In sequencing results, read2 contain the part of KO cassette and adjacent genomic DNA sequences at the insertion site. In this way, the trimming step of the pipeline can differentiate, which part belongs to the KO cassette and where the genomic sequence begins. So, after trimming, the cassette portion of the sequenced read, remaining sequence will be mapped to the reference genome. For an example, see the list of deletion mutants used by Uhse et al. [5] cfg/Uhse_et_al.2018/cassette.fa.

8. For each entry of a deletion mutant, the list must have at least two tags: a unique "Name" and a flag "Neutral". 0 or 1 value of flag will help the pipeline to decide whether a particular deletion strain is included in the reference set of assumed neutral deletions. For an example, see the list of deletion mutants used by Uhse et al. [5] cfg/Uhse_et_al.2018/knockouts.gff.

9. For a single replicate (which consist of input and output library),
   - Four separate compressed FASTQ files, two for input library (e.g., input_read1, input_read2) & two for output library (e.g., input_read1, output_read2).
   - OR, two BAM file (indexed version of the raw sequencing data that contains all the read information e.g., sequences, qualities, and read pair info) one for input library and other for output library.

   When using paired-end sequencing platforms like Illumina, the sequencing output is split into two separate files: one for the first read (sequencing from 5′ end) in each pair and another (sequencing from 3′ end) for the second read. The most common format for raw sequencing data is the FASTQ, and usually files are compressed using the gzip format indicated by the.gz extension, to reduce file size and facilitate storage and transfer. So, the final file names will look like **read1.fastq.gz** and **read2.fastq.gz**. Single BAM file containing unmapped sequencing reads means that BAM file that contains your raw sequencing reads in a single file but not mapped to the *U. maydis* genome yet and mapping will be performed within the iPool-Seq pipeline steps. BAM file are the compressed version that save disk space and efficiently allow random access to reads without reading the whole file. It also maintains pairing information of paired-end reads and widely supported across all bioinformatics tools.

10. If there is an error to download & install environment.tar.gz, then download it manually from http://github.com/Cibiv/ipoolseq-pipeline and place it in the pipeline containing directory. There is no need to install it every time. It will be reactivated (but not recreated), whenever you open a new terminal window.
11. You can reanalyze one of the replicates of Uhse et al. [5] for the confirmation of successful installation of library. To download the raw sequencing data for replicate A1 (data/Uhse et al.2018/expA.r1-in.bam and expA.r1-out.bam) and to run the differential virulence analysis, do

```
snakemake data/Uhse_et_al.2018/expA.r1.dv.tab
```

After successful running of analysis there, you obtain two output files in the directory. One is data/Uhse et al.2018/expA.r1.dv.tab which contains a table listing the differential virulence analysis results for all deletional mutants (see Table 2 for a list of columns), and the other is an accompanying HTML report written to data/Uhse et al.2018/expA.r1.dv.html.
12. For each of library, you can place the sequencing data in anyone of two formats into data/*your design*/:

1. For a single BAM file (unmapped reads):

```
data/your_design/your_lib.bam
```

Table 2
Columns in the differential virulence table

Column title	Data type	Details
knockout	String	Name of the knockout as in the knockout list GFF file
is.neutral	Flag	1 if the knockout is part of the reference set, 0 otherwise
n.out	Integer	Sum of 5′ and 3′ UMI counts after filtering (output pool)
loss.out	Numeric	Average of 5′ and 3′ est. fraction of lost UMIs (output pool)
abundance.out	Numeric	Est. number of genomes, 0.5**n.out/(1-loss.out)* (output pool)
n.in	Integer	Sum of 5′ and 3′ UMI counts after filtering (input pool)
loss.in	Numeric	Average of 5′ and 3′ est. fraction of lost UMIs (input pool)
abundance.in	Numeric	Est. number of genomes, 0.5**n.in/(1-loss.in)* (input pool)
log2fc	Numeric	Virulence log fold change compared to the reference set ($\log_2 \Delta v$)
low.pval	Numeric	*p*-value for *log2fc* being significantly low
high.pval	Numeric	*p*-value for *log2fc* being significantly high
low.qval	Numeric	FDR-corrected *p*-value for log2fc being significantly low
high.qval	Numeric	FDR-corrected *p*-value for log2fc being significantly high

2. Two compressed FASTQ files (paired-end reads):

```
data/your_design/your_lib.1.fq.gz (read1)
data/your_design/your_lib.2.fq.gz (read2)
```

By this naming method of files and folders, the pipeline can relate that which reference genome and KO list belongs to a particular library and (in Step 13) which input and output libraries constitute one replicate. For each library Steps 7–12 must be executed, i.e., twice per replicate, once for *your_replicate-in* and once for *your_replicate-out*.

13. When running the trimming command for either of the read of library, it will automatically run trimming command for other read as well. E.g., if you run it for read1 it will trim read2 as well. If the input was a single unmapped BAM file, the pipeline will automatically convert it into two FASTQ files, with one for each read, during this step. All nongenomic sequences including regions overlapping with UMIs, and the KO cassette are removed from raw paired-end sequencing reads to prevent interference during the mapping with reference genome (Fig. 2A). To retain UMIs throughout downstream processing steps, they are embedded in the read names. Reads that lack a UMI or do not contain a UMI or do not overlap with the KO insertion cassette are removed.
14. After trimming, initial assessment of library preparation and sequencing data can be made by running FastQC on trimmed first and second reads [13]. In the FastQC report, review the following aspects:

    1. Under "BasicStatistics," Ensure that more than 90% of reads have survived after trimming.
    2. Under "Per base sequence quality" and "Per base N content," make sure that base quality is consistently high and contain less or no Ns.
    3. Under "Adaptor content," verify that all sequencing adaptor have been removed from reads.

    If large number of reads are lost during the trimming step, it indicates contamination or the library preparation method did not match the expected library structure. To figure out the potential contaminant, BLAST a few random reads and manually compare the reads to the concept discussed in the (Fig. 2) and to the sequences in data/your_design/cassette.fa. If good percentage of reads is surviving but there is a strong drop-off of base quality score towards ends, add additional quality-based trimming step to Trimmomatic command in cfg/config.yaml [15] If still adaptor sequences are detected in trimmed reads, it is possible that some custom adaptors were used. Additionally, place these custom adaptor sequences to the input adaptor file (cfg/Uhse_et_al.2018.adapters.fa). Trimmomatic will additionally remove these custom adaptors as well. See Trimmomatic manual for additional information.

15. In this step, trimmed reads are mapped to the reference genome by using NextGen Map [16]. On the basis of mapping near the KO cassette, reads are assigned to specific knockout (KO) strains (Fig. 2B). True read pairs are linked to a specific KO cassette flank (5′ or 3′) if one read mapped within the ±10 bp of the respective cassette end and extends away from the cassette. Singleton reads (single read whose mate could not be mapped) must map within 1000 bp (i.e., ≤1000) of the respective end of the cassette and extend towards that cassette. Reads that are not clearly mapped to a specific insertion site are excluded at all. High-confidence reads survived after this filtering are used for downstream analysis of insertional mutations. It is recommended to visually inspect that how reads are mapped to the reference genome and how they are assigned to KO cassette by using a genomic viewer tool, e.g., the Integrated Genomics Viewer [17]. It is expected that most of the reads should map to the flanking regions (3′ and 5′ ends) of the KO cassette insertion sites in the genome and labeled with the KO name and flank in XT tag (metadata tag). These mapped reads will appear extending only up to few base pairs at the genomic regions where KO cassette was inserted. Some reads appeared to be mapped at some unrelated genomic locations and will be ignored by the pipeline and not a concern unless those reads are in abundance. If reads appear to map correctly but are not being assigned to the expected KO cassette insertion site, then it is important to verify that if genomic coordinates listed in your GFF2 file for the KO insertion cassette are correct. If a significant portion of sequencing fragments are longer than 1000 bp or if many reads extend more than a few base pairs into the regions replaced by the KO cassette due to mapping inaccuracy, then you need to adjust the "mapping fuzzyness" or "max fragment length" parameter of the "knockout_assignment" step in cfg/config.yaml file.

16. This step involves phantom removal, error correction, and estimating losses for each unique combination of KO strain and flank. UMI-Tools is used to remove small sequencing errors and to group together similar UMIs if they likely came from the same KO cassette insertion [18]. After correction, merged UMIs are processed further with TRUmiCount, which filter out phantom UMIs [14]. Phantom UMIs are PCR amplification artifacts, which do not represent a real molecule. TRUmiCount counts how many reads support each UMI. The

Table 3
Columns in the knockout abundance table generated by TRUmiCount in Step 10

Column title	Data type	Details
gene	String	KO gene name and flank are combined into a **single string data type**, separated by a colon, "*knockout:flank*"
n.umis	Integer	The **UMIs** after filtering that represent **distinct DNA molecules** detected for that KO:flank pair
n.tot	Numeric	Est. number of total UMI count, *n.umis/(1-loss)*
efficiency	Numeric	Estimated PCR efficiency. Higher values mean better amplification of each original molecule.
depth	Numeric	Avg. number of reads per UMI (including lost UMIs)
loss	Numeric	Estimated fraction of lost (unobserved) UMIs
n.obs	Integer	Same as for *n.umis*
n.reads	Integer	Observed total read count (after filtering)
n.umis.prefilter	Integer	Observed UMI count before read-count filter
n.reads.prefilter	Integer	Observed total read count before read-count filter

UMIs which are supported weakly by reads are most likely false and are removed. Due to technical reasons, some true UMIs are likely lost. TRUmiCount estimates the true missed UMIs and adjust the counts accordingly for each flank of KO cassette insertion. This method minimize the PCR based biasness for KO abundances estimation. The final output includes, for each KO and flank combination: the number of filtered UMIs (representing observed genomes), the estimated loss rate, and the loss-corrected genome count (Table 3).

17. A TRUmiCount phantom is defined as false UMI. It does not represent the real DNA molecule and still shows up in the data. Usually, it is caused by sequencing or PCR artifacts. To set the suitable TRUmiCount's read-count threshold for filtering the reads, it is helpful to check the read-count distribution plot in the TRUmiCount output report data/your design/your lib.count.pdf. TRUmiCount uses a threshold value for read-count filter. The default threshold value is not always suitable for all libraries so you need to adjust it manually on the basis of your data. Increase the threshold to improve the filtering if there is a clear overrepresentation of observed vs. predicted UMIs for read counts slightly greater than the threshold. Conversely decrease the threshold if predicted and observed UMIs agree well even below the current threshold. In this way, you can avoid the unnecessary discard of some true UMIs (Fig. 2C). To set a library-specific threshold *T* for library *your_lib*, add the following lines to the "trumicount" block code in cfg/

config.yaml (While editing the threshold in config.yaml file, match the indentation of the existing lines. Just like the tabs or space define code blocks in Python programming, indentation is crucial for structure of the data in config.yaml file. It tells the program, which settings belong together):

```
file: 'data/your_design/your_lib. *'
opts: '--threshold T'
```

After changing the threshold setting, remove the older output file, i.e., data/your_design/your_lib.count.tab and re-run Step 10 to see the effect of changing the Threshold on filtering of reads.

It may be necessary to adjust the automatically chosen read-count threshold T (Fig. 2C), while filtering the UMIs phantom by TRUmiCount algorithm [14]. This become especially important at higher sequencing depths, where phantom UMIs can represent a large share of UMIs count (but they are supported by lower read count due to their lower abundance as compared to the True UMIs, which reflect real DNA molecules).

If the threshold is set very low (Fig. 2C), phantom UMIs will "survive" more frequently and increase the number of UMIs incorrectly. It will later also affect the mapping and later results will be distorted. If the threshold is set very high, then some of the true UMIs, which are representative of actual genomes in the sequencing pool will be filtered out (Fig. 2C). However, since TRUmiCount algorithm accounts for this loss, but the main consequence is a reduction in precision due to fewer total genome counts, rather than introducing systematic bias. Therefore, when selecting the threshold, it is generally better to choose high threshold rather than a too low one.

18. The differential virulence analysis is based on the KO abundances and loss correction factors from the tables your replicate-in.count.tab and your replicate-out.count.tab, both created in the folder data/your_design during Step 10. Step 13 also produces an accompanying HTML report in data/your_design/your_replicate.dv.html.

    In pooled experiments, the starting abundance of each knockout (KO) can be present at very different levels in the input population. These differences are due to technical factors such as DNA capture bias, rather than biological effects. As a result, underrepresentation of a KO in the output pool might be due to its lower abundance in the input pool. To detect the true virulence effects, it is essential to normalize output abundances based on input levels. The iPool-Seq pipeline accounts for this (see Eq. A) by assuming that a KO's loss-corrected abundance A_{out} in the output pool depends (linearly) on its loss-corrected abundance A_{in} in the input pool, in addition to depending on the virulence factor Δv relative to neutral KOs ($\Delta v = 1$ for neutral KOs). The pipeline also includes a scaling factor λ (which is replicate-specific, but not KO-specific), to account for differences in genome capture efficiency and total genome count between the two libraries. For a KO, the loss-corrected input and output abundances are computed from the (3′ and 5′ summed) filtered UMI counts N_{in} and N_{out}, which are corrected for unobserved UMIs dividing by $1 - \ell_{in}$ and $1 - \ell_{out}$, where ℓ_{in} and ℓ_{out} are the (3′ and 5′ averaged) loss estimates computed by TRUmiCount. The $\log_2$ fold change of a KO's virulence is computed by the pipeline as:

$$\begin{aligned} log_2 \Delta v &= log_2 \frac{A_{out}}{\lambda . A_{in}} \\ &= log_2 \frac{N_{out} \cdot (1 - \ell_{in})}{\lambda . N_{in} . (1 - \ell_{out})} \end{aligned} \tag{A}$$

To test for significant deviations of Δv from neutrality (i.e., Δ v= 1), the pipeline assumes a negative binomial model for N_{out} given N_{in}, which accounts for the uncertainty in A_{out} and A_{in} due to sampling noise, as well as for an additional amount of dispersion d due to random fluctuations of the virulence of neutral KOs:

$$N_{out} | N_{in} \sim \text{NegBin}\left(\mu = \lambda . N_{in} . \frac{(1 - \ell_{out})}{(1 - \ell_{in})}, r = \frac{N_{in}}{1 + d . N_{in}} \right) \tag{B}$$

The two parameters λ and d (see Eq. B) are calculated by setting the model to the set of neutral KOs. Parameter λ measures the relative total size (i.e., the loss-corrected number of UMIs) of the output library relative to the input library, and d the squared coefficient of variation of output abundances due to biological noise (i.e., due to differences in the growth of neutral mutants).

19. To generate plots and to combine data from multiple replicates, load output table data/*your_design*/*your_replicate*.dv.tab from Step 13 into a table-based tool such as Microsoft Excel, GraphPad Prism, R, or SPSS. Then, filter KOs based on fold change, *p*-value, *q*-value, number of genomes, etc. Combine data from multiple replicates either by filtering based on criteria from multiple replicates (e.g., significance) or by calculating a combined *p*-value like Fisher's method. See (Table 2) for the description of the columns of data/your_design/your_replicate.dv.tab, which contains the results of the differential virulence analysis step.

The iPool-Seq results contain several types of information. Firstly, the KO abundance tables (Step 10) enlist the number of genomes per mutant found in the input and output libraries, i.e., they contain information about the absolute abundances of the individual KOs. These tables, in particular, also provide information about which mutants are not detected at all in either input or output (i.e., that show zero detected genomes), which is possibly due to very slow growth or not being viable at all. Gradual changes in a mutant's virulence are detected by comparison of the mutant's input and output abundances to a reference set of neutral mutants and is summarized in the differential virulence report (Step 13).

Step 13 creates the differential virulence report that contains diagnostic statistics and plots, which serve as quality checks and as verification that the assumptions of the statistical model are fulfilled to a reasonable degree. After each step of pipeline, an overview of how many useable reads and UMIs survived can be seen in the "Quality Control/Read and UMI count statistics" report. It is considered normal if one-third of reads during the course of the analysis are discarded. If considerably fewer reads than that remain after the "TRUmiCount" step, then check carefully for the step that is causative for the removing of the largest percentages of reads.

The differential virulence report created in Step 13, contains diagnostic statistics and plots that serve as quality checks and as verification that the assumptions of the statistical model are fulfilled to a reasonable degree. The "Quality Control/Read and UMI count statistics" provide an overview of how many usable reads and UMIs remain after each step of the analysis pipeline. Discarding up to about one-third of reads during the course of the analysis should be considered normal; if considerably fewer reads than that remain after the "TRUmiCount" step, the steps that remove the largest percentages of reads should be checked carefully for problems. For high sequencing depths (on average ~10 reads per UMI or more), TRUmiCount filter all UMIs with a low read count. So, it is normal for the removed percentage to be considerably high.

The precision of the KO abundances determined for the input and output pools is reflected by the correlation (found under "Quality Control/Correlation of 3′ and 5′ Flank Abundances") of the abundances computed for each flank of the KO cassette insertions. After loss correction by TRUmiCount, a correlation of ~0.9 or more can be expected.

Under "Quality Control/Correlation of Input and Output Abundances", can be seen that applied statistical model shows linear relationship between preinfection input abundance and postinfection output abundance of neutral KOs. There is no applicable lower bound for the expected correlation between input and output abundances, as it depends on both the proportion of non-neutral KOs in the experiment and the extent of rate of proliferation of these

KO Mutants. More crucial than the correlation is the qualitative behavior seen in the input vs. output plots. Ideally, the relationship should remain linear across the entire range of input abundances, without showing a plateau at the high end. If such a plateau effect is observed, where output levels stop increasing despite rising input, it is likely that the carrying capacity of the host plants has been reached. In such cases, either the number of mutants used to infect each plant should be reduced or increase the number of infected plants is another effective strategy to deal with the carrying capacity issue. Statistical model adjustment can also help to incorporate the effects of limited host capacity.

References

1. Khan M et al (2024) Tip of the iceberg? Three novel TOPLESS-interacting effectors of the gall-inducing fungus *Ustilago maydis*. New Phytol 244:949. https://doi.org/10.1111/nph.19967
2. Bindics J et al. (2022) Many ways to TOPLESS—manipulation of plant auxin signalling by a cluster of fungal effectors. New Phytol 236(4):1455–1470. https://doi.org/10.1111/nph.18315
3. Saado I et al. (2022) Effector-mediated relocalization of a maize lipoxygenase protein triggers susceptibility to *Ustilago maydis*. Plant Cell 34(7):2785–2805. https://doi.org/10.1093/plcell/koac105
4. Son H et al. (2011) A phenome-based functional analysis of transcription factors in the cereal head blight fungus, *Fusarium graminearum*. PLOS Pathog 7:10. https://doi.org/10.1371/journal.ppat.1002310
5. Uhse S, Pflug FG, Stirnberg A et al (2018) In vivo insertion pool sequencing identifies virulence factors in a complex fungal–host interaction. PLOS Biol 16(4):e2005129. https://doi.org/10.1371/journal.pbio.2005129
6. Uhse S, Pflug FG et al. (2019) Insertion pool sequencing for insertional mutant analysis in complex host-microbe interactions. Curr Protoc Plant Biol 4(3):e20097. https://doi.org/10.1002/cppb.20097
7. Uhse S, Pflug FG, Stirnberg A et al. (2018) In vivo insertion pool sequencing identifies virulence factors in a complex fungal–host interaction. PLOS Biol 16:4. https://doi.org/10.1371/journal.pbio.2005129
8. Köster J, Rahmann S (2012) Snakemake-a scalable bioinformatics workflow engine. Bioinformatics 28(19):2520–2522. https://doi.org/10.1093/bioinformatics/bts480
9. Hoffman CS, Winston F (1987) A ten-minute DNA preparation from yeast efficiently releases autonomous plasmids for transformaion of *Escherichia coli*. Gene 57(2–3):267–272. https://doi.org/10.1016/0378-1119(87)90131-4 ISSN 0378-1119, (https://www.sciencedirect.com/science/article/pii/0378111987901314)
10. Uhse S, Pflug FG, von Haeseler A et al. (2019) Insertion pool sequencing for insertional mutant analysis in complex host-microbe interactions. Curr Protoc Plant Biol 4(3):e20097. https://doi.org/10.1002/cppb.20097
11. Picelli S, Björklund AK, Reinius B, Sagasser S, Winberg G, Sandberg R (2014) Tn5 transposase and tagmentation procedures for massively scaled sequencing projects. Genome Res 24(12):2033–2040. https://doi.org/10.1101/gr.177881.114 Epub 2014 Jul 30. PMID: 25079858; PMCID: PMC4248319
12. Grüning B et al (2018) Bioconda: sustainable and comprehensive software distribution for the life sciences. Nat Methods 15:475. https://doi.org/10.5281/zenodo.1068297
13. Andrews S. A quality control tool for high throughput sequence data. http://www.bioinformatics.babraham.ac.uk/projects/fastqc
14. Pflug FG, von Haeseler A (2018) TRUmiCount: correctly counting absolute numbers of molecules using unique molecular identifiers. Bioinformatics 34(18):3137–3144. https://doi.org/10.1093/bioinformatics/bty283
15. Bolger AM, Lohse M, Usadel B (2014) Trimmomatic: a flexible trimmer for Illumina sequence data. Bioinformatics 30(15):2114–2120. https://doi.org/10.1093/bioinformatics/btu170
16. Sedlazeck FJ, Rescheneder P, Von Haeseler A (2013) NextGenMap: fast and accurate read mapping in highly polymorphic genomes. Bioinformatics 29(21):2790–2791. https://doi.org/10.1093/bioinformatics/btt468
17. Robinson JT et al (2011) Integrative genomics viewer. Nat Biotechnol 29(1):24. https://doi.org/10.1038/nbt.1754

18. Smith T, Heger A, Sudbery I (2017) UMItools: modeling sequencing errors in unique molecular identifiers to improve quantification accuracy. Genome Res 27(3):491–499. https://doi.org/10.1101/gr.209601.116

Chapter 24

Mitotic Heteroallelic and Sister Chromatid Recombination

Jeanette H. Sutherland and William K. Holloman

Abstract

Homologous recombination is a high-fidelity DNA repair mechanism used by mitotic cells to repair double-strand breaks and other types of DNA damage. This process preferentially uses a sister chromatid as a template, ensuring error-free repair, but can also use a homologous sequence on another chromosome. Here, methods are described that can be used to measure recombination between sister chromatids or homologs.

Keywords DNA repair, Homologous recombination, Sister chromatid recombination

1 Introduction

Mitotic cells utilize the homologous recombination system as a high-fidelity DNA repair mechanism to restore genomic integrity after introduction of double-strand breaks and other types of DNA damage. Repair is directed best by a homologous sequence present on the sister chromatid but can also be enabled by a homologous chromosome [1, 2]. Sister chromatids are the ideal and preferred templates for high-fidelity repair, because they are genetically identical and physically more proximal than homologous chromosomes, which are highly similar in sequence although not necessarily identical, and not physically connected. The sister chromatid is also the preferred information donors for another reason, i.e., it is inherently safer. While repair by homologous recombination directed using the homologous chromosome as template serves as a backup mechanism, it does entail the risk of loss of heterozygosity of linked recessive mutations should crossing over occur.

Various assays have been developed to measure recombination between homologs or sister chromatids in a range of experimental systems. In organisms with large chromosomes, mitotic recombination can be monitored by microscopy using fluorescence in situ hybridization [3]. DNA fiber analysis using visual imaging and strand-specific

An N. T. Phan et al. (eds.), *Ustilago maydis*, Methods in Molecular Biology, vol. 3050,
https://doi.org/10.1007/978-1-0716-5364-7_24,

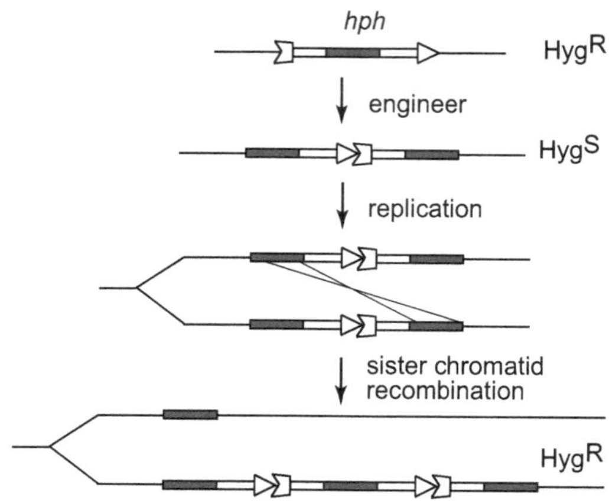

Fig. 1 Sister chromatid recombination assay. Schematic illustrates the construction of the reporter from the *hsp70-hph* gene fusion with a reversed arrangement of proximal and distal regions. Hygromycin resistant recombinants arise from an unequal SCR event

DNA sequencing can be used in organism amenable to in vivo DNA labeling [4, 5]. In microbes, reporter gene systems have been developed such that gene reactivation via recombination can be monitored by growth, color change, or fluorescence [6]. Here, methods are described for measuring interhomolog and intersister mitotic recombination in *U. maydis*. Interhomolog recombination is performed in diploid strains heteroallelic for *nar1*, the structural gene for nitrate reductase, and monitors growth on medium containing nitrate as the sole source of nitrogen. Intersister recombination is performed in haploid strains and monitors reconstruction of overlapping fragments of *hph*, a gene conferring hygromycin resistance (Fig. 1). Interhomolog and intersister recombination methods were developed previously for studies on mitotic gene conversion [7] and DNA repair [8], respectively.

2 Materials

2.1 Strains and Plasmids

1. Parental *U. maydis* test strains UCM350 (*nar1-6 pan1-1 a1 b1*) and UCM520 (*nar1-1 met1-2 a2 b2*) *nar* indicates ability to utilize nitrate as a source of nitrogen; *pan, met* indicate requirements for pantothenic acid, methionine, respectively; *a* and *b* mating type loci.
2. *nar1* (UMAG_03847) encodes 983 codons of the structural gene for nitrate reductase. *nar1-6* is due C1120T transition resulting in Q374* (stop codon). *nar1-1* is due to C2533T transition resulting in P845S mutation.
3. Plasmid pCM54 [9] or other source of *hsp70-hph* transcriptional fusion on a 2.0 kbp *Xho*I-*Sal*I DNA fragment.

4. pCM691 [10] or other pUC19 derivative containing the *U. maydis sdh2* (UMAG_00488) gene variant conferring carboxin resistance.

2.2 Culture Media

1. YEPS medium: 1% yeast extract, 2% peptone, 2% sucrose [+ 2% agar for solid medium] (per liter).
2. Black medium: 1% yeast, 2% peptone, 2% sucrose, 1% charcoal, 1% agar, 1.7% cornmeal agar (per liter).
3. Ammonium or nitrate minimal medium: 1.5 g NH_4Cl or 3 g KNO_3, 16 g KH_2PO_4, 4 g Na_2SO_4, 2 g $MgSO_4$, 1 g $CaCl_2$, 8 mL trace element solution [+2% agar for solid medium]. Adjust to pH 7.0 with HCl. (per liter).
4. Trace element solution: 60 mg H_3BO_3, 140 mg $MnCl_2$-$4H_2O$, 400 mg $ZnCl_2$, 40 mg Na_2MoO_4-$2H_2O$. 100 mg $FeCl_3$-$6H_2O$, 400 mg $CuSO_4$-$5H_2O$ (per liter).

2.3 Laboratory Instruments and Supplies

1. Tabletop microcentrifuge.
2. Standard microbiology laboratory microscope with 40× objective lens.
3. Hausser bright line hemocytometer.
4. Petri dishes (100 × 15 mm) and (150 × 15 mm).
5. 25 mL Erlenmeyer flasks.
6. Gyratory shaker.
7. Sterile transfer loops.
8. Triangular cell spreader.
9. 15 mL Falcon Tubes.
10. Carboxin stock solution 4 mg/mL in ethanol.
11. Hygromycin stock 200 mg/mL in H_2O.

2.4 Genomic DNA Preparation

1. YEPS Sorbitol medium: 1% yeast extract, 2% peptone, 2% sucrose, 182 g sorbitol, 2% agar (per liter).
2. TE −/+ RNaseA: 10 mM Tris-HCl, pH 7.5, 1 mM EDTA ± 10 μg/mL RNaseA.
3. Lysis Buffer: 10 mM Tris-HCl, pH 8.0, 1 mM EDTA, 2% Triton X-100, 1% SDS, 100 mM NaCl.
4. Phenol:chloroform:isoamyl alcohol: 25:24:1.
5. BioSpec 101 Fast Prep Cell disintegrator.
6. 0.5 mm zirconium beads (BioSpec Products).
7. 2 mL XXTuff Microvials (BioSpec Products).
8. GoTaq Green master mix (Promega).
9. TAE electrophoresis buffer: 40 mM Tris-acetate pH 8.3, 1 mM EDTA.

10. Primers: HygSCR forward primer 5′-CGTCTGTCGAGAA GTTTCTG-3′. HygSCR reverse primer 5′-GTCTGCTGCT CCATACAAGC-3′.

3 Methods

3.1 Interhomolog Heteroallelic Recombination

3.1.1 Diploid Construction

1. Using standard gene disruption protocol, knock out gene of interest to be tested in both UCM350 and UCM520 parental strain backgrounds. Grow test strains in 2 mL cultures of YEPS medium in 25 mL Erlenmeyer flasks at 28° overnight in a gyratory shaker at 200 rpm. Cultures generally grow to a density of 2–4 $\times 10^8$ per mL.
2. Remove aliquots (1 mL) to 1.5 mL microcentrifuge tubes, spin at 6000 rpm for 30 s in bench top microcentrifuge at room temperature. Remove spent medium by aspiration and resuspend cells in 1 mL water. Vortex vigorously to resuspend the cells. Repeat the process once, then mix 100 μL aliquots of each strain in a single microcentrifuge tube. Spot aliquots (10 μL) of this mix on Petri plates containing black medium. After drying, seal the plates with parafilm and place in a 28° incubator for 5 days.
3. Using sterile forceps, lift white matted cell pads off the black medium and place in a sterile 15 mL Falcon tube containing 5 mL water. Vortex vigorously for 1 min to disperse the heterokaryons. Plate 100 μL on a Petri dish containing ammonium minimal medium. Spread the cells over the medium with back-and-forth motion using a sterile spreader and place dishes in 28° incubator for 5 days. Diploids are selected due to the forcing auxotrophic markers (*pan1-1 and met1-2*) in the parental strains and grow into colonies. About 100 well-growing colonies should be evident at this time. Restreak colonies onto fresh ammonium minimal plates and incubate 3 days at 28°.

3.1.2 Heteroallelic Recombination

1. Starting with isolated colonies, pick the entire colony with the aid of a sterile transfer loop into 2 mL culture of ammonium minimal medium in 25 mL Erlenmeyer flasks. Grow overnight at 28° in a gyratory shaker (~200 rpm). Split into two aliquots and spin down the cells of each in 1.5 mL microcentrifuge tubes 6000 rpm for 30 s. Wash in water by resuspending in 1 mL sterile water with vigorous vortexing, then recentrifuging. Repeat. After two successive washes, resuspend cells

of each in 0.5 mL sterile water and determine the cell density. Remove 10 μL suspension and add to 1 mL water for a 1/100 dilution. Place 10 μL of the diluted sample in the chamber of a Hausser hemocytometer and determine cell number using the 40× objective of a laboratory light microscope.

2. Prepare nitrate minimal agar plates using large (150 × 15 mm) petri dishes. Dry the plates by removing the lids and placing next to the laminar air flow stream inside a biosafety cabinet equipped with a Hepa filter. After 2 h, remove the plates for use. Add the contents of the entire tube of washed diploid cells and with a back-and-forth motion spread the cell suspension over the surface of the solid medium with the aid of a triangular plate spreader until all the liquid has been absorbed by the agar medium. Place in 28°C incubator for 5–6 days to select for Nar^+ recombinants evident as small white colonies. Total the number of Nar^+ colonies from two paired plates and divide by the number of cells determined by hemocytometer counting. This is the frequency of allelic recombination. This procedure is performed in parallel on seven to nine independently isolated diploids and the median frequency is taken as the representative.

3. Repeat the experiment several times and calculate the average using the median from each experiment. Perform statistical analysis.

3.2 Sister Chromatid Recombination

3.2.1 Reporter Strain Construction

1. *hsp70-hph* transcriptional gene fusion on a 2.0 kbp *Xho*I-*Sal*I fragment derived from pCM54 [9] expresses hygromycin phosphotransferase and confers hygromycin resistance (Hyg^R). Two partially overlapping fragments obtained after digestion with *Xho*I- *Sac*II and *Eco*RI-*Sal*I were ligated together into pBluescript II SK^+ opened with *Eco*RI-*Sac*II to yield a construct, in which the proximal and distal gene fragments were reordered front to back. This construct was moved as a *Hin*dIII-*Sac*I fragment into pCM691 opened with *Hin*dIII-*Sac*I to yield pJS-SCR. pCM691 [10] contains a mutant form of the succinic dehydrogenase 2 iron-sulfur subunit gene *sdh2* that confers carboxin resistance (Cbx^R) in *U. maydis* and is thus allelic to the endogenous *sdh2* gene enabling targeted recombination. Digest 2 μg of pJS-SCR with ten units *Age*I for 4 h at 37°C, then heat inactivate at 70°C for 15 min. Transform protoplasts of the strain of interest with the digested pJS-SCR by standard procedures [9] and spread protoplasts on transformation plates of YEPS sorbitol + 4 μg/mL carboxin. Incubate plates at 28°C 5–7 days until distinct colonies are visible.

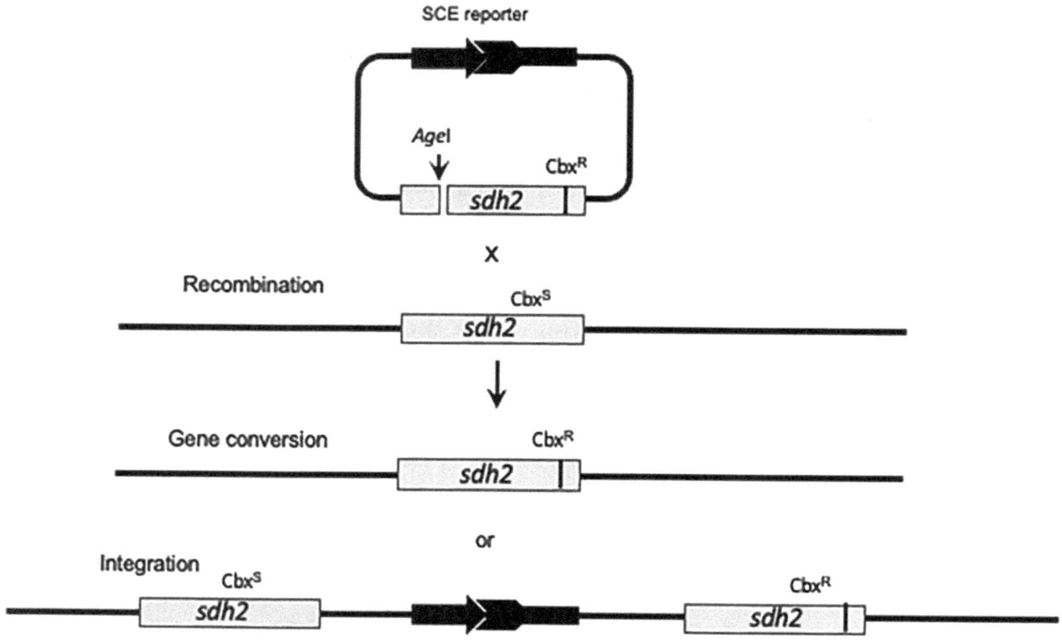

Fig. 2 Schematic view of plasmid vector-by-chromosome recombination. Representation is shown of recombination between the chromosomal allele Cbx^S *sdh2* gene (gray rectangle) and the plasmid harboring the SCE reporter construct (back-to-front black arrow) and containing the homologous *sdh2* mutant gene (Cbx^R) cut with *Age*I within the gene. The mutation resulting in carboxin resistance is indicated (black bar)

2. Restreak transformants to single colonies on YEPS plates without carboxin and incubate 28°C 2 days. Pick isolated colonies, restreak onto plates containing YEPS +2 μg/mL carboxin and incubate 28°C 2 days. This is to ensure that Cbx^R clones are being selected. To distinguish Cbx^R transformants with the integrated SCR reporter gene from Cbx^R transformants without the reporter, genomic DNA of clones must be assayed for the presence of *hph* sequence by PCR (Fig. 2).
3. To prepare genomic DNA, grow up individual transformants overnight at 28°C in 3 mL of YEPS supplemented with 2 μg/mL carboxin. Spin down 2 mL of overnight culture in a XXTuff microvial at 14,000 RPM for 2 min. Remove the supernatant. Using a small spatula, add about 30 μL of 0.5 mm zirconium beads to the tube, as well as 200 μL phenol:chloroform: isoamyl alcohol and 200 μL of Lysis Buffer. Insert tube in the BioSpec Fast Prep cell disintegrator and run for 20 s at a speed setting of 6.5. Remove 200 μL of the top layer and place it in a fresh microcentrifuge tube. Add 200 μL isopropanol to this and mix it by inverting. Leave this to stand on the bench at room temperature for 10 min. Spin the sample for 10 min at 14,000 RPM. Remove and discard the supernatant. Invert over a piece of absorbent paper such as Kimwipe. Let the pellet drain dry

with the cap open for 10 min at room temperature. Add 50 μL of TE + RNase to the pellet and incubate it for 10 min at 70°C.

4. Confirm transformants with integrated SCR reporter substrate by PCR using primers specific for *hph* gene sequence. Perform PCR reactions (25 μL) containing 12.5 μL GoTaqGreen master mix, 1 μL each forward and reverse primers (10 μM), 0.1 μL of genomic DNA sample by cycling 30 times through a regimen of 98°C denaturation, 55°C annealing, and 72°C extension in a thermocycler and analyze products by electrophoresis on a 1% agarose gel in TAE buffer. A 0.7 kbp band indicates clones with the integrated SCR reporter substrate. Save the original cell culture as a glycerol stock for assaying sister chromatid recombination.

3.2.2 Sister Chromatid Recombination Analysis

1. Prepare nine different independent liquid cultures of 2 mL YEPS each in 25 mL Erlenmeyer flasks of the strain to be tested. Do not add carboxin. Grow overnight at 28°C. Determine cell density by counting using a hemocytometer. Remove 1×10^7 cells to separate microcentrifuge tubes and spin down 6000 rpm for 30 s. Remove supernatant and resuspend cells in 1 mL sterile water.
2. Spread 100 μL of each cell suspension on Petri plates containing YEPS solid medium + 50 μg/mL hygromycin. Incubate plates 3–5 days at 28°C. Count hygromycin resistant colonies diploids. The number determined is the frequency per 10^6 cells plated. The median is taken as the representative sister chromatid recombination frequency.
3. Repeat the experiment several times and calculate the average using the median from each experiment. Perform statistical analysis.

Acknowledgments

DNA sequence analysis of the *nar1-1* and *nar1-6* alleles were determined by Dr. Milorad Kojic, University of Belgrade, Serbia. This work was supported in part by a Theresa Patnode Santmann Foundation grant and the Provost Summer Scholarship Grant to JHS.

References

1. Kadyk LC, Hartwell LH (1992) Sister chromatids are preferred over homologs as substrates for recombinational repair in *Saccharomyces cerevisiae*. Genetics 132:387–402. https://doi.org/10.1093/genetics/132.2.387
2. Gonzalez-Barrera S, Cortes-Ledesma F, Wellinger RE et al (2003) Equal sister chromatid exchange is a major mechanism of double-strand break repair in yeast. Mol Cell 11:1661–1671. https://doi.org/10.1016/s1097-2765(03)00183-7
3. Chaganti RS, Schonberg S, German J (1974) A manyfold increase in sister chromatid exchanges in Bloom's syndrome lymphocytes. Proc Natl Acad Sci U S A 71:4508–4512. https://doi.org/10.1073/pnas.71.11.4508

4. Sanders AD, Falconer E, Hills M et al (2017) Single-cell template strand sequencing by Strand-seq enables the characterization of individual homologs. Nat Protoc 12:1151–1176. https://doi.org/10.1038/nprot.2017.029
5. Saxena S, Zou L (2022) Hallmarks of DNA replication stress. Mol Cell 82:2298–2314. https://doi.org/10.1016/j.molcel.2022.05.004
6. Klein HL, Bacinskaja G, Che J et al (2019) Guidelines for DNA recombination and repair studies: cellular assays of DNA repair pathways. Microb Cell 6:1–64. https://doi.org/10.15698/mic2019.01.664
7. Holliday R (1966) Studies on mitotic gene conversion in Ustilago. Genet Res 8:323–337. https://doi.org/10.1017/s0016672300010181
8. Sutherland JH, Holloman WK (2018) Loss of cohesin subunit Rec8 switches Rad51 mediator dependence in resistance to formaldehyde toxicity in *Ustilago maydis*. Genetics 210:559–572. https://doi.org/10.1534/genetics.118.301439
9. Tsukuda T, Carleton S, Fotheringham S et al (1988) Isolation and characterization of an autonomously replicating sequence from *Ustilago maydis*. Mol Cell Biol 8:3703–3709. https://doi.org/10.1128/mcb.8.9.3703-3709.1988
10. Kojic M, Holloman WK (2000) Shuttle vectors for genetic manipulations in *Ustilago maydis*. Can J Microbiol 46:333–338. https://doi.org/10.1139/w00-002

Chapter 25

The Use of Yeast Two-Hybrid Analysis to Study Protein–Protein Interactions for *Ustilago maydis*

Venkata S. Kuppireddy, Shikhi Baruri and Michael H. Perlin

Abstract

Yeast two-hybrid (Y2H) analysis is an important first step in identifying protein–protein interactions and has been used in a variety of biological systems. In fungi, this method has been employed to evaluate proteins that interact within the fungal cell, often as part of signaling pathways. For fungi that infect or are in relationships with other organisms, the method has also been used to identify and characterize host targets for fungal effector proteins, often used in establishing and maintaining the relationship. Here we describe the Y2H method, as it has been applied to work with *Ustilago maydis*. The method is described as used for screening cDNA libraries to identify interactors with a known "bait" protein; further, we describe directed Y2H, where both bait and prey proteins are known. Emphasis is also placed on the need for appropriate controls, since Y2H can be prone to false positives. Finally, we point out how the results of Y2H can be applied to even more biologically relevant experiments that both validate and expand our understanding of fundamental protein–protein interactions.

Keywords Protein–protein interactions, Fungal effectors, Host targets, Reporter genes, DNA binding domain, Transcriptional activation domain

1 Introduction

The study and characterization of protein–protein interactions is important and provides critical insights as to biological role(s) such interactions might play in a variety of processes. For example, in identifying putative interactions within a species or particular cell type, predicted participants in signaling pathways could emerge. Alternatively, interacting partners (as pairs or as part of complexes) for proteins from different cell types or species could yield a clearer picture of how such cell types or species communicate. So, proteins of a pathogen or parasite might interact with target proteins of host cells, and identification of such interactions can be used to better understand how such parasites manipulate their hosts or at least evade the host defense response. Such studies have been used across

An N. T. Phan et al. (eds.), *Ustilago maydis*, Methods in Molecular Biology, vol. 3050,
https://doi.org/10.1007/978-1-0716-5364-7_25,

all five kingdoms of life [1–8], and their overall importance cannot be overstated.

A number of methods have been developed to identify and interrogate protein–protein interactions. These include: (1) yeast 2-hybrid (Y2H) and its varied modifications, (2) Co-immunoprecipitation (CoIP), both in vivo and in vitro, and (3) global cross-linking of proteins in vivo, followed by mass-spectroscopy to identify the interacting partners [9–13]. For Co-IP, the method has been used to discover interactions between a protein from a pathogen, expressed in a heterologous host, absent the pathogen, as well as between the pathogen protein and proteins in a native host(s).

While the Co-IP and global crosslinking methods can be technically challenging, lengthy, and costly, they do potentially provide biologically relevant predictions of interactions, assuming appropriate controls are utilized. Y2H analysis is less expensive, and the rate-limiting steps are those used in development of genetic constructs required for the process. Given the same caveat, that all appropriate controls must be included, this is crucial to obtaining meaningful conclusions from this approach. False positives for interactions are common with Y2H experiments, and such false outcomes need to be eliminated from further consideration. Also, it should be pointed out that Y2H is, by nature, an examination of interactions in a heterologous system (except for *Saccharomyces cerevisiae* proteins); thus, its use and findings should not be overstated but, instead, should be used as a starting point for additional, potentially more biologically relevant approaches.

As was just indicated, Y2H is a very good first step in characterizing protein–protein interactions. The relatively lower cost and reduced technically challenging aspects of the method are some of the reasons for this. In principle, Y2H can be utilized for any organism and any proteins, but the original version is best suited for soluble cytosolic/cytoplasmic proteins. Some versions better serve studies where at least one protein in the interaction is a membrane protein, e.g., both split-ubiquitin and split-YFP adaptations [14–17].

We are primarily interested in fungi, particularly those causing disease on plants. Such fungi can, at least initially, be divided into necrotrophic and biotrophic fungi. Nectrotrophic fungi (e.g., *Botrytis cinerea*, *Sclerotinia sclerotiorum*, *Alternaria species*) require dead or damaged host plant tissue and derive nutrients from the host cells they kill. In contrast, those fungi that require living cells for nutrient acquisition and energy are called biotrophs (e.g., for plant hosts, *Blumeria (Erysiphe) graminis, Uromyces fabae, Ustilago maydis, Cladosporium fulvum, Puccinia graminis*, and the Oomycete, *Phytophthora infestans*; *Pneumocystis jirovecii* (Taphrinomycotina subphylum of the Ascomycota), exclusively a human pathogen, can cause lung diseases). There are also hemibiotrophic fungi, i.e., those that have an initial biotrophic phase for establishing infection, followed by a necrotrophic phase that leads to extensive cell death;

examples include the model, *Magnaporthe oryzae* (rice blast fungus), and other pathogens such as many *Colletotrichum* spp., Southern corn leaf blight (*Bipolaris maydis*), and *Zymoseptoria tritici (syn. Mycosphaerella graminicola*, leaf blotch of wheat).

Our lab studies the smut fungi, biotrophs, in the Basidiomycota. In particular, we study the parasites of maize, *Ustilago maydis* and *Sporisorium reilianum*, as well as the anther smuts, in the genus *Microbotryum*. These fungi follow the typical lifecycle of other smut fungi, whereby haploid cells of opposite/compatible mating types conjugate with their partners to produce an infectious, dikaryotic hypha that penetrates the host tissue, the biomass of the mycelia increases in the plant, and, eventually, there is karyogamy where nuclei fuse and the diploid teliospores are produced in the infected tissue.

Of these fungi, *U. maydis* is unique in that it causes the host plant to produce tumors or galls on any of the above-ground tissues, including leaf, stem, tassels, and cobs. Biotrophic fungi, including *U. maydis*, contain a repertoire of secreted proteins termed effectors. These tend to be small (<250 amino acids) lack matches to the protein database or to PFAM domain databases. Some of these may be cysteine-rich, leading to their protection in the potentially damaging milieu of the apoplastic space of plants [18, 19]. In the following article, we describe the use of Y2H in characterizing protein–protein interactions. This technique has been used in establishing such interactions between specific *U. maydis* proteins (e.g., the ammonium transceptor, Ump2 with Rho1, [15]; the signaling proteins Pdc1 and Rho1, [20, 21]; Ubc2 and Ubc4, [22]; Hsl7 and Smu1, [23]; other signaling components, such as Cdc42, Rho2, etc., [24]). More recently, Y2H has been applied in identifying host protein targets of *U. maydis* effectors (Pit2, [25]; Tin2, [26]; and, a secreted *U. maydis* transcriptional activator, regulating hyperplasia in maize, [27, 28]). Often, as a first step in establishing that a predicted secreted protein may be an effector protein, a method is used initially to confirm that the protein of interest is, in fact, likely secreted from the fungus. This method, the Yeast Secretion Trap [29–30] experimentally tests the putative signal peptide of such proteins for their ability to secrete an invertase from a mutant yeast strain, thereby allowing the strain to grow on sucrose as a sole carbon source. Once such secretion of a putative effector has been established, then Y2H may be pursued to identify, e.g., potential host target proteins.

2 Materials

2.1 Cloning Bait and Prey

- pCR 2.1 TOPO entry vector (Invitrogen/Thermo Fisher, Waltham, MA, USA).
- pGBKT7 bait vector (Fig. 1; Takara Bio).

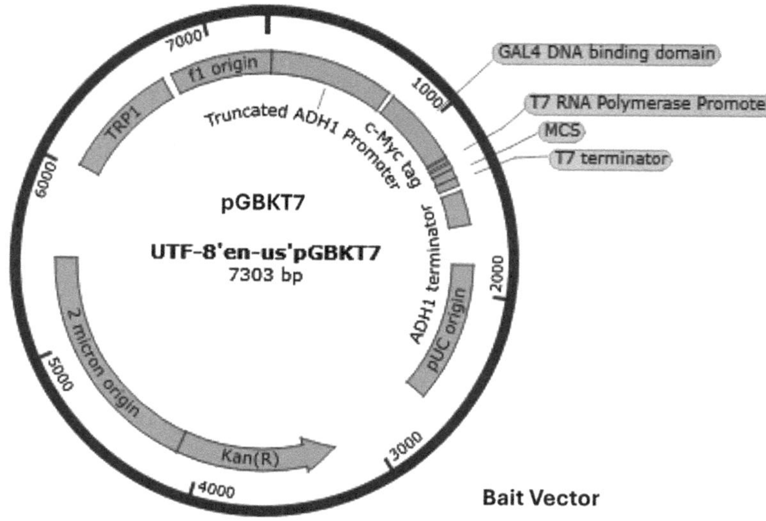

Fig. 1 Bait vector, pGBKT7. This plasmid allows generation of fusion proteins consisting of a "bait" protein, with the Gal4p DNA Binding Domain fused to its N-terminus

- pGADT7 prey vector (Fig. 2; Takara Bio).
- *E. coli* DH5 alpha strain (*fhuA2 lac(del) U169 phoA glnV44 Φ80' lacZ(del)M15 gyrA96 recA1 relA1 endA1 thi-1 hsdR17*).
- Qiagen Plasmid Isolation Kit (Qiagen, Santa Clarita, California) or similar alkaline lysis protocol [31].
- LB Agar plates with antibiotics (200 μg Ampicillin or 50 μg Kanamycin per ml of agar) for selection.
- NEBuilder® HiFi DNA Assembly Master Mix (New England Biolabs).
- RNA isolated from stages of growth for normalized cDNA library
- SuperScript IV First-Strand cDNA synthesis kit (Thermo Fisher).

2.2 Transformation of Yeast Strains

- Bait host (MATa): AH109 (HIS3, ADE2, MEL1/LacZ reporters under Gal4- responsive UAS promoters; Full Genotype: *MATa, trp1-901, leu2-3, 112, ura3-52, his3-200, gal4Δ, gal80Δ, LYS2::GAL1UAS-GAL1TATA-HIS3, MEL1 GAL2UAS-GAL2TATA-ADE2, URA3::MEL1UAS-MEL1TATA-lacZ*; Mating partner).
- *Library host (MATα): Y187 (pretransformed AD/cDNA library). S. cerevisiae* strain Y187 (Full Genotype: *MATα, ura3-52, his3-200, ade2-101, trp1-901, leu2-3, 112, gal4Δ, met–, gal80Δ, URA3::GAL1*$_{UAS}$*-GAL1*$_{TATA}$*-lacZ*; Library host strain).
- pGBKT7-53, control plasmid.

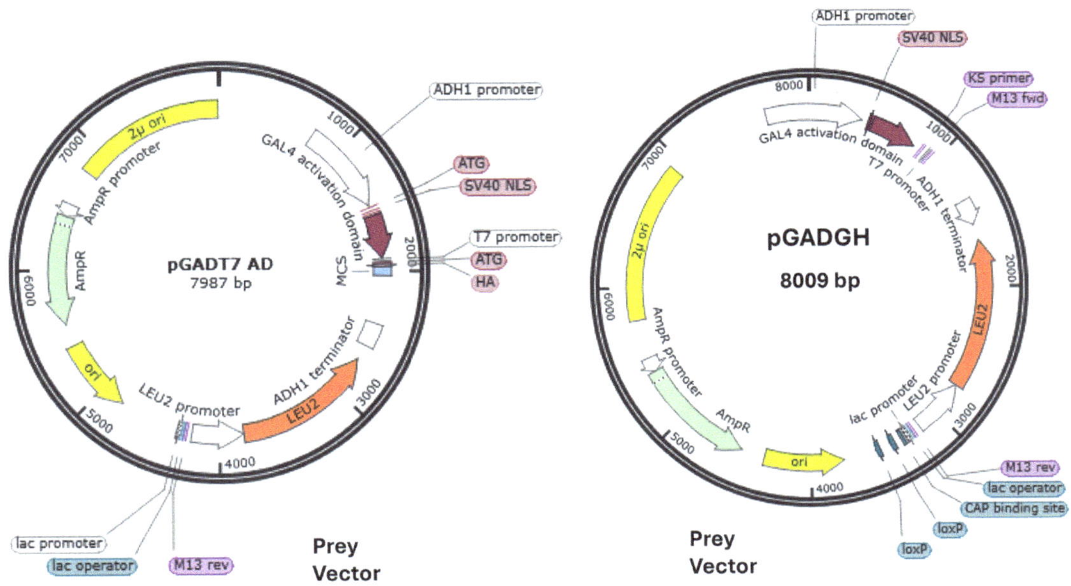

Fig. 2 Prey vectors. Plasmids pGADT7 AD and pGADGH serve as vectors for the cloning of genes encoding "prey" proteins or cDNA libraries of potential prey proteins. In each case, the prey protein is expressed as a fusion to the Gal4p transcriptional Activation Domain (AD). The pGADGH plasmid also provides a nuclear localization signal to assure the captured prey, in successful interactions with a bait protein, is transported to the nucleus

- pGADT7-T, control plasmid.
- Cloned bait gene in pGBTK7 vector (from Sect. 2.1).
- Cloned cDNA library in pGADT7 vector (from Sect. 3.1.1, below)
- Frozen-EZ Yeast Transformation II Kit (Zymo Research, Irvine, California) or,
- Lithium acetate/single-stranded carrier DNA/PEG method [32]: Lithium Acetate (1.0 M)/Polyethylene Glycol 3350 (50%, w/v)/Single-Stranded Carrier DNA (2.0 mg/ml; 200 mg of salmon sperm DNA dissolved in 100 ml of TE buffer (10 mM Tris-HCl, 1 mM Na2 EDTA, pH 8.0). Store 1.0-ml samples at −20° C. Denature the carrier DNA in boiling water bath for 5 min and chill in ice/water before use.)
- YPD medium: 1% Yeast extract, 2% Peptone, 2% Dextrose, (2% Agar, for solid medium).

2.3 Y2H Screening

- Yeast strains, plasmids, and controls:
- Bait Host: AH109.
- Library or Prey Host: Y187.
- Positive controls: AH109[pGBKT7-53] × Y187[pGADT7-T].

- Empty vectors for verification: pGBKT7 (BD; TRP1) and pGADT7-Rec/RecAB (AD; LEU2).
- Flasks 200–250 ml (depends upon the number of experimental groups and controls).
- Flasks 50 ml (depends on the number of co-transformation).
- Incubator (30°C).
- Shaking incubator.
- Cuvettes and Spectrophotometer.
- Centrifuge.
- 2-ml microcentrifuge tubes.
- Media and supplements (see Yeast-Specific Media Recipes, below).
- Rich media: YPDA, 0.5× YPDA, 2× YPDA; optional kanamycin 50 μg/ml to suppress bacteria.
- Minimal selective plates: SD/-Trp (SD drop out medium lacking Trp), SD/-Leu (SD drop out medium lacking Leucine), SD/-Leu/-Trp (DDO, SD medium lacking His and Trp), QDO (SD/-Ade/-His/-Leu/-Trp).
- Chromogenic substrate: X-α-Gal, stock 20 mg/ml in dimethylformamide; final in agar 40 μg/ml (either pre-mix at 2 ml/L to 55°C agar, or spread 200 μl of 4 mg/ml onto 15 cm plates/100 μ l onto 10 cm plates and dry thoroughly). Use X-α-Gal (5-bromo-4-chloro-1H-indol-3-yl α-D-galactopyranoside), <u>not</u> X-Gal (5-bromo-4-chloro-3-indolyl-β-D-galactoside).
- 3AT (3-Amino-1,2,4-triazole).
- Kanamycin: add to cooled media (55°C) to 50 μg/ml.
- QDO agar plates (diameter 150 mm) of media containing SD–
- Ade, Histidine, Tryptophan, Leucine (with containing X-α-
- Gal (DMSO) and 3-amino-1,2,4-triazole (3-AT).

2.4 Yeast-Specific Media Recipes

- YPD: 1% yeast extract, 2% or 10% dextrose, 2% peptone, 2% agar.
- YPDA: 20 g/L Peptone, 10 g/L Yeast Extract, 20 g/L Agar (only for solid medium), 2% Glucose, 0.003% Adenine Sulfate (only for YPDA, recommended for expression of foreign proteins).
- YPS: 1% yeast extract, 2% peptone, 2% sucrose, 1.8% agar.
- PDA: Potato Dextrose Agar may be purchased from a variety of sources (e.g., VWR/Avantar; Thermo Fisher, etc.). Follow the instructions of the manufacturer for reconstitution of the powdered medium with sterile deionized or otherwise purified water.
- Vitamin Concentrations for Minimal Media.

- Vitamin Final conc.:
  - biotin 2 μg/l
  - calcium pantothenate 400 μg/l
  - folic acid 2 μg/l
  - inositol 2 mg/l
  - niacin 400 μg/l
  - p-aminobenzoic acid 200 μg/l
  - pyridoxin HCl 400 μg/l
  - riboflavin 200 μg/l
  - thiamine HCl 400 μg/l

2.5 Making Supplement Mixtures

- There are ten different dropout mixtures. Nine of these lack a single amino acid or base (ADE, ARG, HIS, LEU, LYS, MET, THR, TRP, and URA), and one is missing two amino acids (TYR PHE).
- There is also a complete mixture that contains everything. To make a supplement mixture weigh out the different components (minus 1 or 2) and grind them into a fine powder with a mortar and pestle (~4–5 min).
- One mixture made according to this recipe will make ~40 L of plates.
- Components of Supplement Mixtures:
- adenine 800 mg; arginine 800 mg; histidine 800 mg; leucine 1200 mg; lysine 1200 mg; methionine 800 mg; phenylalanine 2000 mg; threonine 8000 mg; tryptophan 800 mg; tyrosine 1200 mg; uracil 800 mg.
  1. SD Agar Plates: 850 ml distilled water and 20 g microbiological grade agar.
  2. Autoclave.
  3. After autoclaving combine the autoclaved solution and 150 ml of (filter sterilized) Yeast Nitrogen Base, etc.
  4. Mix thoroughly and pour into plates.
- YNB (yeast nitrogen base) + Dextrose + Amino Acids:
  1. Mix 150 ml distilled water, Amino acid supplement mixture, 6.7 g Yeast Nitrogen Base (w/o amino acids, with ammonium sulfate), 20 g. Dextrose.
  2. Filter Sterilize.
  3. After autoclaving combine the autoclaved solution with YNB + Dextrose + Amino Acids.
  4. Mix thoroughly and pour into plates.

- YNB + Dextrose:

  1. Mix 150 ml distilled water + 6.7 g Yeast Nitrogen Base (w/o amino acids, with ammonium sulfate) + 20 g Dextrose.
  2. Filter Sterilize

     - DDO: SD, leaving out 2 amino acids, to select for the bait and prey plasmids.
     - TDO: SD, leaving out 3 amino acids, used in the selection of interactions between bait and prey proteins.
     - QDO: SD, leaving out 4 amino acids, used in the stringent selection of interactions between bait and prey proteins.

3 Methods

3.1 Yeast 2-Hybrid Analyses Using cDNA Libraries

To investigate potential protein–protein interactions, a yeast-based screening approach using *Saccharomyces cerevisiae* serves as an effective method. This method relies on the reconstitution of a transcriptional activator when two hybrid proteins interact within the yeast nucleus (see Fig. 3). The protein of interest, referred to as the "bait," is genetically fused to the Gal4 DNA-binding domain and inserted into an expression vector. Conversely, potential interacting partners, or "prey" proteins, are expressed as fusions with the Gal4 transcriptional activation domain using another vector. When bait and prey proteins form a complex, the separate Gal4 domains are brought into proximity, triggering transcription of specific reporter genes that can be visualized. Typically, prey proteins will be expressed from normalized cDNA libraries of the different stages of the life cycle of the organism of interest during its association with the host partner. The initial identification of potential interactors is based on the yeast strains' ability to grow on progressively more selective auxotrophic media. These strains carry auxotrophic mutations, meaning they either require supplementation with specific nutrients or must exhibit a functional interaction that activates reporter gene transcription. The reporter gene readout is based on the complementation of the growth defect. Additionally, an α-galactosidase reporter gene is employed, which causes a color change to blue-green upon cleavage of 5-bromo-4-chloro-3-indolyl α-d-galactopyranoside (X-α-gal) in the medium. Further investigation will be limited to those interactors that successfully passed all stringent confirmation steps.

3.1.1 Cloning Bait and Prey

1. The coding sequences of each of the effector candidates (Open reading Frame), lacking signal peptides and stop codon, should be PCR amplified using cDNA of the organism of interest.

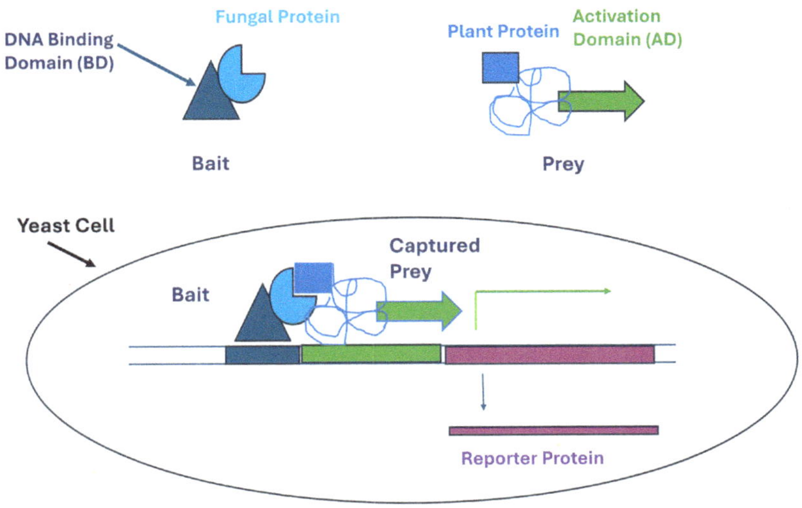

Fig. 3 Overview of Y2H method. In this example, a fungal protein, putatively an effector is expressed as a fusion protein with the Gal4p DNA binding domain (BD); similarly, a host plant protein is expressed as a fusion protein with the Gal4p transcriptional activation domain (AD). When both plasmids are found in the same S. cerevisiae cell, if there is an interaction between the putative effector and the host protein, then the complex can bind upstream of reporter genes to activate their expression, serving as a readout of the interaction

2. Clone the above PCR product into pCR 2.1 TOPO entry vector Invitrogen/Thermo Fisher).
3. Transform the *Escherichia coli* strains (e.g., DH5α) with the above cloned vector with appropriate antibiotic marker for selection.
4. Pick the colonies and re-streak them onto the appropriate antibiotic plates (e.g., ampicillin, 200 μg/ml) overnight at 37°C for growth
5. Culture the resulting bacterial colonies in liquid broth with appropriate antibiotic marker for selection.
6. Isolate the plasmid DNA using a commercial Plasmid DNA isolation kit or a similar alkaline lysis protocol [31].
7. Perform Restriction digestion of the cloned PCR product with appropriate Restriction Enzymes so as to keep the bait and prey coding regions in-frame with the start codon of the gene containing the multiple cloning site for the respective vectors, e.g., the *ADH1* gene.
8. Separate the resulting DNA fragments from the above mix via agarose gel electrophoresis and purify the desired fragment DNA fragment of interest of the appropriate size.
9. Clone the above purified fragment (Open reading frame) into the pGBKT7 destination vector (pGBKT7-GOIΔSP) in frame with GAL4-BD, using appropriate antibiotic selection

plates (50 μg kanamycin per ml agar) for transformation of *E. coli*. (*see* **Note 1**)

10. Incubate overnight at 37°C and examine plates for transformed colonies.
11. Culture the colonies on antibiotic plates to have a stock and cryopreserve the stock in YPD/10% glycerol in a −80°C freezer.
12. Isolate the plasmid DNA from liquid broth cultures using Plasmid DNA isolation kit or similar in-house protocol.
13. Confirm the presence of plasmid and gene through PCR using gene-specific primers and sequencing.
14. As an alternative to the use of restriction enzyme-digested plasmid vectors for cloning into bait vector, Gibson Assembly may be used (New England Biolabs (NEB) sells DNA Assembly kits and master mixes, including NEBuilder HiFi and Gibson Assembly. Gibson Assembly® is licensed to New England Biolabs by TelesisBio; [33]). In this case, primers for amplification of the desired gene of interest (GOI) should contain sequence overlaps with the desired positions in the pGBKT7 vector. Similarly, primers can be designed to amplify the portion of the pGBKT7 vector that will be joined in frame with the GOI.
15. For preparation of normalized cDNA libraries in the prey vector, a cDNA pool should be prepared as indicated in the following example:

RNA for generating the cDNA library should be obtained from axenically grown cultures of haploid strains on nutrient-rich media for 5 days (YPD or YPS or PDA agar media) at 28°C and from the fungal-infected host plant tissue. This latter set of RNAs should then be extracted from desired infected plant tissue stages. The quality of the RNA is then checked by Agilent Bioanalyzer and all the samples indicating highly intact RNA with the RNA integrity scores of at least 7.8 can be used. The total samples should then be pooled equally based on Bioanalyzer quantification to generate a normalized cDNA library. The cDNA library was constructed in a Gal4-based prey vector, pGADT7 (Takara Bio), in a fashion similar to that described above for cloning into bait vector, or the RNAs can be sent to a suitable company (e.g., CD Genomics, Shirley, NY, USA) to generate the cDNA library for later yeast two-hybrid screening. We personally recommend the latter option to minimize variation in library preparation between investigators in one's laboratory.

3.1.2 Transformation of Yeast Strains

1. Retrieve the frozen vials of *S. cerevisiae* library host strain (Y187) and Mating partner strain (AH109) from −80°C freezer and culture them on YPD plates.
2. Allow them to grow at 30°C for 2–3 days.

3. Prepare the SD dropout agar medium plates lacking Leucine (SD/-Leu) and containing chromogenic substrate, X-α-gal.
4. Transform the Y187 with the prey vector (pGADT7-cDNA library interacting partners) and positive control (pGADT7-T) using Frozen-EZ Yeast Transformation II kit (Zymo Research).
5. Calculate the cell density and based on the cell density of the suspension, perform different dilutions to be plated. For example, 10^{-4}, 10^{-5}, 10^{-6}, and 10^{-7} dilutions.
6. Plate the above culture dilutions onto the SD/-Leu + X-α -gal agar plates.
7. Test the bait (autoactivation/toxicity): Transform the AH109 with the bait vector fused to the effector protein (pGBKT7-GOIΔSP) and, separately, positive control (pGBKT7-53) using lithium acetate/single-stranded carrier DNA/PEG method.
8. Plate 100 μl of each type, 1/10 and 1/100 dilutions onto: SD/-Trp/X-α-Gal and TDO/X-α-Gal (SD/-Ade/-His/-Trp) (*see* **Note 2**).
9. Allow the colonies to grow at 30°C for 3–5 days.
10. Confirm the presence of plasmid and gene through PCR using gene-specific primers and sequencing.

3.1.3 Yeast Two Hybrid Screening

1. Perform control mating (strongly recommended).
2. Before screening, mate the positive control strains (AH109[pGBKT7-53] × Y187[pGADT7-T]) and verify expected growth/blue color on DDO/X-α-Gal and QDO/X-α- Gal. These controls validate media performance and selection stringency.
3. Prepare the bait culture.
4. Inoculate a fresh AH109[pGBKT7-bait] colony into 50 ml SD/-Trp. Shake at 250–270 rpm at 30°C to an Absorbance at 600 nm (A_{600}) ≈ 0.8 (16–20 h).
5. Pellet at 1000 × *g* for 5 min and resuspend to >1 × 10^8 cells/ml in 4–5 ml SD/-Trp (count with a hemocytometer).
6. Prepare the library and set up mating.
7. Thaw 1 ml of the Y187 library vial (should contain >2 × 10^7 cells) at room temp. Keep 10 μl for titering on SD/-Leu.
8. In a sterile 2 L flask, combine 1 ml library + 5 ml bait and add 45 ml 2× YPDA + kanamycin (50 μg/ml). Rinse the emptied library vial twice with 1 ml 2× YPDA and add rinses to the flask.

9. Incubate 20–24 h at 30°C with slow shaking at 30–50 rpm. Use the lowest rpm that prevents settling (*see* **Note 3**).
10. Wash and concentrate mated cells.
11. Following mating, pellet the culture by centrifugation at 1000 × *g* for 10 min. Rinse the flask twice with 0.5× YPDA supplemented with kanamycin, pooling the washes to resuspend the pellet.
12. Centrifuge again under the same conditions, then resuspend the final pellet in 10 ml of 0.5× YPDA/Kanamycin.
13. Record the total resuspension volume, including cells and medium (e.g., 11.5 ml).
14. Plate for titering and selection.
15. Titer/coverage plates: From the mated culture, spread 100 μ l of 1/10, 1/100, 1/1000, 1/10,000 dilutions onto SD/-Trp, SD/- Leu, and DDO (SD/-Leu/-Trp) 100 mm plates.
16. Incubate 30°C, 3–5 days. These counts are used to calculate how many diploids are screened.
17. High-stringency selection: Plate the remainder of the mated culture at 200 μl per 150 mm plate onto QDO/X-α-Gal+ 5 mM 3AT (*see* **Note 4**).
18. Incubate 30°C, 3–8 days. Ignore tiny pale colonies that never exceed ~1 mm; true Ade^+/His^+ interactors are robust (> 2 mm) and blue.
19. Calculate library coverage (number of screened diploids).
20. From DDO plate counts after 3–5 days, calculate the following:
    - Screened diploids = (cfu/ml of diploids) × resuspension volume (ml).
    - Example: 70 colonies on the 1/1000 DDO plate with a resuspension volume of 11.5 ml ⇒ 8.05×10^6 screened. Screen $\geq 1 \times 10^6$ diploids to ensure adequate coverage of the library depth.
21. Pick and retest positives (see Fig. 4 for examples).
22. Pick blue colonies from QDO/X-α-Gal+5 mM 3AT and restreak to single colonies onfresh QDO/X-α-Gal+5 mM 3AT to confirm the $Ade^+/His^+/Mel1^+$ phenotype (see Fig. 4).
23. Compare color to the positive and negative controls grown on DDO/X-α-Gal (see Fig. 5). Expect stable blue growth within 2–4 days.
24. Restreak the above transformed colonies onto QDO/X-α-gal plates containing 50 mM 3AT to select strong interactors (*see* **Note 5**).

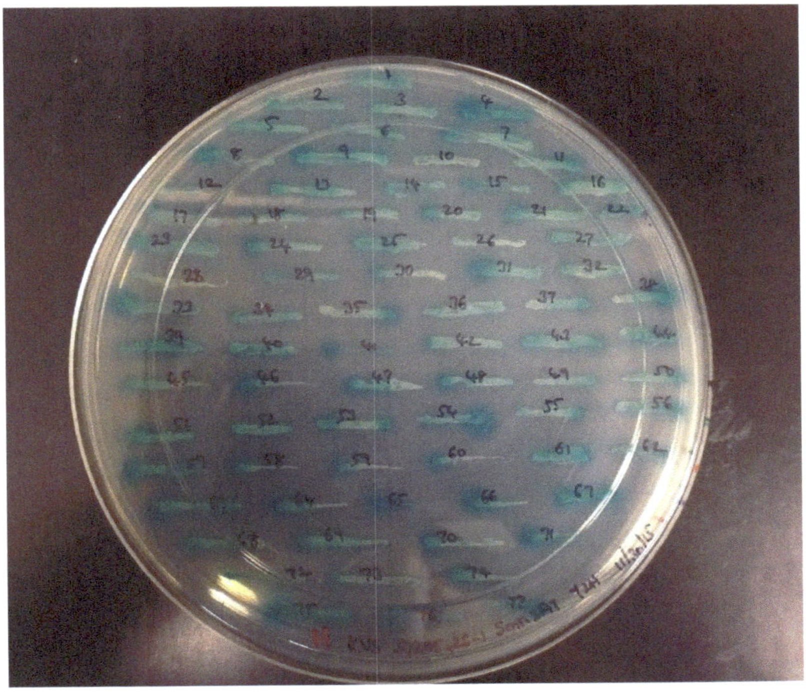

Fig. 4 Picked and re-purified colonies from Y2H initial screening. Colonies were re-purified onto QDO plates

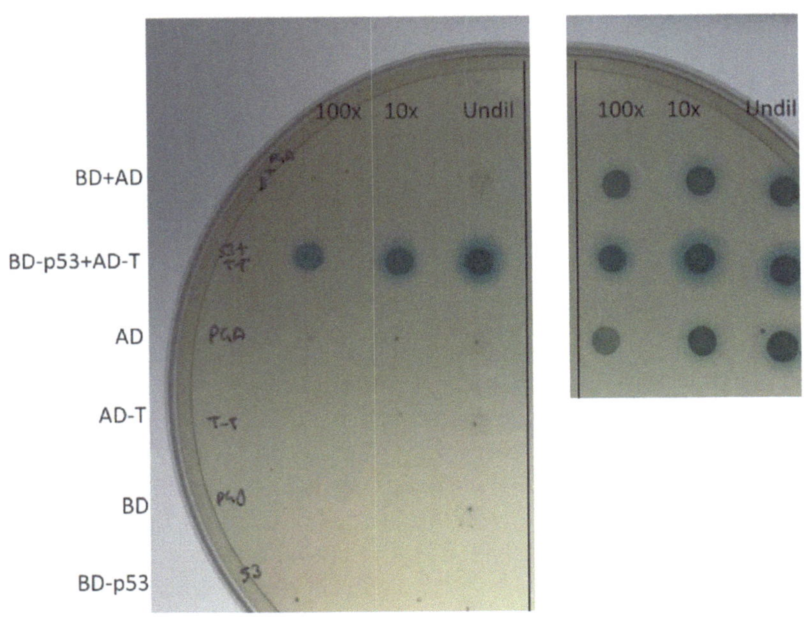

Fig. 5 Comparison of isolated positive colonies from Y2H (right panel) with controls (left panel)

25. Screen only the positive blue-green clones (indicating α -galactosidase activity) that survived on the highest 3AT levels that were used for further screening.
26. Yeast colony PCR to remove duplicates (Optional but recommended).
27. Confirm the positives using yeast colony PCR and analyze products on agarose gels.
28. Rescue the prey plasmid and identify the insert.
29. Segregate plasmids: Purify each QDO-positive diploid by restreaking onto SD/-Leu and SD/-Trp to select for AD (LEU2) and BD (TRP1) plasmids separately.
30. After growth, restreak the LEU2 clones to ensure purity. Then retest segregated LEU2 clones on QDO/X-α-Gal in the presence of the original bait to confirm they carry the interacting prey.
31. Rescue and sequence: Isolate prey and bait plasmid DNA from yeast after transforming into *E. coli* (e.g., DH5α on ampicillin or kanamycin). Prepare plasmid DNA and sequence to identify t cDNA and the bait protein.
32. Verify genuine interactions by co-transformation of yeast with the bait plasmid and the isolated prey plasmid (For more details on such "Spot" testing, see below).
33. Co-transform AH109 with the below and score after 3–5 days at 30°C.

    - (a) GBKT7-bait + candidate prey (pGADT7- Rec/RecAB) and
    - (b) empty pGBKT7 + the same candidate prey. Plate 1/10 and 1/100 dilutions on DDO/X-α-Gal and QDO/X-α -Gal. Genuine positive: (a) grows blue on DDO and QDO; (b) is white on DDO and absent on QDO. False positive: (a) also grows/blue on QDO.

3.2 Y2H Spot Tests

After a high-throughput Y2H library screening process, Y2H spot tests can be used to confirm and validate the protein–protein interactions. While library screens can identify novel interactions, they often require systematic retesting of potential hits. This spot test technique involves direct examination of specific protein pairs for interactions [34, 35]. In the Y2H spot test, defined prey proteins are systematically tested against a bait protein of interest [35].

In a typical Y2H spot test, two proteins of interest are separately cloned into a plasmid with a DNA-binding (BD) and transcriptional activation (AD) domain of a transcription factor, such as Gal4p, similar to what was described above, except that instead of screening a cDNA library in the prey vector, specific individual prey proteins are screened. In the *S. cerevisiae* nucleus, when these two proteins are fused, they reconstitute a functional transcription factor that triggers the expression of reporter genes [36]. The interaction is then visualized by spotting the yeast cells containing both fusion proteins onto selective media and observing growth or color development.

In Y2H interaction assays on selective dropout plates containing α-X-gal and 3-AT, positive interactions would typically result in blue-colored colonies. This color change occurs due to activation of the *MEL1* gene, a *GAL4*-regulated gene used in Y2H screening. *MEL1* encodes the secreted enzyme α-galactosidase reporter gene. When a positive interaction occurs between the bait and prey proteins in the Y2H system, it leads to the activation of reporter genes, including *MEL1*. The α-galactosidase enzyme cleaves α-X-gal resulting in the formation of a blue precipitate [37]. This blue color serves as a visual indicator of protein–protein interactions.

3.2.1 Gene Cloning into Vector

1. The cDNAs that encode the bait and prey proteins should be amplified using a specific set of primers and cloned into the appropriate vector following standard molecular cloning procedures. As indicated above, instead of using restriction endonuclease digestion, we can also design primers with overhangs complementary to the gene of interest (GOI) to amplify the vectors pGBKT7 and pGADT7 or pGADGH, using specific primers under defined high fidelity PCR conditions (using proofreading enzymes such as PrimeStar Max, Takara Bio; or Q5 High Fidelity DNA Polymerase, New England Biolobs), followed by the Gibson assembly method to clone the gene of interest into the plasmid [10, 33, 38].
2. To perform *E. coli* transformation, the ligated mixture of the respective plasmids and genes of interest must be transformed into chemically competent *E. coli* cells using a standard heat-shock transformation protocol [39], or electroporation can be used with electrocompetent cells. Next, select the transformants from LB kanamycin and LB ampicillin agar plates. Following a 24-h incubation period, individual colonies need to be picked and subjected to colony PCR to confirm the successful incorporation of the desired insert [40].

3.2.2 Confirmation Through PCR and Sequencing

1. *E. coli* transformants exhibiting positive results should be utilized to inoculate liquid cultures for plasmid minipreparation.

   The presence of the correct insert in the plasmids must be verified through PCR amplification and DNA sequencing.

3.2.3 Yeast Competent Cells and Transformation

In this study, *S. cerevisiae* strains AH109 and Y187 are used as host strains for the two-hybrid assay. Y187 (MATα genotype) and AH109 (MATa genotype) cells are used as mating partners in Y2H systems because of the fundamental mating process, whereby two alternative mating types, MATa and MATα, regulate mating [41].

For this, we have to prepare two types of competent cells for each strain.

1. First, streak out AH109 and Y187 onto a YPD plate for 24–48 h and incubate at 28–30°C.
2. Inoculate a colony of the yeast strains from the plate into liquid YPD media.
3. Check absorbance at 600 nm (A_{600}) until it reaches 0.8 (approximately 5–6 h). Then competent cells can be prepared using by following the instructions using Frozen-EZ transformation II kit (Zymo research).
4. Next, once the competent cells are ready, add 5 μL (typically, 1–2.5 μg) of the bait plasmid with or without cloned bait cDNA into AH109 competent cell. Likewise, add 5 μL (1–2.5 μg) of the prey plasmid with or without cloned prey cDNA into Y187 competent cells using the solution from Frozen-EZ transformation II kit (Zymo research).
5. Incubate the mixture at 30°C for 30–45 min.
6. After every 15 min, flick the tube.
7. Following the transformation process, the cells are cultured on selective media appropriate for the plasmid-borne selection markers. Specifically, SD-Trp or SD-Leu agar plates are used, as described [42].
8. Wait for 2–4 days for colonies to appear on the plate.
9. Confirm the presence of plasmid and gene through PCRusing gene-specific primers and sequencing.

3.3 Yeast Two-Hybrid Spot Test

1. After confirming the presence of the correct plasmids in the yeast strains, the next step is to perform the spot assay to confirm the interactions between two proteins of interest.
   The controls used previously are very important and need to be included in this co-transformation study, as well.
   Experimental group:
   AH109 with bait plasmid (pGBKT7) with cloned bait protein co-transformed with Y187 with prey plasmid(pGADGH) cloned with prey protein.
2. Inoculate a colony of the yeast strains from the stock working plate into a 200–250 ml flask containing 50 ml of YPDA medium.
3. Place the flasks on the shaker (200 rpm) and incubate at 30°C.
4. Check A_{600} until it reaches 0.6–0.8 (approximately 4–6 h).
5. To co-transform, add 10 μL of each opposite mating type strain in 20 mL of YPD media in a 50 mL flask, allowing the two opposite mating types to fuse and form diploids. This may be checked microscopically.

6. Incubate the inoculated 50 mL flask overnight at 30°C in a shaker at 200 rpm.
7. After overnight incubation, take 2 ml of the culture and centrifuge at 500 × *g* for 4 min.
8. Discard the supernatant and take another 2 ml of overnight culture and repeat the process.
9. Next, the pellet should be resuspended with 2 ml of sterile water.
10. Measure the A_{600} of cell suspension (It will be usually be more than 1.0 after overnight incubation).
11. Next, adjust the cell density of all the 2 ml tubes of stock solutions to A_{600} of 1.0.
12. Then, make tenfold serial dilutions with stock solution/sterile water [10–1, 10–2].
13. Lastly, apply 10 μL aliquots of each dilution onto QDO-SD–Ade, Histidine, tryptophan, leucine, placing samples from the undiluted, 10^{-1}, and 10^{-2} tubes in a linear arrangement.
14. The plates need to be incubated at 30°C for 3–4 days and observed for growth.

Positive interactions of bait and prey proteins would result in blue-colored colonies. However, the intensity of the blue color can vary depending on the strength of the interaction and the level of reporter gene expression. Stronger interactions may result in darker blue colonies, while weaker interactions might produce lighter blue colonies [43]. The presence of 3-AT in the dropout plates helps reduce background growth due to leaky HIS3 gene expression. It is critical for selecting against false positives in two-hybrid assays, and it improves the selection stringency, potentially leading to more reliable results [44].

4 Notes

Y2H can be a very effective first step in identifying potential protein–protein interactions for *U. maydis* and for other fungi, either for interactions with endogenous proteins or with host target proteins. However, since the procedure is prone to false positives, appropriate controls must be provided, including not only those listed above, but also reciprocal experiments where the original bait and prey proteins have their expression switched from the respective vectors, i.e., from the bait vector to the prey vector and vice versa. Once convincing interactions have been identified using this method, then more definitive, but labor-intensive experiments, such as in vivo Co-IP, may be applied towards further testing of a select number of candidate interactors.

1. Upon expression, the fusion protein has the DNA-binding domain of the Gal 4p transcription factor at its N-terminus.
2. Vigorous shaking reduces mating efficiency. Check microscopically for zygotes (typical three-lobed morphology). If absent at 20 h, continue up to 4 h more.
3. In the absence of an interacting partner, bait-expressing colonies should remain white on SD/-Trp/X-α-Gal plates and fail to grow on TDO/X-α-Gal. By contrast, the positive control (AH109[pGBKT7-53]) exhibits robust blue coloration on QDO/X-α-Gal. If the bait construct displays autoactivation under these conditions, this must be corrected prior to proceeding with the library screen. Alternatively, this may suggest that the bait protein acts as a transcription factor.
4. High-stringency QDO/X-α-Gal+5 mM 3AT minimizes false positives by requiring activation of HIS3 and ADE2 and reporting MEL1/LacZ along with overcoming 3AT inhibition.
5. 3AT was used to inhibit the leaky expression that reduces the effectiveness of histidine selection, and to inhibit X-α-gal to allow detection of the MEL1/*lacz* reporter. 3-AT is a competitive inhibitor of the yeast HIS3 protein (His3p), blocking low levels of His3p expression, and thus suppressing background growth on SD medium lacking His.
6. Notes on stringency and troubleshooting

   If you screened $>1 \times 10^6$ diploids and obtained few/no positives on QDO/X-α-Gal, confirm media with controls and consider medium stringency (TDO/X-α-Gal) and, if needed, optimize 3-AT concentration using increasing concentrations to empirically determine the amount needed to reduce or eliminate false positives due to the leaky HIS3 allele [30,38,45].

   If you see too many positives, first ensure your bait does not autoactivate, check QDO media with controls, then sort duplicates by colony PCR before verification. Using a Normalized Library can also help.
7. The following controls (shown in Table 1 and Fig. 5) are very important and need to be included in this experiment, as well as in co-transformation studies (see below) [30,38,44,45]. The combinations of the mating type strains are as follows:

   1. Negative controls:

      Only AH109 containing bait plasmid pGBKT7 empty (without cloned bait cDNA).

      Only Y187 containing prey plasmid pGADGH empty (without cloned prey cDNA; Fig. 2).

      Co-transformation of AH109 pGBKT7empty and Y187pGADGH empty.

      Co-transformation of AH109 pGBKT7 with cloned bait cDNA and Y187 containing prey plasmid pGADGH empty.

Table 1
Controls used in Y2H assay

Survival on selective agar plate						
Construct	**SD/Trp/**		**SD/Trp**		**SD/Trp/Leu/Ade/**	
	SD/Trp	**Leu**	**X-α-Gal**	**/Trp/His**	**SD/Trp/ His/Leu**	**His/X-α -Gal**
pGBKT7-Bait	☺		White	☹	☹	☹
pGBKT7 & pGADT7	☺	☺	White	☹	☹	☹
pGBKT-53 & pGADT7-T	☺	☺	☺	☺	☺	☺

☺, growth; ☹, no growth

Co-transformation of AH109 pGBKT7 empty and Y187 containing prey plasmid pGADGH [10, 42].
The negative controls should not show any growth on the selective media, indicating that the proteins do not interact in the absence of the bait and prey plasmids [10].

2. Positive control:
Co-transformation of any previously known plasmids containing the bait cDNA and another with the prey cDNA (e.g., pGBKT7-53 and pGADT7-T, respectively). These plasmids encode proteins known to interact, allowing researchers to verify that the entire experimental process has been executed properly.

Acknowledgments

This work was supported, in part, by Graduate Assistantships to SB from the University of Louisville and by National Science Foundation (NSF) Award# 2419465, and USDA National Institute of Food and Agriculture (NIFA) Award#2024-67014-43676 to MHP.

References

1. Arévalo-Pinzón G, González-González M, Suárez CF et al. (2018) Self-assembling functional programmable protein array for studying protein-protein interactions in malaria parasites. Malar J 17(1):270. https://doi.org/10.1186/s12936-018-2414-2. PMID: 30016987; PMCID: PMC6050706
2. Christian, N, Perlin, MH (2024) Plant-endophyte communication: Scaling from molecular mechanisms to ecological outcomes. Mycologia 116:227. https://doi.org/10.1080/00275514.2023.2299658
3. Perlin MH, Poulin R, de Bekker C (2025) Invasion of the four kingdoms: the parasite journey across plant and non-plant hosts. Biol Rev Camb Philos Soc 100(2):936–968. https://doi.org/10.1111/brv.13169. Epub 2024 Dec 1. PMID: 39616537
4. Guo J, Guo S, Lu S et al. (2023) The development of proximity labeling technology and its applications in mammals, plants, and microorganisms. Cell Commun Signal 21(1):269. https://doi.org/10.1186/s12964-023-01310-1 Erratum in: Cell Commun Signal.

2024;22(1):403. 10.1186/s12964-024-01750-3. PMID: 37777761; PMCID: PMC10544124
5. Lyu S, Zhang C, Hou X et al. (2022) Tag-based pull-down assay. Methods Mol Biol 2400:105–114. https://doi.org/10.1007/978-1-0716-1835-6_11. PMID: 34905195
6. Soyemi J, Isewon I, Oyelade J et al. (2018) Inter-Species/Host-Parasite Protein Interaction Predictions Reviewed. Curr Bioinform 13(4):396–406. https://doi.org/10.2174/1574893613666180108155851. PMID: 31496926; PMCID: PMC6691774
7. Swapna LS, Stevens GC, Sardinha-Silva A et al. (2024) ToxoNet a high confidence map of protein-protein interactions in Toxoplasma gondii. PLoS Comput Biol 20(6):e1012208. https://doi.org/10.1371/journal.pcbi.1012208. PMID: 38900844; PMCID: PMC11219001
8. Walch P, Selkrig J, Knodler LA et al. (2021) Global mapping of Salmonella enterica-host protein-protein interactions during infection. Cell Host Microbe 29(8):1316–1332.e12. https://doi.org/10.1016/j.chom.2021.06.004. Epub 2021 Jul 7. PMID: 34237247; PMCID: PMC8561747
9. Makuch L (2014) Yeast two-hybrid screen. Methods Enzymol 539:31–51. https://doi.org/10.1016/B978-0-12-420120-0.00003-7. PMID: 24581437
10. Paiano A, Margiotta A, De Luca M et al. (2019) Yeast two-hybrid assay to identify interacting proteins. Curr Protoc Protein Sci 95(1):e70. https://doi.org/10.1002/cpps.70. Epub 2018 Aug 21. PMID: 30133175
11. Lin JS, Ali J, Lai EM (2024) Protein-protein interactions: co-immunoprecipitation. Methods Mol Biol 2715:273–283. https://doi.org/10.1007/978-1-0716-3445-5_18. PMID: 37930535
12. Smits AH, Vermeulen M (2016) Characterizing protein-protein interactions using mass spectrometry: challenges and opportunities. Trends Biotechnol 34(10):825–834. https://doi.org/10.1016/j.tibtech.2016.02.014. Epub 2016 Mar 17. PMID: 26996615
13. de Jong L, Roseboom W, Kramer G (2021) A composite filter for low FDR of protein-protein interactions detected by in vivo cross-linking. J Proteomics 230:103987. https://doi.org/10.1016/j.jprot.2020.103987. Epub 2020 Sep 17. PMID: 32949815
14. Dutta S, Smith MD (2023) Detection of protein-protein interactions utilizing the split-ubiquitin membrane-based yeast two-hybrid system. Methods Mol Biol 2690:37–57. https://doi.org/10.1007/978-1-0716-3327-4_4. PMID: 37450135
15. Paul JA, Barati MT, Cooper M et al. (2014) Physical and genetic interaction between ammonium transporters and the signaling protein Rho1 in the plant pathogen Ustilago maydis. Eukaryot Cell 13(10):1328–1336. https://doi.org/10.1128/EC.00150-14. Epub 2014 Aug 15. PMID: 25128189; PMCID: PMC4187648
16. Denney AS, Weems AD, McMurray MA (2021) Selective functional inhibition of a tumor-derived p53 mutant by cytosolic chaperones identified using split-YFP in budding yeast. G3 (Bethesda) 11(9):jkab230. https://doi.org/10.1093/g3journal/jkab230.
17. Kudla J, Bock R (2016) Lighting the way to protein-protein interactions: recommendations on best practices for bimolecular fluorescence complementation analyses. Plant Cell 28(5):1002–1008. https://doi.org/10.1105/tpc.16.00043. Epub 2016 Apr 20PMID: 27099259; PMCID: PMC4904677
18. Lo Presti L, Lanver D, Schweizer G et al. (2015) Fungal effectors and plant susceptibility. Annu Rev Plant Biol 66:513–545. https://doi.org/10.1146/annurev-arplant-043014-114623. PMID: 25923844
19. Pradhan A, Ghosh S, Sahoo D et al. (2021) Fungal effectors, the double edge sword of phytopathogens. Curr Genet 67:27–40
20. Pham CD, Yu Z, Sandrock B et al. (2009) Ustilago maydis Rho1 and 14-3-3 homologues participate in pathways controlling cell separation and cell polarity. Eukaryot Cell 8(7):977–989. https://doi.org/10.1128/EC.00009-09. Epub 2009 May 1. PMID: 19411618; PMCID: PMC2708450
21. Pham CD, Perlin MH (2010) Possible additional roles in mating for ustilago maydis rho1 and 14-3-3 homologues. Commun Integr Biol 3(1):57–59. https://doi.org/10.4161/cib.3.1.9864
22. Mayorga ME, Gold SE (2001) The ubc2 gene of Ustilago maydis encodes a putative novel adaptor protein required for filamentous growth, pheromone response and virulence. Mol Microbiol 41(6):1365–1379. https://doi.org/10.1046/j.1365-2958.2001.02606.x. PMID: 11580841
23. Lovely CB, Aulakh KB, Perlin MH (2011) Role of Hsl7 in morphology and pathogenicity and its interaction with other signaling components in the plant pathogen Ustilago maydis. Eukaryot Cell 10(7):869–883. https://doi.org/10.1128/EC.00237-10. Epub 2011 May 27PMID: 21622903; PMCID: PMC3147425
24. Weinzierl G, Leveleki L, Hassel A et al. (2002) Regulation of cell separation in the dimorphic fungus Ustilago maydis. Mol Microbiol 45(1):219–231. https://doi.org/10.1046/j.1365-2958.2002.03010.x. PMID: 12100561
25. Mueller AN, Ziemann S, Treitschke S et al. (2013) Compatibility in the Ustilago

maydis-maize interaction requires inhibition of host cysteine proteases by the fungal effector Pit2. PLoS Pathog 9(2):e1003177. https://doi.org/10.1371/journal.ppat.1003177. Epub 2013 Feb 14. PMID: 23459172; PMCID: PMC3573112
26. Tanaka S, Brefort T, Neidig N et al. (2014) A secreted Ustilago maydis effector promotes virulence by targeting anthocyanin biosynthesis in maize. eLife 3:e01355. https://doi.org/10.7554/eLife.01355. Epub 2014 Jan 28. PMID: 24473076; PMCID: PMC3904489
27. Zuo W, Ökmen B, Depotter JRL et al. (2019) Molecular interactions between smut fungi and their host plants. Annu Rev Phytopathol 57:411–430. https://doi.org/10.1146/annurev-phyto-082718-100139. Epub 2019 Jul 23. PMID: 31337276
28. Zuo W, Depotter JRL, Stolze SC et al. (2023) A transcriptional activator effector of Ustilago maydis regulates hyperplasia in maize during pathogen-induced tumor formation. Nat Commun 14(1):6722. https://doi.org/10.1038/s41467-023-42522-w. PMID: 37872143; PMCID: PMC10593772
29. Lee SJ, Rose JK (2012) A yeast secretion trap assay for identification of secreted proteins from eukaryotic phytopathogens and their plant hosts. Methods Mol Biol 835:519–530
30. Tsai MC, Barati MT, Kuppireddy VS et al. (2024) Characterization of *Microbotryum lychnidis-dioicae* secreted effecto r proteins, their potential host targets, and localization in a heterologous host plant. J Fungi (Basel) 10(4):262. https://doi.org/10.3390/jof10040262
31. Birnboim HC, Doly J (1979) A rapid alkaline extraction procedure for screening recombinant plasmid DNA. Nucleic Acids Res 7(6):1513–1523. https://doi.org/10.1093/nar/7.6.1513
32. Gietz RD, Woods RA (2002) Transformation of yeast by lithium acetate/single-stranded carrier DNA/polyethylene glycol method. Methods Enzymol 350:87–96
33. Gibson DG, Young L, Chuang R-Y et al. (2009) Enzymatic assembly of DNA molecules up to several hundred kilobases. Nat Methods 6(5):343. https://doi.org/10.1038/nmeth.1318
34. Koegl M, Uetz P (2008) Improving yeast two-hybrid screening systems. Brief Funct Genomic Proteomic 6(4):302–312. https://doi.org/10.1093/bfgp/elm035
35. Mehla J, Uetz P, Caufield JH (2015) Mapping protein-protein interactions using yeast two-hybrid assays. Cold Spring Harbor Protocols 2015(5):pdb.prot086157. https://doi.org/10.1101/pdb.prot086157
36. Miller J, Stagljar I (2004) Using the yeast two-hybrid system to identify interacting proteins. Methods Mol Biol (Clifton, N.J.) 261:247–262. https://doi.org/10.1385/1-59259-762-9:247
37. Green MR, Sambrook J (2019) Screening bacterial colonies using X-Gal and IPTG: α -complementation. Cold Spring Harb Protoc 2019(12):pdb.prot101329. https://doi.org/10.1101/pdb.prot101329
38. Soellick T-R, Uhrig JF (2001) Development of an optimized interaction-mating protocol for large-scale yeast two-hybrid analyses. Genome Biol 2(12):RESEARCH0052. https://doi.org/10.1186/gb-2001-2-12-research0052
39. Froger A, Hall JE (2007) Transformation of plasmid DNA into E. coli using the heat shock method. J Vis Exp 6:e253
40. Cao M, Pan J, Fu Y et al. (2009) Chlamydomonas (Chlorophyceae) colony PCR. Protoplasma 235:107–110. https://doi.org/10.1007/s00709-009-0036-9
41. Lee C-S, Haber JE (2015) Mating-type gene switching in Saccharomyces cerevisiae. Microbiol Spec 3(2):MDNA3-0013-2014. https://doi.org/10.1128/microbiolspec.mdna3-0013-2014
42. Cloix C, Kaiserli E, Heilmann M et al. (2012) C-terminal region of the UV-B photoreceptor UVR8 initiates signaling through interaction with the COP1 protein. Proc Nat AcadSci USA 109:16366. https://doi.org/10.1073/pnas.1210898109
43. Naz RK, Dhandapani L (2010) Identification of human sperm proteins that interact with human zona pellucida3 (zp3) using yeast two-hybrid system. J Reprod Immunol 84:24–31. https://doi.org/10.1016/j.jri.2009.10.006
44. Uetz P, Giot L, Cagney G et al. (2000) A comprehensive analysis of protein–protein interactions in Saccharomyces cerevisiae. Nature 403:623–627. https://doi.org/10.1038/35001009
45. Kuppireddy VS, Uversky VN, Toh SS et al. (2017) Identification and initial characterization of the effectors of an anther smut fungus and potential host target proteins. Int J Mol Sci 18(11):2489. https://doi.org/10.3390/ijms18112489

Chapter 26

TurboID-Based Proximity Labeling for Exploiting Effector Interacting Proteins in the Fungal Maize Pathogen *Ustilago maydis*

Wei Shi, Hirofumi Nakagami, Isabel M.L. Saur and Gunther Doehlemann

Abstract

TurboID-based proximity labeling (PL) is an advanced method for identifying proteins in the immediate molecular environment of a protein of interest under physiological conditions. TurboID enables the rapid and covalent biotinylation of proteins in proximity to the bait proteins in living cells, capturing both stable and transient interactions with high temporal resolution. Most plant pathogens secrete effectors that modulate host immunity and facilitate infection. The effector-targeted host proteins are commonly involved in plant resistance or susceptibility. Elucidating these proteins is meaningful for crop improvement and design. The corn smut fungus *Ustilago maydis* causes tumor formation on all aerial parts of maize. The *U. maydis*-maize pathosystem is a model system for studying biotrophic fungi and their interactions with plants. To identify *U. maydis* effector targets in maize, a TurboID-based PL approach was developed to explore effector interactors in the *U. maydis*-maize pathosystem. This approach serves as a versatile tool for obtaining interactomes of effector baits in host tissues.

Keywords *Ustilago maydis*, Maize, TurboID, Fungal effectors, Plant–pathogen interaction

1 Introduction

Maize (*Zea mays*) is one of the most important crops in the world, not only for production but also for biofuel production. The basidiomycete smut fungus *Ustilago maydis* infects all aerial parts of maize and causes tumor formation, thereby limiting the yield and quality of maize [1]. *U. maydis* secretes effectors that modulate the biological activity of host proteins to promote infection [2]. However, only a few host targets of *U. maydis* effectors have been functionally characterized. Identifying these targets is essential for understanding pathogenic mechanisms and crop resistance.

Conventional methods for identifying protein–protein interactions include yeast two-hybrid (Y2H) library screening and affinity purification followed by mass spectrometry (AP-MS) [3, 4]. In the

An N. T. Phan et al. (eds.), *Ustilago maydis*, Methods in Molecular Biology, vol. 3050,
https://doi.org/10.1007/978-1-0716-5364-7_26,

Y2H system, interaction detection relies on the reconstitution of the transcription factor GAL4, where the DNA-binding domain (BD) is fused to one protein and the activation domain (AD) to another [5]. The interaction between the two proteins brings the BD and AD together, activating transcription of a reporter gene. However, this assay relies on artificial localization and may fail to detect interactions when proteins misfold or mislocalize in yeast. Additionally, as a heterologous expression system, certain proteins can independently activate transcription or engage in non-specific, low-affinity interactions, potentially leading to false-positive and false-negative results.

Affinity purification methods such as co-immunoprecipitation (Co-IP) are effective for isolating stable protein complexes, but they often fail to capture weak or transient interactions. Moreover, these techniques typically require cell lysis under non-physiological conditions and depend heavily on binding affinity, which can bias recovery toward abundant or tightly bound partners and may lead to artifacts or loss of compartment-specific interactions [6].

To overcome these limitations, enzyme-catalyzed proximity labeling (PL) techniques have emerged as powerful alternatives for mapping protein interaction networks. These methods enable in vivo labeling of proteins in close spatial proximity to a bait protein under near-physiological conditions, improving the detection of dynamic, low-affinity, or spatially restricted interactors.

PL captures weak or transient protein interactions by biotinylating neighboring proteins. The enzymes used for proximity labeling are engineered ascorbate peroxidase (APEX) and *Escherichia coli* biotin ligase BirA [7, 8]. The APEX-based proximity labeling relies on the toxic H_2O_2 for rapid labeling. The utilization of H_2O_2 and high endogenous peroxidase activity limits its applicability. BioID uses a point-mutated form of BirA (R118G), which generates reactive biotin intermediates that covalently bind to exposed lysines on nearby proteins. This process requires extended incubation times and elevated temperatures. The 35 kDa TurboID has 15 mutations of BirA resulting in high catalytic efficiency at lower temperatures, which makes it particularly suitable for rapid biotin labeling using significantly less biotin than BioID in plant systems [9]. The binding of biotinylated proteins to streptavidin for pull-down and further MS analysis does not depend on their native conformation. This allows extraction and binding under denaturing conditions, including the use of high concentrations of detergents, facilitating isolating and identifying membrane proteins [10]. The TurboID-based PL approach has been successfully applied in yeast, plants, and animals [11]. Considering the high proximity labeling activity of TurboID, TurboID-based PL may be an efficient tool to identify specific fungal effector-host protein interactions for the effector interaction network. This PL approach was developed to identify effector-target interaction networks in the *U. maydis*-maize pathosystem [12]. This chapter provides a comprehensive step-by-step protocol for performing TurboID-based PL from maize leaves during *U. maydis* infection.

2 Materials

All solutions should be prepared using Milli-Q-grade distilled or deionized water unless specified otherwise.

2.1 Biological Materials

Escherichia coli and *Ustilago maydis* strains were stored at −80°C in 25% glycerol.

Zea mays cv. Early Golden Bantam or Golden Bantam (Demeter International, Germany) was used in this study.

2.2 Cloning for Vectors

1. Q5 High-Fidelity DNA Polymerase.
2. dNTPs.
3. GeneRuler 1 kb DNA Ladder.
4. 1× TAE: Dissolve 4.84 g Tris–HCl in 500 mL H_2O, add 2 mL 0.5 M Na_2EDTA pH 8.0 and 1.14 mL Acetic acid; adjust to 1 L with H_2O.
5. 1% TAE-agarose gel: Weigh out 1 g of universal agarose into a flask and add 100 mL of 1× TAE. Heat until agarose is completely dissolved. Cool down to 60°C and add 3.5 μL 1% Ethidium bromide solution and pour into gel tray.
6. NEBuilder HiFi DNA Assembly Master Mix.
7. NucleoSpin Gel and PCR clean-up kit.
8. NucleoSpin Mini kit for plasmid DNA.
9. dYT: Dissolve 16 g Tryptone—Peptone, 10 g yeast extract, and 5 g NaCl in 1 L H_2O. Sterilize by autoclaving.

2.3 Generation of Fungal Strains

1. Novozyme 234 solution: dissolve 7 mg Novozyme 234 in 1 mL SCS solution (*see* **Note 17**).
2. Carboxin: 2.5 μg/mL.
3. Heparin solution (1 mg/mL).
4. SCS solution: 20 mM Na-Citrate, pH 5.8, 1 M Sorbitol, sterile filtered.
5. STC solution: 10 mM Tris–HCl, pH 7.5, 100 mM $CaCl_2$, 1 M Sorbitol, sterile filtered.
6. STC/PEG solution: dissolve 10 g PEG 3350 in 15 mL STC.
7. Regeneration of agar plates: Dissolve 20 g Peptone, 10 g yeast extract, 20 g Sucrose, 182.2 g Sorbitol, and 15 g Agar in 1 L H_2O. Sterilize by autoclaving.
8. PD plates: Potato dextrose agar plates (39 g/L).

2.4 Infected Maize Sample Preparation

1. YEPS light liquid medium: Dissolve 4 g Peptone, 10 g yeast extract, and 4 g Sucrose in 1 L H_2O. Sterilize by autoclaving.
2. Charcoal-Platte agar plate: 1 % Activated charcoal in potato dextrose agar plates.
3. Tween 20.
4. Biotin solution (0.5 M biotin): Dissolve 122 mg of biotin in Dimethyl sulfoxide (DMSO).

2.5 Protein Extraction and Enrichment

1. Protein-lysis buffer: 50 mM Tris–HCl, pH 7.5, 150 mM NaCl, 10 % glycerol, 2 mM EDTA, 10 mM DTT, 0.5 % IGEPAL, 1 mM PMSF with cOmplete™ Protease Inhibitor Cocktail.
2. Wash buffer: 50 mM Tris–HCl pH 7.5, 150 mM NaCl, 10% glycerol, 2 mM EDTA.
3. SDT-lysis buffer: 10 mL 10% SDS, 2.5 mL 1 M DTT, 2.5 mL 1 M Tris–HCl pH 7.5, 10 mL ddH_2O.
4. Methanol (MeOH).
5. Chloroform.
6. PBS buffer: 144 mL 0.5 M Na_2HPO_4, 56 mL 0.5 M NaH_2PO_4, 100 mL 1.5 M NaCl, to 1 L water.

2.6 Laboratory Equipment

1. 1 mL syringe and 22-gauge needle.
2. PCR thermal cycler.
3. Agarose gel electrophoresis.
4. Tabletop centrifuge.
5. Nanodrop spectrophotometer.
6. Incubator (37°C and 28°C).
7. Ultrasonic bath.
8. Magnet rack (DynaMag™-2).
9. Eppendorf Thermomixer (Heat block).
10. Tube rotator (rolling wheel).
11. Polyacrylamide gel electrophoresis (SDS-PAGE).
12. ChemiDoc™ MP imaging system.

3 Method

3.1 Cloning of Ustilago maydis Sequences into p123-Ppit2-TurboID-3HA Vector

1. Select the coding sequence of the *U. maydis* effector gene of interest (*see* **Note 1**).
2. Use NEBuilder® Assembly Tool (https://nebuilder.neb.com/#!/) to design PCR primers for the adjacent DNA

fragment. The primers for HiFi DNA assembly should contain overlap sequences corresponding to the cloning vector region (including *Hin*dIII or *Bam*HI restriction sites) and the gene-specific sequence.

3. Amplify the fragment in a total volume of 50 μL with Q5 High-Fidelity DNA Polymerase.
4. Digest approximately 1 μg of p123-Ppit2-TurboID-3HA vector (Fig. 1A) with *Hin*dIII and *Bam*HI (1 μL each) restriction enzymes in a total volume of 100 μL at 37°C overnight (*see* **Note 1**).
5. Separate the digested vector and PCR product by electrophoresis on a 1% TAE-agarose gel and excise the relevant bands and purify the DNA fragments.
6. Purify both the digested vector and PCR product with NucleoSpin Gel and PCR clean-up kit according to the manufacturer's protocol, and measure the DNA concentration using a Nanodrop spectrophotometer.
7. Use the NEBuilder® Protocol Calculator (https://nebuildercalculator.neb.com/) to calculate optimal insert-to-vector ratios, and set up the HiFi DNA assembly reaction on ice

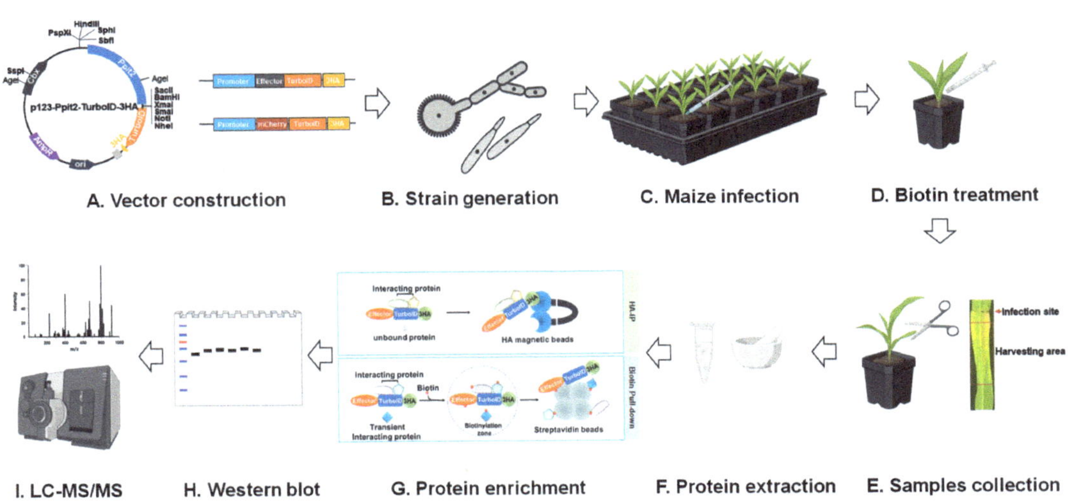

Fig. 1 Experimental design for identifying effector-interacting proteins in maize using TurboID-based proximity labeling during *Ustilago maydis* infection. The process begins with vector construction **(A)**, depicted by a circular plasmid map annotated with restriction sites. Generation of *U. maydis* transformants **(B)** is illustrated using a mechanical tool symbolizing transformation. Maize infection **(C)** is represented by seedlings arranged in trays. This is followed by biotin treatment **(D)**, shown as a syringe injecting the plants. Infected tissues are harvested from maize, indicated by highlighted infection and sampling sites on a plant **(E)**. Protein extraction **(F)** is depicted with a mortar and a test tube. Protein enrichment **(G)** is represented by interaction diagrams and magnetic beads. Western blot analysis **(H)** is shown using an electrophoresis gel. Finally, LC-MS/MS analysis **(I)** is visualized with a mass spectrometer and a representative output graph

according to NEBuilder HiFi DNA Assembly Master Mix protocol.

8. Incubate the assembly reaction samples in a Thermocycler at 50°C for 15–60 min.
9. Transform 50 µL competent *E. coli* DH5α cells with the assembly reaction product by mixing gently.
10. Incubate the cells on ice for 30 min, heat shock at 42°C for 1 min, then immediately add 300 µL of dYT medium, incubate 1 h at 37°C by shaking at 200 rpm. Plate the cells onto YT-agar plates supplemented with ampicillin, and incubate for 16 h at 37°C.
11. Pick single colonies for colony PCR screening and confirm positive clones by sequencing.

3.2 Generation of Fungal Strains

1. Linearize the constructed plasmid with SspI or AgeI to enable homologous recombination.
2. Prepare *U. maydis* protoplasts using Novozyme 234 as described below [13] (*see* **Note 17**).
3. *U. maydis* strains were freshly grown in YEPS light liquid medium until the culture reached an OD600 of 0.6–0.8.
4. The culture was then centrifuged at 2000 × *g* for 10 min, and the pellets were resuspended in 10 mL SCS and centrifuged again for 5 min at 2000 × *g*.
5. The pellets were resuspended in 2 mL of SCS containing 7 mg/mL of Novozyme 234 and incubated at room temperature for 10 min.
6. 40%–50% of the cells were observed to form protoplasts under a microscope, then added 10 mL of ice-cold SCS and centrifuged at 2000 × *g* for 10 min at 4°C.
7. The pellets were washed carefully twice with 10 mL of ice-cold SCS.
8. The pellets were washed carefully with 10 mL of ice-cold STC.
9. The pellets were resuspended in 500 µl of ice-cold STC and divided into 50 µl portions in pre-chilled Eppendorf tubes as *U. maydis* protoplasts.
10. Add up to 10 µL of linearized plasmid DNA into 50 µL prepared protoplasts.
11. Add 1 µL of heparin solution (1 mg/mL) into the mixture and incubate on ice for 10 min.
12. Gently mix the protoplast suspension with 500 µL of STC/PEG solution by five to eight times pipetting up and down.
13. Incubate the mixture on ice for an additional 15 min.

14. Plate the protoplasts onto freshly prepared regeneration agar plates with a 10 mL bottom layer containing the selective antibiotic (carboxin: 2.5 μg/mL) and a 10 mL top layer without antibiotic selection.
15. Incubate the plates at 28°C and flip the transformation plate upside down until the next day.
16. Single out colonies after 3–6 days at 28°C from regeneration agar to PD plates with carboxin selection for an additional 2–3 days.
17. Isolate genomic DNA from the resulting fungal strains for further analysis by PCR and Southern blotting.

3.3 Maize Infection for Sample Preparation

1. Inoculate *U. maydis* strains from PD-agar plate into 10 mL YEPS light liquid medium and incubate at 28°C with shaking at 200 rpm overnight (*see* **Notes 2 and 3**).
2. Dilute the *U. maydis* overnight culture to OD_{600}=0.2 in fresh YEPS light liquid medium, and continue incubation at 28°C with 200 rpm shaking until the culture reaches an OD_{600} of 0.8.
3. Transfer the culture to a 50 mL Falcon and centrifuge at 2000 × *g* for 10 min.
4. Discard the supernatant and wash the pellet with 30 mL sterile distilled water.
5. Centrifuge again at 2000 × *g* for 10 min and remove the supernatant.
6. Resuspend the pellet in sterile water and adjust the suspension to OD_{600}=3.0, and add 0.1% [v/v] Tween 20 to facilitate maize infiltration.
7. Spot 5 μl of the prepared cell suspension onto Charcoal-Platte agar plate and incubate at room temperature (RT) for 3 days to check the filamentation of *U. maydis* (Fig. 2B).
8. Seven-day-old maize seedlings of Early Golden Bantam (EGB) are used for *U. maydis* infection by syringe with needle (Fig. 2A). Three biological replicates for each construct are performed, using at least 20 maize seedlings per replicate.
9. Infected maize plants are grown in a controlled phytochamber under 16 h light at 28°C and 8 h dark at 22°C.

3.4 Biotin Treatment for TurboID-Based PL Sample Preparation

1. Prepare a biotin infiltration solution by diluting a high concentration of biotin (0.5 M biotin in DMSO) with sterile water to a final concentration of 100 μM (*see* **Note 5**).
2. At 2 days post-infection (dpi), infiltrate the biotin solution directly into the infected regions of maize leaves using a syringe without a needle.

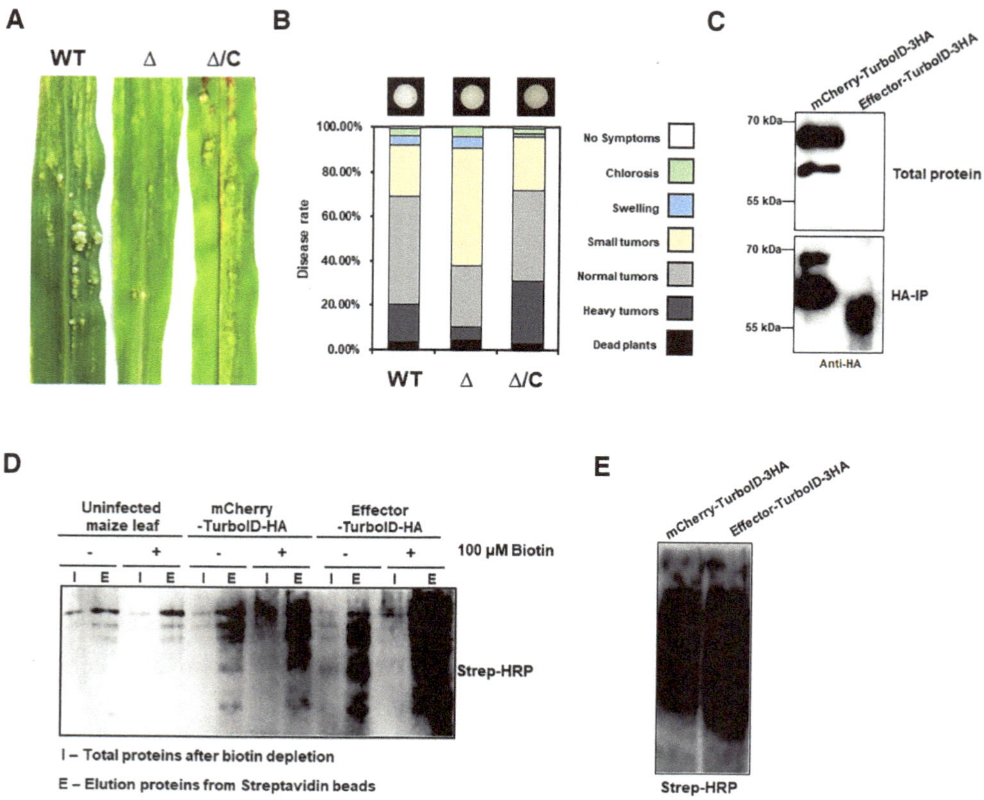

Fig. 2 TurboID-based proximity labeling in the *Ustilago maydis* –maize pathosystem. (**A**) Phenotype of infection symptoms on maize seedlings infected with *U. maydis* strains (wild-type [**WT**], effector deletion mutant [Δ], and effector-TurboID-3HA complementation [**Δ/C**]) at 12 dpi. (**B**) Quantification of infection symptoms on maize seedlings infected with *U. maydis* strains as indicated in A at 12 dpi. (**C**) Western blot analysis detecting mCherry-TurboID-3HA and effector-TurboID-3HA in infected maize leaves upon delivery by *U. maydis*. from total protein and anti-HA immunoprecipitation. Maize leaves were harvested at 3 days post-infection. (**D**) Western blot analysis detecting total and eluted proteins after biotin treatment for TurboID-based proximity labeling. 7-day-old maize seedlings were infected with *U. maydis* strains expressing mCherry-TurboID-3HA or effector-TurboID-3HA. Non-infected leaves were used as a negative control. 2 days post-infection non-infected and infected leaves were treated with (+) or without (–) 100 μM biotin. Infected leaves were harvested 1-day post biotin treatment for protein extraction and Streptavidin-based precipitation of biotinylated proteins. (**E**) Biotinylated proteins from *U. maydis*-infected maize leaves were enriched using streptavidin beads and analyzed by Western blot for further LC-MS/MS analysis

3. After biotin treatment, maize plants are maintained in a controlled phytochamber under the same growth conditions: 16 h light at 28°C and 8 h dark at 22°C
4. Harvest leaf samples by excising 3 cm leaf segments approximately 1 cm below the infection site at 3 dpi (Fig. 1E), immediately freeze the samples in liquid nitrogen.
5. Grind the frozen plant material into a fine powder using liquid nitrogen, and aliquot approximately 300 mg of the ground

tissue into 2 mL pre-chilled Eppendorf tubes for further protein extraction.

3.5 Protein Extraction and Immunoprecipitation for Expression Verification

1. Add 1.6 mL ice-cold protein-lysis buffer (50 mM Tris–HCl pH 7.5, 150 mM NaCl, 10 % glycerol, 2 mM EDTA, 10 mM DTT, 0.5 % IGEPAL, 1 mM PMSF) with cOmplete™ Protease Inhibitor Cocktail (Roche) to a tube containing frozen plant powder. Vortex thoroughly and incubate on ice for 30 min.
2. Centrifuge the tubes at 13,300 × *g* for 10 min at 4°C.
3. Carefully transfer the supernatant (soluble protein fraction) to a new tube and centrifuge again at 13,300 × *g* for 20 min at 4°C.
4. Transfer the resulting clear supernatant to a fresh tube as the total protein extract.
5. Aliquot 50 μL of the total protein extract for Western blot analysis and store at −20°C (*see* **Note 4**).
6. Using a pipette tip with the end cut off, aliquot 10 μL aliquot of HA magnetic beads (Pierce Anti-HA magnetic beads, Thermo Fisher Scientific) per sample into a chilled tube at 4°C and mix well (*see* **Notes 9 and 13**).
7. Separate the HA magnetic beads on a magnetic separator until the supernatant is clear, and remove the supernatant.
8. Wash the beads by adding 1 mL ice-cold wash buffer (50 mM Tris–HCl, pH 7.5, 150 mM NaCl, 10% glycerol, 2 mM EDTA) to the beads and mixing well, then magnetically separating and discarding the supernatant.
9. Repeat the wash step three times.
10. Add the total protein extracts to the washed beads and incubate the mixture on a rolling wheel at 4°C for 3 h or overnight.
11. Magnetically separate the HA magnetic beads and remove the supernatant.
12. Wash the beads once with 1 mL ice-cold wash buffer and magnetically separate.
13. Remove the supernatant and repeat the wash step three times.
14. Resuspend the beads in 500 μL wash buffer and transfer the mix to a fresh tube.
15. Aliquot 25 μL aliquots of the mix for Western blot analysis and store at −20°C.
16. Magnetically separate the remaining mix and remove the supernatant completely.

17. Store the remaining beads at −80 °C until further use (*see* **Note 16**).
18. Use all aliquots for immunoblotting with mouse anti-HA primary antibody and anti-mouse IgG secondary antibody (Fig. 2C).

3.6 Protein Isolation from TurboID-Based PL

1. Prepare SDT-lysis buffer (10 mL 10% SDS, 2.5 mL 1 M DTT, 2.5 mL 1 M Tris–HCl pH 7.5, 10 mL ddH_2O) and heat the buffer to 95°C using an Eppendorf Thermomixer (*see* **Note 5**).
2. Add 500–750 μL of the hot SDT-lysis buffer to a tube containing frozen powder. Vortex to mix thoroughly and briefly centrifuge.
3. Incubate the samples at 95°C for 5 min in the Eppendorf Thermomixer with shaking.
4. Sonicate the samples for 10 min, then centrifuge at 13,300 × *g* for 10 min at room temperature.
5. Transfer the supernatant to a fresh tube and centrifuge again at 13,300 × *g* for 10 min.
6. Collect the resulting clear supernatant into a fresh tube as the total protein extract.
7. Aliquot 50 μL of the total protein extract for Western blot analysis and store at −20°C.
8. Snap-frozen the remaining samples in liquid nitrogen and stored at −80 °C until further use.

3.7 Biotin Depletion for Biotinylated Proteins Enrichment

Use the methanol-chloroform-precipitation method [14].

1. Add 666 μL methanol and 166 μL chloroform to 500 μL total protein extracts and mix thoroughly with 300 μL ddH_2O.
2. Centrifuge the mixture at 1500 × *g* for 10 min and discard the supernatant.
3. Wash the protein pellets with 600 μL methanol and centrifuge at 1500 × *g* for 10 min.
4. Discard the supernatant, and repeat the washing step two or more times until the pellet turns white (*see* **Note 6**).
5. Air-dry the protein pellet for 5 min (*see* **Note 7**).
6. Resuspend the protein pellets in 500 μL SDT-lysis buffer and sonicate for 10 min (*see* **Note 8**).
7. Incubate the mixture in an Eppendorf Thermomixer for at 1250 rpm 30 min while shaking.

8. Centrifuge the mixture for 10 min at 1500 × *g* and transfer the supernatant to a fresh tube as the biotin-depleted samples.
9. Take 50 μL aliquots of the biotin-depleted samples for Western blot analysis and store at −20°C (Fig. 2C).
10. Snap-frozen the remaining samples in liquid nitrogen and stored at −80 °C until further use.

3.8 Biotinylated Proteins Enrichment

1. Use a cut-off pipette tip to take 100 μL aliquot per sample from a 50% Streptavidin-agarose beads slurry at 4°C and mix well (*see* **Note 9**).
2. Add 5 mL PBS (0.1 M phosphate, 0.15 M NaCl, pH 7.2) into a 15 mL Falcon tube to wash the beads and centrifuge for 2 min at 2000 × *g* and 22°.
3. Discard the supernatant carefully, and repeat the wash step three times with PBS buffer.
4. Resuspend the washed beads 100 μL PBS with 0.5% SDS per sample.
5. Dilute the biotin-depleted samples with PBS buffer to a final concentration of 0.5% SDS.
6. Add 100 μl beads to each sample and incubate the samples on a rolling wheel at 22°C for 20 h.
7. Centrifuge the beads for 3 min at 2000 × *g* at 22°C and remove the supernatant carefully.
8. Wash once with 2 mL PBS buffer containing 2% SDS, and centrifuge for 3 min at 2000 × *g* at 22°C and discard the supernatant.
9. Add 10 mL PBS buffer to wash the beads and centrifuge for 3 min at 2000 × *g* at 22°C.
10. Discard the supernatant carefully and repeat the wash step six times with 10 mL PBS buffer (*see* **Note 11**).
11. Add 500 μL PBS buffer to the beads and transfer the mix to a fresh tube (*see* **Note 12**).
12. Take 25 μL aliquots of the mix for Western blot analysis and store at −20°C (*see* **Note 14**).
13. Centrifuge the remaining mix for 3 min at 2000 × *g* at 22°C and remove the supernatant completely (*see* **Note 10**).
14. Store the remaining beads at −80 °C until further use.
15. All the aliquots are used for immunoblotting with the Strep-HRP antibody (Fig. 2D) (*see* **Notes 15 and 16**).

4 Notes

1. The vector backbone includes the UmPit2 (UMAG_01375) promoter (high expression throughout infection) for overexpression effector proteins during infection [15, 16]. If this promoter is not required, it can be removed using suitable restriction enzymes and replaced with the promoter and coding sequence of your effector.
2. TurboID fusion may affect effector functionality. It is recommended to introduce the construct into the corresponding effector knockout strain and select transformants that restore the wild-type phenotype through complementation for further experiment (Fig. 2A).
3. Select appropriate controls to account for endogenous biotinylated proteins in maize plants. It is recommended to introduce the uninfected plants as a negative control and the control strain-infected plants as an additional control.
4. Detecting the target protein expression in total protein extracts by Western blot is often difficult. A co-immunoprecipitation step is performed using the total protein lysate, and approximately 5%–10% of the beads is then analyzed by Western blot to assess target protein expression. If a detectable signal is obtained, then conduct Streptavidin-based enrichment. otherwise, either prepare new infection samples or generate new strain expressing higher levels of the fusion protein.
5. Freshly prepare the biotin solution, protein lysis buffer, and SDT lysis buffer to use.
6. During biotin depletion, carefully remove supernatants without disturbing the protein pellets. Wash the pellets repeatedly with methanol until the pellets appear white, as residual plant pigments may compromise downstream analyses.
7. After biotin depletion, the protein pellets should not be overdried, as this may hinder subsequent resuspension.
8. Ensure that the protein pellets are fully dissolved in SDT lysis buffer. If necessary, increase the volume of SDT lysis buffer accordingly.
9. Mix well the beads before transferring, and use a pipette tip with the end removed (cut approximately 3–5 mm from the tip) to prevent bead shearing and ensure complete transfer, as some beads may be relatively large.
10. When removing supernatant from Streptavidin-agarose beads, attach a 20 μl gel-loading tip to the end of a blue pipette tip. This setup minimizes the risk of aspirating beads

while allowing for efficient volume removal, thereby reducing the risk of contamination. Never touch the beads with the pipette tip when removing the supernatant, and do not centrifuge the beads at speeds exceeding 2000 × *g*.

11. Thoroughly wash the enriched Streptavidin-agarose beads with PBS to ensure complete removal of SDS contaminants. Residual SDS may adversely affect trypsin digestion and ionization efficiency during downstream mass spectrometry analysis.
12. Use low-protein-binding tubes and sterile, unopened pipette tips. Avoid using autoclaved tubes and tips, as they may affect subsequent mass spectrometry analysis.
13. While this protocol describes the use of HA-tagged magnetic beads and streptavidin-conjugated agarose beads as representative examples, other comparable affinity resins may be used following proper optimization.
14. To detect biotinylated proteins enriched by Streptavidin-agarose beads, resuspend the beads in SDS sample buffer supplemented with 20 mM biotin (prepared in 4× SDS sample buffer). Boil the samples for 10 min at 95°C with shaking. Afterward, centrifuge briefly and load as much of the supernatant as possible onto the gel for electrophoresis.
15. For immunoblotting with Strep-HRP antibody, membranes should be blocked with 5% BSA in PBST (PBS with 0.1% Tween 20). After blocking, wash membranes 3×5 min in PBST before antibody incubation to minimize non-specific binding.
16. Store the remaining beads at −80 °C and avoid repeated freeze-thaw cycles. After Western blot analysis of all bead samples, the remaining beads can be used for mass spectrometry (MS) analysis.
17. Since Novozyme 234 is out of production, alternative lysing enzymes might be used for protoplastation.

Acknowledgments

We would like to thank Dr. Katharina Melkonian, Dr. Sara Christina Stolze, and Anne Harzen for helpful practical advice on the TurboID-based PL approach. This project has received funding from the European Research Council (ERC) under the European Union's Horizon 2020 research and innovation programme (Grant Agreement No 771035), as well as funding by the Deutsche Forschungsgemeinschaft (DFG, German Research Foundation) under Germany's Excellence Strategy- EXC-2048/1- Project ID:

390686111, Research Grant DFG-Az: DO 1421/3-3 and DFG Emmy Noether Programme (SA 4093/1-1).

References

1. Doehlemann G, Wahl R, Horst RJ et al (2008) Reprogramming a maize plant: transcriptional and metabolic changes induced by the fungal biotroph *Ustilago maydis*. Plant J 56(2):181–195
2. Zuo W, Depotter JRL, Stolze SC et al (2023) A transcriptional activator effector of *Ustilago maydis* regulates hyperplasia in maize during pathogen-induced tumor formation. Nat Commun 14(1):6722
3. Paraoan LI, Sharif U, Kelly L et al (2020) Intracellular RPE interactome analysis of protein variants associated with AMD by affinity purification mass spectrometry. Invest Ophthalmol Vis Sci 61(7):3688–3688
4. Woloschuk RM, Reed PMM, McDonald S et al (2020) Yeast two-hybrid screening of photoswitchable protein-protein interaction libraries. J Mol Biol 432(10):3113–3126
5. Cowell IG (1997) Yeast two-hybrid library screening. cDNA library protocols. Humana Press, Totowa, NJ, pp 185–202
6. Sciuto MR, Coppola V, Iannolo G et al (2019) Two-step co-immunoprecipitation (TIP). Curr Protoc Mol Biol 125(1):e80
7. Chen C-L, Hu Y, Udeshi ND et al (2015) Proteomic mapping in live Drosophila tissues using an engineered ascorbate peroxidase. Proc Natl Acad Sci U S A 112(39):12093–12098
8. Kim DI, Jensen SC, Noble KA et al (2016) An improved smaller biotin ligase for BioID proximity labeling. Mol Biol Cell 27(8):1188–1196
9. Branon TC, Bosch JA, Sanchez AD et al (2018) Efficient proximity labeling in living cells and organisms with TurboID. Nat Biotechnol 36(9):880–887
10. Melkonian K, Stolze SC, Harzen A et al (2022) miniTurbo-based interactomics of two plasma membrane-localized SNARE proteins in *Marchantia polymorpha*. New Phytol 235(2):786–800
11. Zhang Y, Song G, Lal NK et al (2019) TurboID-based proximity labeling reveals that UBR7 is a regulator of N NLR immune receptor-mediated immunity. Nat Commun 10(1):1–17.
12. Shi W, Stolze SC, Nakagami H et al (2023) Combination of in vivo proximity labeling and co-immunoprecipitation identifies the host target network of a tumor-inducing effector in the fungal maize pathogen *Ustilago maydis*. J Exp Bot 74(15):4736–4750
13. Kämper J (2004) A PCR-based system for highly efficient generation of gene replacement mutants in *Ustilago maydis*. Mol Genet Genomics 271(1):103–110
14. Melkonian K, Stolze SC, Harzen A et al (2023) Proximity-dependent in vivo biotin labeling for interactome mapping in *Marchantia polymorpha*. Plant proteostasis: methods and protocols. Springer US, New York, NY, pp 295–308
15. Mueller AN, Ziemann S, Treitschke S et al (2013) Compatibility in the *Ustilago maydis*–maize interaction requires inhibition of host cysteine proteases by the fungal effector Pit2. PLoS Pathog 9(2):e1003177
16. Misas Villamil JC, Mueller AN, Demir F et al (2019) A fungal substrate mimicking molecule suppresses plant immunity via an inter-kingdom conserved motif. Nat Commun 10(1):1576

Check for updates

Chapter 27

Immunodetection of Cell-Surface-Associated Proteins in *Ustilago maydis* and Infected Maize Leaves

Chibbhi Bhaskar, Minh-Quang Chau and Lay-Sun Ma

Abstract

Immunolocalization relies on the sequential binding of a specific primary antibody to its target antigen, followed by detection with a fluorophore-conjugated secondary antibody to visualize the subcellular localization of proteins. This technique enables precise and high-resolution mapping of protein localization and distribution patterns after secretion. This chapter presents a detailed protocol for the immunolocalization of secreted proteins from *Ustilago maydis*, both in vitro and in infected maize leaves.

Keywords Immunolocalization, Cell-wall-associated protein, Effectors, Antibody-based detection, *Ustilago maydis*, Maize

1 Introduction

Ustilago maydis deploys a repertoire of secreted proteins to establish biotrophic growth within its host [1]. These proteins can associate with fungal or plant cell surfaces, accumulate in the plant apoplast, or translocate into plant cells and localize to different organelles. Determining their subcellular localizations is essential for identifying host targets and elucidating their roles in fungal virulence, a critical step toward defining their mechanisms of action.

Fluorescence-based immunolocalization is widely used to detect protein localization, especially for detecting fungal cell surface-associated proteins [2, 3]. This technique relies on the specific binding of a primary antibody to its target epitope, such as an HA tag, His tag, or a specific amino acid sequence within the protein of interest, followed by binding of a fluorophore-conjugated secondary antibody. The resulting fluorescence signals, visible on the surface of both yeast-like and filamentous cells, can be detected using a fluorescence microscope, enabling high-specificity localization of proteins in live cells without the need for fixation.

An N. T. Phan et al. (eds.), *Ustilago maydis*, Methods in Molecular Biology, vol. 3050,
https://doi.org/10.1007/978-1-0716-5364-7_27,

In contrast, immunodetection in infected maize leaves is challenging, as the epidermal layer restricts antibody access to fungal surfaces. This limitation can be addressed by partially removing the epidermis through mechanical means, followed by treatment with plant cell wall-degrading enzymes to loosen surrounding tissues. This combined approach facilitates antibody access to the epitope, enabling effective immunodetection of secreted proteins on fungal hyphal surface or associated with the plant cell wall.

This chapter presents an immunostaining protocol for examining the surface localization of secreted proteins in *U. maydis*. Using Rsp3 [4]—a cell surface-associated effector tagged with HA—as a case study, a detailed procedure for immunodetection using an anti-HA primary antibody and a fluorophore-conjugated secondary antibody is described. The resulting fluorescence signal enables the visualization of HA-tagged protein distribution in sporidial cells, in vitro-induced filaments, and fungal hyphae within infected maize tissues using confocal microscopy.

2 Materials

All nutrient media and buffers prepared with Milli-Q water are sterilized by autoclaving at 121°C and 15 psi for 15–20 min, and stored at room temperature until use.

2.1 Fungal Strains

The solopathogenic *U. maydis* strain SG200 (a1::mfa2 bE1bW2), derived from FB1 wildtype *U. maydis* strain [5], and its derivative SG200Δrsp3-Rsp3HA, which expresses C-terminal HA-tagged Rsp3 under the constitutive *otef* promoter generated via homologous recombination [4], were used for immunolocalization experiments in this study.

2.2 Culture Media and Growth Conditions

1. Potato dextrose agar (PDA): Prepare agar medium containing 10 mM Tris-HCl (pH 8.0), 2.4% potato dextrose, and 2% agar in Milli-Q water.
2. Yeast extract peptone sucrose liquid (YEPSL) medium: Prepare liquid medium containing 0.4% yeast extract, 0.4% peptone, and 2% sucrose in Milli-Q water (*see* **Note 1**).
3. Autoclaved glass test tubes and 500-mL conical flasks (*see* **Note 2**).
4. Orbital shaker incubator maintained at 28°C.
5. Visible-light spectrophotometer.

2.3 Chemicals and Reagents for Filament Induction

1. 2% YEPSL medium: Dilute YEPSL medium 50-fold with sterile Milli-Q water.

2. Hydroxy fatty acids: Prepare a solution containing 10 mM of 16-hydroxyhexadecanoic acid in absolute ethanol. Aliquot them and store at −20°C until use (*see* **Note 3**).
3. 1-oz spray bottles: Sterilize with 100% ethanol, followed by rinsing with sterile Milli-Q water (Fig. 1A; *see* **Note 4**).
4. Glass microscope slides: size 1″ x 3″ with a thickness of 1–1.2 mm.
5. PARAFILM M: Prepare several 1″ x 2″ pieces of PARAFILM, each slightly shorter than a microscope slide, and tape them onto glass slides.
6. Sterile Petri dishes with moistened paper towels: Cut several paper towels to the same size as the Petri dishes, moisten them with sterile Milli-Q water, and place them inside the dishes.

2.4 Maize Infection and Reagents for Plant Cell Wall Digestion

1. Maize seedlings for infection: Seven-day-old maize seedlings are grown under controlled conditions (15 h/9 h light/dark cycle; 28°C day/25°C night), used for *U. maydis* infection.
2. Sterile 1-mL disposable syringes with 26 G x 1/2″ needles.
3. 2-(N-morpholino)ethanesulfonic acid (MES) buffer: Prepare a buffer containing 10 mM MES in Milli-Q water. Adjust the pH to 5.7 with 1 N NaOH, and sterilize the buffer by autoclaving.
4. Cell wall-digestion enzyme solution: Prepare a solution containing 1.5% cellulase from *Trichoderma* sp., 0.3% Macerozyme™ R-10, and 0.6 M mannitol in 10 mM MES buffer (pH 5.7). Heat the mixture at 50°C for 10 min, then add $CaCl_2$ and BSA

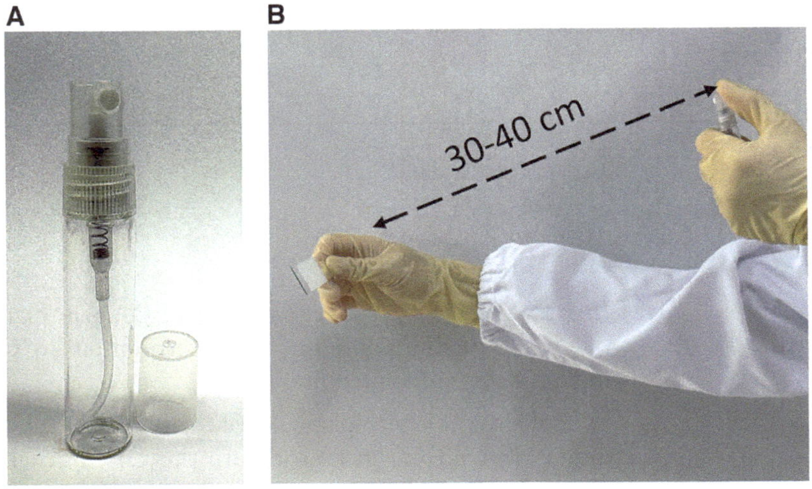

Fig. 1 Panel figure with two images labeled (**A**) and (**B**). Image (**A**) A clear spray bottle with a removable cap, positioned upright on a light background. Image (**B**) A person wearing gloves and a lab coat, holding a small glass slide in one hand and a spray bottle in the other. A dashed line with the text "30–40 cm" indicates the distance between the spray bottle and the slide

to final concentrations of 1 mM and 0.1%, respectively. Filter-sterilize the solution through a 0.22 μm polyethersulfone (PES) membrane before use.

2.5 Buffers and Reagents for Immunostaining

1. 10x Phosphate-buffered saline (PBS) buffer: Prepare a solution containing 1.37 M NaCl, 27 mM KCl, 100 mM Na2HPO4, and 18 mM KH2PO4 in Milli-Q water, adjust the pH to 6.9 or 7.4 as required, and sterilize the solution by autoclaving.
2. 1x PBS buffer: Dilute 10x PBS stock solution 10-fold with sterile Milli-Q water.
3. 4% formaldehyde (Fixation reagent, pH 6.9; freshly prepared): Prepare 100 mL of solution by dissolving 4 g of paraformaldehyde powder in heated 1x PBS (pH 6.9). Adjust the pH if necessary.
4. 0.1 M glycine (Quenching reagent): Prepare 100 mL of solution by dissolving 0.757 g of glycine in 1x PBS (pH 7.4).
5. 3% Blocking solution: Prepare 10 mL of solution by dissolving 0.3 g of bovine serum albumin (BSA) in 1x PBS (pH 7.4). Filter-sterilize the solution through a 0.22 μm PES membrane.
6. Wash buffer: Prepare 1x PBS (pH 7.4) containing 0.02% Tween-20. For leaf sample immunostaining, omit Tween-20 from the wash buffer.
7. Primary antibody solution: Dilute the mouse anti-HA monoclonal antibody to a 1:2000 ratio in the 3% blocking solution.
8. Secondary antibody solution: Dilute the goat anti-Mouse IgG (H+L) Cross-Adsorbed Secondary Antibody, Alexa Fluor™ 488 to a 1:4000 ratio in the 3% blocking solution.
9. A commercially available mounting medium.

3 Methods

3.1 Sample Preparation

3.1.1 Preparation of U. maydis Sporidial Cells

1. Inoculate *U. maydis* cells grown on a PDA agar plate into 5 mL of YEPSL medium and incubate overnight at 28°C with shaking at 240 rpm in an orbital shaker.
2. Subculture the overnight culture to an initial OD_{600} of 0.2 and incubate at 28°C for 4 h, or until the OD_{600} reaches approximately 0.8.
3. Harvest 1 mL of the culture by centrifugation at 1300 x*g* for 5 min at room temperature.

4. Wash the pellet three times with 1 mL of 1x PBS each time (*see* **Note 5**).
5. Resuspend the cell pellet in 1 mL of fixation reagent and incubate for 10 min at room temperature.
6. Harvest the cells and wash the pellet three times with 1x PBS before proceeding to immunostaining as described in Sect. 3.2.

3.1.2 Preparation of Hydroxy Fatty Acid-Induced U. maydis Filaments on a Parafilm

1. The *U. maydis* cell pellet obtained from Steps 1–4 in Sect. 3.1.1, without fixation, is resuspended in 2 mL of 2% YEPLS medium.
2. Add 20 μL of 10 mM 16-hydroxy hexadecanoic acid to the 2 mL cell suspension to obtain a final concentration of 0.1 mM.
3. Incubate the mixture at 28°C for 10 min, then transfer the cell suspensions into a glass sprayer.
4. Spray the culture evenly onto Parafilm M, which is taped on a glass slide, from a distance of 30–40 cm, allowing fine mist droplets to form on the Parafilm surface (Fig. 1B) (*see* **Note 6**).
5. After spraying, place these glass slides with droplets facing upward into a Petri dish lined with wet paper.
6. Seal the Petri dish with Parafilm and incubate at 28°C for 15–16 h.
7. Examine the Parafilm under a microscope to confirm filament formation before proceeding.
8. Remove the Parafilm from the glass slide and place it in a Petri dish containing 1x PBS (pH 7.4), with the sprayed side facing the buffer.
9. Gently wash the Parafilm three times on a shaker at 70 rpm to remove sporidial cells.
10. Filaments attached on the Parafilm are ready for immunostaining (Sect. 3.2).

3.1.3 Preparation of U. maydis-Infected Maize Leaf Samples

1. Select the third leaf from *U. maydis*-infected maize seedlings displaying clear symptoms. Excise a 2-cm segment just below the infection sites (marked by the injection holes) within the chlorotic region (*see* **Note 7**).
2. Briefly rinse the leaf segment in sterile Milli-Q water to remove surface-attached cells, and gently blot dry using a paper towel.
3. Using a sterile razor blade, make a shallow cut without cutting through the leaf, and carefully peel the epidermal layer from one side of the cut (*see* **Note 8**).

4. Incubate the peeled segment in 3 mL of cell wall-digestion enzyme solution in a 5-mL Eppendorf tube at 28°C with gentle shaking (80 rpm) for 1 h (*see* **Note 9**).
5. Wash the cell wall-enzyme-digested leaf segments three times in 4 mL of 1x PBS (pH 7.4) for 10 min each at room temperature with gentle rotation.
6. Fix the segments in 4 mL of fixation reagent for 30 min at room temperature with gentle rotation.
7. Wash the segments twice with 1x PBS (pH 7.4), then quench any residual formaldehyde by incubating them in 0.1 M glycine prepared in 1x PBS for 15 min at room temperature with rotation.
8. Wash the segments three times with 1x PBS for 10 min at room temperature before proceeding to immunostaining (Sect. 3.2).

3.2 Immunostaining

1. Block non-specific binding by incubating samples in the blocking solution for 1 h at room temperature.
2. Remove the blocking solution and incubate the samples in the primary antibody solution for 1 h at room temperature or overnight at 4°C, with gentle shaking at 70 rpm.
3. Wash samples three times with the wash buffer.
4. Incubate with the secondary antibody solution for 1 h at room temperature in the dark (wrapped in aluminum foil) with gentle shaking (*see* **Note 10**).
5. Wash samples three times with the wash buffer for 15 min each (*see* **Note 11**).
6. Mounting the samples as instructed below:

   - For sporidial cells: Resuspend the cell pellet in 40 μL mounting medium, add 0.5 μL of the suspension onto a glass slide, and gently place a coverslip over it, tapping lightly to spread the cells evenly (*see* **Note 12**).
   - For filaments: Place the Parafilm on the slide with the filament side facing up. Add a drop of mounting medium onto the filaments and carefully cover with a coverslip.
   - For maize leaf samples: Place the leaf on a microscope slide with the peeled epidermal side facing up. Gently place a coverslip over the sample and then add a drop of sterile Milli-Q water to adhere the sample to the slide.

7. Image the mounted samples using a confocal microscope (*see* **Note 13**). Representative confocal images of immunostained Rsp3-HA in *U. maydis* sporidial cells and filaments are shown under both in vitro (Fig. 2A-B) and *in planta* (Fig. 2C) conditions.

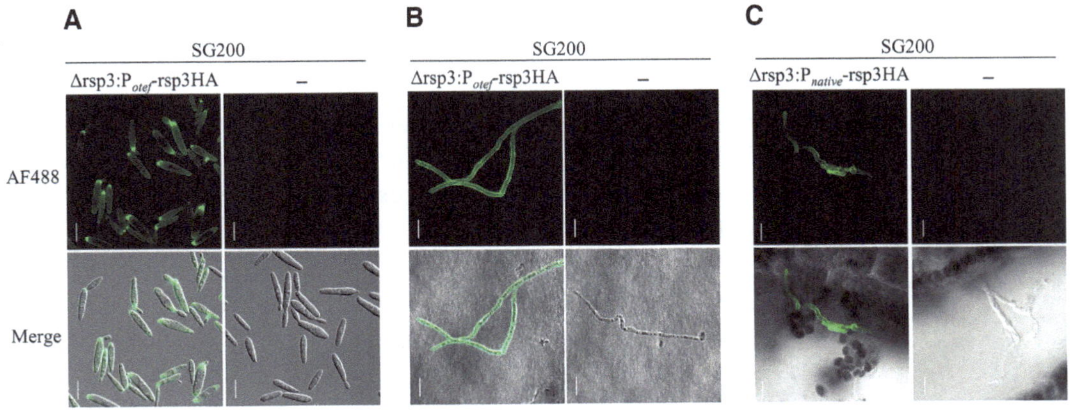

Fig. 2 Fluorescence microscopy images showing immunolocalization of HA-tagged Rsp3 proteins in **U. maydis**. Panel (**A**) shows sporidial cells with Rsp3 green fluorescence localized at the bud neck, while the control cells display no green fluorescence. Panel (**B**) shows filamentous structures with Rsp3 along filaments induced by hydroxy fatty acids. Panel (**C**) illustrates Rsp3 localization on the hyphal surface in infected maize leaves. Images include epifluorescence (488 nm) and merged differential interference contrast (DIC) views. Scale bars, 10 μm, are present in each image

4 Notes

1. The final pH of the YEPSL medium typically ranges between 6.0 and 7.0, which supports the sporidial growth of *Ustilago maydis*. Ensure the pH does not become acidic, as lower pH may induce filamentation, probably reflecting a stress response.
2. The *U. maydis* cultures require sufficient aeration; therefore, 500-mL conical flasks are strictly preferred for 50 mL of cell culture.
3. Avoid more than two freeze-thaw cycles of hydroxy fatty acids, as repeated freeze-thawing may reduce filamentation efficiency. Therefore, after preparing the hydroxy fatty acid stock, aliquot 100 μL into 1.5-mL Eppendorf tubes and store at −20°C.
4. Spray bottles are reusable. To avoid contamination, fill them with 100% ethanol overnight, then wash thoroughly with sterile Milli-Q water.
5. To prevent contamination, all procedures involving sporidia cells should be performed in a laminar flow cabinet.
6. Droplet size determines the success of filament formation, so it is important to produce a fine mist. Adjust the spraying distance or use a sprayer capable of generating a consistent mist.

7. The yellowed chlorotic region caused by fungal colonization is a suitable area for observing fungal structures under a microscope.
8. Excessive peeling of leaf epidermal cells can damage both host tissue and fungal structures, potentially compromising staining quality and microscopic observations. Therefore, peel just enough to allow antibody access while preserving tissue integrity.
9. Gentle shaking at 28°C has been found to enhance the efficiency of cell wall loosening, as evidenced by increased separation of leaf cells.
10. Avoid prolonging the incubation with the secondary antibody, as this may increase non-specific binding, resulting high background signal.
11. If necessary, increase the number of PBS washes or extend the washing time to improve specificity and reduce background.
12. Handle the samples gently when tapping, as excessive force may damage the cells and increase background signals.
13. Fungal structures are often more visible at cut-edge regions, where the enzyme solution has better access during digestion, or in areas where mesophyll cells have loosened or separated due to enzymatic activity.

References

1. Lanver D, Muller AN, Happel P et al (2018) The biotrophic development of *Ustilago maydis* studied by RNA-seq analysis. Plant Cell 30(2):300–323. https://doi.org/10.1105/tpc.17.00764
2. Lin YH, Xu MY, Hsu CC et al (2023) *Ustilago maydis* PR-1-like protein has evolved two distinct domains for dual virulence activities. Nat Commun 14(1):5755. https://doi.org/10.1038/s41467-023-41459-4
3. Ma LS, Tsai WL, Damei FA et al (2023) Maize antifungal protein AFP1 elevates fungal chitin levels by targeting chitin deacetylases and other glycoproteins. mBio 14(2):e0009323. https://doi.org/10.1128/mbio.00093-23
4. Ma LS, Wang L, Trippel C et al (2018) The *Ustilago maydis* repetitive effector Rsp3 blocks the antifungal activity of mannose-binding maize proteins. Nat Commun 9(1):1711. https://doi.org/10.1038/s41467-018-04149-0
5. Kamper J, Kahmann R, Bolker M et al (2006) Insights from the genome of the biotrophic fungal plant pathogen *Ustilago maydis*. Nature 444(7115):97–101. https://doi.org/10.1038/nature05248

Chapter 28

Protein Extraction Methods for *Ustilago maydis*

María Dolores Pejenaute Ochoa, José Ignacio Ibeas Corcelles and Ramón Ramos Barrales

Abstract

Proteins perform essential structural, enzymatic, and regulatory functions, many of which depend on their subcellular localization. In plant–pathogen interactions, they are crucial in establishing virulence. The phytopathogenic fungus *Ustilago maydis* employs protein-based strategies, including effector secretion, cell wall remodeling, membrane-associated transport, and regulatory signaling mechanisms to infect the maize plant. To facilitate the study of these proteins, a set of optimized and reproducible protocols has been developed for isolating cytosolic, membrane-bound, secreted, and glycosylated protein fractions. These methods combine cryogenic disruption, specific lysis buffers, TCA–DOC precipitation, and affinity enrichment, providing a robust workflow for comprehensive protein analysis in fungal pathogenesis.

Keywords Protein extraction, Secreted proteins, Membrane proteins, Glycosylation, Plant-pathogenic fungi, *Ustilago maydis*.

1 Introduction

Proteins are essential components of all cellular organisms, performing a wide range of structural, enzymatic, and regulatory functions [1]. Their activity is intrinsically linked to their location within specific cellular compartments [2], such as the extracellular space, cellular membranes, or the cytosol. To investigate these spatially defined functions, the separation of cellular components into distinct protein fractions is essential. This technique not only reveals a protein's native location but is also a crucial first step for its purification and functional analysis.

The maize pathogen *U. maydis* has become a widely used model for studying plant–pathogen interaction [3]. Significant progress has been made in identifying proteins crucial for pathogenic development. To establish infection, *U. maydis* secretes numerous effector proteins [4] such as Tin2 [5] and Cmu1 [6], which redirect host metabolism, and Rsp3, which modifies the cell wall to mask the fungus from the plant immune system [7]. Membrane-associated

An N. T. Phan et al. (eds.), *Ustilago maydis*, Methods in Molecular Biology, vol. 3050,
https://doi.org/10.1007/978-1-0716-5364-7_28,

proteins are also critical for pathogenesis. For example, Yup1, a soluble SNARE protein involved in early endocytic vesicle fusion, is necessary for proper pheromone perception and conjugation hyphae formation [8]. The protein Row1, found in membranes and secretory vesicles, is required for cell wall remodeling during appressoria formation [9]. Additionally, the transmembrane proteins Vmp1 and Vmp2 are required for full virulence, though their precise functions remain to be fully elucidated [10]. Cytoplasmic proteins, especially involved in signaling regulatory cascades, are also essential for pathogenic development in *U. maydis*. Mating and filamentous development are coregulated by the MAP kinase (MAPK) and the PKA pathways, converging on one of the master regulators of infection, Prf1 [11–13]. Many of these proteins, especially secreted, membrane, and cell-wall proteins, undergo post-translational modifications such as glycosylation [14], which is critical for infection by ensuring protein stability, localization, and function [15, 16]. Examples include the β-1,3-glucanase Erc1, which facilitates cell-to-cell extension within the plant bundle sheath [17]; the mucin-like transmembrane protein Msb2, which plays a role in signaling during pathogenic development [18]; the effector Cmu1, which modifies host metabolism [6]; and the protein disulfide isomerase Pdi1, whose N-glycosylation is essential for full virulence [19].

While proteins are key mediators of pathogenic development, their study presents significant hurdles, including inherent instability and complex subcellular localization. To overcome these obstacles, this chapter provides reproducible protocols for isolating distinct protein fractions, including total, cytosolic, membrane-bound, and secreted, as well as enriching for glycoproteins from *U. maydis*. These methods combine cryogenic cell disruption, specific lysis buffers, TCA-DOC precipitation for secreted proteins, and affinity-based enrichment for glycoproteins. This toolkit provides a solid foundation for dissecting the subcellular proteome of *U. maydis* and advancing the study of its pathogenesis.

2 Materials

2.1 Media and Buffer Solutions

2.1.1 Strain Growth Media

1. **YEPSL (Yeast Extract Peptone Sucrose Liquid medium, rich medium) rich medium:** 1% yeast extract, 0.4% bactopeptone, and 0.4% saccharose as carbon source. Preparation: mix all components in distilled water. Sterilize by autoclaving.
2. **YPDU (Yeast Extract Peptone Dextrose Uridine medium) rich medium:** 1% yeast extract, 2% bactopeptone, and 2% agar. 1% D-glucose as carbon source. Preparation: mix all components in distilled water except agar and D-glucose, adjust pH to 7.0, and add agar and autoclave. After cooling to ~50°C, add filter-sterilized D-glucose.

3. **CMD(complete medium D-glucose):** 0.25% casaminoacids, 0.1% yeast extract, 1% vitamin solution, 6.25% "Holliday'74" salts, 0.15% NH_4NO_3, pH 7.0 adjusted, 1% D-glucose as carbon source. Preparation: mix all components in distilled water except D-glucose, adjust pH to 7.0, and autoclave. Before using, add filter-sterilized D-glucose.
4. **MM-NH$_4$(Minimal Medium with Ammonium as nitrogen source):** 6.25% "Holliday, 74" salts solution, 0.3% ammonium sulfate ($(NH_4)_2SO_4$) as nitrogen source, 2% D-glucose as carbon source, adjusted to pH 7.0 with NaOH. Preparation: mix all components in distilled water except D-glucose, adjust pH to 7.0, and autoclave. Before using, add filter-sterilized D-glucose.
5. **Holliday'74" salts solution:** 16% KH_2PO_4, 4% Na_2SO_4, 8% KCl, 4.09% $MgSO_4 \times 7H_2O$, 1.32% $CaCl_2 \times 2H_2O$ and 8% trace elements.
6. **Vitamin Solution:** 0.1% thiamine, 0.05% riboflavin, 0.05% pyridoxine, 0.2% calcium pantothenate, 0.05% p-aminobenzoic acid, 0.2% nicotinic acid, 0.2% choline chloride, and 1% myo-inositol. All compounds were obtained from Sigma-Aldrich.

2.1.2 Buffer Solutions

All buffers must be prepared using ultrapure water and pre-chilled before use. All centrifugations should be performed at 4°C, and both the centrifuges and rotors should be pre-chilled. Proteins must be kept on ice throughout the process.

2.2 Total Protein Extraction

Strain Growth Media

1. YPDU Plates.
2. YEPSL.

Buffers and Solutions

1. Wash buffer (Tris–HCl 20 mM, pH 8.8).
2. RIPA lysis buffer (50 mM Tris/HCl, pH 8, 150 mM NaCl, 1%Nonidet P-40, 0.5% sodium deoxycholate, 0.1% SDS).

Reagents

1. RC-DC™ Protein Assay (BioRad).
2. Inhibitor Cocktail cOmplete Tablets EDTA-Free™ (Roche) (*see* **Note 1**).

Equipment

1. 1.5 mL tubes (low-protein binding recommended).
2. 1.5 mL screw-cap microcentrifuge tubes.
3. 50 mL polypropylene tubes.
4. Mortar and pestle.

5. Small scoop.
6. Ice bucket.
7. Liquid nitrogen (LN_2).
8. Liquid nitrogen container.
9. Refrigerated centrifuge.
10. Refrigerated microcentrifuge.
11. Spectrophotometer.

2.3 Soluble and Membrane Protein Extraction

Strain Growth Media

1. YPDU Plates.
2. YEPSL.

Buffer and Solutions

1. Wash buffer (Tris–HCl 20 mM, pH 8.8).
2. NaCl lysis buffer (Tris–HCl 20 mM, NaCl 0.5 M pH 7.4).
3. Urea lysis buffer (8 M Urea, 0.5 M NaCl, 2% SDS, 20 mM Tris, pH 7.5). Store this buffer at room temperature, as urea tends to become cloudy when stored cold.

Reagents

1. RC-DC™ Protein Assay (BioRad).
2. Inhibitor Cocktail cOmplete Tablets EDTA-Free™ (Roche).

Equipment

1. 1.5 mL tubes (low-protein binding recommended).
2. 1.5 mL screw-cap microcentrifuge tubes.
3. 50 mL polypropylene tubes.
4. Mortar and pestle.
5. Small scoop.
6. Ice bucket.
7. Liquid nitrogen (LN_2).
8. Liquid nitrogen container.
9. Refrigerated centrifuge.
10. Refrigerated microcentrifuge.
11. Spectrophotometer.
12. Sonicator.

2.4 Secreted Protein Extraction

Yeast Growth Media

1. YPDU Plates.
2. MM-NH_4.
3. YEPSL.

Buffer and Solutions (*see* **Note 2**)

1. 2% Deoxycholate (DOC) solution.
2. TS buffer (Urea 7 M, Thiourea 2 M, CHAPS 4%).

Reagents

1. Acetone stored at −20°C.
2. Trichloroacetic acid (TCA).
3. RC-DC™ Protein Assay (BioRad).

Equipment

1. 1.5 mL tubes (low-protein binding recommended).
2. High-speed polypropylene copolymer (PPCO) centrifuge tubes (250 mL).
3. Ice bucket.
4. Liquid nitrogen (LN_2).
5. Liquid nitrogen container.
6. Refrigerated superspeed centrifuge.
7. Refrigerated microcentrifuge.
8. Thermomixer.
9. Spectrophotometer.
10. Vortex.
11. Vacuum pump.

2.5 Glycosylated Protein Extraction

Yeast Growth Media

1. YPDU Plates.
2. CMD.
3. YEPSL.

Buffer and Solutions (*see* **Note 3**)

1. Wash buffer: Tris–HCl 20 mM, pH 8.8.
2. Binding buffer: 20 mM Tris–HCl, 0.5 M NaCl, 2 mM $MnCl_2$, 2 mM $CaCl_2$, pH 7.4.
3. Elution buffer: 20 mM Tris–HCl, 0.5 M NaCl, and 0.5 M methyl-α-O-glucopyranoside, pH 7.4.
4. 5X SDS-PAGE buffer: SDS 10% (w/v), Glycerol 50% (v/v), DTT 5% (v/v), Bromophenol 0.05% (w/v).

Reagents

1. HiTrap® Con A 4B columns 1 mL (Cytiva).
2. Inhibitor Cocktail cOmplete Tablets EDTA-Free (Roche).
3. Trichloroacetic acid (TCA).

4. Acetone stored at −20°C.
5. RC-DC™ Protein Assay (BioRad).

Equipment

1. 1.5 mL tubes (low-protein binding recommended).
2. High-speed polypropylene copolymer (PPCO) centrifuge tubes (50 mL and 250 mL).
3. 50 mL polypropylene tubes.
4. 15 mL polypropylene tubes.
5. 5–10 mL pipette.
6. 0.45 μm filters.
7. Ice bucket.
8. Liquid nitrogen (LN_2).
9. Liquid nitrogen container.
10. Cryogenic grinder.
11. Refrigerated superspeed centrifuge.
12. Refrigerated microcentrifuge.
13. Refrigerated centrifuge.
14. Fast Protein Liquid Chromatography System (FPLC).
15. Vortex.
16. Vacuum pump.
17. Spectrophotometer.

3 Methods

3.1 Total Protein Extraction

This protocol allows the extraction of total proteins, including cytosolic, nuclear, lumenal organelles, and partially membrane-associated proteins. However, highly hydrophobic integral membrane proteins as multi-spanning membrane proteins, certain proteins embedded within membranes, and proteins very strongly bound to DNA may not be fully solubilized. Protein extraction is developed as indicated in [20] with some modifications. The workflow for total protein extraction is schematically illustrated in Fig. 1A.

3.1.1 Cultures Growth

Once the culture is harvested for centrifugation, all subsequent steps, including centrifugations, should be performed on ice or at 4°C to prevent protein degradation and preserve protein integrity.

1. Inoculate 5 mL of *U. maydis* pre-culture in YEPSL medium overnight, using a mixture of colonies previously streaked on a

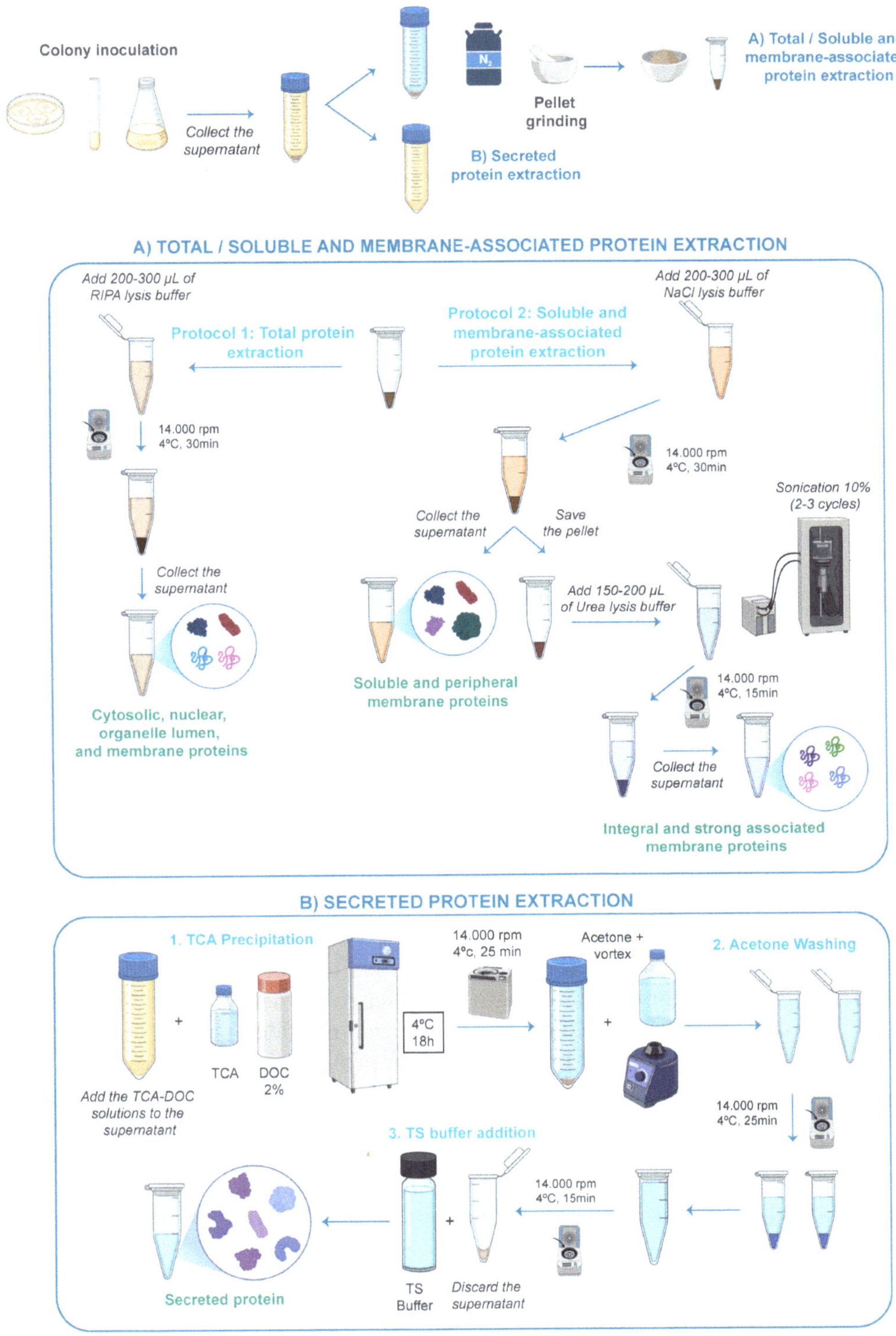

Fig. 1 Overview of total, soluble/membrane-associated, and secreted protein extraction: schematic representation of the general workflow, including *U. maydis* culture inoculation and scale-up to exponential phase. (**A**) Total proteins are extracted through mechanical disruption via cryogenic grinding and cell lysis (1. Total protein extraction); soluble and membrane fractions are obtained using specific buffers and treatments (2. Soluble and membrane-associated proteins). (**B**) Secreted proteins are recovered by DOC–TCA precipitation of the culture supernatant, followed by acetone washing and pellet resuspension

YPDU plate from a frozen glycerol stock and incubated for 2–3 days. Scale up by transferring the culture to 50 mL. Grow the culture at 28°C to the exponential phase (OD_{595} = 0.6–0.8).

2. Collect and centrifuge the culture at 3400 × *g* for 5 min at 4°C in a refrigerated centrifuge. Carefully discard the supernatant (*see* **Note 4**). Wash the resulting cell pellet twice with cold Wash buffer and transfer it to a 1.5 mL screw-cap microcentrifuge tube. Immerse the tube in liquid nitrogen to ensure rapid freezing (*see* **Note 5**).
3. Grind the pellet under liquid nitrogen using a pre-chilled mortar and pestle. Collect the powder with a pre-chilled small scoop and transfer it to a new screw-cap tube. Immerse the tube again in liquid nitrogen (*see* **Note 6**).

3.1.2 Protein Extraction

1. Resuspend the resulting cell powder in 200–300 μL of RIPA Lysis buffer supplemented with 2× protease Inhibitor Cocktail (cOmplete™, EDTA-free).
2. Incubate on ice for 10–15 min with occasional mixing by inversion or pipetting to ensure complete lysis (*see* **Note 7**).
3. Centrifuge the lysate at 18.000 × *g* for 30 min at 4°C in a refrigerated microcentrifuge, carefully collect the supernatant and transfer it to a new 1.5 mL tube.
4. Centrifuge the supernatant again for 15 min under the same conditions to further clarify it and recover the final supernatant.
5. Quantify the protein concentration using the RC-DC™ Protein Assay, following the manufacturer's instructions and store the samples at −80°C for long-term preservation (*see* **Note 8**).

3.2 Soluble and Membrane Protein Extraction

This protocol describes a two-step extraction protocol based on the method described in [9]. The first fraction is obtained using a non-denaturing lysis buffer to primarily enrich in soluble proteins, including cytosolic, proteins inside organelles, and peripheral membrane proteins. Nuclear proteins tightly associated with DNA, integral and strongly membrane-associated proteins are retained in the second fraction, which is then solubilized under denaturing conditions, using urea and sonication. The workflow for soluble and membrane-associated protein extraction, including key steps, buffers, and volumes used, is schematically illustrated in Fig. 1A.

3.2.1 Cultures Growth

Once the culture is harvested for centrifugation, all subsequent steps, including centrifugations, should be performed on ice or at 4°C to prevent protein degradation and preserve protein integrity.

1. Inoculate 5 mL of *U. maydis* pre-culture in YEPSL medium overnight using a mixture of colonies previously streaked on a YPDU plate from a frozen glycerol stock and incubated for 2–3

days. Scale up by transferring the culture to 50 mL. Grow the culture at 28°C to the exponential phase (OD595 = 0.6–0.8).

2. Collect and centrifuge the culture at 3400 × *g* for 5 min at 4°C in a refrigerated centrifuge. Carefully discard the supernatant. Wash the resulting cell pellet twice with cold Wash buffer and transfer to a 1.5 mL screw-cap microcentrifuge tube. Immerse the tube in liquid nitrogen to ensure rapid freezing (*see* **Note 5**).
3. Grind the pellet under liquid nitrogen using a pre-chilled mortar and pestle. Collect the powder with a pre-chilled small scoop and transfer it to a new 1.5 mL screw-cap microcentrifuge tube. Immerse the tube again in liquid nitrogen (*see* **Note 6**).

3.2.2 Soluble Protein Extraction

1. Resuspend the resulting cell powder in 200–300 μL of NaCl lysis buffer supplemented with 2× protease inhibitor cocktail (cOmplete™, EDTA-free, Roche).
2. Centrifuge the lysate at 18.000 × *g* for 30 min at 4°C in a refrigerated microcentrifuge and carefully collect the supernatant and transfer it to a new 1.5 mL tube. Save the insoluble pellet for membrane fraction extraction (see Sect. 3.2.3).
3. Centrifuge the supernatant again at 18.000 × *g* for 15 min at 4°C to further clarify it and recover the final supernatant.
4. Determine the protein concentration using the RC-DC™ Protein Assay, following the manufacturer's instructions and store the samples at −80°C for long-term preservation (*see* **Note 8**).

3.2.3 Membrane Protein Extraction

1. Resuspend the pellet previously obtained from Step 3.2.2 (Soluble protein extraction) in 150–200 μL of urea lysis buffer.
2. Sonicate at 10% amplitude in cycles of 5–10 s followed by 1 min on ice. Repeat this Step for two to three cycles (*see* **Note 9**).
3. Centrifuge for 15 min at 18.000 × *g* at 4°C in a refrigerated microcentrifuge. Collect the supernatant and transfer it to a new 1.5 mL tube. This fraction contains the previously insoluble proteins that have now been solubilized. The remaining pellet contains proteins that could not be solubilized.
4. Quantify the protein concentration using the RC-DC™ Protein Assay, following the manufacturer's instructions and store the samples at −80°C for long-term preservation (*see* **Note 8**).

3.3 Secreted Protein Extraction

The protocol for the precipitation of secreted proteins is based on the method described in [21] with some modifications. This method incorporates modifications and detailed steps that enhance protein recovery and reproducibility, addressing limitations of previously reported procedures. The workflow for secreted protein precipitation is schematically illustrated in Fig. 1B.

3.3.1 Cultures Growth

Once the culture is harvested for centrifugation, all subsequent steps, including centrifugations, should be performed on ice or at 4°C to prevent protein degradation and preserve protein integrity.

1. Inoculate 5 mL of *U. maydis* pre-culture in YEPSL medium using a mixture of colonies previously streaked on a YPDU plate from a frozen glycerol stock and incubated for 2–3 days. Scale up to 25 mL of YEPSL and incubate until the culture reaches the exponential phase (OD_{595} 0.6–0.8). Use this culture to inoculate 200 mL of MM-NH_4 medium supplemented with 2% D-glucose (*see* **Note 10**) and grow under exponential conditions (*see* **Note 11**) for protein collection.
2. Collect and centrifuge the culture at 3100 × *g* for 7 min at 4°C in a refrigerated superspeed centrifuge and carefully collect the supernatant (*see* **Note 4**). Repeat the centrifugation of the recovered supernatant to ensure complete removal of any residual pellet from the secreted protein fraction. After the final collection of the supernatant, proceed to TCA-DOC precipitation.

3.3.2 TCA-DOC Precipitation

1. Add 0.02% of sodium deoxycholate (DOC) to the secreted culture supernatant. For a 200 mL volume, this can be achieved by adding 2 mL of a 2% DOC stock solution. Gently invert the tubes three to four times to mix. Incubate the mixture on ice or at 4°C for 30 min. During this incubation, mix the tubes three to four times at 10-min intervals to ensure homogeneous mixing (*see* **Note 12**).
2. Add 20 mL of TCA to 200 mL of culture supernatant to reach a final TCA concentration of 10% and mix thoroughly by gently inverting the tubes to ensure complete contact between TCA and proteins, facilitating efficient precipitation. Incubate the mixture at 4°C for at least 18 h (*see* **Note 13**).
3. Centrifuge the protein precipitate at 30.000 × *g* for 25 min at 4°C in a refrigerated superspeed centrifuge. After centrifugation, quickly discard the supernatant by inverting the tube and remove any remaining liquid using a vacuum pump (*see* **Note 14**).
4. Resuspend the pellet in 2 mL of cold acetone, using a vortex to detach the pellet completely from the tube walls. Perform cycles of vortexing and incubation on ice to ensure proper resuspension while minimizing protein degradation. Transfer the suspension into two pre-chilled 1.5 mL tubes.
5. Centrifuge for 25 min at 18.000 × *g* at 4°C in a refrigerated microcentrifuge. Discard the supernatant and resuspend the resulting

pellets in 500 μL of cold acetone and combine the samples into a single tube. Centrifuge again at 18.000 × *g* for 15 min at 4°C.

6. Repeat the acetone wash step three times, washing the pellet a total of five times with cold acetone. Discard the supernatant and remove any residual liquid using a vacuum pump. Air-dry the pellet at room temperature until all residual acetone has completely evaporated (*see* **Note 15**).

3.3.3 Solubilization of Secreted Protein Pellet in TS Buffer (see Note 16)

1. Resuspend the dried protein pellet in 30 μL of TS buffer. Mix thoroughly by pipetting and vortex in a thermomixer for 5 min at 1200 rpm at room temperature.
2. Centrifuge for 2 min at 18.000 × *g*. Transfer the supernatant to a new 1.5 mL tube.
3. Resuspend the remaining pellet in an additional 30 μL of TS buffer. Mix by pipetting and vortex in the thermomixer for 5 min at 1200 rpm at room temperature.
4. Centrifuge for 2 min at 18.000 × *g* and combine the supernatant with the previously collected 30 μL.
5. Use the combined 60 μL protein extract to resuspend the remaining pellet once again. Mix thoroughly by pipetting and vortex in the thermomixer for 10 min at 1200 rpm at room temperature.
6. Transfer the final supernatant to a new 1.5 mL tube (*see* **Note 17**).
7. Quantify the protein concentration using the RC-DC™ Protein Assay, following the manufacturer's instructions and store the samples at −80°C for long-term preservation (*see* **Note 8**).

3.4 Glycosylated Protein Extraction

This protocol outlines a comprehensive and reproducible method for the isolation and purification of glycoproteins using concanavalin A (ConA) affinity chromatography. The purified samples are suitable for downstream applications such as SDS-PAGE, Western blotting, and LC–MS/MS analysis.

3.4.1 Cultures Growth

1. Inoculate 10 mL of *U. maydis* pre-culture in YEPSL medium using a mixture of colonies previously streaked on a YPDU plate from a frozen glycerol stock and incubated for 2–3 days. Scale up by transferring the culture to 1 L of CMD medium supplemented with 1% D-glucose. Grow the culture at 28°C to the exponential phase (OD595 0.6–0.8).
2. Centrifuge the culture at 3100 × *g* in a superspeed centrifuge for 5 min at 4°C. Carefully discard the supernatant.
3. Wash the resulting cell pellet twice with cold Wash buffer. Add 40 mL of cold Wash buffer to the cell pellet and gently resuspend. Transfer the suspension to a sterile 50 mL tube and centrifuge at 3400 × *g* for 7 min at 4°C in a refrigerated centrifuge. Carefully discard the supernatant.

4. Resuspend the resulting pellet in 5 mL of cold Wash Buffer by pipetting with a big volume pipette (*see* **Note 18**). Avoid generating foam or introducing air bubbles during resuspension (*see* **Note 19**).
5. Dispense approximately 20–30 mL of liquid nitrogen into a sterile 50 mL tube. Gradually add the 5 mL of resuspended culture into the liquid nitrogen, avoiding overly slow dripping to minimize nitrogen evaporation (*see* **Note 20**). Immediately transfer the resulting frozen sample to −80°C for long-term preservation or immerse the tube back into liquid nitrogen for pellet disruption.
6. Disrupt the frozen cell pellets using four cycles of 1 min using a cryogenic grinder and transfer the powder to a 15 mL tube (*see* **Note 21**).

3.4.2 Protein Extraction

1. Resuspend the resulting cell powder in 1.5 volumes of binding buffer supplemented with 2× protease inhibitor cocktail (cOmplete™, EDTA-free). Distribute the suspension into 1.5 mL microcentrifuge tubes.
2. Centrifuge at 18.000 × *g* for 30 min at 4°C in a refrigerated microcentrifuge. If the supernatant remains viscous or unclear, perform an additional centrifugation for 15 min under the same conditions.
3. Carefully collect and combine the supernatants in a sterile 15 mL tube. Filter the volume collected through a 0.45 μm membrane filter to remove any remaining debris and transfer it to FPLC equipment using a ConA column.

3.4.3 Isolation of Glycosylated Proteins Using Affinity Chromatography (see Note 22)

1. Column Preparation: equilibrate the ConA column by sequentially washing with bidistilled water followed by at least 5 column volumes of binding buffer.
2. Sample Loading: load the clarified protein lysate onto the column at a low flow rate (~0.2 mL/min) to allow efficient binding of glycoproteins to the matrix.
3. Washing: wash the column with 25 column volumes of binding buffer at a flow rate of 1 mL/min to remove non-specifically bound proteins and other contaminants.
4. Elution: elute glycosylated proteins using the elution buffer at a flow rate of 0.1 mL/min. Collect a total of 20 mL, fractionated into five 15 mL tubes of 4 mL each.

3.4.4 TCA Glycoprotein Precipitation (see Note 2)

1. Mix the collected fraction in a 50 mL high-speed centrifuge tube and add 10% (v/v) of cold trichloroacetic acid (TCA). Mix thoroughly by gentle inversion. Incubate the sample overnight at 4°C to allow complete protein precipitation.
2. Centrifuge the sample at 25.200 × *g* for 25 min at 4°C to pellet the precipitated proteins. Carefully discard the supernatant and remove residual liquid using a vacuum pump.

3. Resuspend the pellet in 1 mL of cold acetone, using vortex. Perform several vortex–ice–vortex cycles to ensure complete resuspension while avoiding protein denaturation (*see* **Note 23**).
4. Transfer the sample to a 1.5 mL microcentrifuge tube and centrifuge at 18.000 × *g* for 15 min at 4°C. Repeat the acetone wash step a total of four times to ensure removal of TCA and contaminants.
5. Discard the supernatant and air-dry the pellet at room temperature or under vacuum until all residual acetone has completely evaporated (*see* **Note 15**).
6. Resuspend the dried protein pellet in 30 μL of SDS loading buffer. Ensure thorough resuspension by gently pipetting up and down. If the pellet is difficult to dissolve, use a thermomixer set at 1200 rpm for 10 min to facilitate complete resuspension.
7. Quantify the protein concentration using the RC-DC™ Protein Assay according to the manufacturer's instructions. Store the resuspended samples at −80°C until further use.

4 Notes

1. When extracting labile proteins such as histones, it is recommended to supplement the lysis or extraction buffer with protease inhibitors such as 1 μg/mL Bestatin and 1 mM PMSF to improve protein stability and prevent degradation.
2. Nitrile gloves should be worn throughout all procedures to ensure personal protection. Particular caution is required when handling trichloroacetic acid (TCA), sodium deoxycholate (DOC), and acetone; always consult the corresponding safety data sheets (SDS) prior to use. TCA is a corrosive agent; DOC is an irritant; and acetone is a highly volatile and flammable solvent that can also cause skin and eye irritation. All steps involving TCA and DOC should be performed in a fume hood.
3. All buffers should be prepared using ultrapure water and filtered through a 0.45 μm membrane filter. Prior to use, all solutions should be pre-chilled and maintained at 4°C to prevent protein degradation.
4. This step should be performed rapidly to prevent the cell pellet from dispersing into the supernatant.
5. Use screw-cap cryovials (1.5 mL) with a conical or round bottom to store samples in liquid nitrogen. Standard 1.5 mL microcentrifuge tubes may pop open or crack, causing sample loss. Conical-bottom tubes are easier to handle when

retrieving and grinding frozen samples. The pellet can be stored at −80°C before protein extraction.

6. It is essential to keep the pellet completely frozen during grinding to prevent protein degradation. Work quickly and avoid thawing. Grinding the pellet into a fine, homogeneous powder disrupts the cell structures facilitating subsequent protein extraction step. The powder obtained can be stored at −80°C for long-term storage until further extraction is performed.
7. This step is optional but enhances cell lysis.
8. It is recommended to aliquot the samples in volumes of approximately 30 μL. Samples stored at −80°C remain stable for up to three freeze–thaw cycles without significant loss of protein integrity.
9. To avoid foam formation or air bubbles during sonication, ensure that the sonicator probe is placed just below the surface of the liquid without touching the tube walls or bottom. Use short pulses (5–10 s) followed by cooling on ice, and avoid excessive amplitudes. Do not shake the tube before or during sonication. Use a tube with sufficient volume to minimize splashing and turbulence.
10. MM-NH_4 medium is used for the collection of secreted proteins due to its ability to enhance effector secretion compared to rich media. While some proteins can also be recovered from YEPSL, MM-NH_4 is generally preferred for effector enrichment.
11. Higher optical density (OD) correlates with better protein yield. An OD of 0.8 is optimal for protein extraction. Cultures with an OD below 0.5 should be avoided, as they yield insufficient protein for effective recovery.
12. Sodium deoxycholate (DOC) enhances protein precipitation by altering protein solubility and facilitating aggregation when combined with TCA, resulting in improved protein yield.
13. Before precipitation, small white protein particles may be observed. However, the absence of these particles does not indicate the lack of protein precipitation. The incubation time can be extended up to 24 h if needed for better precipitation.
14. This step must be performed quickly and carefully, as the pellet resuspends rapidly. If the pellet begins to mix, centrifuge again to recover the pellet.
15. Complete evaporation is essential to ensure later efficient resuspension of the pellet in TS buffer. 30 min to 1 h should be sufficient for complete evaporation.

16. Complete resuspension of the protein pellet in TS buffer can be challenging due to its compact and adhesive nature. Multiple resuspension steps may be needed to maximize protein recovery from the pellet.
17. Although full solubilization of the pellet may not be achieved due to its compact nature, the recovered protein yield is adequate for downstream applications, including SDS-PAGE followed by Western blotting or secretome profiling via mass spectrometry.
18. Due to the high viscosity and density of the pellet suspension, it is recommended to resuspend it using a high-volume pipette (e.g., 5, 10, or 25 mL), as these allow smooth handling without introducing bubbles. Standard pipette tips are not suitable, as they tend to clog easily and make resuspension inefficient.
19. If bubbles appear or the pellet does not dissolve completely, an additional 2 mL of cold Wash Buffer may be added to facilitate the process. The suspension should appear homogeneous, with no visible clumps or undissolved material, before proceeding to the next step.
20. Carefully drop the resuspended cell sample into the liquid nitrogen, one drop at a time. This step facilitates subsequent disruption of the pellet, improving handling and processing in downstream applications.
21. Ensure the system remains at cryogenic temperatures throughout the process to avoid thawing, and make sure the pellet is fully ground into a fine powder after the cryogenic grinding step. Complete pulverization is essential to maximize protein extraction efficiency.
22. The described procedure may require adjustments depending on the specific column and FPLC system employed. It is essential to optimize volumes, flow rates, and washing and elution conditions in accordance with the manufacturer's guidelines and the characteristics of the equipment to ensure optimal binding efficiency and protein purity.
23. If the pellet is not completely detached from the wall of the tube, use a pipette to help dislodge it.

Acknowledgments

We thank Ismael Moreno-Sánchez and Miriam Marín-Menguiano for kindly providing the initial protocols on which the present methods were based, and Laura Tomás-Gallardo for her continuous support and valuable expertise throughout this work. This work was supported by MCIN/AEI/10.13039/501100011033/Grant No. PID2019-110477GB-I00 to JII and AEI/10.13039/501100011033/FEDER, UE, Grant No. PID2022-142783NB-I00 to JII and RRB.

References

1. Alberts B et al (2002) Molecular biology of the cell, 4th edn. Garland Science
2. Lundberg E, Borner GHH (2019) Spatial proteomics: a powerful discovery tool for cell biology. Nat Rev Mol Cell Biol 20:285–302. https://doi.org/10.1038/s41580-018-0094-y
3. Dean R et al (2012) The top 10 fungal pathogens in molecular plant pathology. Mol Plant Pathol 13:414–430. https://doi.org/10.1111/j.1364-3703.2011.00783.x
4. Lanver D et al (2017) Ustilago maydis effectors and their impact on virulence. Nat Rev Microbiol 15:409–421. https://doi.org/10.1038/nrmicro.2017.33
5. Tanaka S et al (2019) Neofunctionalization of the secreted Tin2 effector in the fungal pathogen Ustilago maydis. Nat Microbiol 4:251–257. https://doi.org/10.1038/s41564-018-0304-6
6. Djamei A et al (2011) Metabolic priming by a secreted fungal effector. Nature 478:395–398. https://doi.org/10.1038/nature10454
7. Ma L-S et al (2018) The Ustilago maydis repetitive effector Rsp3 blocks the antifungal activity of mannose-binding maize proteins. Nat Commun 9: 1711. https://doi.org/10.1038/s41467-018-04149-0
8. Fuchs U et al (2006) Endocytosis is essential for pathogenic development in the corn smut fungus Ustilago maydis. Plant Cell 18:2066–2081. https://doi.org/10.1105/tpc.105.039388
9. Pejenaute-Ochoa MD et al (2024) Row1, a member of a new family of conserved fungal proteins involved in infection, is required for appressoria functionality in Ustilago maydis. New Phytol 243:1101–1122. https://doi.org/10.1111/nph.19798
10. Weiland P, Altegoer F (2021) Identification and characterization of two transmembrane proteins required for virulence of Ustilago maydis. Front Plant Sci 12:669835. https://doi.org/10.3389/fpls.2021.669835
11. Brefort T et al (2009) Ustilago maydis as a pathogen. Annu Rev Phytopathol 47:423–445. https://doi.org/10.1146/annurev-phyto-080508-081923
12. Kaffarnik F (2003) PKA and MAPK phosphorylation of Prf1 allows promoter discrimination in Ustilago maydis. EMBO J 22:5817–5826. https://doi.org/10.1093/emboj/cdg554
13. Zhao X et al (2007) Mitogen-activated protein kinase pathways and fungal pathogenesis. Eukaryot Cell 6:1701–1714. https://doi.org/10.1128/EC.00216-07
14. Fernández-Álvarez et al (2010) Protein glycosylation in the phytopathogen Ustilago maydis: from core oligosaccharide synthesis to the ER glycoprotein quality control system, a genomic analysis. Fungal Genet Biol 47:727–735. https://doi.org/10.1016/j.fgb.2010.06.004
15. Fernández-Álvarez A et al (2009) The *O* -Mannosyltransferase PMT4 is essential for normal appressorium formation and penetration in *Ustilago maydis*. Plant Cell 21:3397–3412. https://doi.org/10.1105/tpc.109.065839
16. Pejenaute-Ochoa MD et al (2021) Structural, evolutionary, and functional analysis of the protein O-Mannosyltransferase family in pathogenic fungi. JoF 7:328. https://doi.org/10.3390/jof7050328
17. Ökmen B et al. (2022) A conserved enzyme of smut fungi facilitates cell-to-cell extension in the plant bundle sheath. Nat Commun 13: 6003. https://doi.org/10.1038/s41467-022-33815-7
18. Lanver D et al (2010) Sho1 and Msb2-related proteins regulate appressorium development in the smut fungus Ustilago maydis. Plant Cell 22:2085–2101. https://doi.org/10.1105/tpc.109.073734
19. Marín-Menguiano M et al (2019) N-glycosyla tion of the protein disulfide isomerase Pdi1 ensures full Ustilago maydis virulence. PLoS Pathog 15:e1007687. https://doi.org/10.1371/journal.ppat.1007687
20. Navarrete B et al. (2023) Systematic characterization of Ustilago maydis sirtuins shows Sir2 as a modulator of pathogenic gene expression. Front Microbiol 14:1157990. https://doi.org/10.3389/fmicb.2023.1157990
21. Fernández-Álvarez A et al (2012) Identification of O-mannosylated Virulence Factors in Ustilago maydis. PLoS Pathog 8:e1002563. https://doi.org/10.1371/journal.ppat.1002563

Chapter 29

Studying Cell Wall-Associated Effectors in *U. maydis*

Shigeyuki Tanaka

Abstract

Fungal cell wall-binding secreted effector proteins play pivotal roles in plant–microbe interactions and cell wall remodeling in *Ustilago maydis*. Accurate assessment of their localization and binding mode is essential for understanding their biological function. Here, a protocol combining immunolocalization with fluorescence microscopy and immunoprecipitation (IP) under both native and denaturing (SDS-containing) conditions is described. Immunolocalization provides direct visualization of protein distribution at the cell wall, while comparative IP experiments reveal the strength and nature of protein–cell wall associations. This combined approach offers a reproducible and adaptable framework for analyzing secreted proteins that associate with cell walls.

Keywords Fungal cell wall, Immunocytochemistry, Effector proteins, Fluorescent microscopy, Immunoprecipitation

1 Introduction

Plant pathogenic fungi rely on secreted effector proteins to successfully colonize their host plants [1, 2]. Most research has traditionally focused on effectors that target apoplastic or intracellular components of the plant, where they suppress immune responses or manipulate metabolic pathways to favor pathogen survival [1, 3, 4]. These studies have provided detailed insights into the molecular strategies that pathogens employ to overcome plant defenses.

More recently, works in the corn smut fungus *Ustilago maydis* have highlighted a distinct group of effectors that do not primarily interact with plant proteins but instead associate with the fungal cell wall [5–9]. These cell wall-binding effectors appear to play critical roles in maintaining fungal structure, modulating interactions with the host surface, and protecting the pathogen from plant-derived antimicrobial compounds [6, 8].

Biochemical analysis of these proteins is often challenging, as their tight association with polysaccharides and the extracellular matrix complicates extraction and purification, limiting the applicability of

An N. T. Phan et al. (eds.), *Ustilago maydis*, Methods in Molecular Biology, vol. 3050,
https://doi.org/10.1007/978-1-0716-5364-7_29,

conventional assays [10, 11]. To overcome these limitations, immunological approaches provide practical alternatives. Immunolocalization enables visualization of protein distribution directly in situ, whereas comparative immunoprecipitation under non-denaturing and denaturing conditions can indicate whether a secreted protein is tightly bound to the fungal cell wall.

In this chapter, experimental strategies for investigating cell wall-binding effectors are presented. Specifically, protocols are provided for analyzing the localization of secreted proteins using antibody-based fluorescence microscopy, as previously described [8, 12], and for comparing non-denaturing and denaturing immunoprecipitation to assess the stability of cell wall association. Together, these approaches offer a framework for the functional analysis of secreted proteins that bind to the fungal cell wall, providing new insights into their roles in infection and host colonization.

2 Materials

2.1 Culture Medium for *U. maydis*

The medium to be prepared will be autoclaved at 121°C for 5 min before use.

Potato dextrose agar (PD plate): 3.9% (w/v) potato dextrose agar (Difco), 10 mM Tris-HCl (pH 8.0).

YEPS light liquid medium: 1.0% (w/v) yeast extract, 0.4% (w/v) peptone, 0.4% (w/v) sucrose.

CM liquid medium with glucose (CM-glucose): 0.25% (w/v) casamino acids, 0.1% (w/v) yeast extract, 1.0% (v/v) vitamin solution by Holliday '74, 6.25% (v/v) salt solution by Holliday '74, 0.05% (w/v) deoxyribonucleic acid from herring sperm, 0.15% (w/v) NH_4NO_3, 10 mM Tris-HCl (pH 8.0). Adjust the pH to 7.0 with 5 M NaOH. After autoclave, add 2% (w/v) glucose.

Nitrate minimal liquid medium with glucose (NM-glucose): 0.3% (w/v) KNO_3, 6.25% (v/v) salt solution by Holliday '74. Adjust the pH to 7.0 with 5 M NaOH . After autoclave, add 2% (w/v) glucose.

Salt solution by Holliday '74: 1.6% (w/v) KH_2PO_4, 0.4% (w/v) Na_2SO_4, 0.8% (w/v) KCl, 0.4% (w/v) $MgSO_4 \cdot 7H_2O$, 0.13% (w/v) $CaCl_2 \cdot 2H_2O$, 0.8% (v/v) trace elements by Holliday '74.

Trace elements by Holliday '74: 0.006% (w/v) H_3BO_3, 0.014% (w/v) $MnCl \cdot 4H_2O$, 0.04% (w/v) $ZnCl_2$, 0.04% (w/v) $Na_2MoO_4 \cdot 2H_2O$, 0.01% (w/v) $FeCl_3 \cdot 6H_2O$, 0.003% (w/v) $CuSO_4$.

Vitamin solution by Holliday '74: 0.01% (w/v) thiamine hydrochloride (Sigma T4625), 0.005% (w/v) riboflavin, 0.005% (w/v) pyridoxine, 0.02% (w/v) calcium pantothenate, 0.005% (w/v) aminobenzoic acid, 0.02% (w/v) nicotinic acid, 0.02% (w/v) choline chloride, 0.1% (w/v) myo-inositol.

2.2 Protein Extraction, Sodium Dodecyl Sulfate (SDS)-Polyacrylamide Gel Electrophoresis and Western Blot

Protein extraction buffer: 50 mM Tris-HCl (pH 7.5), 150 mM NaCl, 1 mM EDTA, 10% glycerol, and a protease inhibitor cocktail. Optionally, 0.1% SDS, 0.5% Triton X-100, or 0.5% NP-40 can be added if stronger lysis is required to release proteins associated with the cell wall.

12% SDS-polyacrylamide gel: For the resolving (separating) gel, use 12% acrylamide/bis, 375 mM Tris–HCl (pH 8.8), and 0.1% SDS. For the stacking gel, use 4% acrylamide/bis, 125 mM Tris-HCl (pH 6.8), and 0.1% SDS. To initiate polymerization and solidify the gel, add 0.05% ammonium persulfate (APS) and 0.05% N,N,N′,N′-tetramethylethylenediamine (TEMED).

SDS running buffer: 25 mM Tris, 192 mM glycine, 0.1% SDS.

Tris-buffered saline with Tween 20 (TBST): 20 mM Tris-HCl (pH 7.5), 150 mM NaCl, 0.1% (v/v) Tween 20.

Towbin buffer: 25 mM Tris, 192 mM glycine, 20% (v/v) methanol. For high-molecular-weight proteins, 0.1% SDS is added to improve transfer.

2.3 Antibodies

Anti-HA magnetic beads.
Monoclonal anti-HA antibody (mouse).
Polyclonal anti-HA antibody (rabbit).
HRP-conjugated anti-rabbit IgG antibody.
Alexa Fluor 488-conjugated anti-mouse IgG antibody.

3 Methods

Prior to performing the following experiments, fungal strains constitutively expressing the target secreted protein fused to an appropriate epitope tag (e.g., hemagglutinin [HA]) should be generated.

3.1 Immunocytochemistry of Cell Wall-Binding Secreted Proteins

For immunocytochemistry, comparison with a negative control strains that lack expression of the target protein or express the diffusely secreted protein carrying the same epitope tag is indispensable. Because the method of filament induction differs depending on whether the available transformant is based on the *U. maydis* SG200 strain or the AB33 strain, the respective protocols are described separately.

3.1.1 Observation of Filaments from U. maydis SG200-Derived Strains

If transformants derived from *U. maydis* solopathogenic strain SG200 [13] are available, apply this protocol.

1. Grow *U. maydis* SG200-derived transformants constitutively expressing the putative cell wall-binding secreted protein in YEPS light liquid medium at 28°C overnight with shaking at 180 rpm (*see* **Note 1**).

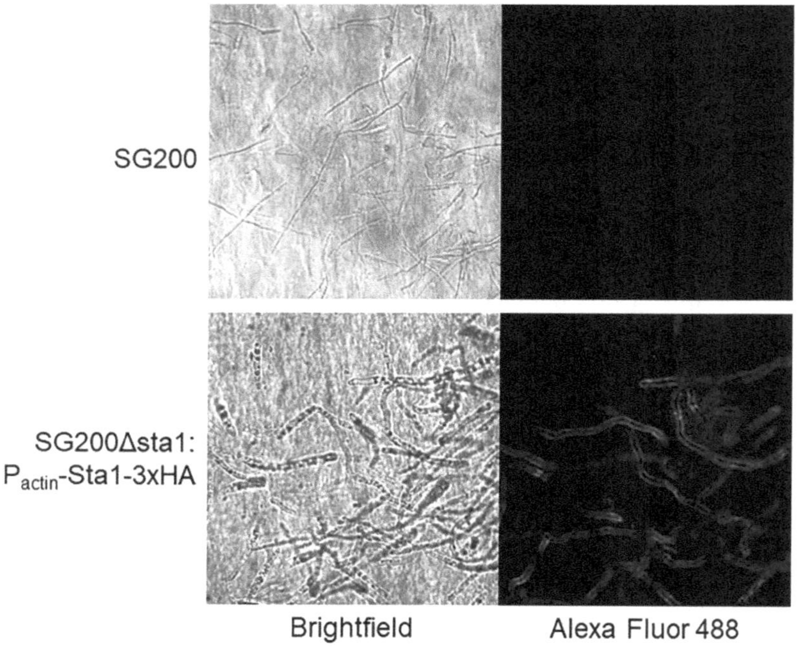

Fig. 1 Immunocytochemistry of Sta1-3xHA protein in filamentous cells from *U. maydis* SG200 and SG200Δ sta1:P_{actin}-Sta1-3xHA on parafilm. Green signal of Alexa Fluor 488 indicates the localization of Sta1-3xHA

2. Dilute the overnight culture in 2% YEPS light medium to an OD_{600} of 0.5 and supplement with 0.1 mM 16-hydroxyhexadecanoic acid [14].
3. Spray the sporidial suspension onto Parafilm and incubate at 28°C for 16 h under saturated humidity to induce filamentation (*see* **Note 2**).
4. Wash the Parafilm by gently applying phosphate-buffered saline (PBS) to remove unbound sporidia. Then, apply 500 μL of PBS containing mouse anti-HA antibody (1:1500 dilution) and 3% bovine serum albumin to Parafilm. Incubate the samples at 4°C overnight under saturated humidity (*see* **Note 3**).
5. Wash the Parafilm three times with 500 μL of PBS by gentle application and removal. Apply PBS containing Alexa Fluor 488–conjugated goat anti-mouse IgG secondary antibody (1:1500 dilution) and incubate for 1 h at room temperature.
6. Wash the Parafilm three times with PBS as above. Visualize filaments on the Parafilm using a confocal laser-scanning microscope under the following conditions: excitation at 488 nm and detection at 500–540 nm. Expected result after microscopic observation is shown in Fig. 1, which illustrates the localization of Sta1 effector of *U. maydis* on the filaments.

3.1.2 Observation of Filaments from *U. maydis* AB33-Derived Strains

If transformants derived from *U. maydis* AB33 strain [15] are available, the following protocol can be applied.

1. Grow *U. maydis* AB33-derived transformants constitutively expressing the putative cell wall-binding secreted protein in CM-glucose liquid medium at 28°C overnight with shaking at 180 rpm.
2. Dilute the overnight culture in NM-glucose liquid medium to an OD_{600} of 0.5 and induce filamentation at 28°C for 4–6 h with shaking at 180 rpm.
3. Harvest cells by centrifugation at 10,000 × *g* for 1 min and discard the supernatant.
4. Resuspend the cell pellet in PBS and fix with 3% formaldehyde for 30 min at room temperature (*see* **Note 4**).
5. Centrifuge at 10,000 × *g* for 1 min and remove the supernatant.
6. Wash the cell pellet by resuspending in PBS through gentle pipetting and centrifuging at 10,000 × *g* for 1 min; discard the supernatant. Repeat this wash step three times.
7. Incubate cells in PBS containing mouse anti-HA antibody (1:500 dilution) and 3% bovine serum albumin for 1 h at room temperature.
8. Wash the cells three times with PBS as described above.
9. Incubate in PBS containing Alexa Fluor 488–conjugated goat anti-mouse IgG secondary antibody (1:500 dilution) for 1 h at room temperature.
10. Wash the cells three times with PBS as described above.
11. Resuspend the final pellet in PBS, mount on a glass slide, and visualize AB33 filaments using a confocal laser-scanning microscope under the following conditions: excitation at 488 nm and detection at 500–540 nm.

3.2 Immunoprecipitation and Western Blot

When immunoprecipitating secreted proteins from infected leaves, strong detergents are generally not required for proteins that do not associate with the cell wall. In contrast, for cell wall–binding proteins, inclusion of a relatively strong detergent such as 0.1% (w/v) SDS in the extraction buffer can improve recovery, enabling detection by Western blotting. The epitope recognized by the antibody must tolerate the presence of SDS. Alternatively, milder detergents (e.g., 0.5% Triton X-100 or NP-40) can be used initially, with SDS added only if extraction efficiency is insufficient. The presence or absence of a detectable signal under these different extraction conditions can serve as an empirical indicator of whether the protein associates with the cell wall. However, it should be interpreted cautiously, as detection also depends on extraction efficiency and antibody recognition.

1. Harvest leaf tissue inoculated with SG200-derived transformants at the appropriate infection stage (*see* **Note 5**). Immediately freeze the tissue in liquid nitrogen and grind thoroughly with a pre-chilled mortar and pestle until a fine powder is obtained. Resuspend the powdered tissue in ice-cold protein extraction buffer (with or without detergents such as 0.1% SDS, 0.5% Triton X-100, or NP-40, depending on the protein of interest) at a ratio of ~2 mL buffer per gram of tissue. Mix briefly to ensure complete solubilization.
2. Incubate the lysate on a rotary shaker at 4°C for 20 min to promote protein extraction. Centrifuge at 14,000 × *g* for 15 min at 4°C. Carefully transfer the clarified supernatant to a fresh pre-chilled tube, avoiding contamination with insoluble debris.
3. Wash 5 μL of anti-HA magnetic beads three times with 500 μL protein extraction buffer to remove storage solution. Add the equilibrated beads to the clarified protein extract. Incubate the mixture with gentle rotation for 2 h at 4°C to allow specific binding of HA-tagged proteins.
4. Collect the beads using a magnetic stand and discard the supernatant. Wash the beads three times with 500 μL protein extraction buffer to minimize nonspecific binding. Perform each wash by gentle resuspension followed by magnetic separation.
5. Resuspend the washed beads in 20–30 μL of protein extraction buffer. Add 2%–5% (v/v) β-mercaptoethanol, then heat the suspension at 95°C for 10 min to denature the protein complexes and release bound proteins into the supernatant.
6. Place the sample on a magnetic stand and transfer the supernatant (the immunoprecipitated sample) to a new pre-chilled microcentrifuge tube. Keep on ice until use.
7. Mix immunoprecipitated proteins and input samples with appropriate loading buffer, and load onto a 12% SDS-polyacrylamide gel. Perform electrophoresis in SDS running buffer until the dye front reaches the bottom of the gel (*see* **Note 6**).
8. Transfer proteins from the gel to a PVDF membrane using Towbin transfer buffer. For semi-dry transfer, apply 10–15 V constant voltage, ensuring that the recommended current capacity of the blotting apparatus is not exceeded (*see* **Note 7**).
9. Block nonspecific binding sites on the membrane by incubating in 5% (w/v) skim milk prepared in TBST for 1 h at room temperature with gentle shaking.
10. Dilute rabbit polyclonal anti-HA antibody 1:10,000 directly in the blocking solution. Incubate the membrane overnight at 4°C with gentle shaking to ensure efficient binding.

11. Wash the membrane with TBST for 5 min at room temperature with shaking. Repeat the washing step three times to remove unbound antibody.
12. Incubate the membrane with horseradish peroxidase (HRP)-conjugated anti-rabbit IgG secondary antibody diluted 1:10,000 in 5% (w/v) skim milk in TBST. Incubate for 1 h at room temperature with gentle agitation.
13. Wash the membrane again three times with TBST (5 min each) to remove excess secondary antibody.
14. Incubate the membrane with enhanced chemiluminescence reagent according to the manufacturer's instructions. Visualize chemiluminescence using X-ray film exposure or a CCD-based imaging system.

4 Notes

1. Prior to cultivation in liquid medium, the frozen stock should be revived on a PD plate, and colonies derived from this plate should be used to inoculate the liquid culture. Direct inoculation of frozen stocks into liquid medium often results in poor growth.
2. In practice, Parafilm is cut to match the size of a glass slide, placed in a humidified chamber, and the sample is sprayed onto the Parafilm within this chamber.
3. If necessary, an additional piece of Parafilm of the same size may be placed on top of the PBS solution to prevent evaporation and to ensure that the antibody solution is evenly distributed across the entire surface of the Parafilm.
4. For fixation and the following antibody incubation steps, the samples should be placed on a rotating mixer (rotator) to provide continuous gentle agitation throughout the incubation.
5. For extraction from infected leaves, the symptomatic regions should be excised to ensure that the fungal biomass can be as large as possible compared with the leaf tissue volume.
6. Running the gel at a constant current of 30–40 mA for approximately 30 min can be acceptable.
7. For semi-dry transfer, the constant voltage to 10–15 V for 60–90 min can be applied following the manufacturer's recommendation not to exceed 25 V in case of a Bio-Rad system.

References

1. Lo Presti L et al (2015) Fungal effectors and plant susceptibility. Annu Rev Plant Biol 66: 513–545. https://doi.org/10.1146/annurev-arplant-043014-114623
2. Li G et al (2024) Fungal effectors: past, present, and future. Curr Opin Microbiol 81:102526. https://doi.org/10.1016/j.mib.2024.102526
3. Tanaka S et al (2015) Microbial effectors target multiple steps in the salicylic acid production and signaling pathway. Front Plant Sci 6:349. https://doi.org/10.3389/fpls.2015.00349
4. Franceschetti M et al (2017) Effectors of filamentous plant pathogens: commonalities amid diversity. Microbiol Mol Biol Rev 81(2). https://doi.org/10.1128/MMBR.00066-16
5. Tanaka S, Kahmann R (2021) Cell wall-associated effectors of plant-colonizing fungi. Mycologia 113(2):247–260. https://doi.org/10.1080/00275514.2020.1831293
6. Ma LS et al (2018) The *Ustilago maydis* repetitive effector Rsp3 blocks the antifungal activity of mannose-binding maize proteins. Nat Commun 9(1):1711. https://doi.org/10.1038/s41467-018-04149-0
7. Christ M et al (2025) Structural and functional analysis of the lectin-like protein Llp1 secreted by *Ustilago maydis* upon infection of maize. J Fungi (Basel) 11(2):164. https://doi.org/10.3390/jof11020164
8. Tanaka S et al (2020) The functionally conserved effector Sta1 is a fungal cell wall protein required for virulence in *Ustilago maydis*. New Phytol 227(1):185–199. https://doi.org/10.1111/nph.16508
9. Weiland P et al (2023) Structural and functional analysis of the cerato-platanin-like protein Cpl1 suggests diverging functions in smut fungi. Mol Plant Pathol 24(7):768–787. https://doi.org/10.1111/mpp.13349
10. Jamet E et al (2006) Cell wall proteins: a new insight through proteomics. Trends Plant Sci 11(1):33–39. https://doi.org/10.1016/j.tplants.2005.11.006
11. Gow NAR et al (2017) The fungal cell wall: structure, biosynthesis, and function. Microbiol Spectr 5(3). https://doi.org/10.1128/microbiolspec.FUNK-0035-2016
12. Yoshimoto R et al (2022) The production and secretion of tRNA-derived RNA fragments in the corn smut fungus *Ustilago maydis*. Front Fungal Biol 3:958798. https://doi.org/10.3389/ffunb.2022.958798
13. Kamper J et al (2006) Insights from the genome of the biotrophic fungal plant pathogen *Ustilago maydis*. Nature 444(7115):97–101. https://doi.org/10.1038/nature05248
14. Mendoza-Mendoza A et al (2009) Physical-chemical plant-derived signals induce differentiation in *Ustilago maydis*. Mol Microbiol 71(4):895–911. https://doi.org/10.1111/j.1365-2958.2008.06567.x
15. Brachmann A et al (2001) Identification of genes in the bW/bE regulatory cascade in *Ustilago maydis*. Mol Microbiol 42(4):1047–1063. https://doi.org/10.1046/j.1365-2958.2001.02699.x

Chapter 30

Computational Identification and Structural Analysis of *Ustilago maydis* Effector Proteins

Max Heinen and Florian Altegoer

Abstract

To successfully colonize their hosts, fungi secrete specialized proteins known as effectors that manipulate host physiology, for example, by suppressing immune responses or altering host cellular processes. Effector proteins are central to fungal virulence, yet they frequently lack recognizable domains and show little to no sequence conservation, reflecting their rapid adaptation during host–pathogen coevolution. As a result, annotation is limited and structure–function relationships remain poorly understood. In this study, a systematic computational workflow is presented for the identification and structural analysis of putative effector proteins in *Ustilago maydis*, a biotrophic model fungus. The protocol begins with retrieval of genome-derived protein sequences, prediction of secretion signals, and conversion of identifiers into standardized formats. Structural models are then obtained from the AlphaFold Protein Structure Database or generated de novo using AlphaFold 3/ColabFold. To facilitate comparative analysis, predicted structures can subsequently be clustered and explored using tools such as FoldMason and Foldseek, enabling inference of potential functional relationships. This approach integrates widely available resources with accessible computational steps, lowering the barrier for researchers without extensive bioinformatics expertise. It provides a framework to prioritize effector candidates for experimental validation and can be readily applied to diverse fungal pathogens.

Keywords Structure prediction, Structural bioinformatics, Effector proteins, Structure-based clustering, Host–pathogen interactions

1 Introduction

Effector proteins are typically small, secreted proteins used by fungi to manipulate their hosts, partner organisms, and surrounding microbial communities [1, 2]. These proteins are tailored to specific functions in diverse niches and accordingly often share limited sequence similarities, besides being short in sequence and mostly lacking functional domains [3]. Consequently, traditional sequence-based annotation methods often fail to recognize them. Structural approaches overcome many of these limitations. Since the protein fold space is more confined than the sequence space, three-dimensional models can

An N. T. Phan et al. (eds.), *Ustilago maydis*, Methods in Molecular Biology, vol. 3050,
https://doi.org/10.1007/978-1-0716-5364-7_30,

reveal distant evolutionary relationships or functional motifs that are missed by sequence-based searches [3, 4]. AlphaFold 2's breakthrough demonstrated that deep learning can predict structures with near-experimental accuracy, even in the absence of structures from similar proteins [5]. This breakthrough has enabled genome-wide structural inference across diverse organisms. Complementary tools such as Foldseek and FoldMason translate backbone geometry into a "structural alphabet," allowing rapid comparisons and clustering [6] of entire predicted structural datasets in a way comparable to sequence searches, such as BLAST [7]. Using these approaches, fungal effectors with unrelated sequences have been shown to converge on similar folds, forming so-called sequence-unrelated structurally similar (SUSS) families [4]. Such fold-based clustering enables functional inference from structural similarity, even when sequence conservation is absent [3, 6]. This means that an effector with a novel sequence can still be linked to a known enzyme or toxin fold based on its structure. Similar analyses have already revealed new connections, such as fungal RNase-, hydrophobin-, and toxin-like effector families, which were invisible to BLAST but form tight clusters by fold [3, 4].

The workflow (Fig. 1) presented here applies these advances to *Ustilago maydis*, but is widely applicable. First, the proteome is retrieved from Ensembl Fungi, and candidate secreted proteins

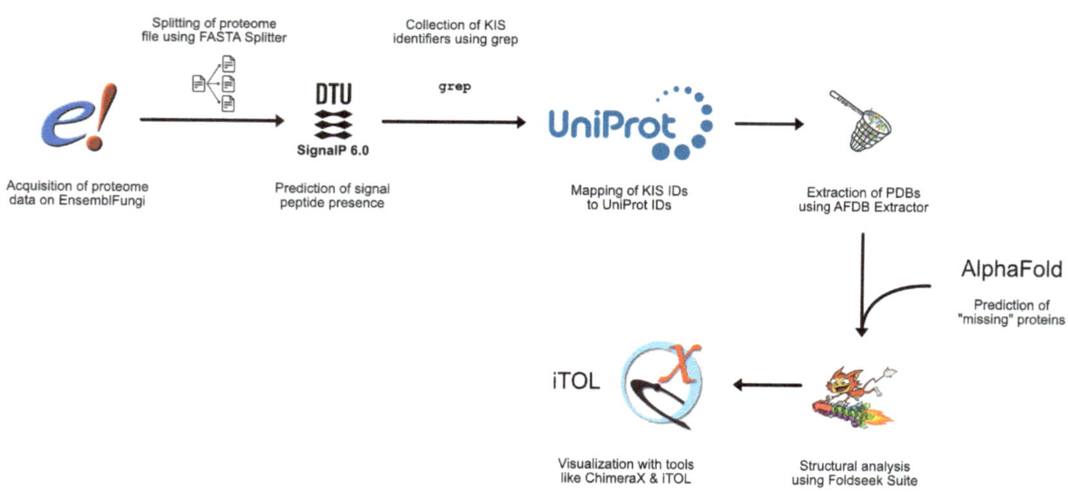

Fig. 1 Workflow for the structure-based identification of fungal effector proteins. (left to right) Ensembl Fungi as the source of protein-coding sequences. After splitting proteome data into FASTA files containing max. 1000 sequences each, SignalP 6.0 is used to predict N-terminal signal peptides and cleavage sites. With usage of the grep function, the KIS identifiers are extracted from the processed_entries.fasta file output by SignalP 6.0. Mapping the KIS IDs to UniProt IDs using the UniProt mapping tool allows to use the latter as input for the AFDB Extractor, which is used to retrieve protein structures from the AlphaFold Database. Protein structures not found on AFDB can be predicted online utilizing AlphaFold web server to get a complete dataset of structural files. This then serves as input for structure-based analysis using the Foldseek suite, taking advantage of the ultrafast structure-to-structure similarity searches and alignments. Usage of logos is for illustrative purposes only and does not imply endorsement

are predicted using SignalP [8]. These candidates are mapped to UniProt identifiers, and their structural models are obtained from the AlphaFold Protein Structure Database or generated de novo using AlphaFold. These predicted structures are then fed into Fold-Mason for multiple-structure alignment and into Foldseek for rapid pairwise comparisons. Clustering similar folds and inspecting the results enables users to select representative effectors and generate hypotheses (e.g.,, identifying a candidate as a protease inhibitor or a chaperone-like effector based on structural similarity to a known protein). Importantly, all steps rely on accessible resources, such as web servers (e.g., SignalP, AlphaFold DB, and Foldseek) and simple scripts or spreadsheets for handling IDs. Thus, the protocol is designed for bench scientists with only basic command-line experience.

Researchers can use the protocol to rapidly prioritize targets (e.g., grouping effectors into families or highlighting those with familiar folds). Although the protocol is focused on a model smut fungus, the same steps can be applied to any fungal genome, making it a general framework for effector biology. In summary, this user-friendly workflow uses modern predictive models and fast structure searches to convert annotated genome data into insights about effector function and evolution.

2 Materials

The conduction of the described protocol requires the following computational equipment and software.

2.1 Computational Equipment

A computer with UNIX accessibility is required. Windows supports Linux through the Windows Subsystem for Linux (WSL), while macOS has native support.

2.2 Software

1. Text editor (e.g., Notepad, Nano, Vim) and spreadsheet software (Excel, LibreOffice, R).
2. FastaSplitter (Perl script; https://kirill-kryukov.com/study/tools/fasta-splitter/).
3. SignalP 6.0 (online or local installation).
4. PyMOL or ChimeraX.
5. AlphaFold 3 web server (https://alphafoldserver.com).

3 Methods

3.1 Identification of Secreted Protein Candidates

The first step in analyzing effector proteins is to generate a list of *U. maydis* proteins predicted to be secreted.

3.1.1 Retrieval of *Ustilago maydis* Protein Sequences

1. To start the analysis, a genome file containing the protein sequence data is required (*see* **Note 1**). Navigate to the index page for all genomes (https://fungi.ensembl.org/species.html) and select *Ustilago maydis.*
2. Download the translated coding sequences (CDS) in FASTA format from the pep/ directory (e.g., `Ustilago_maydis.Umaydis521_2.0.pep.all.fa.gz`).
3. Extract the .gz archive to obtain the proteome file in plain FASTA format (*see* **Note 2**).

3.1.2 Splitting the Protein Sequences for Downstream Analysis

1. The file containing all protein-coding sequences is in single-letter amino acid code. Since the genome contains several thousand sequences, split the FASTA file into smaller files as web servers are mostly restricted in the maximum amount of sequences that can be used (max. 1000 sequences each) or run a local installation of SignalP 6.0 [8] for the next step (refer to material/software prerequisites chapter for this).
2. A convenient way of splitting the sequences is using the FastaSplitter Perl script [9]. Split the *U. maydis* genome FASTA file using following command in a Linux terminal (*see* **Note 3**): `perl fasta-splitter.pl --part-size 1000 --measure count path/to/Ustilago_maydis.Umaydis521_2.0.pep.all.fa --out-dir path/to/save_files`
3. For the *U. maydis* genome, this should provide seven separate FASTA files, each containing a maximum of 1000 sequences.

3.1.3 Prediction of Secreted Proteins

1. In the next step, the presence of signal peptides has to be predicted to identify potential effectors. This is done using the program SignalP 6.0 (https://services.healthtech.dtu.dk/services/SignalP-6.0/).
2. Here, the individual split FASTA files can be uploaded. Select Eukarya under Organism, Short output under Output format and Fast under Model mode (*see* **Note 4**).
3. Submit all files (each in a separate browser tab, if desired). After clicking on Submit, wait until the prediction is finished.
4. Download at least the `processed_entries.fasta` file from the output files generated by SignalP (*see* **Note 5**).

3.1.4 Extraction of Protein Identifiers

For the following steps, the structure predictions of the secreted proteins are required. As protein prediction via online tools is limited by the number of predictions (e.g., AlphaFold 3: 30 predictions every 24 h) or computational power (https://github.com/sokrypton/ColabFold), predicted structures will be obtained from the AlphaFold Protein Structure Database. For many reference genomes, such as

U. maydis, most of the proteins have already been predicted and are available on the AlphaFold Protein Structure Database (https://alphafold.ebi.ac.uk/).

1. Extract Ensembl Fungi protein identifiers (KIS IDs) from the `processed_entries.fasta` file generated by SignalP 6.0.
2. Use a terminal using following command: `grep "^>" path/to/processed_entries.fasta | cut -d' ' -f1 | sed 's/^>//' > output_KIS_IDs.txt`
3. Merge all identifier lists into a single .txt file. In the case of *U. maydis*, this will result in a total of ~ 580 sequences (*see* **Note 6**).

3.1.5 Mapping to UniProt Identifiers

Mapping the acquired KIS identifiers to more commonly used UniProt IDs enables further processing of the sequences, as well as consolidation of information in one central mapping table.

1. Open the UniProt ID Mapping Tool (https://www.uniprot.org/id-mapping) [10].
2. Paste all KIS IDs into the input field.
3. Select Ensembl Protein as input database, and UniProtKB as output.
4. Submit and download the mapping result (TSV format).

The following page will be displayed as soon as the mapping job is finished (https://www.uniprot.org/tool-dashboard). Here, clicking on the job will bring up a new context menu, where the download format can be selected (*see* **Note 7**).

3.2 Structure Prediction of Effector Proteins

3.2.1 Retrieval of Structural Models from the AlphaFold Database

1. From the TSV file, copy all UniProt IDs.
2. Upload these IDs to the AlphaFold Database Structure Extractor (https://project.iith.ac.in/sharmaglab/alphafoldextractor/index.html) [11].
3. Select output format (PDB or mmCIF) and provide an email for notification (*see* **Note 8**).
4. Download the resulting directory containing structural models, summary CSV, and a `Missing_IDs.txt` file (*see* **Note 8**).

3.2.2 Handling Proteins Without AlphaFold Models

1. To predict the missing structures, take the identifiers listed in `Missing_IDs.txt` file to search for the KIS ID in the ID mapping TSV and write out the KIS IDs. These can then be used to search the `processed_entries.fasta` files output

by SignalP 6.0 to get the sequences needed to predict the missing proteins.

2. Open `Missing_IDs.txt`.
3. Map missing UniProt IDs back to KIS IDs using the mapping TSV.
4. Retrieve the corresponding amino acid sequences from the SignalP `processed_entries.fasta` file.
5. Save all missing sequences into a new FASTA file.
6. These sequences can be submitted to ColabFold (https://colab.research.google.com/github/sokrypton/ColabFold/blob/main/AlphaFold2.ipynb), AlphaFold 3 (https://alphafoldserver.com/welcome), or a locally installed structural prediction program for de novo structure prediction (*see* **Note 9**).

3.2.3 Optional: Using AlphaFold 3

1. Head to https://alphafoldserver.com/welcome and sign in using a Google account.
2. Click on **Server** in the top left-hand corner.
3. Clear the input mask by pressing the **Clear** button on the right-hand side.
4. Enter the amino acid sequence, ensuring there are no spaces or special characters, as these will cause an error.
5. Press **Continue and preview job**. Name the job (using the corresponding ID is highly recommended) and press **Confirm and submit job**.
6. Wait till the job is finished, then click on it to bring up the overview page.
7. Different metrics, plots, and a 3D structure preview of the predicted protein are shown here. Clicking on **Download** will prompt to choose a storage location for the zipped output file.
8. The folder contains several JSON and CIF files. Per default, AlphaFold 3 will predict five models and rank them according to model quality (models 0–4), with 0 being the best model. The full data JSON file contains all quality metrics, while the summary confidence file contains a summarized quality score (chain pTM) (*see* **Note 10**). The CIF files contain the coordinates of the predicted protein structures.

3.3 Analysis of Effector Protein Candidates

3.3.1 Phylogenetic Tree Generation Based on Structural Information

1. To get a first idea of structural similarity relationships, the FoldMason web server can be used (https://search.foldseek.com/foldmason) (*see* **Note 11**).

2. The option to upload PDB files that have been downloaded or computed is available under the Load Accession menu.
3. Pressing Align translates the structural information of the PDB files into an own 3Di alphabet, reflecting the relative amino acid positions of the proteins in the structure. Once translated, FoldMason will perform a multiple sequence alignment (MSA) with these 3Di sequences, providing a large-scale structure-based alignment.
4. On the results page, an interactive guide tree is shown, a structure viewer window of the sequences selected in the guide tree as well as the actual 3Di MSA.
5. By clicking on the file names in the guide tree, the corresponding structural files can be viewed in the structure viewer. A reference structure (blue) and an aligned structure (yellow) have to be selected. This allows to select representative structures of closely related protein families.
6. To download the results, press the Downloads section in the context menu on the left-hand side. Downloading MSTA will provide a zipped file containing the MSAs in FASTA format, as well as the guide tree in Newick (NW) format. This can be used for further analysis and figure generation. Downloading all data provides a JSON file, which can be uploaded to FoldMason to get back into the web-based analysis.

3.3.2 Finding Representative Structures for Further Analysis

1. Representative structures can be identified using the FoldMason web server, which generates a guide tree based on structural similarity among predicted proteins (Fig. 2a).
2. Within this tree, proteins that occupy central positions in well-supported clusters often represent the shared core fold of that group. Selecting such representatives provides good starting points for downstream analyses.
3. Representatives can be flagged in the UniProt/Ensembl mapping table or collected into a separate FASTA/PDB list for convenience.
4. As an example, the structure **AF-A0A0D1C2J4-F1.pdb** shows a conserved core fold highly similar to its neighboring proteins in the tree (Fig. 2b, c), marking it as a suitable candidate for further exploration.

3.3.3 Using Representatives for Further Analysis with the Foldseek Search Suite

1. Identified representatives can be compared to structural databases using the Foldseek search suite (https://search.foldseek.com/searchIt). It is recommended to select databases of interest under the DBs & search settings before starting the search to keep the results more concise. This is also the place to choose the structural alignment algorithm.

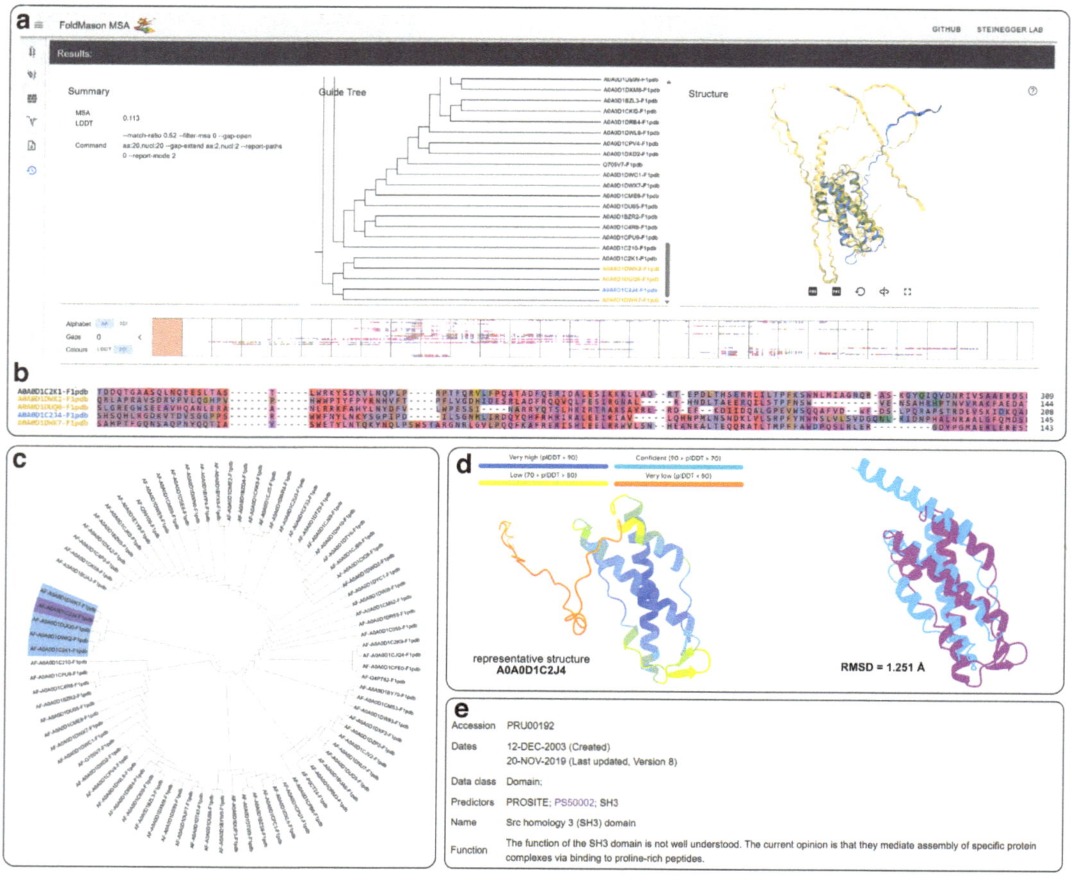

Fig. 2 Outputs of protein family characterization using FoldMason MSTA and subsequent analysis using free-to-use tools and databases. (**a**) The FoldMason MSTA interface displays a summary of the alignment parameters, a guide tree showing the grouped sequences, and a predicted structural superposition of the aligned proteins. (**b**) A sequence alignment excerpt of selected protein family members is shown, highlighting structurally conserved regions (shaded in similar coloring), although sequence diversion is observable. (**c**) Phylogenetic clustering of protein sequences based on MSTA results. The NW file provided by FoldMason was imported into the iTOL web application for figure adaptation. The radial tree shows structurally grouped protein family members together, with the highlighted clade indicating the proteins that have been selected for further structural comparison. (**d**) Structural analysis of a representative protein model (A0A0D1C2J4). Left: predicted structure colored by confidence (average pLDDT score of 72.56) with coloring guide shown above. Right: superposition of representative structure (purple) aligned to protein family member (blue), showing structural conservation (RMSD = 1.251 Å). (**e**) Domain annotation from InterPro/PROSITE suggesting the highly conserved structural fold being an SH3 domain, with excerpt shown here. The SH3 domain (PRU00192) and the associated functional description suggest that it plays a role in mediating the assembly of protein complexes by binding to proline-rich peptides. Usage of screenshots and elements from web applications is just for illustrative purposes and does not imply endorsement. (All copyrights remain with the original authors. Screenshots taken from FoldMason [12], AlphaFold Protein Structure Database [13, 14], and PROSITE [15]. Illustrations generated using iTOL [16] and ChimeraX [17])

2. Before launching the search, select relevant target databases under DBs & search settings. This step also allows choosing the preferred structural alignment algorithm.
3. Submitting the query initiates a structural similarity search across the selected databases. Results are displayed in an overview table, with top hits ranked by quality metrics (e.g., alignment score, coverage), similar to BLAST output.
4. As an example, searching with **A0A0D1C2J4** yields top hits showing nearly complete coverage of the query structure. These alignments can be visualized under the Alignment tab (indicated by three horizontal lines).
5. Clicking on a hit's accession links to external resources such as UniProt, providing additional annotation. In this case, the representative matches N-terminal SH3 domains from both the ascomycete *Histoplasma capsulatum* and the mammal *Sus scrofa*, suggesting strong structural conservation (Fig. 2d). SH3 domains are known to mediate protein–protein interactions via binding to proline-rich peptides, giving a functional hypothesis for the query (Fig. 2e).
6. This example illustrates how Foldseek enables rapid functional inference from structure alone. By linking representatives to known folds, users can prioritize candidates for deeper bioinformatic analysis or for experimental testing in the laboratory.

3.3.4 Tips for Visualization of Data Generated

1. Protein structures: Three-dimensional models can be explored using molecular visualization tools such as PyMOL (https://pymol.org/) or ChimeraX (https://www.rbvi.ucsf.edu/chimerax/download.html). ChimeraX is particularly user-friendly for beginners, with an extensive online documentation and tutorials (https://www.cgl.ucsf.edu/chimerax/docs/index.html). These programs allow detailed inspection of folds, surface properties, and potential functional motifs.
2. Phylogenetic/structural trees: Tree files produced by FoldMason (e.g., Newick format, .nw) can be uploaded to the interactive tree viewer iTOL (https://itol.embl.de/). iTOL provides extensive options for annotation, coloring, and interactive exploration of clusters, making structural relationships easier to interpret.
3. Multiple sequence alignments (MSAs): If representative sequences are aligned (e.g., for comparison across clusters), visualization and annotation can be done with JalView (https://www.jalview.org/). JalView supports color schemes by residue property, secondary structure overlays, and links to structural viewers.

3.4 Summary and Perspectives

This protocol outlines a streamlined, structure-based workflow for discovering and analyzing effector proteins in *U. maydis*. Starting with the retrieval of the genome from Ensembl Fungi, candidate secreted proteins are identified using SignalP 6.0 and mapped to UniProt identifiers. They are then linked to structural predictions from the AlphaFold database, with de novo modeling via AlphaFold 3 being employed as required. These structures are then clustered using FoldMason to identify representative folds and group them into structurally similar families. Meanwhile, Foldseek provides rapid, database-wide comparisons to infer potential functions, even in the absence of sequence similarity. Integrating high-quality predictive tools with web-accessible resources means that the method requires only minimal computational expertise. Beyond *U. maydis*, this framework provides a broadly applicable pipeline for fungal effector biology, transforming genome data into functional hypotheses and laying the groundwork for accelerating effector discovery and mechanistic insights across diverse plant–pathogen systems.

4 Notes

1. **Procuring genomic files**. While Ensembl Fungi is used for this particular example, other genomic databases such as NCBI Genome (https://www.ncbi.nlm.nih.gov/datasets/genome/), the JGI Data Portal (https://data.jgi.doe.gov/), or fungal species-specific databases such as the *Saccharomyces* Genome Database (https://www.yeastgenome.org/) can be used to acquire fungal genomes. To be compatible with this workflow, the filetype used should always be FASTA.
2. **Unzipping compressed files**. Due to the size of biological datasets, a lot of files will be compressed. The most common forms of compression files encountered are gzip (.gz) and ZIP (.zip). Both Windows and Mac can unzip those filetypes natively. For Windows, just right-click on the compressed file, select Extract All and go through the pop-up menu. For Mac, just double-click on the compressed file. The UNIX command for unzipping is `gunzip name_of_file.gz` or `unzip name_of_file.zip` respectively. Using this command on Linux for the first time might prompt you to install gunzip first, just follow the instruction displayed.
3. **Installing scripts**. Scripts are singular files, which can simply be downloaded and run directly without further installation. In the case of the FASTA splitter, download the newest version from the link (https://kirill-kryukov.com/study/tools/fasta-splitter/) provided, unzip it, and open a terminal/console in the same file path as the Perl script using the command shown.

4. **Choosing SignalP 6.0 settings**. The choice of *Fast* vs. *Slow* mode in SignalP trades speed for accuracy. For a quick overview, *Fast* is sufficient, but for high-confidence results, *Slow* mode may be preferable.
5. **Data management and naming**. Each step generates multiple intermediate files (split FASTAs, SignalP outputs, identifier lists, mapping TSVs). Organize them in a clear directory structure to avoid confusion, particularly when multiple users contribute to the project. Good practice is using a shared suffix for all files in one project (e.g., `UM_eff_pred_`).
6. **Example dataset used**. For demonstration purposes, only the first set of 82 potential effector sequences from the first split FASTA file will be used in this protocol.
7. **Choosing settings for ID mapping**. Once the IDs have been mapped, clicking on **Download** will bring up a context menu asking for the download format. Selecting a tabular format, such as TSV, will expand the context menu and display a wide range of options that can be added as columns to the table. These include options such as known Pfam or Interpro categorization. If the purpose is simply to remap IDs, deselect everything so that only the KIS ID and the UniProtKB name it was mapped to remain.
8. **Differences between mmCIF and PDB**. While the PDB file format uses a fixed-width, position-based structure with limitations on the number of atoms and chains it can represent, the mmCIF format is more flexible and uses named fields. This makes it better suited to modern structural data, as it supports larger and more complex structures and provides better data standardization and validation through a defined dictionary system. Consequently, mmCIF has become the standard format for depositing macromolecular structures.
9. **Differences between different prediction models**. AlphaFold 3 can predict the structures of proteins, nucleic acids, and ligands in complexes, which is an improvement on the protein-focused AlphaFold 2. The ColabFold web server offers a fast, easy-to-use, cloud-based tool making AlphaFold2 accessible, but with size limits. Running AlphaFold 3 locally provides full control and higher accuracy for large or custom complexes, although this requires significant computational resources and technical setup. In short, ColabFold is quick and accessible, while local AlphaFold 3 is powerful and flexible, but more resource-intensive.
10. **Understanding AlphaFold prediction metrics**. To gain a deeper understanding of AlphaFold, how it works, and the metrics it produces to classify the quality of the structural prediction, visit: https://www.ebi.ac.uk/training/online/courses/alphafold/inputs-and-outputs/

evaluating-alphafolds-predicted-structures-using-confidence-scores/

11. **Structural comparison using FoldMason**. FoldMason is a fast, open-source tool that creates multiple structural alignments of large sets of protein structures. It does this by converting the structures into string representations using its own 3Di structural alphabet, and then performing progressive alignment guided by a minimum spanning tree. Leveraging efficient string-based comparisons, it achieves high speed and accuracy, provides confidence scores via LDDT and offers interactive visualizations for exploring structural conservation.

Acknowledgments & Statement

FA and MH acknowledge funding from the German Research Foundation (Deutsche Forschungsgemeinschaft, DFG) (Project ID 458090666 / CRC1535/B02). All materials are available from the corresponding author upon request.

Author contributions MH designed the study with input from FA. FA and MH wrote the paper. FA contributed funding and resources.

References

1. Mesny F, Bauer M, Zhu J et al. (2024) Meddling with the microbiota: fungal tricks to infect plant hosts. Curr Opin Plant Biol 82:102622. https://doi.org/10.1016/j.pbi.2024.102622
2. Snelders NC, Rovenich H, Thomma BPHJ (2022) Microbiota manipulation through the secretion of effector proteins is fundamental to the wealth of lifestyles in the fungal kingdom. FEMS Microbiol Rev 46:fuac022. https://doi.org/10.1093/femsre/fuac022
3. Franceschetti M, Maqbool A, Jiménez-Dalmaroni MJ et al (2017) Effectors of filamentous plant pathogens: commonalities amid diversity. Microbiol Mol Biol Rev 81:e00066-16. https://doi.org/10.1128/mmbr.00066-16
4. Seong K, Krasileva KV (2023) Prediction of effector protein structures from fungal phytopathogens enables evolutionary analyses. Nat Microbiol 8:174–187. https://doi.org/10.1038/s41564-022-01287-6
5. Jumper J, Evans R, Pritzel A et al. (2021) Highly accurate protein structure prediction with AlphaFold. Nature 596:583–589. https://doi.org/10.1038/s41586-021-03819-2
6. van Kempen M, Kim SS, Tumescheit C et al. (2024) Fast and accurate protein structure search with Foldseek. Nat Biotechnol 42:243–246. https://doi.org/10.1038/s41587-023-01773-0
7. Altschul SF, Gish W, Miller W et al. (1990) Basic local alignment search tool. J Mol Biol 215:403–410. https://doi.org/10.1016/S0022-2836(05)80360-2
8. Teufel F, Almagro Armenteros JJ, Johansen AR et al. (2022) SignalP 6.0 predicts all five types of signal peptides using protein language models. Nat Biotechnol 40:1023–1025. https://doi.org/10.1038/s41587-021-01156-3
9. FASTA Splitter. https://kirill-kryukov.com/study/tools/fasta-splitter/. Accessed 27 Aug 2025
10. Huang H, McGarvey PB, Suzek BE et al. (2011) A comprehensive protein-centric ID mapping service for molecular data integration. Bioinformatics 27:1190–1191. https://doi.org/10.1093/bioinformatics/btr101
11. Saraf N, Karthick V, Sharma G (2025) AlphaFold Database Structure Extractor: a web server and API to download AlphaFold

structures using common protein accessions. BMC Bioinform 26:305

12. Gilchrist CLM, Mirdita M, Steinegger M (2026) Multiple protein structure alignment at scale with FoldMason. Science 391:485–488
13. Abramson J, Adler J, Dunger J et al. (2024) Accurate structure prediction of biomolecular interactions with AlphaFold 3. Nature 630:493–500. https://doi.org/10.1038/s41586-024-07487-w
14. Varadi M, Bertoni D, Magana P et al. (2024) AlphaFold Protein Structure Database in 2024: providing structure coverage for over 214 million protein sequences. Nucleic Acids Res 52:D368–D375. https://doi.org/10.1093/nar/gkad1011
15. ProRule PRU00192. https://purl.expasy.org/prosite/rule/PRU00192. Accessed 1 Sept 2025
16. Letunic I, Bork P (2007) Interactive Tree Of Life (iTOL): an online tool for phylogenetic tree display and annotation. Bioinformatics 23:127–128. https://doi.org/10.1093/bioinformatics/btl529
17. Meng EC, Goddard TD, Pettersen EF et al. (2023) UCSF ChimeraX: tools for structure building and analysis. Protein Sci 32:e4792. https://doi.org/10.1002/pro.4792

Chapter 31

Chromatin Immunoprecipitation (ChIP) to Investigate Histone-Modifications in Smut Fungi

Lukas Dorian Dittiger, Juliane Kühne, Christian Müller and Jan Schirawski

Abstract

Post-translational modification of histone tails can change the complete gene expression profile of an organism. Chromatin immunoprecipitation (ChIP) is a technique that can be used to identify genomic regions wrapped around modified histones. Specifically, histones and their bound DNA are co-immunoprecipitated with antibodies targeting histones carrying a defined modification. Subsequent analysis of the associated genomic regions allows the identification of genes and regulatory elements marked by histones carrying this specific modification thereby providing insights into a genome-wide regulatory mechanism of gene expression. This chapter describes the experimental setup and provides a detailed protocol for performing chromatin immunoprecipitation in *Ustilago maydis* or related smut fungi.

Keywords Epigenetics, ChIP, Histone modifications, Chromatin, Immunoprecipitation, Antibody, Nuclear DNA

1 Introduction

Since the early 2000s, the field of epigenetic research has been rising in importance. Literally, epigenetics means "on top of genetics." It describes changes in gene functions and gene regulation that are heritable and do not involve a change of DNA sequence [1]. The study of epigenetics includes, for example, chemical modification of the DNA structure, like DNA methylation, or processes altering the structure of chromatin, called chromatin remodeling processes. The post-translational modification of histone proteins is one major process affecting the chromatin structure. Histone modifications usually occur at the easily accessible N-terminal histone tails that are sticking out of the assembled nucleosome complex. Known modifications of histone tails include acetylation, phosphorylation,

These authors Lukas Dorian Dittiger and Juliane Kühne contributed equally to this work.

An N. T. Phan et al. (eds.), *Ustilago maydis*, Methods in Molecular Biology, vol. 3050,
https://doi.org/10.1007/978-1-0716-5364-7_31,

ubiquitination, sumoylation, ADP-ribosylation, glycosylation, biotinylation, and methylation [2]. Some of these modifications (e.g., acetylation, phosphorylation, ADP-ribosylation, glycosylation, and biotinylation) have a direct effect by physically changing chromatin accessibility, while processes like methylation, ubiquitination, or sumoylation have indirect effects and are first recognized by other (regulatory) proteins that subsequently change the chromatin structure [2].

Chromatin Immunoprecipitation (ChIP) allows identification of genomic regions associated with specific histone modifications by co-immunoprecipitating modified histones along with their bound DNA [3]. The only requirement is a specific, high-quality antibody that targets the histone modification to be investigated. A ChIP experiment can roughly be divided into four steps: Firstly, the cells are fixed by treating with formaldehyde. This leads to the covalent crosslinking of DNA to attached proteins. Secondly, the fixed cells are lysed, the chromatin is isolated followed by its fragmentation by either sonication or enzymatic digestion. The third defining step of a ChIP experiment is the following immunoprecipitation. The fragmented chromatin is incubated with an antibody targeting the histone modification to be investigated. Using protein A or protein G beads, the crosslinked DNA-protein complex can be immunoprecipitated. Lastly, the formaldehyde crosslinking is reversed to release the DNA from the proteins, and the DNA is purified for subsequent analysis [4]. Quantitative PCR (ChIP-qPCR) can be used to investigate the association of known genomic regions with the investigated histone modification. For an unbiased genome-wide analysis, the immunoprecipitated DNA can be sequenced (ChIP-seq), and histone modification-associated genomic regions can be identified by mapping the sequencing reads to a reference genome followed by peak calling and further downstream analyses [5, 6].

In the biotrophic maize pathogen *Ustilago maydis* histone modifications, i.e., histone acetylation and de-acetylation, were shown to be involved in the regulation of genes important for key processes affecting pathogenicity. The histone deacetylase Hda1 was shown to control repression of the *maize-induced gene 1* (*mig1*) gene, one of the most highly expressed effector genes during the biotrophic phase. However, deletion of *HDA1* did not affect plant penetration or colonization. Instead, the gene seems to be important for the maturation of fungal teliospores [7, 8]. Two more histone deacetylases of *U. maydis*, Clr3 and Hos2, were found to be important for virulence and the completion of the fungal cycle. Deletion of both genes affected the cAMP-dependent expression of mating genes and since mating is a prerequisite for plant colonization the deletion also impacted the virulence of the phytopathogen [9]. In case of histone acetylation, it could be shown that deletion of the histone acetylase gene *GCN5* also affected *U. maydis* virulence, likely because GCN5 positively regulating the expression of effector genes [10]. The results of studying histone acetylation and deacetylation in *U. maydis* suggest that histone modifications are regulating central processes required for pathogenicity. However, even though other histone modifications, like histone methylation, are known to be involved in the virulence of other fungal plant pathogens they have not yet been studied in smut fungi [11]. This chapter discusses the experimental setup and provides a detailed protocol to perform a ChIP experiment in smut fungi like *U. maydis*.

2 Materials

2.1 Cultivation of Fungal Strains

1. Potato Dextrose (PD) agar: 2.4% PD broth, 10 mM Tris-HCl at pH 8, 2% agar. Autoclave for 20 min at 121°C.
2. YEPS-light: 1% yeast extract, 0.4% Tryptone, 0.4% Saccharose. Autoclave for 20 min at 121°C.

2.2 Cell Fixation/ Crosslinking

1. 23.5% Formaldehyde, methanol-free.
2. 2.5 M Glycine.
3. 10× PBS at pH 7.4: 79 mM Na2HPO4, 145 mM KH2PO4, 5 mM MgCl2 × 6 H2O, 27 mM KCl, 1.37 M NaCl. Adjust pH to pH 7.4 using NaOH.

2.3 Chromatin Fragmentation by Sonication

1. 0.1 M Phenylmethylsulfonyl fluoride (PMSF).
2. Protease Inhibitor.
3. LB140: 50 mM HEPES/ KOH at pH 7.4, 140 mM NaCl, 1 mM EDTA at pH 8.0, 1% Triton X-100, 0.1% sodium deoxycholate.
4. Sterile glassbeads (0.25–0.5 mm) (*see* **Note 1**).
5. IKA VXR basic Vibrax® (IKA-Werke GmbH & Co. KG, Cat. #: 0002819000).
6. SS-DNA: 1% salmon sperm DNA. Dissolve for 2–4 h at room temperature, aliquot, store at −20°C, denature before use at 99°C for 5 min.
7. CHIP dilution buffer: 0.01 % SDS, 1.1 % Triton X-100, 1.2 mM EDTA at pH 8.0, 16.7 mM Tris–HCl at pH 8.1, 167 mM NaCl. Filter-sterilize and store at 4°C.
8. 1.5 ml Bioruptor® Pico Microtubes with Caps (Diagenode, Cat. #: C30010016).
9. Bioruptor® Pico (Diagenode, Cat. #: B01080010).

2.4 Quality Control of Sonicated Chromatin

1. 1× TE buffer at pH 8: 1 mM EDTA at pH 8.0, 10 mM Tris–HCl at pH 8.0.
2. 2% Glycogen: Dissolve in water, prepare freshly.

3. Proteinase K solution: 50 mM Tris–HCl at pH 8.0, 1 mM CaCl2, 2 % Proteinase K.
4. Phenol/Chloroform/Isoamyl alcohol (25:24:1).
5. Chloroform/Isoamyl alcohol (24:1).
6. IKA VXR basic Vibrax® (IKA-Werke GmbH & Co. KG, Cat. #: 0002819000).
7. 5 M NaCl.

2.5 Chromatin Immunoprecipitation

1. Antibody of choice (*see* **Note 2**).
2. Immunoglobulin G (IgG) (*see* **Note 3**).
3. SS-DNA: 1% salmon sperm DNA. Dissolve for 2–4 h at RT, aliquot, store at −20°C, denature before use at 99°C for 5 min.
4. Test Tube Rotator.
5. μMACS Protein A/G Starting Kit (Miltenyi Biotec, Cat. #: 130-042-601) (*see* **Note 4**).
6. ChIP dilution buffer: 0.01% SDS, 1.1% Triton X-100, 1.2 mM EDTA at pH 8.0, 16.7 mM Tris-HCl at pH 8.1, 167 mM NaCl. Filter-sterilize and store at 4°C.
7. 5% blocking buffer: 5% non-fat dry milk in 1xPBS at pH 7.4.
8. Low salt immune complex wash buffer: 0.1 % SDS, 1% Triton X-100, 2 mM EDTA at pH 8.0, 20 mM Tris-HCl at pH 8.1, 150 mM NaCl.
9. High salt immune complex wash buffer: 0.1% SDS, 1% Triton X-100, 2 mM EDTA at pH 8.0, 20 mM Tris-HCl at pH 8.1, 500 mM NaCl.
10. LiCl immune complex wash buffer: 0,25% LiCl, 1 % IGEPAL–CA 630 (NP-40), 1% sodium deoxycholate, 1 mM EDTA at pH 8.0, 10 mM Tris-HCl at pH 8.1.
11. 1× TE buffer at pH 8: 1 mM EDTA at pH 8.0, 10 mM Tris-HCl at pH 8.0.
12. Elution buffer: 1% SDS, 0.1 M NaHCO3. Dissolve NaHCO3 in ddH$_2$O, add SDS, aliquot á 1 mL, heat to 95°C before use and avoid foaming.

2.6 Decrosslinking and DNA Purification

1. 5 M NaCl.
2. Proteinase K solution: 50 mM Tris-HCl pH 8.0, 1 mM CaCl2, 2% Proteinase K.
3. NucleoSpin® Gel and PCR Clean-up Kit (Macherey-Nagel, Cat. #: 740609.50).
4. Binding buffer NTB (Macherey-Nagel, Cat. #: 740595.150).
5. Qubit™ Flex Fluorometer (Thermo Fisher Scientific, Invitrogen).

3 Methods

3.1 Experimental Design

To use ChIP to investigate genomic regions associated with a specific histone modification, appropriate controls and enough replicates are required. Commonly, one of two controls is used for downstream analysis: (1) An "Input control," for which DNA isolated from fixed cells is used. Here, no immunoprecipitation is done, and the total amount of DNA is represented allowing fold-enrichment analysis and normalization. It corrects for a technical bias, e.g., uneven sonication, however it neglects non-specific binding to the antibody or beads used during immunoprecipitation. (2) An "IgG control" can be used by performing an IP using purified non-specific immunoglobulin G. This mimics the processing of the samples more closely and accounts for non-specific binding. However, it does not allow normalization to the total DNA amount. The protocol provided here uses the latter, however, both controls have their own advantages and disadvantages [12]. It is recommended to use at least three biological replicates for all sample conditions to allow subsequent statistical analysis following qPCR or sequencing. To save sequencing costs, one replicate for the input or the IgG control per investigated strain is sufficient.

3.2 Cultivation of Fungal Strains

1. Freshly streak your fungal strains from their -80°C stock on PD-agar plates and incubate under appropriate conditions until you observe single colonies (e.g., incubate the plates for 2 days at 28°C).
2. Prepare a liquid culture of the strains by inoculating 5 ml of YEPS-light medium with a small amount (preferably a single colony) of fungal cell material and incubate at 28°C overnight.
3. Measure the optical density at 600 nm (OD) of the overnight culture and check absence of microbial contamination using microscopy. For each pre-culture, inoculate one 250 ml Erlenmeyer flask containing 50 ml of YEPS-light medium using a calculated amount of inoculum (*see* **Note 5**). Cultures should reach an OD of 0.7–0.8 on the next day.
4. Measure the optical density (OD) of the overnight culture and check absence of microbial contamination using microscopy. Proceed with the fixation of the cells.

3.3 Cell Fixation/Crosslinking

1. Add formaldehyde to a final concentration of 1% to each flask.
2. Incubate for 15 min at 28°C and 180 rpm.
3. To quench the reaction, add 2.5 M Glycine to a final concentration of 0.25 M to each flask.
4. Incubate for 10 min at room temperature.
5. Transfer the cultures into 50 ml-centrifugation tubes, harvest the cells by centrifugation at 3500 × *g* at 4°C for 8 min and discard the supernatant completely.

6. Wash the pellet twice using 10 ml PBS at pH 7.4 before resuspending it in 1 ml PBS at pH 7.4 and transferring the suspension into a fresh 2 ml screwcap microcentrifugation tube.
7. Centrifuge at 3500 × *g* and 4°C and completely remove the supernatant.
8. Shock-freeze the pellets in liquid nitrogen. Store the samples at −80°C or directly proceed with sonication.

3.4 Chromatin Fragmentation by Sonication

1. Thaw the samples on ice. Add 700 µl LB140 (freshly add protease inhibitor and PMSF) and ~100 µl glassbeads.
2. Vibrax three times for 30 min at 4°C. Let the samples cool down for 10 min on ice after each vibraxing step.
3. Use a needle to puncture a small hole in the lid of each sample and place the microcentrifugation tube bottom-up in a 15 ml centrifugation tube.
4. Centrifuge for 15 min at 1250 × *g* and 4°C. The liquid should be transferred from the screwcap microcentrifugation tube into the 15-ml centrifugation tube. Check if all liquid was completely transferred. If not, repeat the centrifugation step.
5. Centrifuge for 5 min at 16,000 × *g* and 4°C. Proceed with the pellet.
6. Wash the pellet once with 700 µl of Chip Dilution Buffer (CDB), by resuspending the pellet followed by centrifugation for 5 min at 16,000 × *g* at 4°C. Discard the supernatant.
7. Resuspend the pellet in 700 µl CDB. Transfer 50 µl into a labelled, precooled 1.5 ml tube and store at −20°C ("unsonicated control sample").
8. Aliquot 300 µl of each sample into two Bioruptor® Microcentrifugation tubes and proceed with sonification with five cycles of sonification for 30 s followed by 30 s rest using the Bioruptor® Pico (*see* **Note 6**).
9. Centrifuge the sonicated samples for 10 min at 16.000 × *g* and 4°C and combine the supernatant of both 300-µl aliquots in a labeled precooled 1.5-ml centrifugation tube. Aliquot two times 15 µl of each sample into two labeled, precooled 1.5-ml tube to serve as "input control" for the following immunoprecipitation as well as to check protein integrity using SDS-polyacrylamide gel electrophoresis (SDS-PAGE) and Coomassie Brilliant Blue staining. The sonicated samples can be stored at −20°C.
10. Proceed with Sect. 3.6 when using *U. maydis* SG200. The sonication conditions may have to be adjusted when using other strains or other smuts. In this case, ensure proper sonication of chromatin and proceed with Sect. 3.5 (Quality control for sonicated chromatin) before continuing with chromatin immunoprecipitation.

3.5 Quality Control of Sonicated Chromatin

1. Transfer 50 µl of sonicated sample into a fresh 1.5 ml tube. For decrosslinking, incubate the sample for 16 h at 65°C before adding a master mix consisting of 62.5 µl TE-buffer, 0.25 µl 20 mg/ml Glycogen and 1.75 µl Proteinase K solution to each sample. Incubate the mixture at 37°C for 2 h.
2. Add 450 µl of phenol/ chloroform/ isoamyl alcohol (25:24:1), vibrax for 20 s and centrifuge for 5 min at maximum speed and 14°C.
3. Transfer the supernatant into a labeled, precooled 1.5-ml tube and add 450 µl chloroform/isoamyl alcohol (24:1). Vibrax for 5 s and centrifuge for 2 min at maximum speed at 14°C.
4. Transfer the supernatant into a labelled, precooled 1.5-ml tube. Add 12.5 µl 5 M NaCl and 1 ml ice-cold ethanol, invert several times and incubate for 2 h at −80°C.
5. Centrifuge the samples for 10 min at maximum speed and 4°C. Discard the supernatant and wash the pellet once by adding 1 ml of ice-cold 70% ethanol and centrifugation for 3 min at maximum speed and 4°C.
6. Discard the supernatant completely and let the pellet dry at room temperature for 5 min.
7. Resolve the pellet in 50 µl TE buffer. Check chromatin shearing by agarose gel electrophoresis, use the collected "unsonicated control sample" as negative control (*see* **Note 7**).

3.6 Chromatin Immunoprecipitation (ChIP)

1. Immunoprecipitation is performed using the µMACS™ Protein A/G system, which enables the use of magnetic beads with an antibody of choice. Four samples can be processed at the same time using the µMACS™ magnetics separator. If you have more samples, it is recommended to process three biological replicates and their respective control samples at the same time.
2. Thaw your sonicated samples on ice. Add 4 µg of antibody or IgG (for control samples) and incubate rotating for 16–20 h at 4°C.
3. To pre-block the magnetic beads, add 42 µl of µMACS™ Protein A/G beads as well as 0.42 µl salmon sperm DNA (freshly denatured at 99°C for 5 min) per sample to a fresh pre-cooled 1.5-ml tube and incubate for 1 h 40 min on ice.
4. Pipet 40 µl of pre-blocked Protein A/G beads with a cut-off pipette tip to your IP and control samples and incubate at 4°C while rotating.
5. Place the µColumns into the magnetic field of the separator (*see* **Note 8**) and prepare a suitable waste container below the columns.
6. Equilibrate the columns using 200 µl of CDB (freshly add protease inhibitor and SS-DNA). The columns are flow-stop and do not run dry.

7. Add 1 ml of 5% blocking buffer to your samples, mix by pipetting and directly transfer the mixture to the μColumns (*see* **Note 9**)
8. Wash the column once with 1 ml low salt immune complex wash buffer.
9. Wash the column once with 1 ml high salt immune complex wash buffer.
10. Wash the column once with 1 ml LiCl immune complex wash buffer.
11. Wash the column once with 1 ml TE buffer at pH 8.
12. Remove remaining droplets at the tip of the μColumn using a sterile pipette tip.
13. Place a 1.5 ml micro centrifugation tube on ice below each column.
14. Add 50 μl of pre-heated (95°C) elution buffer to each column and collect the eluate. Remove remaining droplets at the tip of the μColumns using a sterile pipette tip and add to eluate.
15. Repeat Step 14 three times to collect a total of 200 μl eluate.
16. Thaw your 15-μl "input control" samples (Sect. 3.4, Step 9) on ice and add 185 μl of hot elution buffer to a total volume of 200 μl. Proceed with ChIP and input samples for decrosslinking and DNA purification (Sect. 3.7).

3.7 Decrosslinking and DNA Purification

1. Add 8 μl 5 M NaCl and 1 μl Proteinase K solution to each IP- (all samples with antibody or IgG treatment) and input sample and incubate for 2 h at 65°C with shaking at 500 rpm.
2. Spin the tubes shortly to collect the liquid at the bottom of the tube and proceed with DNA purification using the NucleoSpin Extract II Kit (Macherey Nagel). In this kit, exchange the NT buffer with NTB buffer. Follow the manufacturer's instructions.
3. Measure the purity and concentration of the extracted DNA using Qubit Flex Fluorometer (Invitrogen). DNA can be used for downstream analysis like qPCR or ChIP-sequencing to identify genomic regions and genes associated with the histone modification targeted by your antibody of choice.

4 Notes

1. Glass beads should be free of DNA and DNase. Sterile certified DNase and RNase-free glass beads can be bought, or unsterile glass beads can be freed of DNA and DNase by first soaking

in 25% HCl for 60 min with occasional shaking, followed by several washing steps with sterile water until the pH of the washing fluid is neutral. Sterilize and dry clean glass beads in an oven at 180°C for at least 6 h.

2. The antibody is the key factor determining the success of immunoprecipitation and determines which histone modification is investigated. The choice of antibody should be carefully considered regarding the specificity of the antibody as well as the suitability for immunoprecipitation. Ideally, use an antibody that was previously validated for ChIP and test whether the antibody can detect your targeted histone modification by performing a Western blot on a histone extract under your specific experimental conditions.

3. There are two possible ways to generate control samples for downstream analysis of a ChIP experiment. You can either use the DNA isolated from fixed cells without performing the following immunoprecipitation, which is considered as an "input" control (see Sect. 3.4) or perform an immunoprecipitation adding purified IgG instead of antibody targeting your histone modification ("IgG" control). The latter is recommended as it imitates the conditions for your samples more closely. For the IgG control, choose the same source organism as the one in which the primary antibody was raised (e.g., when using a primary IgG antibody generated in a mouse, use purified IgG from a mouse).

4. Other protein A/G bead systems should also work. We recommend using magnetic beads to reduce non-specific binding to the bead surfaces as well as to facilitate the procedure. Whether Protein A or Protein G beads work best depends on the specific antibody you are using. Check the beads' manufacturing instructions for the recommended choice.

5. You can calculate the amount of inoculum using the following formula:

$$Amount\ of\ inoculum\ \ [ml] = \frac{(OD600\ \ of\ main\ culture)*(Volume\ of\ main\ culture\ \ [ml])}{(OD600\ \ of\ preculture)*2^{\frac{Incubation\ hours}{Doubling\ time}}}$$

6. To ensure even sonification of all samples always make sure all six sample slots of the Bioruptor® Pico are occupied.

7. The DNA is properly fragmented if a smear between 200 bp and 500 bp can be observed on the agarose gel.

8. Avoid touching the tip of the μColumn while placing it in the separator.

9. To ensure proper mixing of blocking buffer and sample, immediately pipette each sample on the μColumn before proceeding with the next step.

References

1. Wu Ct MorrisJR (2001) Genes, genetics, and epigenetics: a correspondence. Science 293(5532):1103–1105. https://doi.org/10.1126/science.293.5532.1103
2. Bannister AJ, Kouzarides T (2011) Regulation of chromatin by histone modifications. Cell Res 21(3):381–395. https://doi.org/10.1038/cr.2011.22
3. Hebbes TR, Thorne AW, Crane-Robinson C (1988) A direct link between core histone acetylation and transcriptionally active chromatin. EMBO J 7(5):1395–1402. https://doi.org/10.1002/j.1460-2075.1988.tb02956.x
4. Saleh A, Alvarez-Venegas R, Avramova Z (2008) An efficient chromatin immunoprecipitation (ChIP) protocol for studying histone modifications in Arabidopsis plants. Nat Protoc 3(6):1018–1025. https://doi.org/10.1038/nprot.2008.66
5. Oba GM, Nakato R (2025) A beginner's guide to ChIP-seq analysis. Biochem (Lond) 47(1):21–26. https://doi.org/10.1042/bio_2024_920
6. Nakato R, Sakata T (2021) Methods for ChIP-seq analysis: a practical workflow and advanced applications. Methods 187:44–53. https://doi.org/10.1016/j.ymeth.2020.03.005
7. Torreblanca J, Stumpferl S, Basse CW (2003) Histone deacetylase Hda1 acts as repressor of the *Ustilago maydis* biotrophic marker gene *mig1*. Fungal Genet Biol 38(1):22–32. https://doi.org/10.1016/s1087-1845(02)00505-4
8. Reichmann M, Jamnischek A, Weinzierl G et al (2002) The histone deacetylase Hda1 from *Ustilago maydis* is essential for teliospore development. Mol Microbiol 46(4):1169–1182. https://doi.org/10.1046/j.1365-2958.2002.03238.x
9. Elías-Villalobos A, Fernández-Álvarez A, Moreno-Sánchez I et al (2015) The Hos2 histone deacetylase controls *Ustilago maydis* virulence through direct regulation of mating-type genes. PLOS Path 11(8):e1005134. https://doi.org/10.1371/journal.ppat.1005134
10. Martínez-Soto D, González-Prieto JM, Ruiz-Herrera J (2015) Transcriptomic analysis of the GCN5 gene reveals mechanisms of the epigenetic regulation of virulence and morphogenesis in *Ustilago maydis*. FEMS Yeast Res 15(6):fov055. https://doi.org/10.1093/femsyr/fov055
11. Lai Y, Wang L, Zheng W, et al (2022) Regulatory roles of histone modifications in filamentous fungal pathogens. J Fungi (Basel) 8(6):565. https://doi.org/10.3390/jof8060565
12. Xu J, Kudron MM, Victorsen A et al (2021) To mock or not: a comprehensive comparison of mock IP and DNA input for ChIP-seq. Nucleic Acids Res 49(3):e17. https://doi.org/10.1093/nar/gkaa1155

Chapter 32

Expression and Detection of Heterologous Proteins in *Ustilago maydis*

Simon Wegmann, Lesley Plücker, Thomas Drepper, Florian Finkernagel and Kerstin Schipper

Abstract

Expression of transgenes of both bacterial and eukaryotic origin is critical for the versatility of model species. A range of heterologous genes has been successfully transplanted to and cognate proteins functionally synthesized in *Ustilago maydis*, ranging from cytosolic reporters like the enhanced green fluorescent protein and its derivatives to cytoplasmic or secreted enzymes from bacterial, plant, or filamentous ascomycete sources. Here, a brief summary of the most important considerations for expression of heterologous genes, including expression strain selection, codon optimization, the choice of insertion loci, as well as potential protein fusions for targeting, localization, purification, and detection, is provided. A detailed protocol for the prevalent constitutive expression of heterologous proteins and subsequent verification via Western blot analysis is provided. All steps are described in detail using the example of the secreted endo-glucuronoxylanase Xyn30A, a carbohydrate-active enzyme from *Thermothelomyces thermophilus*. In essence, recombinant protein synthesis is central for the broad exploitation of *U. maydis* as a fungal model, and applications range from fundamental research to biotechnological applications.

Keywords Insertion locus, Heterologous protein, Secretion, Unconventional secretion, Codon-optimization, Protein tag, Reporter

1 Introduction

The insertion and functional expression of heterologous genes is very important for potent model microorganisms. Efficient transgene expression ensures the applicability of reporters and markers and allows for the introduction of foreign enzymes or transplantation of complete biosynthetic pathways. Multiple heterologous genes have been successfully inserted and functionally expressed in *U. maydis*. These include genes for fluorescent proteins like eGfp, mCherry, mKate2, and others, used for example for protein localization studies [1]. Heterologous enzymes that have been implemented in *U. maydis* encompass, among others, bacterial or fungal carbohydrate-active enzymes [2], luciferases like *Photinus pyralis* (Firefly) FLuc, *Gaussia*

An N. T. Phan et al. (eds.), *Ustilago maydis*, Methods in Molecular Biology, vol. 3050,
https://doi.org/10.1007/978-1-0716-5364-7_32,

princeps GLuc or *Renilla reniformis* RLuc [3, 4], alternative antibody formats [5, 6] or the bacterial reporter enzymes β-galactosidase and β-glucuronidase [7, 8]. Even new biosynthetic routes have been functionally inserted, exemplified by the expression of plant or fungal sesquiterpene synthases yielding α-cuprenene or (+)-valencene [9].

Prior to a typical protein expression experiment as described here, the precise planning and selection of the expression strain and the contained expression construct are critical. The focus lies on the most important points to consider:

a) Expression strain: Different strain backgrounds are available, providing a solid basis for the expression of transgenes depending on the experimental goals. These include the solopathogenic strain SG200 for efficient infection studies, strains AB33 and AB34 for synchronous induction of hyphae formation independent of the infection context [10]. AB33P8Δ is an engineered expression strain for the production of secreted proteins that lacks 8 secreted proteases leading to a strong reduction in extracellular protease activity to stabilize secreted proteins in the supernatant [11]. MB215 is a wild-type isolate especially valuable for production of biotechnologically relevant secondary metabolites [12].

b) Insertion loci for chromosomal integration: Different loci are frequently used for insertion of recombinant expression cassettes. These include, for example, the *ip* locus (*cbx*) [8] with the advantage of possible multi-copy insertion, and the protease-encoding loci *upp3* and *pep4* [1, 7].

c) Di-codon bias: Due to differences in the codon utilization, expression of heterologous genes can eventually lead to premature polyadenylation of the transcript [13]. As a consequence, insertion of non-optimized genes can result in the lack of detectable protein amounts as described for *tetR* [13]. Similar observations were made for bacterial flavin-binding fluorescent protein (FbFP) [14], for which heterologous protein could only be detected after di-codon optimization (unpublished observation). An automated, web-based tool is available for *U. maydis*, speeding up and guiding the design of di-codon optimized sequences (see Sect. 3).

d) Promoter selection: Mostly, heterologous genes are expressed constitutively in *U. maydis*. Promoters with differing expression strength are available. The strong promoters P_{rpl10}, P_{tef}, P_{otef}, and the very strong P_{oma} promoter are the most frequently used genetic elements. Alternatively, inducible promoters exist, including two nutrient-regulated promoters P_{crg} (carbon source regulated) and P_{nar} (nitrogen source regulated) [10], as well as synthetic tetracycline (doxycycline) controlled systems [13, 15].

e) Protein targeting: Heterologous proteins can reside in the cytoplasm, targeted to distinct organelles or destined for secretion via corresponding targeting peptides. The repertoire of available targeting sequences for *U. maydis* allows, for example, the transport of proteins into mitochondria, peroxisomes, endosomes, the nucleus, or the ER. In addition, signal sequences can be fused to the target protein for conventional secretion via the endomembrane system (i.e., the signal peptide of Cmu1) [16]. Eventually, in the case of heterologous eukaryotic and especially fungal proteins, the native signal sequence may be kept as their function is usually conserved (unpublished observation). Protein fusions with Cts1 or Jps1 can be used to mediate unconventional secretion via the fragmentation zone [4, 8].

f) Protein fusions: Multiple protein appendices, including reporters for detection, quantification, localization, and purification, have successfully been implemented. The most frequently used protein fusions encompass Gfp, hemagglutinin-tag (HA), 3x myc tag, taptag, strep tag, and his-tag [9, 17, 18]. Caution: The function of fusion proteins that rely on externally provided substrates needs to be carefully verified, because the substrates often fail to enter the fungal cells due to the rigid cell wall.

Based on these important considerations, genetic constructs and strains for heterologous expression can be designed. Valuable plasmid collections compatible with modern cloning techniques are available for efficient genetic modification of *U. maydis*, and also strain generation and confirmation are well-established [19–25]. Next, protein expression can be conducted, including cultivation, isolation of proteins, and detection of successful biosynthesis at the level of Western blot analysis. Here, a brief summary of the steps required for expression strain generation based on existing protocols is provided (3.1), followed by a detailed description of a standard protein expression strategy exemplified by the expression and secretion of a heterologous endo-glucuronoxylanase (3.2–3.6).

2 Materials

Solutions, media, and buffers should be prepared with high-purity water, obtained from respective purification devices. All chemicals should be purchased in the *per analysis* (p.a.) grade.

2.1 Cultivation and Storage

All solutions are prepared using ultrapure water.

1. Cultivation: Complete medium supplemented with glucose (CM-glc) is used for cultivation (*see* **Note 1**). Prepare complete medium (CM) using stock solutions mentioned below.

For solid CM-glc plates, add 2% (w/v) agar-agar. Autoclave the solution for 20 min at 121°C. Add 1% (v/v) glucose after autoclaving.

Complete Medium (CM): 0.25% (w/v) casaminoacids (Difco), 0.1% (w/v) yeast extract, 1% (w/v) Holliday vitamins, 6.25% (w/v) Holliday salt solution, 0.05% (w/v) Herring sperm DNA (Sigma, D-3159), 0.15% (w/v) NH_4NO_3. Note that Holliday vitamins and Holliday salt solution are complex mixtures that need to be prepared beforehand. Adjust the pH to 7.0 with 1 M KOH. Sterilize by autoclaving at 121°C for 5 min. Store at room temperature (RT). Caution: The medium needs to be supplemented with the appropriate carbon source prior to use.

Holliday vitamins: 0.1% (w/v) thiamine hydrochloride (Serva 36020), 0.05% (w/v) riboflavin (Sigma R-4500), 0.05‰ (w/v) pyridoxine hydrochloride (Sigma p–9755), 0.2‰ (w/v) D-(+)-pantothenic acid calcium salt, 0.05‰ (w/v) aminobenzoic acid (Free Acid, Sigma A 9878), 0.2‰ (w/v) nicotinic acid (free acid, Sigma N–4126), 0.2‰ (w/v) choline chloride (Sigma C–1879), 1‰ (w/v) myo-inositol (Sigma I–5125). Store aliquots at −20°C.

Holliday salt solution: 16% (w/v) KH2PO4, 4% (w/v) Na-2SO4, 4% (w/v) KCI, 8% (w/v) CaCl2*2H2O, 1.32% (w/v) Holliday trace elements, 2% (w/v) MgSO4. Note that the Holliday trace elements are a separate solution. Sterilize by filtering using a 0.22 µm filter. Store at RT.

Holliday trace elements: 0.6% (w/v) H_3BO_3, 0.14% (w/v) MnCl*$4H_2O$, 0.4% (w/v) $ZnCl_2$, 0.4% (w/v) Na_2MoO_4*$2H_2O$, 0.1% (w/v) $FeCl_3$*$6H_2O$, 0.04% (w/v) $CuSO_4$*$5H_2O$. Store at RT.

Glucose solution (medium supplement): 50% (w/v) glucose. The solution needs to be filter sterilized (0.22 µm).

2. Long term storage: *U. maydis* strains are conserved in NSY-glycerol at −70° or − 80°C. Mix an exponential culture with NSY-glycerol in a ratio of 1:1. Caution: Do not freeze in liquid nitrogen. It is recommended to check for the absence of contamination by microscopy.

   NSY-glycerol for storage: 0.8% (w/v) nutrient broth (Difco), 0.1% yeast extract, 0.5% sucrose, 80% (v/v) glycerin. Prepare with H_2O dest and autoclave 20 min 121°C.

2.2 Extraction of Intracellular Proteins

1. 500 mM sodium phosphate buffer, pH 8: Final volume of 1 L. Weigh 85.5 g of Na_2HPO_4 x 2 H_2O (MW 178 g/mol) (final concentration of 480.2 mM) and 2.736 g NaH_2PO_4 x H_2O (MW 138 g/mol) (final concentration of 19.82 mM) and mix in 1 L of water.

2. 1 M Tris/HCl buffer pH 8.0: Final volume of 500 ml. Weigh 121.14 g of Tris (MW 121.14 g/mol) and resuspend in 500 ml water. Adjust pH using 37% HCl.
3. Lysis buffer: 8 M urea, 100 mM sodium phosphate buffer pH 8.0, 10 mM Tris/HCL buffer pH 8.0, 1 mM dithiothreitol, 1 mM benzamidine hydrochloride, 2×cOmplete protease Protease inhibitor cocktail tablet (Roche,*see* **Note 2**). Resuspend in water, store at RT. Prepare freshly prior to purification.
4. Glass beads: 0.40–0.60 mm Ø, B. Braun Biotech International GmbH (BBI-8541701)

2.3 Trichloroacetic Acid Precipitation of Proteins

1. 50% (w/v) Trichloroacetic acid (TCA, AppliChem PanReac, Darmstadt): Weigh 500 g TCA and resuspend H_2O_{bidest} ad 1 L.
2. Ice cold acetone. Stored at −20°C.
3. 2x Laemmli buffer: Mix in respective final volume 100 mM Tris/HCL (pH 6.8), 4% (w/v) SDS, 20% (v/v) glycerol, 10% (v/v) β-mercaptoethanol and 0.002% bromophenol blue. Fill up to the desired volume with H_20_{bidest}.
4. 2x Laemmli buffer containing 0.12 M NaOH: Mix in respective final volume 100 mM Tris/HCL (pH 6.8), 4% (w/v) SDS, 20% (v/v) glycerol, 10% (v/v) β-mercaptoethanol, 0.002% bromophenol blue, and 0.12 M NaOH. Fill up to the desired volume with H_2O_{bidest}.

2.4 Bradford Assay

1. Bradford reagent: BIORAD #500–0006, BioRad, Hercules.
2. Protein standards: Dilute laboratory-grade bovine serum albumin (BSA) either in water or in lysis buffer to fixed concentrations of 25 μg/ml, 50 μg/ml, 100 μg/ml, 250 μg/ml, and 500 μg/ml.

2.5 SDS Polyacrylamide Electrophoresis

1. Preparation of the gel. Caution: Toxic solutions are used. Use appropriate personal protective equipment (gloves, lab coat, safety glasses) and prepare within a fume hood. Pre-mix the given solutions for the separating and stacking gel. Start with the separating gel. TEMED and APS should be added freshly prior to pouring the gel, as this starts the polymerization reaction. Mix carefully to avoid extensive formation of bubbles.
   Separating gel: 0.375 M Tris-HCl pH 8.8, 12% (v/v) acrylamide/bis-solution (30%), 1.25% (v/v) glycerin, 0.1% (v/v) sodium dodecyl sulfate (SDS), 0.05% (v/v) ammonium persulfate (APS; $NH_4)_2S_2O_8$, 0.1% (v/v) N,N,N′,N′-tetramethylethylenediamine (TEMED). The given values can be scaled.
   Stacking gel: 0.125 M Tris-HCl pH 6.8, 5% (v/v) acrylamide/bis-solution (30%), 0.1% (v/v) sodium dodecyl sulfate (SDS), 0.05% (v/v) ammonium persulfate (APS; $(NH_4)_2S_2O_8$, 0.1% (v/v) N,N,N′,N′-tetramethylethylenediamine (TEMED). The given values can be scaled.

2. Gel run. Caution: Take careful safety measures if you handle sodium dodecylsulfate. It is advised to use pellets.
   10× Running buffer: 25 mM Tris pH 8.4, 192 mM glycine, 0.1% (v/v) sodium dodecyl sulfate (SDS). Running buffer is prepared in a 10× concentrated manner and diluted prior to work using water.

2.6 Immunoblotting

1. Polyvinylidene fluoride (PVDF) membrane, pore size 0.45 μm (*see* **Note 3**).
2. Blotting solutions (discontinuous system; **Note 4**): For this blotting technique, three buffers are needed. Caution: Methanol needs to be handled with utmost care as it's highly toxic. Strictly avoid skin contact and inhalation. Apply careful safety measures, including lab coat, safety glasses, and special gloves.
   Anode buffer 1: 0.3 M Tris–HCl, pH 10.4, 15% (v/v) methanol.
   Anode buffer 2: 30 mM Tris–HCl, pH 10.4, 15% (v/v) methanol.
   Cathode buffer: 25 mM Tris–HCl, pH 9.4, 15% (v/v) methanol, 40 mM ε -aminocaproic acid.
3. Solutions for immunoblot development. Include TBS, TBS-T, blocking buffer, and examples for needed antibodies in their individual dilutions.
   TBS: 20 mM Tris–HCl, pH 7.6, 136 mM NaCl.
   TBS-T: 100% TBS buffer, 0.05% Tween 20.
   Blocking buffer: 100% TBS-T buffer, 3% (w/v) skimmed milk powder (*see* **Note 5**).
   Standard antibodies (examples): anti-Ha from mouse igG (primary), diluted 1:3000; anti-mouse IgG HRP conjugate from goat (secondary), diluted 1:3000. Antibodies are diluted in blocking buffer.
4. Solutions for PVDF membrane staining after detection (Tab. 17 and Tab. 18). Coomassie Brilliant Blue staining will visualize the bound proteins (loading control, transfer control).
   Coomassie Brilliant Blue (CBB) staining solution: 20% (v/v) ethanol, 0.05% (w/v) Coomassie® Brilliant Blue G-250, 5% (v/v) acetic acid.
   Destainer solution for PVDF membrane: 20% (v/v) ethanol, 10% (v/v) acetic acid.

2.7 Equipment

1. Centrifuges to host 1.5 ml and 50 ml reaction tubes, and the possibility to adjust temperature.
2. Fluorimeter (e.g., Tecan plate reader).
3. Photometer.
4. Retch mill with metal containers to host 1.5–2 ml reaction tubes.

5. SDS-PAGE system, i.e., Mini-Protean III.
6. Thermomixer, i.e., Eppendorf Thermomix.
7. (Semi-dry) blotting device, i.e., Biometra electroblot chamber.
8. Fume hood.
9. Clean bench.
10. Chemiluminescence detection machine, i.e., ImageQuant 800 (Amersham).
11. Shaking device for bigger boxes or tubes holding blots or gels.
12. Cultivation equipment: 250 ml (baffled) flasks, glass tubes, 28°C cupboards or rooms, rotary wheel, and culture shaker or incubator.

3 Methods

3.1 Strain Generation

Use virtual cloning to carefully plan and verify your genetic construct carrying the heterologous gene. There is freeware like Benchling that provides all the required tools (https://www.benchling.com/).

1. Eventually generate a (di-)codon optimized version of your heterologous gene of interest (Fig. 1) (*see* **Note 6**). A helpful program is available for automated dicodon-optimization: https://tyberiusprime.github.io/dicodon_optimizer_wasm/index.html. Include tag, targeting, or reporter sequences for protein fusions if required. In the present example of Xyn30a, the native signal sequence was kept and a c-Myc tag was added during cloning.
2. Generate a plasmid carrying the expression unit for your heterologous gene (promoter, (optimized) transgene, terminator) and a resistance cassette (HygR, G418R, NatR or CbxR) encased by about 1 kb flanking regions (at least 500 bp) for genome targeting by homologous recombination using standard molecular cloning techniques like restriction-ligation, Golden Gate or Gibson cloning [20, 21, 27] (Fig. 2). Different insertion loci are available (*see* **Note 7**).
3. Linearize the plasmid for transformation using an appropriate restriction enzyme to obtain a single fragment encompassing the genetic construct (*see* **Note 8**).
4. Transform a suitable *U. maydis* strain using published protocols [19].
5. Verify transformants using published protocols [19] (*see* **Note 9**).

A

```
Original : ATG TAT TCT CTG CTC ATT GCC CTC CTC TGC GCC GGC ACT GCC GTC      45
Optimized: --- --C --C --C --G --- --T --- --- --- --- --- --C --G --G      45
AA       : M   Y   S   L   L   I   A   L   L   C   A   G   T   A   V

Original : GAC GCC CAG GCT CTC CAG CAG CGA CAG GCC GGG ACA ACT CTC ACC      90
Optimized: --T --G --A --C --G --A --A --- --- --T --A --T --G --- --G      90
AA       : D   A   Q   A   L   Q   Q   R   Q   A   G   T   T   L   T

Original : GTC GAC CTG TCC ACG ACG TAC CAG AGG ATC GAC GGG TTC GGT ACC      135
Optimized: --- --T --C AG- --T --A --- --- C-C --- --T --T --T --A --A      135
AA       : V   D   L   S   T   T   Y   Q   R   I   D   G   F   G   T

Original : TCG GAG GCC TTC CAG CGC GCG GTG CAG ATG AGC AGG CTG CCC GAG      180
Optimized: --C --A --- --- --- --- --- --- --- --- TC- C-A --T --A --A      180
AA       : S   E   A   F   Q   R   A   V   Q   M   S   R   L   P   E

Original : GAA GGA CAG CGG CGC GCT CTC GAC GTC CTG TTT AGT ACC ACC AAC      225
Optimized: --- --C --- --T --- --- T-G --T --- --T --C --C --G --T --T      225
AA       : E   G   Q   R   R   A   L   D   V   L   F   S   T   T   N

Original : GGC GCC GGG CTC TCC ATC CTT CGC AAC GGC ATC GGG TCC TCA CCC      270
Optimized: --- --- --A --T --- --T T-G --A --T --T --- --C --T --T --T      270
AA       : G   A   G   L   S   I   L   R   N   G   I   G   S   S   P

Original : GAC ATG AGC TCC GAC CAC ATG GTC TCG ATC GCG CCC AAG AGC CCC      315
Optimized: --T --- --- AG- --T --T --- --T --C --- --- --- --A TCT ---      315
AA       : D   M   S   S   D   H   M   V   S   I   A   P   K   S   P

                                  ...

Original : ATG ACC ACG GTG GTG CTG TAT CCG GCG GCC GAC GCG                  1431
Optimized: --- --G --A --C --T --- --- --- --T --- --- --T                  1431
AA       : M   T   T   V   V   L   Y   P   A   A   D   A
```

B

	xyn30A	
	Original	Optimized
A	263	310
C	526	401
G	420	381
T	222	339
%GC	*66.1*	*54.6*

Fig. 1 Example for di-codon optimization of heterologous genes. (**A**) Dicodon-optimized amino acid sequence (partial) of the enzyme Xyn30A in direct comparison to the native sequence from the filamentous fungus *Thermothelomyces thermophilus*. The sequence includes the native signal peptide for secretion (position 1–19). The web tool was used to generate the optimized sequence. Original gene sequence as found in *Thermothelomyces thermophilus*. Optimized gene sequence after dicodon-optimization. Diverging bases are displayed while identical bases are indicated by dashes. AA, encoded amino acids. (**B**) Amount of the 4 different bases in the original and di-codon optimized gene. The GC content is reduced in the optimized gene

Fig. 2 Typical architecture of an expression cassette for heterologous genes for genomic insertion at a distinct genetic locus, e.g. *upp3*. UF/DF, upstream and downstream flank homologous to insertion locus of about 1 kb; Grey scales, resistance cassette consisting of constitutive promoter (P), resistance gene, terminator (T); expression cassette for the transgene consisting of promoter (P), (optimized) heterologous gene sequence (transgene), terminator (T) (Figure not to scale)

6. Store verified strains at −70 or −80°C: Inoculate a 3 ml culture in a glass tube using CM-glc and incubate overnight at 28°C in a rotating wheel. Next morning, verify that the culture is not contaminated using a simple microscope with 40x magnification, mix 900 µl of the culture and 900 µl of NSY-glycerol in a 2 ml cryo tube with screw cap, mix carefully and freeze at −70–80°C (*see* **Note 10**).

3.2 Cultivation and Harvesting

The following protocol describes an expression culture for constitutively expressed transgenes (*see* **Note 11**) with harvest during late exponential growth. Follow the steps provided in Fig. 3 to verify production (and secretion) of your heterologous target protein. In the present example, a secreted enzyme, Xyn30A, is produced.

1. Preculture: Inoculate 3 ml CM-glc (*see* **Note 12**) with a small amount of the respective confirmed strain (*see* **Note 13**) using sterile equipment like a pipette tip and incubate for at least 16 h at 28°C on a rotating wheel.
2. Main culture: Use the preculture to inoculate 50–200 ml of CM plus 1% glucose to reach an $OD_{600\ nm}$ between 0.8 and 1.0 after one night of growth at 28°C (expected doubling time 2 h) (*see* **Note 14**).

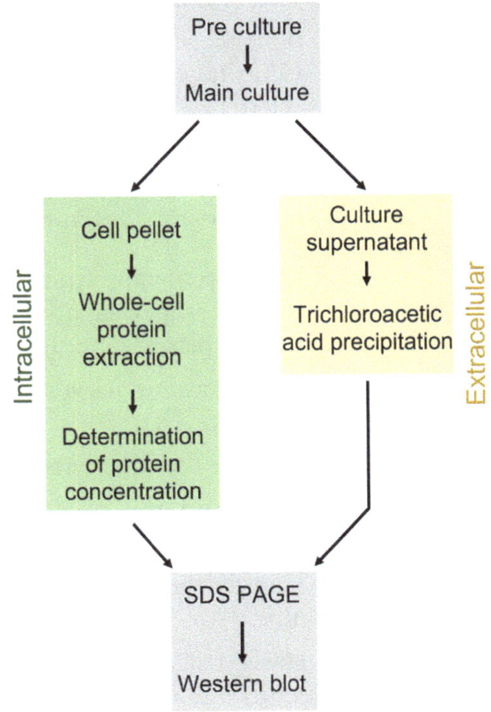

Fig. 3 Schematic overview of the experimental pipeline for verifying the production of intra- and extracellular proteins

3. Determine optical density ($OD_{600\,nm}$) in the morning (should be between 0.8 and 1) using a standard lab photometer. Prepare appropriate dilutions of your culture to ensure that the OD measurement is within the linear range of your photometer.
4. Harvest main cultures at 3,500 x *g* for 3 min in 50 ml Falcon tubes. Working at 4°C is recommended to improve protein stability.
5. Transfer the supernatant into a fresh tube and place both the cell pellet and the supernatant on ice. Then proceed either with the preparation of whole-cell protein extracts (Sect. 3.3) or isolation of proteins from culture supernatant (Sect. 3.4). Keep the pellet on ice for isolation of intracellular proteins to minimize degradation.

3.3 Preparation of Whole Cell Protein Extracts (Intracellular Proteins)

1. Resuspend the cell pellet by adding 1 ml fresh lysis buffer (store at room temperature). This can be done by vortexing, pipetting up and down, or shaking (*see* **Note 15**).
2. Add 200 μl glass beads into 2 ml reaction tubes for each sample. Tubes can also be prepared beforehand and cooled down for unstable proteins.
3. Transfer the resuspended cells from the 50 ml Falcon tube into the reaction tubes containing the glass beads (*see* **Note 16**). From here on, continue with all steps at 4°C.
4. Place the reaction tubes into safe-lock cups for an oscillating mill and crush the cells at 30 Hz for 15 min.
5. Centrifuge the samples at 16,000 x*g* for 30 min to pellet the cell debris.
6. Transfer the supernatant containing the cell extracts into fresh 1.5 ml reaction tubes and store the extracts on ice or at 4°C until protein concentration determination (*see* **Note 17**).

3.4 Isolation of Proteins From Culture Supernatant (TCA Precipitation)

1. Use supernatant from 3.2 Step 6. Gently mix 10 ml of the resulting supernatant with 2 ml of 50% (w/v) TCA solution in a 15 ml plastic tube. Incubate at 4°C overnight.
2. Next morning, harvest proteins at 16,000 x*g* at 4°C for 30 min (*see* **Note 18**).
3. Remove the supernatant.
4. Resuspend the pellet in 1 ml of −20°C cold acetone (*see* **Note 19**).
5. Transfer the resulting solution into a fresh cold 1.5 ml reaction tube.
6. Harvest again at 16,000 x*g* at 4°C for 30 min in a suitable centrifuge.
7. Repeat Steps 4 and 6 (skip Step 5).

8. Carefully remove all acetone using a micro pipette and let the sample dry.
9. Resuspend the pellet in 20–30 µl of the 2× Laemmli containing NaOH (*see* **Note 20**).
10. Heat the sample at 95°C for 10 min and use directly or store at −20°C until further use (*see* **Note 21**).

3.5 Protein Concentration Determination by Bradford Assay

Bradford protein determination can be applied to whole-cell protein extracts. This is important to ensure equal loading for SDS-PAGE analysis and subsequent immunoblot. TCA precipitates cannot be used in Bradford assays (*see* **Note 22**).

1. Prepare protein standard (see Sect. 2.4).
2. Dilute the staining solution (Bradford reagent BIORAD #500–0006, BioRad, Hercules) 1:5 in H_2O_{bidest} (*see* **Note 23**).
3. Prepare samples: When high protein concentrations are expected 1:10 or 1:100 dilutions should be used. Eventually, include all three concentrations. Use the lysis buffer for the dilution of the samples.
4. Load triplicates of 10 µl of each sample and each standard onto a transparent 96-well plate.
5. Add 200 µl of the diluted staining solution to each sample (*see* **Note 24**)
6. Incubate the plate for 5 min at room temperature.
7. Use a fluorescence spectrometer to determine the absorbance of the samples at 595 nm (*see* **Note 25**). A 3 s shaking step prior to the measurement in the device is recommended to ensure an equal mixture.
8. Generate a graph by entering the protein amounts of the protein standards (x–axis) versus the determined average absorbance at 595 nm (y–axis) in an appropriate computer program, introduce a trend line, and calculate its function. Using this formula, determine the protein concentration of your cell lysates. Determine the required sample volume for the next experimental step, in the present case 10 µg total protein for SDS PAGE.
9. Transfer the desired amount of protein into a fresh safe-lock tube and add 2x Laemmli buffer to a final concentration of 1x (*see* **Note 26**).
10. Boil samples at 95–99°C for 5 min and subsequently cool them down to room temperature. Centrifuge 5 min 16,000 x*g* in a tabletop centrifuge to pellet insoluble matter. The samples can be use directly or store at −20°C (*see* **Note 27**).

3.6 SDS Polyacrylamide Gel Electrophoresis (SDS-PAGE)

Caution: Due to the use of toxic substances, SDS gels should be poured and run inside a fume hood.

1. Assemble the glass plates (short plate + spacer plate) into a gel cassette sandwich in the casting frame and put them into the casting stand equipped with gaskets (*see* **Note 28**).
2. Pour the separation gel using the basic solution with freshly added TEMED and APS, and overlay the solution with 2-propanol (*see* **Notes 29 and 30**). Let it solidify for about 20 min.
3. Remove 2-propanol completely (*see* **Note 31**).
4. Pour the stacking gel with freshly added TEMED and APS until the upper front of the short plate, then carefully apply the comp. Let solidify for about 20 min (*see* **Note 32**). Caution, do not splash, wear eye protection.
5. Remove solidified gel from the casting stand and frame and assemble in the electrode assembly. Use a dummy in case you aim to run only one gel (*see* **Note 33**).
6. Insert the electrode assembly with the gel into the chamber and fill with 1× SDS running buffer, and remove the comp.
7. Load your samples. Include an appropriate protein size marker (*see* **Note 34**).
8. Add the lid and connect cables to a power supply. Run the gel with appropriate settings, e.g., 120–150 V for 1–2 h.
9. Stop the run at an appropriate time. Usually, you can track the loading dye front and stop once it has just left the lower front of the glass plates (*see* **Note 35**).
10. Carefully unpack the SDS gel from the glass plates, remove and discard the stacking gel, keep the separation gel in water for short time and quickly proceed with 4 (*see* **Note 36**).

4 Immunoblot

The following protocol describes the procedure of a semi-dry Western blot using PVDF membranes (*see* **Note 37**). Caution: Due to the use of toxic compounds, especially methanol, careful safety measures should be taken, and the procedure should be conducted in a fume hood. Wear a lab coat, safety glasses, and protective gloves.

1. Preparative steps: During the SDS-PAGE cut 6 Whatman papers (3 mm) to the size of the separation gel. Cut a PVDF membrane of similar size without removing the protective cover. Prepare 3 containers below the fume hood filled with anode buffer 1, anode buffer 2, and cathode buffer. Use just

enough buffer to soak up to 3 Whatman papers. Activate the PVDF membrane in methanol for 1 min, remove methanol, and then store in H_2O_{bidest}. Caution: Buffers contain methanol, which is toxic. Handle with careful safety measures, do not inhale, and use forceps to handle wet papers.

2. Assemble the blot according to Fig. 4. Use forceps to handle the wet papers.
3. Apply the lid carefully. Connect the blotting device to a power supply and start the protein transfer at about 5 mA/cm^2 for approximately one hour.
4. Stop the power supply. Remove the lid and carefully disassemble the Whatman papers. Use forceps to put the gel into a small box (protein side up) with blocking solution (*see* **Note 38**). Put on an orbital shaker for agitation at room temperature for at least 30 min (*see* **Note 39**).
5. Discard the blocking solution.
6. Add the primary antibody directed against the protein tag/protein at the established concentration in the blocking solution. Incubate for at least 1 h at RT with agitation using an orbital shaker (*see* **Note 40**).
7. Discard antibody solution and wash three times for 10 min with TBS-T.
8. Add the secondary antibody used for detection, i.e., a horseradish peroxidase conjugate. Incubate for 1 h at RT with agitation using an orbital shaker (*see* **Note 41**).
9. Discard antibody solution and wash with TBS-T three times for 10 min.
10. Wash with TBS once.
11. Develop the blot according to the manual provided with the developing reagents, i.e., for HRP conjugates, you can use Cytiva Amersham™ ECL Select™ Western Blotting Detection Reagents (Fisher Scientific).
12. Use a suitable device for signal detection, i.e., ImageQuant 800 (Amersham) (*see* **Note 42**). An example is provided in Fig. 5.

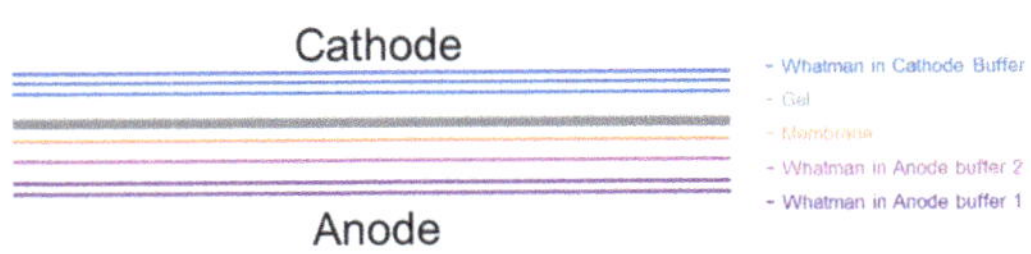

Fig. 4 Western blot assembly for semi-dry machines using a discontinuous buffer system

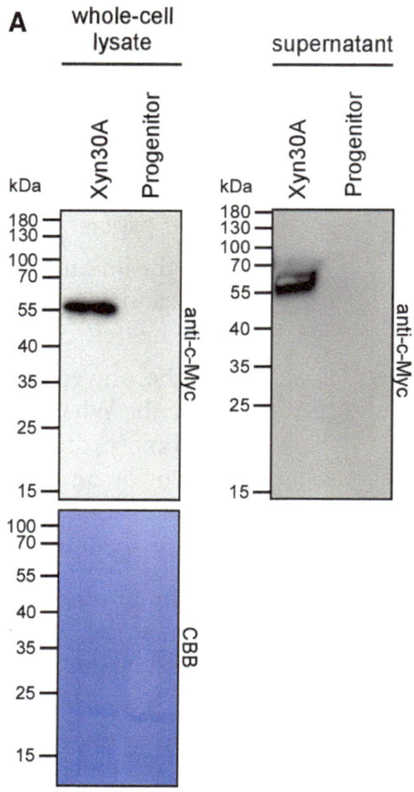

Fig. 5 Immunoblot verification of the production of heterologous endo-glucuronoxylanase Xyn30A from *Thermothelomyces thermophilus* in *U. maydis*. The protein carries a c-Myc tag for detection. Expected protein size after signal peptide cleavage, 53 kDa. 10 μg whole-cell protein extract (left blot) and TCA precipitated protein of 10 mL culture supernatant were applied (right blot, the white spot inside the band results from complete consumption of the ECL substrate due to high protein amounts). Progenitor, expression strain without the recombinant expression cassette. A primary mouse anti-c-Myc antibody was used in combination with an anti-mouse-horseradishperoxidase (HRP) conjugate for detection. CBB, Coomassie brilliant blue staining of the PVDF membrane after detection

13. Optionally: Stain the PVDF membrane with Coomassie brilliant blue after detection by placing the membrane into Coomassie brilliant blue (CBB) solution for 10 min on a rotary shaker at room temperature (*see* **Note 43**).
14. Remove the CBB solution and rinse the membrane with VE water.
15. Destain the membrane in Destainer solution on a rotary shaker. Replace the Destainer regularly as long as it turns blue.
16. Remove the Destainer solution and rinse the membrane with VE water.
17. Dry membrane on a paper towel in a fume hood (*see* **Note 44**).

5 Notes

1. Different media might be required or could be tested, depending on the used strain background. For example, SG200 is preferentially grown in rich medium like YepsLight (0.4% (w/v) yeast extract, 0.4% (w/v) peptone, 2.0% (w/v) sucrose, pH 7.0). Verduyn [26], a defined minimal medium, is preferred in biotechnologically relevant strains.
2. Scale the protocol for higher volumes if required. Urea precipitates at low temperatures, so the lysis buffer should be stored and used at RT.
3. For low-molecular-weight proteins (<15 kDa) 0.2 μm pore size membranes should be used. Alternatively, nitrocellulose membranes can be used (no activation step with methanol). In comparison, nitrocellulose membranes are more prone to tearing and are not suitable for stripping and reprobing, but show less autofluorescence than PVDF membranes. The most versatile membrane option is low fluorescence PVDF membranes, which show strongly reduced autofluorescence compared to standard PVDF membranes. Sticking to the established protocols of your laboratory is recommended. Success of the experiment can depend on the membrane type. Depending on its chemical properties, your protein of interest may bind better to one of the two membrane types. In case of trouble with detecting your protein of interest, it might be worth to switch the membrane type.
4. Suitable alternative protocols exist and can be exploited according to the local laboratory standard.
5. Alternatives for the blocking reaction, like BSA, exist and can be used according to the local laboratory standards.
6. Utilization of commercial gene synthesis is recommended to generate the heterologous sequence if it diverges from the native sequence or if DNA from the native organism is not available. Make sure that for fusing gene sequences encoding C-terminal tags the stop codon of the target gene is removed. Any tag must be fused in frame.
7. A large collection of published plasmids is available at the Feldbrügge lab, suitable for traditional cloning using restriction ligation and Golden Gate cloning [20, 21]. For example, use a pRabX1 or pRabX2 backbone for insertion of expression cassettes for heterologous proteins using integrative plasmids in the *ip* locus [6, 8] or pUMa4500 derivatives for insertion of expression cassettes in *upp3* locus [9]. Flanking regions should not be shorter than 500 bp [21]. Avoid using phleomycin-resistance cassettes (PhleoR), as this antibiotic

can cause side mutations. Caution: Repetition of genetic elements, e.g., using identical terminators for the expression cassette and resistance cassette, is possible but might reduce the efficiency of strain generation and strain stability under stress conditions.

8. It is essential that the genetic construct for transformation is linear and complete. For homologous recombination via 1 kb flanks, 100% homologous flank ends are advantageous, and strategies for their generation exist [21].
9. The gold standard is Southern blot, but diagnostic PCRs can also be sufficient.
10. Eventually, on the next day, streak the strain on CM-glc agar plates to verify that the stock is functional.
11. Alternatively, nutrient-regulated inducible promoters exist [10].
12. Depending on the heterologous target protein, the medium components may be varied, for example, if a cofactor is critical for function.
13. Use cell material grown on agar plates not older than two weeks for inoculation.
14. The culture volume should not exceed one-fifth of the flask volume.
15. A denaturing buffer for generating protein extracts is used in the present example. If you aim to isolate proteins in their native, active form, a non-denaturing buffer needs to be used that is, in the best case, compatible with downstream biochemical assays or applications.
16. Fill the 2 ml tubes with a maximum of 1 ml to guarantee efficient cell disruption. In case of low yields, the protocol can be scaled up. Cell pellets of larger culture volumes can also be disrupted using metal beads in a metal container compatible for use in the cryo mill.
17. Longer storage at 4°C is not recommended. Place on −20 or −80°C for extended storage.
18. Remember to balance the centrifuge tubes carefully.
19. No complete resuspension can be reached here as the sample is a precipitate, and this step mainly has the purpose to allow transfer of the precipitate into a new tube.
20. The resulting solution may be sticky. Foam formation might occur but should be avoided by slow pipetting. Eventually, residuals of the precipitate might remain. Color of the sample should be blue; if not, 1M NaOH can be added in small amounts (μl).

21. After storage at −20°C, the samples should eventually be boiled again.
22. Certain substances are not compatible with the Bradford reagent. A list of compatible substance concentrations of common buffer ingredients is available from BIORAD. It is advisable to test if the extraction buffer of choice is compatible with the Bradford assay.
23. You will need 200 μl of diluted staining solution for each sample.
24. Bradford solution is difficult to handle. Pipette carefully to prevent cross-contamination of your protein samples.
25. The color of the Bradford Reagent is pH dependent. Thus, the reagent may turn green under neutral conditions. This does not influence protein detection of the sample.
26. Laemmli buffer can be prepared at least up to 6 x concentrated, which can be useful for low sample concentrations. Typical yields for whole cell extracts generated with the present protocol are 3–6 mg/ml in a volume of 700–800 μl.
27. If the samples were stored at −20°C, Step 10 should be repeated.
28. Make sure that the glass plates are clean without solid gel residuals and dry. In the frame, they should be very well aligned. Ensure that the lower edges are not broken. It is important to fit the cassettes into the stand carefully to ensure tightness.
29. Make sure to leave space for the comp. If the remaining gel solution is kept in a closed tube and is solidified, you can safely proceed to the next step.
30. TEMED solution is prone to oxidization. The formation of yellow droplets in the TEMED solution indicates oxidization and will result in increased polymerization time.
31. Residual liquid can be removed by the insertion of paper towels into the empty space between the two plates.
32. If the remaining gel solution is kept in a closed tube and is solidified, you can safely proceed to the next step. Gels can be stored for several days at 4°C in wet paper towels if disassembled from the cast frame.
33. It is important to insert the gaskets carefully to ensure chamber tightness.
34. Make sure not to spill the samples into neighboring pockets.
35. Make sure to know the expected size of your protein and eventually adapt the gel run. Using a pre-stained protein ladder can aid to judge the best time for stopping the run.

Eventually, gradient gels or gels with differing percentages may be needed to resolve the target protein.

36. Make sure the gel is not exposed to the air for longer periods as it will quickly dry out and break. As an alternative to Western blotting, SDS PAGE gels can be stained with Coomassie brilliant blue or by silver staining to visualize separated proteins.

37. Other buffer systems and blotting methods are available. Adapt the procedure according to the established methodology in your laboratory.

38. Prior to the blocking step, the membrane can be temporarily stained with Ponceau S to visualize protein loading and as a control for the transfer to the membrane. To make a 100 mL Ponceau S staining solution, dissolve 0.1 g (100 mg) of Ponceau S powder in 95 mL H_20_{bidest}, then add 5 mL of glacial acetic acid and mix well to obtain a final concentration of 0.1% (w/v) Ponceau S and 5% (v/v) glacial acetic acid. Store the solution at 4°C or room temperature, protected from light. After complete destaining in tap water, blocking solution can be applied as described.

39. The blocking step can be extended. Overnight blocking should be conducted at 4°C due to the danger of contamination.

40. This step can be prolonged up to overnight incubation at 4°C. Other antibody systems established for the protein target or tag can be used, for example, fluorescent conjugates. The following steps need to be adapted accordingly.

41. Prolonging this step can result in nonspecific signals.

42. The production of the protein, as verified by immunoblotting, does not essentially mean that the protein is also functional and targeted to the correct cellular location. If available, perform activity assays or other experiments to proof protein functionality. For secreted proteins, analyze the culture supernatant (Note that secreted proteins may be *N*-glycosylated and thus appear larger in SDS PAGE). To confirm the localization of a heterologous protein, attaching a fluorescence protein tag like eGfp might be informative.

43. Alternatively, the membrane can be stained with Ponceau solution after the transfer as indicated in Sect. 3.6, Step 4.

44. The background of the membrane is still stained even after incubation with the Destainer. Drying at room temperature will reduce that background thus the protein bands become visible.

Acknowledgments

We acknowledge the contribution of Theresa Heidrich on strain generation and Western blot analysis of secreted heterologous xylanase and Jörg Kämper for promoting the development of the di-codon optimization tool. The project was funded in part by the Deutsche Forschungsgemeinschaft (DFG, German Research Foundation) – SFB1535 - Project ID 458090666 (projects B03 and B01). The scientific activities of the Bioeconomy Science Center were financially supported by the Ministry of Culture and Science within the framework of the NRW Strategieprojekt BioSC (No. 005-2012-0107).

References

1. Müntjes K, Philipp M, Hüsemann L et al (2020) Establishing polycistronic expression in the model microorganism *Ustilago maydis*. Front Microbiol 11:1384
2. Stoffels P, Müller MJ, Stachurski S et al (2020) Complementing the intrinsic repertoire of *Ustilago maydis* for degradation of the pectin backbone polygalacturonic acid. J Biotechnol 307:148–163
3. Heucken N, Tang K, Hüsemann L et al (2023) Engineering and implementation of synthetic molecular tools in the basidiomycete fungus *Ustilago maydis*. J Fungi (Basel) 9:480
4. Philipp M, Hussnaetter KP, Reindl M et al (2021) A novel potent carrier for unconventional protein export in *Ustilago maydis*. Front Cell Dev Biol 9:816335
5. Philipp M, Müller L, Andree M et al (2023) Efficient virus detection utilizing chitin-immobilized nanobodies synthesized in *Ustilago maydis*. J Biotechnol 366:72–84
6. Sarkari P, Reindl M, Stock J et al (2014) Improved expression of single-chain antibodies in *Ustilago maydis*. J Biotechnol 191:165–175
7. Reindl M, Stock J, Hussnaetter KP et al (2020) A novel factor essential for unconventional secretion of chitinase Cts1. Front Microbiol 11:1529
8. Stock J, Sarkari P, Kreibich S et al (2012) Applying unconventional secretion of the endochitinase Cts1 to export heterologous proteins in *Ustilago maydis*. J Biotechnol 161:80–91
9. Lee J, Hilgers F, Loeschke A et al (2020) *Ustilago maydis* serves as a novel production host for the synthesis of plant and fungal sesquiterpenoids. Front Microbiol 11:1655
10. Brachmann A, Weinzierl G, Kämper J et al (2001) Identification of genes in the bW/bE regulatory cascade in *Ustilago maydis*. Mol Microbiol 42:1047–1063
11. Terfrüchte M, Wewetzer S, Sarkari P et al (2018) Tackling destructive proteolysis of unconventionally secreted heterologous proteins in *Ustilago maydis*. J Biotechnol 284:37–51
12. Hewald S, Josephs K, Bölker M (2005) Genetic analysis of biosurfactant production in *Ustilago maydis*. Appl Environ Microbiol 71:3033–3040
13. Zarnack K, Maurer S, Kaffarnik F et al (2006) Tetracycline-regulated gene expression in the pathogen *Ustilago maydis*. Fungal Genet Biol 43:727–738
14. Wingen M, Potzkei J, Endres S et al (2014) The photophysics of LOV-based fluorescent proteins - new tools for cell biology. Photochem Photobiol Sci 13(6):875–883
15. Ingole KD, Nagarajan N, Uhse S et al (2022) Tetracycline-controlled (TetON) gene expression system for the smut fungus *Ustilago maydis*. Front Fungal Biol 3:1029114
16. Lo Presti L, Zechmann B, Kumlehn J et al (2017) An assay for entry of secreted fungal effectors into plant cells. New Phytol 213:956–964
17. Terfrüchte M, Reindl M, Jankowski S et al (2017) Applying unconventional secretion in *Ustilago maydis* for the export of functional nanobodies. Int J Mol Sci 18:937
18. Stoffel NK, Sankaranarayanan S, Müntjes K et al (2025) Microbial iCLIP2: enhanced mapping of RNA-protein interaction by promoting protein and RNA stability. RNA 31:258–272
19. Bösch K, Frantzeskakis L, Vranes M et al (2016) Genetic manipulation of the plant pathogen *Ustilago maydis* to study fungal biology and plant microbe interactions. J Vis Exp 125:54522
20. Brachmann A, König J, Julius C et al (2004) A reverse genetic approach for generating gene replacement mutants in *Ustilago maydis*. Mol Genet Genomics 272:216–226

21. Terfrüchte M, Joehnk B, Fajardo-Somera R et al (2014) Establishing a versatile Golden Gate cloning system for genetic engineering in fungi. Fungal Genet Biol 62:1–10
22. Schuster M, Schweizer G, Reissmann S et al (2016) Genome editing in *Ustilago maydis* using the CRISPR-Cas system. Fungal Genet Biol 89:3–9
23. Schuster M, Trippel C, Happel P et al (2018) Single and multiplexed gene editing in *Ustilago maydis* using CRISPR-Cas9. Bio Protoc 8:e2928
24. Wege SM, Gejer K, Becker F et al (2021) Versatile CRISPR/Cas9 systems for genome editing in *Ustilago maydis*. J Fungi (Basel) 7:149
25. Zuo W, Depotter JR, Doehlemann G (2020) Cas9HF1 enhanced specificity in *Ustilago maydis*. Fungal Biol 124:228–234
26. Müller MJ, Stachurski S, Stoffels P et al (2018) Online evaluation of the metabolic activity of *Ustilago maydis* on (poly)galacturonic acid. J Biol Eng 12: 34
27. Kämper J (2004) A PCR-based system for highly efficient generation of gene replacement mutants in *Ustilago maydis*. Mol Genet Genomics 271:103–110

Chapter 33

Mannosylerythritol Lipid and Ustilagic Acid Analysis Using Thin-Layer Chromatography

Jessica Tiefenbacher and Björn Sandrock

ABSTRACT

Secondary metabolites produced by *Ustilago maydis*, like mannosylerythritol lipids and ustilagic acid, are usually secreted together, leading to a glycolipid mixture. This makes it difficult to distinguish which compounds are present in the mixture or not. Here, a simple and fast approach for analysis of mannosylerythritol lipids and ustilagic acid using thin-layer chromatography is shown, enabling the separation and visualization of those from the mixture. A step-by-step protocol from the induction of such glycolipid synthesis in *U. maydis* over the isolation and their separation via thin-layer chromatography is presented.

Keywords Secondary metabolites, *Ustilago maydis* , Mannosylerythritol lipids, Ustilagic acid, Thin-layer chromatography

1 Introduction

Secondary metabolites can be naturally synthesized by *Ustilago maydis* and many other fungi. Some of these secondary metabolites can be produced under lab conditions as well and are proposed to find application in various industrial fields [1]. The glycolipids ustilagic acid (UA) and mannosylerythritol lipid (MEL) are such secondary metabolites. They are synthesized and secreted under nitrogen starvation conditions by *U. maydis* [2]. Other basidiomycetes of the family of Ustilaginaceae solely synthesize MELs [3]. The natural function of such glycolipids has not been understood completely, although they are predicted to function as external carbon storage and be involved in the pathogenicity process of *U. maydis*, as a defense mechanism against competing bacteria during plant infection, for example [4, 5].

As biosurfactants, UAs and MELs exhibit many interesting properties, like biological activity or emulsifying activity. This makes them potential alternatives to synthetic-derived surfactants, due to lower toxicity and better biodegradability [6]. UAs have been shown to

An N. T. Phan et al. (eds.), *Ustilago maydis*, Methods in Molecular Biology, vol. 3050,
https://doi.org/10.1007/978-1-0716-5364-7_33,

exhibit antifungal activity, while MELs have been demonstrated to show hemolytic activity, antibacterial activity against Gram-positive bacteria and emulsifying activity [6–9].

In order to analyze and separate the secreted MELs and UAs in the glycolipid mixture, thin-layer chromatography (TLC) can be used as a fast and low-cost method. It allows separation of different compounds, applied to a silica gel plate, according to their hydrophobicity [10]. The silica gel plate hereby functions as a solid phase, while the solvent consisting of chloroform and methanol acts as a mobile phase and leads to the separation of the glycolipid mixture. More hydrophobic compounds (MELs) migrate faster than less hydrophobic compounds (UA), leading to a distinct pattern after separation. The separation of the compounds can be visualized on the plate, making it a fast and simple approach for the separation of the glycolipid mixture. Although TLC might not be as efficient in the separation of the compounds and seems more labor-intensive than high-performance liquid chromatography (HPLC), it allows high-throughput screening due to the ability to simultaneously analyze multiple samples at once [10]. Another advantage of TLC is the lack of expensive equipment necessary for a simple analysis of the presence or absence of different MEL-variants and UAs in a mixture. Knowledge about the composition of the glycolipid mixture can be helpful to quickly verify deletions in the UA- or MEL-cluster.

For this reason, analysis of MELs and UAs via thin-layer chromatography can be seen as crucial to get an overview of the glycolipids synthesized by *U. maydis*, before testing them further. In the following, a step-by-step protocol from the induction of glycolipid synthesis under nitrogen starvation conditions, over the isolation using ethyl acetate and the analysis using TLC is described.

2 Materials

2.1 Media

YEPS-light medium for overnight cultures: 1% yeast extract, 0.4% bacto peptone, 0.4% sucrose dissolved in ddH_2O.

PD-plates: 2.4% potato dextrose, 2% bacto-agar dissolved in ddH_2O.

YNB-medium for glycolipid induction: 0.17% yeast nitrogen base w/o amino acids and ammonium sulfate dissolved in ddH_2O, addition of 1%–2% (w/v) sterile filtered glucose (50% w/v, dissolved in ddH_2O) after autoclaving.

2.2 Solutions

Solvents for thin-layer chromatography: chloroform:methanol (9:1, v/v), chloroform:methanol:water (65:25:4, v/v/v).

Solvent for visualization of sugar-compounds: ethanol:sulfuric acid:p-anisaldehyde (18:1:1, v/v/v).

2.3 Consumables

2 ml reaction tubes (*see* **Note 1**), Falcon tubes of suitable size (50 ml), pipette tips, polystyrene semi-micro cuvettes, TLC silica gel 60 F_{254} plates, sterile syringe with sterile syringe filter (0.45 μm), ethyl acetate, methanol (HPLC-grade, *see* **Note 2**).

2.4 Equipment

Minifuge centrifuge for 50 ml falcon-tubes, centrifuge for 1.5 ml/2 ml reaction-tubes, baffle flasks for incubation, culture tubes, vibrax shaker, rocker shaker, thermocycler for 70°C, two glass tanks with lid, fume hood for evaporation and working with organic solvents, shaking incubators for incubation at 23°C and 28°C, photometer for OD_{600}, sprayer, oven at 160°C, autoclave for sterile media preparation and waste inactivation.

3 Methods

3.1 Induction of MEL and Ustilagic Acid Synthesis

1. Preparation of *U. maydis* overnight culture: Inoculation of *U. maydis* grown on PD-agar plates in 3 ml YEPS-light media and incubation at 28°C shaking overnight.
2. Determination of OD_{600} of the overnight culture using a photometer. 1 ml of the overnight culture is diluted 1:20 in YEPS-light, before determination of the OD_{600}.
3. Induction of *U. maydis* under nitrogen starvation conditions: Synthesis of MELs and UA can be achieved in YNB media supplemented with glucose as a carbon source. *U. maydis* cells are added to 3 ml (or up to 50 ml) YNB supplemented with 1 or 2% glucose (*see* **Note 3**) at a final optical density of $OD_{600}=0.1$.

   MEL-synthesis: Culture is incubated at 23°C, shaking for 6 days. UA-synthesis: Culture is incubated at 28°C, shaking for 3 days.

3.2 Isolation of Glycolipids

Isolation of glycolipids from the culture can be performed 6 days (3 days) after induction and incubation at 23°C (28°C).

1. In 2 ml reaction tubes or 50 ml Falcon tubes (depending on the culture volume) mix equal amounts of culture with ethyl acetate (*see* **Note 4**) and incubate for 10 min shaking on a vibrax shaker or rocker shaker.
2. Centrifuge at room temperature for 10 min at 16,000 × *g* for reaction tubes and 2500 × *g* for Falcon tubes. Then, two phases (organic ethyl acetate phase and aqueous culture phase) should be clearly distinguishable.
3. Transfer the upper phase consisting of ethyl acetate and the glycolipids into new 2 ml reaction tubes. Sometimes multiple tubes are necessary if larger amounts of culture and ethyl acetate are used (*see* **Note 5**). The ethyl acetate mixture containing the glycolipids is evaporated under a fume hood on a

thermocycler at 70°C slowly shaking with the lid open until all liquid is completely evaporated. The evaporation of 2 ml ethyl acetate containing glycolipids usually takes between 1 ½ and 2 h.

4. After all liquid is evaporated the glycolipids can be dissolved in methanol (*see* **Note 2**). The addition of 50 μL methanol per 2 ml reaction tube is suitable to dissolve the glycolipids in methanol; the tube can be vortexed briefly. If multiple reaction tubes were used during evaporation, the glycolipids dissolved in methanol can be pooled together.
5. Storage of glycolipids dissolved in methanol at 4°C is suitable for short-term storage. For long-term storage, −20°C is recommended, as well as the use of safe-lock reaction tubes (*see* **Note 1**).

3.3 Thin-Layer Chromatography

In order to analyze the MELs and UAs in the glycolipid mixture thin-layer chromatography can be performed. All following steps should be performed wearing gloves since the plates are sensitive to direct skin contact.

1. Preparation of a silica gel plate for separation of the glycolipid mixture (Fig. 1A). With a pencil mark the spots for application of the glycolipid mixture on the plate. Leave 1 cm space from the bottom to draw a line and around 1.5 cm space from each side. Mark 0.5 cm above the line on which the spots of the glycolipid mixture samples will be spotted. The spacing between different sample spots is ideally around 1 cm and 1.2 cm. The sketch of the plate design in Fig. 1A can be used for reference on how an ideal plate design can look like (*see* **Note 6**).
2. The glycolipid mixture can then be spotted (4 × 3 μl) on the previously marked spots on the silica gel plate. Hereby, multiple spotting steps with low volume and drying of the spots are ideal to concentrate the samples.
3. For the separation of the glycolipid mixture according to its hydrophobicity, two different solvent systems are used. Chloroform:water:methanol (65:25:4) can be used for the separation of UA and MELs; chloroform:methanol (9:1) can be used for further separation of MELs into the different MEL-variants. The glass tanks containing the solvents should be closed with a lid to avoid evaporation and should be filled at least 15 min prior to the TLC to allow settling of the solvents (*see* **Note 7**). The filling level of the solvent in the glass tanks should not exceed 1 cm to allow the solvent to move as a mobile phase and enable proper separation of UAs and MELs. The silica gel plate is carefully placed into the tank with the spotted samples at the bottom of the plate, so the applied

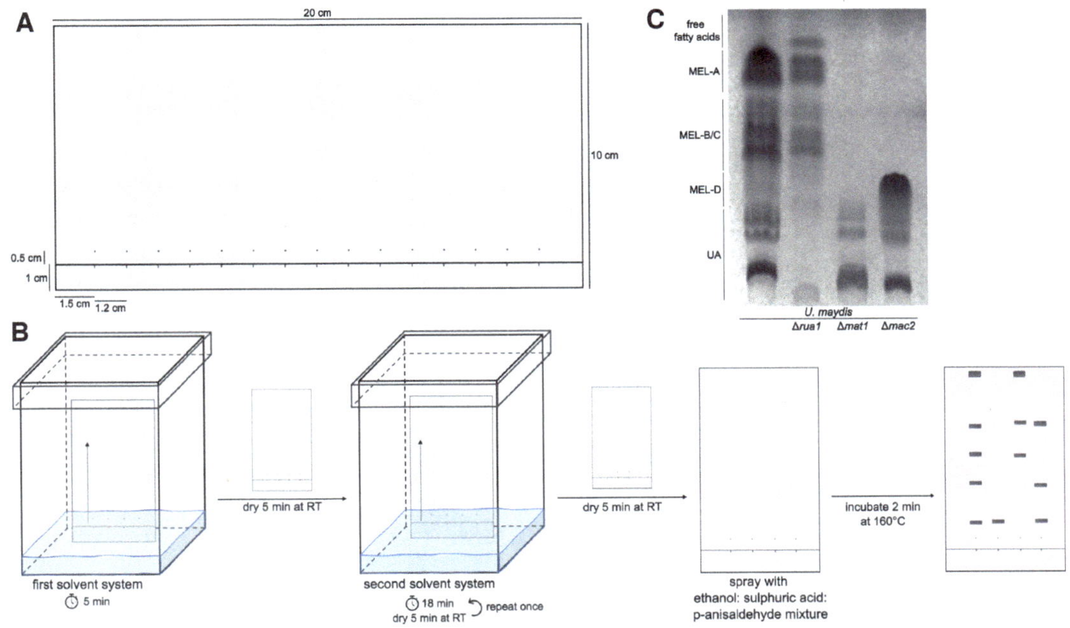

Fig. 1 Analysis via thin-layer chromatography. (**A**) Silica gel plate design for TLC. (**B**) Set up thin-layer chromatography. Silica gel plate with glycolipid mixture is developed in different solvent systems, then sprayed and incubated at 160°C for visualization of UAs and MELs. (**C**) TLC of glycolipids synthesized by different *U. maydis* strains. *U. maydis* WT synthesizes UAs and MELs, *U. maydis* Δ*rua1* synthesizes solely MELs, *U. maydis* Δ*mac2* solely produces UAs, and *U. maydis* Δ*mat1* synthesizes UAs and the non-acetylated MEL-D-variant.

sample can migrate towards the top of the plate during the development (Fig. 1B).

(a) Incubation of the silica gel plate for 5 min in chloroform:water:methanol (65:25:4) (first solvent system), followed by drying of the plate for 5 min at room temperature (*see* **Note 8**).

(b) Incubation of the plate twice for 17 min each in chloroform:methanol (9:1) (second solvent system). After each of the two incubation steps in the second solvent system, the plate is dried for 5 min at room temperature.

4. For visualization of UAs and MELs, the sugar compound of these glycolipids can be visualized using an ethanol:sulfuric acid:p-anisaldehyde mixture. The plate is evenly sprayed with the mixture and dried for 2 min. To visualize the sugar compounds, the silica gel plate is then incubated for 2 min in an oven at 160°C (*see* **Note 9**). Documentation of the visualized glycolipids on the silica gel plate should be performed immediately after the incubation and can be performed using a simple office scanner or just by taking a photo using a phone or camera (*see* **Note 10**).

3.4 Analysis of the TLC

MELs and UAs in the glycolipid mixture were separated according to their hydrophobicity on the silica gel plate. Less hydrophobic UAs migrate slower on the plate during TLC, while MELs migrate faster. The different MEL-variants (MEL-A, MEL-B/C and MEL-D), also migrate at different speed depending on the degree of acetylation, the di-acetylated MEL-A migrates faster than the non-acetylated MEL-D. This leads to the separation of the glycolipid mixture into distinguishable bands on the silica plate after visualization (Fig. 1C) (*see* **Notes 11 and 12**). From bottom to top, the glycolipid mixture is separated into the following:

1. UAs (*see* **Note 13**).
2. MEL-D.
3. MEL-B/C.
4. MEL-A.
5. Free fatty acids (*see* **Note 14**).

4 Notes

1. 2 ml reaction tubes for storage: 2 ml safe-lock tubes are better for long-term storage of glycolipids in methanol, since they avoid evaporation of methanol over time.
2. Use of methanol (HPLC-grade): For downstream applications, it is convenient to dissolve the glycolipids in methanol suitable for HPLC analysis. This is a standard step to avoid using the wrong methanol during HPLC analysis by forgetting to switch to the HPLC-grade methanol. This avoids an additional evaporation step of the methanol and dissolving in HPLC-grade methanol in case HPLC needs to be performed.
3. Volume of YNB+glucose can vary depending on the desired follow-up experiments. If only small amounts of MELs are needed, e.g., checking for MEL-synthesis of the strain via TLC, 3 ml YNB+glucose is suitable. If higher amounts of MELs are needed, e.g., for different downstream applications of MELs, higher volumes of the initial starting volume are recommended (50 ml YNB + glucose).
4. The amount of ethyl acetate used for the isolation of glycolipids can be reduced from equal amounts (1:1) to about 800 µl ethyl acetate per 1 ml culture. This allows for faster evaporation and less consumption of ethyl acetate during the isolation step.
5. The evaporation of the ethyl acetate mixture containing glycolipids can be performed in multiple 2 ml reaction tubes for

higher volumes of ethyl acetate. Alternatively, to decrease the waste of reaction tubes and for limited space on the thermocycler, the ethyl acetate phase containing the glycolipids can be added step-wise to the same tube, with evaporation in between. For this, it is recommended to transfer the upper phase after centrifugation into a new Falcon tube and transfer it step-wise into the evaporation tube at 70°C to avoid mixing of the two phases over time.

6. The silica gel plate design can be used as a reference for easy spotting. 1–1.2 cm spacing in between spots ensures uniform separation of the glycolipid mixture, but can be increased or reduced, if necessary, less than 0.8 cm spacing is not recommended since the bands between different samples might become less distinguishable then. Most commercially available silica gel plates are 20 cm×20 cm, using half of the plate (20 cm × 10 cm) for one thin-layer chromatography assay is recommended since 10 cm migration pathway is sufficient for separation of the UAs and MELs. Depending on the number of samples, the plate can also be shortened to fit the number of samples, e.g., 9 cm × 10 cm for five samples.
7. Filling and preparing the tanks in advance allows settling of the solvents giving better results. Additionally, the lid prevents evaporation of the solvents and therefore ensures correct ratios of the solvents, which could impact migration of the glycolipid mixture on the silica gel plate. To ensure correct ratios in the solvent systems, exchange through fresh solvents can become necessary after some time, if the separation of the glycolipids on the plate is not as expected anymore. Exchanging the solvents at least every 2 months is recommended.
8. Drying of the silica gel plate in between the incubation steps in the solvent systems can be varied, as long as the plate completely dried before the next incubation step. 5 min drying time ensures complete drying of the plate.
9. Heating the plate using an oven at 160°C allows for even "baking" of the plate, enabling uniform coloring and visualization of the glycolipids on the silica gel plates. Alternatively, a hot fan can be used for visualization of the silica gel plate, if no oven is available, although results might not look as nice, since the heat most of the time is not distributed as even as in the oven.
10. Documentation of the visualized glycolipids should be performed immediately since the coloring on the silica gel plates faints over time. If visualization is necessary again the plate can simply be sprayed again with the ethanol:sulfuric

acid:p-anisaldehyde mixture and incubated again for 2 min at 160°C.

11. Separation of UAs and MELs during TLC strongly depends on the solvent systems used. A third solvent system consisting of chloroform:methanol (4:1) allows for separation of MELs only. This third solvent system could also be used for a fast TLC approach, since one incubation step for 18 min is sufficient for separation of the MELs, if simply MEL-synthesis should be checked.
12. For further analysis of specific bands appearing during TLC, TLC coupled to mass spectrometry (TLC/MS) can be used [3]. First, each sample is spotted in duplicates in direct proximity. Next, the TLC protocol is carried out as described above, except for Step 4. After separation, the duplicate samples are split by cutting the plate and one duplicate is sprayed for visualization. After heating the sprayed plate, the desired bands are marked on the untreated plate by placing each duplicate next to each other. TLC/MS can then be performed with the marked area.
13. UAs mainly migrate in the first solvent system. If separation of UA is not necessary, the incubation time in the first solvent system can also be reduced to 3 min instead of 5 min. For better separation of UAs instead of the MELs, the silica gel plate can be incubated for 17 min in the first solvent system only.
14. Free fatty acids might appear as a top band on the plate after TLC and can be ignored for the MEL and UA analysis.

Additional note: This method can also be used for analysis of UA and MELs from other basidiomycetes. Induction and isolation of UA and MELs work under the same conditions for other basidiomycetes as well [11, 12].

Acknowledgments

This work was supported by the grant SA 1018/5-1 from the Deutsche Forschungsgemeinschaft (DFG) to B.S.

References

1. Bölker M, Basse CW, Schirawski J (2008) *Ustilago maydis* secondary metabolism—from genomics to biochemistry. Fungal Genet Biol 45:S88–S93
2. Haskins RH, Thorn JA, Boothroyd B (1955) Biochemistry of the ustilaginales: XI. Metabolic products of *Ustilago zeae* in submerged culture. Can J Microbiol 1:749–756
3. Beck A, Haitz F, Grunwald S et al. (2019) Influence of microorganism and plant oils on the structure of mannosylerythritol lipid (MEL) biosurfactants revealed by a novel thin layer chromatography mass spectrometry method. J Ind Microbiol Biotechnol 46:1191–1204
4. Becker F, Stehlik T, Linne U et al. (2021) Engineering *Ustilago maydis* for production of

tailor-made mannosylerythritol lipids. Metabol Eng Commun 12:e00165
5. Yu C, Qi J, Han H et al. (2023) Progress in pathogenesis research of *Ustilago maydis*, and the metabolites involved along with their biosynthesis. Mol Plant Pathol 24:495–509
6. Kitamoto D, Morita T, Fukuoka T et al. (2009) Self-assembling properties of glycolipid biosurfactants and their potential applications. Curr Opin Colloid Interface Sci 14:315–328
7. Teichmann B, Linne U, Hewald S et al. (2007) A biosynthetic gene cluster for a secreted cellobiose lipid with antifungal activity from *Ustilago maydis*. Mol Microbiol 66:525–533
8. Hewald S, Josephs K, Bölker M (2005) Genetic analysis of biosurfactant production in *Ustilago maydis*. Appl Environ Microbiol 71:3033–3040
9. Ceresa C, Hutton S, Lajarin-Cuesta M et al. (2020) Production of Mannosylerythritol Lipids (MELs) to be used as antimicrobial agents against *S. aureus* ATCC 6538. Curr Microbiol 77:1373–1380
10. Santiago M, Strobel S (2013) Thin layer chromatography. Methods in enzymology. Elsevier, pp 303–324
11. Deinzer H-T, Linne U, Xie X et al. (2019) Elucidation of substrate specificities of decorating enzymes involved in mannosylerythritol lipid production by cross-species complementation. Fungal Genet Biol 130:91–97
12. Tiefenbacher J, Linne U, Freitag J et al (2025) Two gene clusters are required for mannosylerythritol lipid biosynthesis in *Sporisorium reilianum*. mBio 16(9):e0089925

Chapter 34

Studying Peroxisomes in *Ustilago maydis*: A Defined Workflow

Nils Bäcker, Damon Freitag, Gert Bange, Judith M. Klatt and Johannes Freitag

Abstract

Peroxisomes are highly conserved oxidative organelles essential for the breakdown of fatty acids and the detoxification of hydrogen peroxide (H_2O_2), a byproduct of this process. Beyond these central roles, peroxisomes contribute to diverse cellular functions, including amino acid biosynthesis and secondary metabolism. In recent years, the basidiomycete *Ustilago maydis* has emerged as a valuable model to uncover novel aspects of peroxisome biology. Here, we present a workflow to investigate the subcellular and suborganellar localization of peroxisomal proteins of interest and to assess their functional relevance. The outlined experimental strategies are broadly applicable to research questions beyond peroxisome biology and can be adapted to other systems.

Keywords Peroxisomes, Protein import, *Ustilago maydis*, *Mycosarkoma maydis*, PTS1, Fatty acid oxidation, Organelle purification, Micro sensors

1 Introduction

Peroxisomes are vesicular, single-membrane-bound eukaryotic organelles discovered and characterized by Christian de Duve and colleagues in the 1960s [1]. They harbor a variety of oxidative enzymes, including those involved in the β-oxidation of fatty acids, the detoxification of H_2O_2, and the oxidation of uric acid, a waste product of purine metabolism [2–4]. Peroxisomes are highly dynamic organelles that change their shape, number, and protein content depending on the developmental stage or the environmental conditions and can fulfill organism-specific functions. In plants, they are required for a process termed photorespiration to compensate for the oxidase function of Ribulose-1,5-bisphosphate carboxylase/oxygenase; in mammals, peroxisomes are essential and are relevant

The authors Nils Bäcker and Damon Freitag contributed equally to this work.

An N. T. Phan et al. (eds.), *Ustilago maydis*, Methods in Molecular Biology, vol. 3050,
https://doi.org/10.1007/978-1-0716-5364-7_34,

for the biosynthesis of plasmalogens [5, 6]. In fungi, peroxisomes contain a variety of anabolic metabolic pathways, e.g., for production of secondary metabolites or several amino acids [7, 8]. Peroxisomes lack their own DNA and therefore rely on the import of proteins from the cytosol. This process requires specific targeting signals—either a C-terminal peroxisomal targeting signal type I (PTS1) or an N-terminal peroxisomal targeting signal type II (PTS2), which are recognized by soluble receptor proteins in the cytosol [9].

The basidiomycete *Ustilago maydis* has emerged as a valuable model system to study peroxisome biology, including metabolism [10, 11], protein targeting [12–14], and organelle sub-functionalization [15]. Critical steps in the biosynthesis of mannosylerythritol lipids (MELs)—glycolipids with strong surface-active properties that reduce surface and interfacial tension between liquids (e.g., oil and water), thereby allowing them to mix more easily—take place inside the peroxisomes of *U. maydis*. Two of the enzymes essential for this process accumulate within detergent-resistant core structures in the peroxisomal lumen and are further segregated into distinct peroxisome subpopulations [11, 15, 16]. Furthermore, peroxisomes and peroxisomal fatty acid degradation are relevant for infection of the host plant *Zea mays* [13, 17, 18]. *U. maydis* also serves as a model to decipher the spatial organization of peroxisomes inside of cells, e.g., it has been proven useful to generate insights into long-distance trafficking and formation of organellar contact sites [19, 20, 14].

Here we present a workflow to analyze both the subcellular and intraperoxisomal localization of putative peroxisomal proteins, and to assess peroxisome functionality through growth assays and measurements of respiration rates under defined conditions. Establishing such a workflow is essential for linking protein localization to organelle function, thereby enabling a deeper understanding of peroxisome biology and its impact on cellular physiology. This protocol will enable a standard characterization of any predicted peroxisomal protein of interest in *U. maydis* but in principle can be adapted other organisms and parts of the protocol can be followed autonomously for more specific questions. First, we present a simple test to study the functionality of PTS1 motifs, which are characteristic features of the majority of peroxisomal matrix proteins. Secondly, we describe two strategies to analyze the peroxisomal and sub-peroxisomal localization of full-length proteins either by a microscopic approach or a biochemical fractionation experiment followed by fluorescence microscopy. Finally, we introduce two different ways to analyze growth and oxygen consumption in medium containing oleic acid ((9Z)-octadec-9-enoic acid)—a process that requires peroxisome function [10, 13]. Please note that the workflow described here requires transgenic *U. maydis* strains containing insertions in the *ip*-locus [21], deletion of genes [22, 23] and molecular cloning [24]. The equipment and procedures to generate these strains are not detailed in this chapter, but are thoroughly described in the chapter from Jessica Tiefenbacher and Björn Sandrock on CrispR-Cas9 gene editing and are also specified elsewhere.

2 Materials and Equipment

All described buffers, solutions, and media should be prepared with double-distilled (ddH_2O) or MilliQ water. Note that for preparation of competent *U. maydis* cells, MilliQ water is strongly recommended as the quality of the water drastically affects transformation efficiency. For all presented protocols standard laboratory equipment such as benchtop centrifuges, centrifuges covering larger volumes (up to 50 mL) (ideally with a cooling function), autoclaves, tubes (50 mL, 15 mL, and 1.5 mL), pipettes, tips, disposable plastic cuvettes, and incubation chambers with adjustable temperature is required. Other standard laboratory equipment, such as gel systems for DNA, a Nanodrop device, and a PCR machine is required as well if you prepare transgenic strains yourself.

2.1 Growth Media for *Ustilago maydis*

1. YEPS-light 10 g/L Yeast extract (Chemsolute, TH Geyer, Germany) 4 g/L Peptone (Gibco, Thermo Scientific, USA) 0.4% (w/v) Sucrose.
2. YNB-Media (liquid) 1.7 g/L Yeast Nitrogen Base (Difco™, Becton Dickinson, USA) 0.2% (w/v) ammonium sulfate pH = 5.6 (adjusted with NaOH).
3. Additives:
   0.1% (v/v) oleic acid (Roth, Germany).
   0.1% (v/v) TWEEN40 (Sigma Aldrich, USA).
4. YNB-Media (solid).
   1.7 g/L Yeast Nitrogen Base 0.2% (w/v) ammonium sulfate.
   pH = 5.6 (adjusted with NaOH).
   Additives:
   0.2% (v/v) oleic acid.
   0.2% (v/v) TWEEN40.
   1.5% (w/v) agar (BD Difco™, Becton Dickinson, USA).
   or
   2% (w/v) D(+)-glucose.
   1.5% (w/v) agar.

2.2 Fluorescence Microscopy

1. 1.5% (w/v) agarose (Roth, Germany) in ddH_2O (for immobilization of cells on microscopy slides)
   Dissolve 1.5% (w/v) agarose in sterile ddH_2O.
2. 1.5% (w/v) agarose in Lysis Buffer (for immobilization of organelles on microscopy slides)
   Dissolve 1.5% (w/v) agarose in Lysis Buffer
3. Coverslips
4. Microscopy slides

5. Photometer for determination of optical density at a wavelength of 600 nanometers (OD_{600})
6. Fluorescence microscope (Structured illumination microscopy [25] is a benefit to resolve intraperoxisomal features)

2.3 Isolation of Crude Organelles

1. SCS (Sodium citrate, Sorbitol)-solution (for stabilization of protoplast).
   SCS solution 1: 0.6% (w/v) Sodium citrate × H_2O (final concentration (f.c.) 20 mM), 18.2% (w/v) Sorbitol (final concentration 1 M) in ddH_2O.
   SCS solution 2: 0.4% (w/v) Citric acid × H2O (f.c. 20 mM), 18.2% (w/v) Sorbitol (1 M) in ddH_2O.
   Titrate solution 2 into solution 1 to reach pH 5.8 and autoclave.
2. *Trichoderma* Lysing Enzymes (Sigma-Aldrich, USA) in SCS solution for cell wall degradation.
   Dissolve Trichoderma lysing enzymes in SCS Solution to a final concentration of 10 mg/mL; filtered sterile.
3. Lysis Buffer (for lysis of protoplasts).
   5 mM 2-(N-morpholino)ethanesulfonic acid (MOPS), 0.5 mM EDTA, 1 mM KCl, 0.6 M Sorbitol.
   Protease inhibitors: 1 μg/mL Antipain, 1 μg/mL Aprotinin, 0.2 mg/mL Benzamidine hydrochloride, 1 μg/mL Bestatin, 1 μ g/mL Chymostatin, 1 μg/mL Leupeptin, 0.2 mg/mL NaF, 1 μ g/mL Pepstatin, 10 μg/mL PMSF.
   - Combine MOPS, EDTA, KCL and Sorbitol in MilliQ water, then add the protease inhibitors and adjust to pH 6.0 with 1 M KOH at room temperature.
   - Protease inhibitor cocktail for yeast and fungal extracts (Sigma-Aldrich, USA) can be used as alternative.
4. Potter-Elvehjem homogenizer (make sure that it fits tightly enough to induce pressure.
5. Photometer for determination of OD_{600}.

2.4 Oxygen Optode Respiration Measurements

1. YNB medium (liquid/solid) as in Sect. 2.1 without additives
   Additive stocks: D-Glucose: 40% (w/v) stock for short-term injections. 2% (w/v) in YNB media for long-term experiments
   Oleic acid emulsion: 1:1 (v/v) oleic acid:TWEEN40 emulsion for short-term injections. 0.1% (v/v) oleic acid/TWEEN40 in YNB (liquid) for long-term experiments.
2. Sterile YNB (without carbon source) for washing equipment.
3. Sterile ddH_2O for preparation of all buffers & solutions.
4. ~3 mL septum vials (e.g., Exetainers, Labco, UK) with butyl rubber septa (pierceable multiple times).

5. Glass-coated magnetic stir bars, 4 mm.
6. Optical oxygen sensor spots (e.g., PyroScience).
7. Silicone adhesive.
8. Bent tweezers with rounded tips.
9. Hamilton syringes (250 μL volume) for injecting substrate through septa.
10. Syringe needles for substrate injections & flushing media with gas.
11. Sterile filters (0.2 μm, luer-lock) to place inline on gas needles/tubing for sterile air or nitrogen flushing.
12. Oxygen optode meter with fiber-optic readout and holders (e.g., FireStingO_2, PyroScience) plus manufacturer software on a laptop for acquisition.
13. Fiber-optic sensor spot holders/adapters (commercial or custom made).
14. Magnetic stirrer (multi-position recommended) with vial rack.
15. Autoclave (for vials with glued spots and stir bars; lids sterilized separately).
16. UV source/biosafety cabinet (for septum/cap surface sterilization).
17. Gas supply for calibration: sterile-filtered air and sterile-filtered nitrogen (flushed via syringe/needle + 0.2 μm filter).

3 Methods

3.1 Assessment of PTS1 Functionality via a Fluorescence Microscopic Approach

1. The efficiency of PTS1-dependent import is affected by the nature of the last 10–12 amino acids of a PTS1 containing protein [26, 27]. To obtain an initial indication of the functionality of putative motifs, the PTS1 predictor (https://mendel.imp.ac.at/pts1/) can be used [28] (Fig. 1A). To analyze functionality, insert the DNA sequences encoding the last 12 amino acids of the protein of interest, followed by a stop codon downstream of the GFP open reading frame. This results in a fusion construct coding for GFP at its N-terminus. A suitable vector is P_{etef}-GFP-Ala_6-MMXN ([29, 30]; available from the authors) that allows constitutive expression. It can be integrated into the *ip*-locus of *U. maydis* cells via homologous recombination resulting in resistance towards the antibiotic carboxin [21]. Constructs should be verified by sequencing. Petef-GFP-Ala_6-MMXN is also suitable to test functionality of putative PTS2 motifs. To this end, motifs or

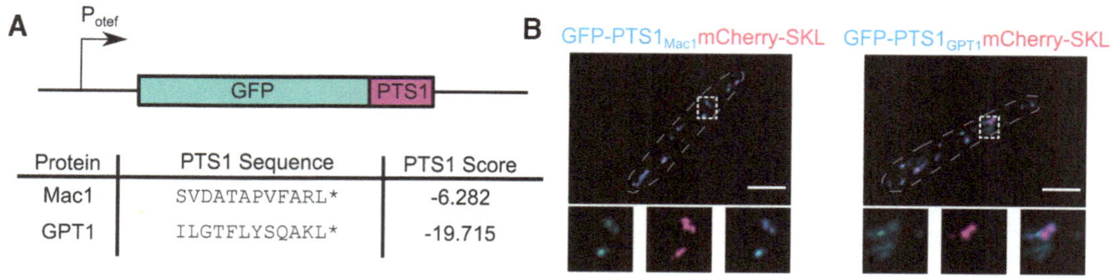

Fig. 1 Evaluation of PTS1 motifs. (**A**) Schematic drawing of the construct used (top). C-terminal dodecamers were tested using the PTS1PTS1 prediction tool [28]. Values larger than 0 suggest peroxisomal targeting (bottom). (**B**) Experimental validation: *U. maydis* strains expressing N-terminally GFP-tagged to the depicted dodecamers (cyan) and the peroxisomal marker protein mCherry-SKL (magenta) were inspected by epifluorescence microscopy. Full images are shown as overlays of two channels. For insets single channels and merged channels are provided. Scale bars: 5 μm. Please note that the prediction tool is often correct. We chose these examples to show that testing can be useful for sequences with negative prediction. Mac1, acyltransferase involved in MEL production [43]; GPT1, glucose-6-phosphate transporter from *Arabidopsis thaliana* [44]

full-length proteins containing a predicted PTS2 are inserted upstream of the GFP open reading frame.

2. Prepare competent cells from a strain expressing a peroxisomal marker protein, such as mCherry-SKL and follow a standard transformation protocol [31] to generate strains expressing a GFP-PTS1 construct.
3. Inoculate individual colonies into YEPS-light medium. Incubate overnight at 30°C with shaking (~180 rpm).
4. Measure the OD_{600} of overnight cultures. Dilute each culture to an OD_{600} of 0.2 in fresh YEPS-light medium and incubate for an additional 4 h at 30°C to reach mid-log phase.
5. Centrifuge cultures at 1920 x*g* for 10 min to pellet the cells.
6. Wash the cell pellets three times with 5 mL sterile ddH_2O, centrifuging at 1920 x*g* for 10 min between washes.
7. Resuspend the cells in YNB medium supplemented with 0.2% (w/v) ammonium sulfate (serving as the nitrogen source) and 0.1% (v/v) TWEEN40 and 0.1% (v/v) oleic acid (1:1 ratio) as the carbon sources. First premix TWEEN40 and oleic acid (1:1) and then add 0.2% (v/v) to the medium.
8. Incubate the cultures for 4 h at 30°C with shaking to induce peroxisome proliferation. Note: Validation of peroxisome targeting can also be performed in D-glucose-grown cells, for assessment of defined organelle subpopulations or luminal subdomains incubation with oleic acid is beneficial as it induces peroxisome proliferation [15].
9. Harvest cells equivalent to 1 OD_{600} unit by centrifugation. Wash the pellet three times with ddH_2O as described above.

10. Prepare microscope slides using 1.5% (w/v) agarose dissolved in ddH_2O to form agarose pads. Prepare your agarose cushions 30–45 min in advance. This ensures that they are dry but not brittle.
11. Resuspend the washed cell pellet in 1 mL of sterile ddH_2O water. Spot 4 μL of the suspension onto the agarose pad and gently place a coverslip on top. Addition of Carbonyl cyanide m-chlorophenyl hydrazone (CCCP) to imaging samples can be beneficial as it blocks movement of peroxisomes [15].
12. Using a fluorescence microscope, capture a minimum of five images per biological replicate. Each image should include at least 10 cells. Perform imaging for at least three independent replicates per strain. Representative images for two PTS1 motifs are shown in Fig. 1B.
13. Analyze the images using appropriate image analysis software (e.g., Fiji/ImageJ) [32]. Determine the Pearson correlation coefficient between GFP and mCherry fluorescence for each image. Compute the mean Pearson correlation across the five images for each biological replicate.
14. **High Pearson correlation values** (positive, +1) indicate a positive linear relationship of both signals pointing to colocalization of GFP with the peroxisomal marker.
15. **Low or negative Pearson correlation values** indicate no or even a negative relationship and colocalization is unlikely.

3.2 Identification of Peroxisomal Subdomains and Subpopulations by Fluorescence Microscopy

1. Use a strain expressing a suitable peroxisomal marker such as mCherry-SKL [13] to prepare competent *U. maydis* cells [31].
2. Generate N-terminal GFP fusion constructs encoding your protein of interest, verify by sequencing, and transform into the competent *U. maydis* cells prepared in Step 1. The vector Petef-GFP-Ala_6-MMXN (3.1) is suitable and might therefore be used.
3. Select single colonies on selection media containing carboxin.
4. Inoculate positive clones into YEPS-light medium and grow overnight.
5. The next morning, determine the OD_{600} of your cultures, dilute the cells in fresh YEPS-light media to an OD_{600} of 0.2, and allow them to grow for an additional 4 h.
6. Harvest the cells by centrifugation at 1920 x*g* for 10 min.
7. Wash the cells three times with 5 mL of sterile ddH_2O, centrifuging at 1920 x*g* for 10 min each time.

8. Transfer the cells to YNB media containing ammonium sulfate as the nitrogen source and a 1:1 mixture of TWEEN40 and oleic acid as the carbon sources.
9. Incubate the cells for an additional 4 h.
10. Harvest 1 mL of a culture with OD_{600} of 1 by centrifugation as in Step 6, then wash three times with ddH_2O.
11. Prepare slides with agarose cushions.
12. Resuspend the cell pellet in 1 mL of sterile ddH_2O.
13. Spot 4 μL of the resuspended cells onto the agarose cushion and cover with a coverslip.
14. Using a fluorescence microscope, capture at least 5 images per replicate, with each image containing a minimum of 10 cells. Aim for at least 3 replicates per strain.
15. Analyze the images using appropriate software and calculate the Pearson correlation coefficients. Compute the mean Pearson correlation across the 5 images to represent each replicate.
16. Interpretation:
17. Pearson correlation coefficients below 0.3 suggest the presence of peroxisomal subdomains if your protein of interest colocalizes with peroxisomes per se (Fig. 2A, right panel; Fig. 2B). Coefficients above 0.7 indicate strong co-localization between signals (Fig. 2A, left panel; Fig. 2B).

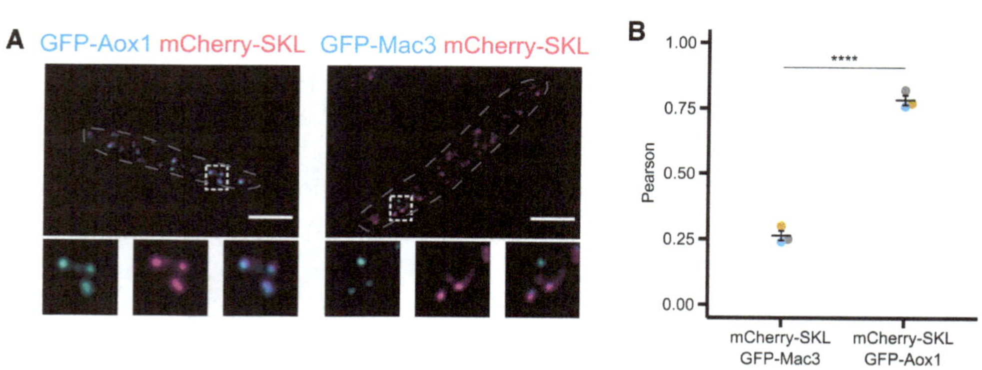

Fig. 2 Imaging of full-length peroxisomal proteins. (**A**) *U. maydis* strains expressing N-terminally GFP-tagged versions of Aox1 and Mac3 (cyan) and the peroxisomal marker protein mCherry-SKL (magenta) were inspected by epifluorescence microscopy. Full images are shown as overlays of two channels. For insets single channels and merged channels are provided. Scale bars: 5 μm. (**B**) Quantifications show Pearson's correlation coefficients of GFP and mCherry signals for indicated strains. Each dot represents one biological replicate. Center line, mean; error bars, standard error of the mean. Significance assessed by an unpaired, two-sided Student's *t*-test. **** refers to a $p \leq 0.0001$

3.3 Isolation of Peroxisomal Core Structures by Cell Fractionation and Detergent Treatment

To prepare crude organelle extracts, you can use this protocol adapted from [33].

1. Inoculate your strains of interest in YEPS-light media and grow overnight.
2. Dilute your cells to an OD_{600} of 0.2 in fresh your cells to an OD_{600} of 0.2 in fresh YEPS-light media and let them grow for 4 h. You need at least 100 mL of logarithmic culture.
3. After 4 h harvest your cells by centrifugation (i.e., 1920 x*g*, 10 min).
4. Subsequently, wash your cells two times with 20 mL of deionized and sterile water (1920 x*g*, 10 min).
5. After that wash your cells twice with 20 mL of SCS Solution (1920 x*g*, 10 min).
6. Achieve formation of protoplasts by incubating cells in SCS solution containing *Trichoderma* lysing enzymes (10 mg/mL) for 30 to 45 min. Degradation of the cell wall was followed microscopically. Monitor the formation of protoplasts after 20 min, and continue checking at 5-min intervals thereafter.
7. Afterward, keep the protoplasts on ice and gently wash them twice with SCS solution, centrifuging each time for 10 min at 1000 x*g* and 4°C.
8. Wash cells one time with 20 mL of Lysis Buffer (1000 x*g*, 10 min at 4°C).
9. Resuspend cells in 15 mL of lysis Buffer and freeze them overnight at −80°C.
10. Let your cell suspension thaw on ice.
11. Homogenization is achieved through 2 × 10 strokes with a Potter-Elvehjem homogenizer interrupted by chilling the samples on ice for 2 min.
12. Remove nuclei and cell debris via centrifugation at 1000 x*g* for 10 min at 4°C.
13. Repeat Step 12.
14. Subsequently, the post-nuclear supernatant (PNS) obtained in Step 13 is diluted to an OD_{600} of 1, aliquoted, and frozen at −80°C.
15. Thaw a PNS Aliquot on ice and concentrate 10 times via centrifugation (10,000 x*g*, 10 mins at 4°C), and resuspend in Lysis Buffer.
16. Split the sample into two halves. To one half, add Triton X-100 to a final concentration of 1% (v/v) and incubate on ice for 45 min.
17. Prepare slides with agarose cushions. For this use, 1.5% (w/v) agarose dissolved in Lysis Buffer.

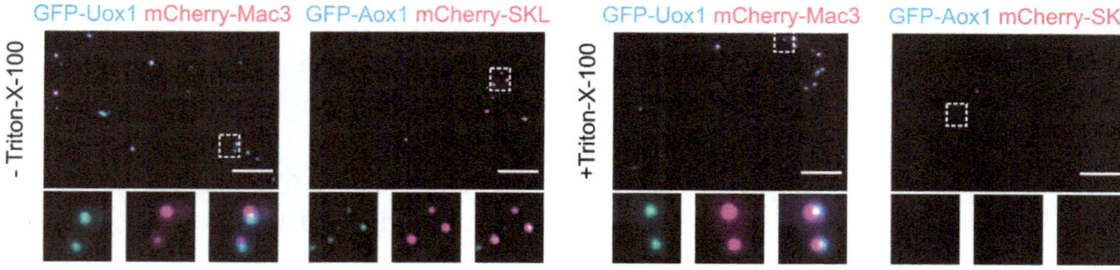

Fig. 3 Example of an organelle preparation experiment to detect peroxisomal cores. Crude organelle preparations of indicated strains were analyzed by epifluorescence microscopy after incubation in lysis buffer (left) or lysis buffer supplemented with Triton X-100 (right) for 45 min. Full images are shown as overlays of two channels. For insets single channels and merged channels are depicted. Scale bars: 5 μm

18. Analyze the two samples using a fluorescence microscope. Spot 4 μL of each aliquot onto an agarose cushion and cover with a coverslip.
19. Proteins residing in peroxisomal subdomains/core structures are resistant to treatment with Triton X-100 [15]. A corresponding fluorescent signal should be detectable in both samples (Fig. 3). Proteins that are not localized to peroxisomal subdomains are sensitive to detergent treatment; therefore, a signal is expected only in the untreated control sample (Fig. 3).

3.4 Identification of Growth Phenotypes via Spot Assays on Oleic Acid Medium

1. Inoculate strains of interest, along with appropriate control strains (e.g., the Δ*pex5*-deletion strain), into YEPS-light medium and incubate overnight to allow growth.
2. The following morning, measure the OD_{600} of each culture. Dilute the cells in fresh YEPS-light medium to an OD_{600} of 0.2 and continue incubation for an additional 4 h.
3. Harvest cells equivalent to 1 OD_{600} unit by centrifugation at 1920 x*g* for 10 min.
4. Wash the cell pellet three times with 5 mL of sterile ddH_2O, centrifuging at 1920 x*g* for 10 min each time to remove residual media.
5. Resuspend cells in 1 mL of sterile ddH_2O.
6. Prepare six additional tubes per strain, each containing 900 μL of ddH_2O.
7. Perform serial dilutions by transferring 100 μL of the resuspended cells into the first tube, vortex briefly, and then transfer 100 μL from that tube into the next, repeating this process until reaching a 10^{-6} dilution.

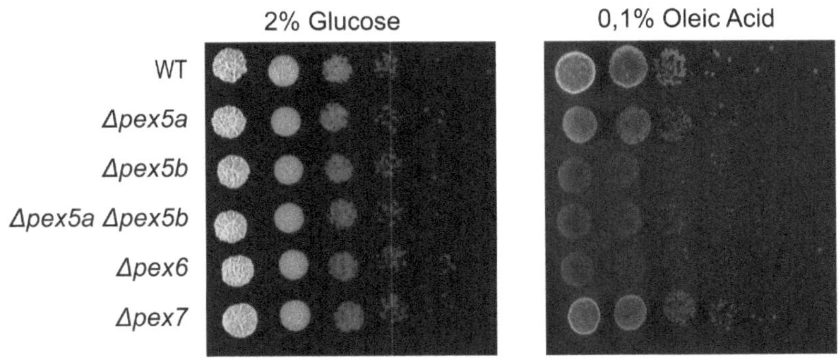

Fig. 4 Exemplary growth assay. Serial dilutions of indicated strains were spotted on minimal media containing either D-glucose or oleic acid as carbon source. *Pex* genes encode proteins required for peroxisome biogenesis or protein import [45]

8. Prepare your agar plates:
   a. Control plates: YNB media with ammonium sulfate as the nitrogen source, 2% (w/v) D(+)-glucose as the carbon source, and 1.5% (w/v) agar.
   b. Test plates: YNB with ammonium sulfate and 1.5% (w/v) agar, supplemented with a 1:1 mixture of TWEEN 40 and oleic acid (final concentration 0.1% (v/v) each), were used as the carbon source. Oleic acid plates should be dried under a clean bench before use; otherwise, drops of cell suspension may merge.
   c. Spot aliquots of the serial dilutions onto both sets of plates and allow them to dry for at least 30 min before incubation. Avoid moving the plates prior to incubation at 28°C for 48–72 h. For longer incubation times, sealing of the plates with Parafilm M is recommended. Plates should be dried under a clean bench to prevent spot merging.
9. After incubation, photograph the plates and analyze colony number, size, and morphology to assess growth differences and potential deficiencies (Fig. 4). Taking magnified photographs of colonies through a binocular microscope can help you to judge altered colony morphology [15].

3.5 *U. maydis* Respiration Methods

Molecular oxygen serves as the terminal electron acceptor of the mitochondrial electron transport chain, making oxygen consumption a direct proxy for respiratory activity. This approach is particularly valuable in *U. maydis* when oleic acid is used as a carbon source, since emulsified fatty acids make cultures turbid and prevent reliable OD measurements. An oxygen optode spot is a sensor coated with an oxygen-sensitive dye that allows non-invasive, optical measurement of dissolved oxygen concentrations in liquids or on surfaces [34].

These sensors allow non-invasive, continuous monitoring of dissolved oxygen in small culture volumes for respiratory activity—a schematic drawing of the experimental setup is shown in Fig. 5A. Below, we first describe the common preparation and calibration steps, then outline two strategies: (i) short-term assays probing acute substrate responses (Fig. 5B), and (ii) long-term monitoring to approximate growth (Fig. 5C).

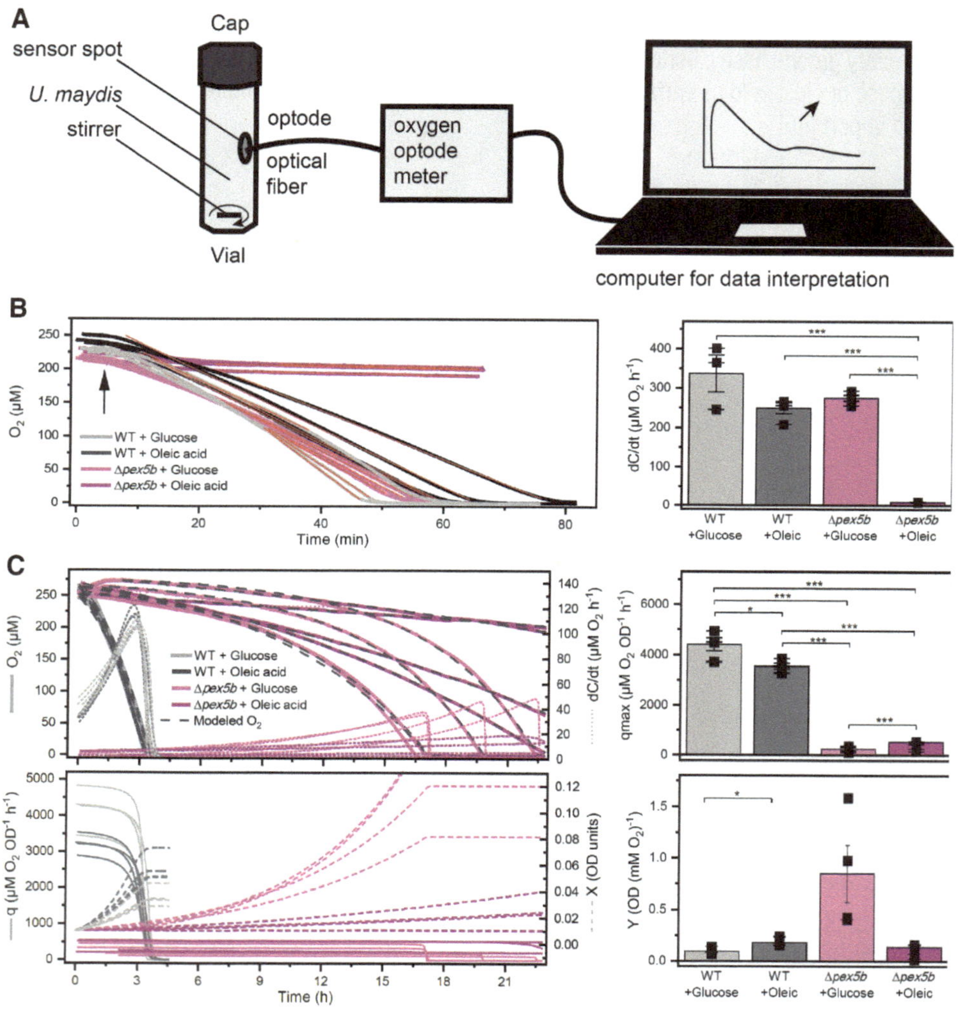

Fig. 5 Exemplary oxygen optode measurements in *U. maydis* wild type and Δ*pex5b* mutant on D-glucose and oleic acid. (**A**) Schematic drawing depicting a typical optode setup in the septum vial and carefully sealed without headspace. The septum vial is shown with the optode spot affixed to the glass in a position accessible to the fiber optics cable. Using a commercial or makeshift holder (not shown), contact between the fiber optics cable and the optode spot is maintained as the optical response is read into the meter and subsequently saved by the data logging software. Constant mixing is achieved by the glass stirrer at the bottom of the vial to maintain a homogenous mixture and prevention of oxygen gradient formation. Substrates may be injected through the septum in the cap. (**B**) Short-term incubations: oxygen concentration trajectories after substrate

3.5.1 General Procedure

1. Prepare small septum vials of ~3 mL volume by affixing an oxygen optode spot to the inner wall with ELASTOSIL® E43 silicone glue and adding a 4 mm Pyrex glass stir bar. These sealed vials function as miniature batch reactors: once closed without headspace, the contained oxygen pool is consumed only by the cells making it possible to monitor respiration dynamics until depletion.
   a. Glass stir bars are preferred as they do not absorb or store gases; in contrast, Teflon-coated stir bars can trap and release oxygen, introducing variability in small-volume assays.
   b. To affix the optode spot, dip the readout side of the spot into a drop of silicone glue. Use bent tweezers with rounded tips to avoid damaging the sensor coating. Press the spot firmly against the inner wall of the glass vial until the glue begins to solidify. Pressing is necessary as the vial wall is slightly curved; this prevents the formation of air bubbles and ensures close contact between spot and glass.
   c. We recommend exetainers (Labco, UK) because they can be easily filled and sealed without leaving headspace, and their septa can be pierced multiple times for substrate additions without significant leakage. Their cylindrical shape, which does not taper at the top, also provides a flat surface for attaching sensor spots and ensures consistent alignment of the fiber-optic readout (Fig. 5A).
   d. We used oxygen sensor spots from PyroScience GmbH (Germany), which we favor for their flat geometry, robustness, ease of fixation, and reusability (including tolerance to autoclaving). Alternatives are available from other vendors or can be custom-made (e.g., [35]).

Fig. 5 (continued) addition (indicated by the arrow), with regression fits as described in Sect. 3.5.2. Bar charts summarize mean respiration rates (± SEM) across biological triplicates. (**C**) Long-term incubations: oxygen concentration over time with model fits (Sect. 3.5.3). The coupled biomass–respiration model is shown, including derived trajectories of respiration rate dC/dt, biomass X, and biomass-specific respiration q. Bar charts compare parameter estimates across strains and substrates, highlighting differences in Y (biomass yield per oxygen) and qmax (maximum specific respiration rate). These parameters make the method highly effective for detecting phenotypic differences, but their physiological interpretation requires caution. Oxygen consumption can be shaped by multiple processes, and growth may be limited by factors other than oxygen. In the Δ*pex5b* mutant on D(+)-glucose, for instance, the initial rate of oxygen consumption is very low, forcing the model to assign a low qmax. As rates increase later in the incubation, the model explains this by a disproportionately high yield Y. Rather than reflecting truly high growth efficiency, this pattern more likely results from physiological effects such as oxygen toxicity, relief from ROS stress, or metabolic rewiring after a prolonged adaptation phase. The script used to analyze both short-term and long-term data is provided at https://doi.org/10.17617/3.VPN9ND

2. Autoclave the vials containing glued optodes, stir bars, and aluminum foil covers. Septum lids should not be autoclaved, as they deform; sterilize separately by UV irradiation for a minimum of 20 min in a sterile environment (e.g., under a clean bench).
3. Connect the fiber-optic readout cables from the oxygen meter (e.g., FireStingO_2, PyroScience GmbH) to the sensor spots. These cables transmit excitation and emission light between the meter and the spot, enabling non-invasive oxygen readout. The oxygen meter is controlled by a laptop running the manufacturer's software for data recording (Pyroscience GmbH) (Fig. 5A).
4. Secure the fiber-optic cables in place using custom-made or commercially available holders to maintain a fixed distance at a 90° angle to the spot. This ensures stable signals over long measurements (Fig. 5A).
5. Place the vials on a magnetic stirrer (200–750 rpm) and keep them continuously stirred throughout calibration and subsequent measurements. Continuous mixing ensures homogeneous distribution of oxygen and prevents gradients forming near the optode spot.
6. Calibrate the sensor spots directly in the assay medium, which should be added to the vials under sterile conditions. For air-saturated calibration, flush the medium with sterile-filtered air through a needle fitted with a sterile filter. For anoxic calibration, flush with sterile-filtered nitrogen. Keep the aluminum foil cover in place during calibration to minimize contamination risk. Calibrating in the actual assay medium (including oleic acid/TWEEN40 when used) avoids systematic offsets. Note that the sensor detects only dissolved oxygen in the aqueous phase; emulsified droplets act as a secondary reservoir and can buffer oxygen levels, but this effect is minor at the concentrations used here.
7. Keep calibration and measurement at a stable temperature. Even brief transfers between the incubator and bench can cause apparent fluctuations in dissolved oxygen concentration.
8. Empty the vials after calibration and wash them with sterile media without a carbon substrate.

3.5.2 Short-Term Response Studies

1. Grow *U. maydis* according to Sect. 3.1, Steps 3 and 4.
2. For each condition, inoculate as biological replicates starting from individual colonies. For example, for each strain tested D(+)-glucose and oleic acid in triplicate require six flasks grown in YEPS-light.
3. Harvest and wash cells according to Sect. 3.1, Steps 5 & 6.

4. Resuspend in YNB medium without a carbon source. Adjust an OD_{600} of 0.1 (high enough for measurable consumption but not so high that oxygen is depleted too rapidly). Different physiological parameters or variables may require different cell densities and preliminary testing to determine the ideal density is encouraged.
5. To minimize the influence of intracellular storage compounds, pre-incubate cultures in YNB media for ~24 h in Erlenmeyer flasks on a shaker without carbon substrate. This ensures that subsequent respiration primarily reflects the supplied carbon source rather than mobilization of pre-existing storage.
6. For each flask, transfer 5 mL of the starved culture into a 15 mL conical tube and fully aerate by inverting ~20 times.
7. Immediately load 3 mL into the pre-assembled septum vials (containing the calibrated sensor spot, stir bar, and foil cover), and pre-positioned for measurement.
8. Multiple vials can be run in parallel on a multi-position stir plate, enabling simultaneous measurements across conditions. For consistency, each vial should remain dedicated to one substrate type throughout its use, as residues (especially from fatty acid emulsions) may adsorb to the glass and bias subsequent assays.
9. Rapidly seal vials tightly without leaving headspace, avoiding bubbles that act as oxygen reservoirs.
10. Start stirring and maintain continuous stirring throughout recording to prevent oxygen gradients near the spot.
11. Immediately begin acquiring dissolved oxygen at intervals of 5 s.
12. Allow ~5 min to stabilize without external carbon to establish the baseline (endogenous) respiration for each replicate.
13. Add your substrate through the septum using a Hamilton syringe (Hamilton, Switzerland) fitted with a 23-gauge (0.6 mm) Sterican needle (Braun, Germany). Below are typical final concentrations for each substrate:
    a. Oleic acid: inject 150 μL of a predilution of 1:50 oleic acid/TWEEN40 emulsion for a final concentration 0.1% (v/v) oleic acid.
    b. Glucose: inject 150 μL of 40% (w/v) D-glucose for a final concentration of 2% (w/v) D-glucose.
14. Continue recording until oxygen is depleted (Fig. 5B).
15. For each vial, model oxygen concentration c(t) (μM) over a post-injection window where the decline is approximately linear:
    $c(t) = \beta_0 + \beta_1 t + \varepsilon,$

where β_0 is the intercept (oxygen concentration at the start of the fit), β_1 is the slope (change per unit time), and ε is the residual error. The respiration rate is defined as: $r = -\beta_1$.

16. Choose one of two complementary approaches for comparing respiration rates across treatments:
    a. Option A (two-stage comparison of vial-level slopes).
        i. Compute a slope estimate r_v for each vial v.
            1. Compare distributions of r_v across treatments:
            2. For two treatments, use Welch's t-test [36].
            3. For more than two conditions or multiple strains, use one-way ANOVA (with Welch variant if variances differ), followed by Tukey HSD or Games–Howell post hoc tests [37, 38].
        ii. Report mean ± standard error mean (SEM) of r_v, exact *p*-values, and effect sizes (e.g., mean difference in μM min^{-1}). Apply Holm correction or another multiple-testing adjustment when necessary.
    b. Option B (one-stage mixed-effects model)
        i. Fit a linear mixed-effects model to the full time series across all vials [39]: $c\_iv(t) = \beta_0 + \beta_1 t + \beta_2 X + \beta_3 (t \times X) + b_{0v} + b_{1v} t + \varepsilon\_iv$, where X represents treatment, and b_{0v}, b_{1v} are vial-level random effects (intercept and slope).
        ii. Use the interaction coefficient β_3 to test for slope differences between treatments.
        iii. Assess significance with two-sided tests. Approximate your degrees of freedom according to Satterthwaite's method [40].
    c. When performing multiple pairwise slope comparisons, Holm-adjusted *p*-values should be reported.
17. Practical guidance (for either option)
    a. Define the window: pick a post-injection segment that is visually linear and within sensor range; exclude the first ~10–20 s after injection (mixing transient).
    b. Baseline vs stimulated: If "delta rate" is desired, compute $\Delta r = r\text{post} - r\text{base}$ per vial (fit the same model on a pre-injection window), then compare Δr across treatments (Welch t-test or mixed model with t and X).
    c. Autocorrelation/heteroscedasticity: Mixed models largely handle correlation via random effects. When modeling two-stage, your test is on vial-level slopes (independent across vials) without issue. If you must work directly on timepoints without mixed models, use GLS

with AR(1) errors. However, the mixed-model route produces cleaner results.

d. *p*-values: State your threshold (commonly 0.05) and report exact *p* values and confidence intervals for the slope(s) or differences. Correct for multiple comparisons when applicable (Holm).

e. Diagnostics: Check for linearity in the chosen window (e.g., $R^2 > 0.95$ on per-vial fits), inspect residual plots, and verify the absence leaks/bubbles (non-monotonic oxygen).

3.5.3 Respiration Rates as a Growth Estimate

1. Grow *U. maydis* according to Sect. 3.1, Steps 3 and 4.
2. For each tested condition, prepare biological replicates as outlined in Sect. 3.5.2, Step 2.
3. Harvest and wash cells according to Sect. 3.1, Steps 5 and 6.
4. Resuspend in YNB medium containing the desired carbon source (e.g., 0.1% (v/v) oleic acid or 2% (w/v) D-glucose). Adjust to an OD_{600} of 0.01.
5. Proceed with vial preparation, calibration, and measurement as described in Sect. 3.5.2, starting from Step 6. In contrast to the short-term protocol, oxygen concentrations can be recorded at longer intervals (e.g., every 60 s).
6. Measurements are continued until oxygen is depleted, which typically requires 4–24 h for *U. maydis*.
7. Analysis depends strongly on the resultant data. We outline two possible analysis approaches that have proven to be useful for experiments involving (1) D-glucose and oleic acid and (2) only oleic acid.
8. When grown on oleic acid, *U. maydis* respiration rates typically increase linearly suggesting that cell growth is not exponential, possibly due to substrate accessibility constraints (e.g., micelle partitioning) [15] (Fig. 5C). Thus, a typical exponential growth model cannot be used to fit the oxygen respiration data. Instead, we suggest a model in which biomass increase is limited by respiration rate and governed by a growth yield factor such as the one described in the following (see supplementary code). This model can be used for cells grown on either oleic acid or **D**-glucose.
9. Define the respiration rate of the vial as
   $r(t) = dC/dt$
   with units μM (unit time)$^{-1}$.
10. Model the respiration rate per biomass unit, q(C), as oxygen-dependent:
    $q(C) = qmax * C / (Kc + C)$

where C is the concentration of oxygen, qmax is the maximum specific respiration rate (μM O_2 (unit biomass)$^{-1}$ (unit time)$^{-1}$), and Kc is the half-saturation constant for oxygen [41].

11. Relate vial respiration to biomass as Δ
r(t)=v q(C) * X(t)
where X(t) is biomass.
12. Define the biomass growth rate as proportional to the specific respiration:
μ(C)=Y * q(C)
where Y is the maximum growth yield per oxygen consumed (unit biomass (μM O_2)$^{-1}$).
13. The coupled dynamics are therefore:
dC/dt=−q(C) * X & dX/dt=μ(C) * X
14. Fit the parameters Y, qmax, Kc, X0 (and optionally the initial oxygen concentration C_0) by integrating the ODE system and minimizing squared errors between predicted and observed oxygen concentrations [37].
15. Select fitting windows to include:
    a. the stable post-equilibration baseline
    b. the monotonic decline segment
    c. the terminal zero-oxygen plateau. Including the plateau constrains the time-to-depletion and prevents overestimation of μmax.
16. Use simple bounds for identifiability (e.g., all parameters >0; maximum Kc constrained to air saturated O_2 levels). Optionally seed initial values for qmax*X0 from the early slope of oxygen decline, with X0 representing the initial OD_{600} or cell number.
17. Focus on qmax and Y as the primary comparison metric, whereas Kc describes respiratory physiology.
18. Apply statistical testing of replicate qmax and Y estimates with Welch's *t*-test for two groups or ANOVA for more than two groups [36].
19. For ANOVA, apply suitable post hoc procedures such as Tukey's HSD (for equal variances) or Games–Howell (for unequal variances or sample sizes) [38]. Correct for multiple testing across comparisons using procedures such as Holm adjustment [42].
20. Interpret results based on *p*-values (typically $p<0.05$) and report adjusted values when corrections are applied. Summarize as mean ± SEM and visualize data as replicate scatter plots or boxplots overlaid with model fits.

21. Confirm assumptions of normality and homogeneity of variances; apply nonparametric alternatives (e.g., Mann–Whitney U, Kruskal–Wallis) if assumptions are not met [37].
22. Parameters Y (biomass yield *per* oxygen) and qmax (maximum specific respiration rate) should be interpreted with caution. While they are efficient for detecting phenotypic differences, they do not provide definitive physiological descriptions. Within the model, Y reflects the coupling of oxygen consumption to biomass formation, and qmax defines the upper bound of respiration per biomass unit. In practice, oxygen consumption can be influenced by multiple processes, and growth may be limited by factors other than oxygen. Nevertheless, these parameters are highly useful for hypothesis generation and should be validated by independent measurements, such as substrate and product turnover rates or direct measures of biomass growth (e.g., cell counts or dry weight measurement).
23. When *U. maydis* grows on **oleic acid only**, the analysis can also be simplified:
24. Compute instantaneous respiration rates $r(t) = dC/dt$ from the oxygen time series, applying mild smoothing if necessary.
25. Fit a simple linear model to the respiration rates:

    a. $r(t) = \beta 0 + \beta 1\ t + \varepsilon$,

    b. where $\beta 1$ is the acceleration of respiration ($\mu M\ min^{-2}$).

    Compare replicate-level estimates of $\beta 1$ between conditions. The statistical framework for regression analysis, replicate handling, and hypothesis testing is identical to that described for short-term incubations in Sect. 3.5.2.

Acknowledgments

We thank Jessica Tiefenbacher and Björn Sandrock for kindly sharing their chapter prior to publication. G.B. gratefully acknowledges support from the European Research Council (ERC) through the Advanced Grant KIWIsome (Grant No. 101019765). G.B. and J.M.K. further acknowledge the Max Planck Society and the LOEWE program of the State of Hesse for their continuous support. J.F. was supported by the Deutsche Forschungsgemeinschaft (DFG, grant ID FR-3586/2-1).

References

1. De Duve C, Baudhuin P (1966) Peroxisomes (microbodies and related particles). Physiol Rev 46:323–357
2. Islinger M, Grille S, Fahimi HD, Schrader M (2012) The peroxisome: an update on mysteries. Histochem Cell Biol 137:547–574
3. Kumar R, Islinger M, Worthy H, Carmichael R, Schrader M (2024) The peroxisome: an update on mysteries 3.0. Histochem Cell Biol 161:99–132
4. Smith JJ, Aitchison JD (2013) Peroxisomes take shape. Nat Rev Mol Cell Biol 14:803–817
5. Hu J, Baker A, Bartel B, Linka N, Mullen RT, Reumann S, Zolman BK (2012) Plant peroxisomes: biogenesis and function. Plant Cell 24:2279–2303
6. Wanders RJA, Baes M, Ribeiro D, Ferdinandusse S, Waterham HR (2023) The physiological functions of human peroxisomes. Physiol Rev 103:957–1024
7. Bartoszewska M, Opalinski L, Veenhuis M, van der Klei IJ (2011) The significance of peroxisomes in secondary metabolite biosynthesis in filamentous fungi. Biotechnol Lett 33:1921–1931
8. Stehlik T, Sandrock B, Ast J, Freitag J (2014) Fungal peroxisomes as biosynthetic organelles. Curr Opin Microbiol 22:8–14
9. Walter T, Erdmann R (2019) Current advances in protein import into peroxisomes. Protein J 38:351–362
10. Camões F, Islinger M, Guimarães SC, Kilaru S, Schuster M, Godinho LF, Steinberg G, Schrader M (2015) New insights into the peroxisomal protein inventory: acyl-CoA oxidases and-dehydrogenases are an ancient feature of peroxisomes. Biochim Biophys Acta (BBA)-Molecular Cell Res 1853:111–125
11. Freitag J, Ast J, Linne U, Stehlik T, Martorana D, Bölker M, Sandrock B (2014) Peroxisomes contribute to biosynthesis of extracellular glycolipids in fungi. Mol Microbiol 93:24–36
12. Ast J, Bäcker N, Bittner E, Martorana D, Ahmad H, Bölker M, Freitag J (2022) Two Pex5 proteins with different cargo specificity are critical for peroxisome function in *Ustilago maydis*. Front Cell Dev Biol 10:858084
13. Freitag J, Ast J, Bölker M (2012) Cryptic peroxisomal targeting via alternative splicing and stop codon read-through in fungi. Nature 485:522–525
14. Bittner E, Stehlik T, Lam J, Dimitrov L, Heimerl T, Schöck I, Harberding J, Dornes A, Heymons N, Bange G et al. (2024) Proteins that carry dual targeting signals can act as tethers between peroxisomes and partner organelles. PLOS Biol 22:1–36
15. Bäcker N, Ast J, Martorana D, Renicke C, Berger J, Mais C-N, Christ M, Stehlik T, Heimerl T, Wernet V et al. (2025) Peroxisomal core structures segregate diverse metabolic pathways. Nat. Commun. 16: 1802
16. Becker F, Linne U, Xie X, Hemer A-L, Bölker M, Freitag J, Sandrock B (2022) Import and export of mannosylerythritol lipids by *Ustilago maydis*. MBio 13(5):e0212322
17. Klose J, Kronstad JW (2006) The multifunctional β-oxidation enzyme is required for full symptom development by the biotrophic maize pathogen *Ustilago maydis*. Eukaryot Cell 5:2047–2061
18. Kretschmer M, Klose J, Kronstad JW (2012) Defects in mitochondrial and peroxisomal β -oxidation influence virulence in the maize pathogen *Ustilago maydis*. Eukaryot Cell 11:1055–66
19. Guimaraes SC, Schuster M, Bielska E, Dagdas G, Kilaru S, Meadows BRA, Schrader M, Steinberg G (2015) Peroxisomes, lipid droplets, and endoplasmic reticulum "hitchhike" on motile early endosomes. J Cell Biol 211:945–995
20. Kors S, Schuster M, Maddison DC, Kilaru S, Schrader TA, Costello JL, Islinger M, Smith GA, Schrader M (2024) New insights into the functions of ACBD4/5-like proteins using a combined phylogenetic and experimental approach across model organisms. Biochim Biophys Acta - Mol Cell Res 1871:119843
21. Broomfield PLE, Hargreaves JA (1992) A single amino-acid change in the iron-sulphur protein subunit of succinate dehydrogenase confers resistance to carboxin in *Ustilago maydis*. Curr Genet 22:117–121
22. Kämper J (2004) A PCR-based system for highly efficient generation of gene replacement mutants in *Ustilago maydis*. Mol Genet Genomics 271:103–110
23. Brachmann A, König J, Julius C, Feldbrügge M (2004) A reverse genetic approach for generating gene replacement mutants in *Ustilago maydis*. Mol Genet Genomics 272:216–226
24. Sambrook J, Fritsch EF, and Maniatis T (1989) Molecular cloning: a laboratory manual. Cold Spring Harbor Laboratory Press
25. Gustafsson MGL (2000) Surpassing the lateral resolution limit by a factor of two using structured illumination microscopy. J Microsc 198:82–87
26. Nötzel C, Lingner T, Klingenberg H, Thoms S (2016) Identification of new fungal peroxisomal

matrix proteins and revision of the PTS1 consensus. Traffic 17:1110–1124
27. Brocard C, Hartig A (2006) Peroxisome targeting signal 1: is it really a simple tripeptide?. Biochim Biophys Acta 1763:1565–1573
28. Neuberger G, Maurer-Stroh S, Eisenhaber B, Hartig A, Eisenhaber F (2003) . Prediction of peroxisomal targeting signal 1 containing proteins from amino acid sequence. J Mol Biol 328:581–592
29. Freitag J, Lanver D, Böhmer C, Schink KO, Bölker M, Sandrock B (2011) Septation of infectious hyphae is critical for appressoria formation and virulence in the smut fungus *Ustilago maydis*. PLOS Pathog 7:e1002044
30. Böhmer C, Ripp C, Bölker M (2009) The germinal centre kinase Don3 triggers the dynamic rearrangement of higher-order septin structures during cytokinesis in *Ustilago maydis*. Mol Microbiol 74:1484–1496
31. Schulz B, Banuett F, Dahl M, Schlesinger R, Schäfer W, Martin T, Herskowitz I, Kahmann R (1990) The b alleles of *U. maydis*, whose combinations program pathogenic development, code for polypeptides containing a homeodomain-related motif. Cell 60:295–306
32. Schneider CA, Rasband WS, Eliceiri KW (2012) NIH Image to ImageJ: 25 years of image analysis. Nat Methods 9:671–675
33. Cramer J, Effelsberg D, Girzalsky W, Erdmann R (2015) Small-scale purification of peroxisomes for analytical applications. Cold Spring Harb Protoc 2015(9):pdb.prot083717
34. Kautsky H (1939) Quenching of luminescence by oxygen. Trans Faraday Soc 35:216–219
35. Kühl M, Polerecky L (2008) Functional and structural imaging of phototrophic microbial communities and symbioses. Aquat Microb Ecol 53:99–118
36. Ruxton GD (2006) The unequal variance t-test is an underused alternative to Student's t-test and the Mann–Whitney U test. Behav Ecol 17:688–690
37. Field A, Miles J, Field Z (2012) Discovering statistics using R. SAGE Publications Ltd, London
38. Games PA, Howell JF (1976) Pairwise Multiple Comparison Procedures with Unequal N's and/or Variances: A Monte Carlo. Study. J. Educ. Stat. 1:113–125
39. Pinheiro, JC, Bates, DM, 2000. Mixed-effects models in S and S-PLUS. Springer, New York
40. Kuznetsova A, Brockhoff PB, Christensen RHB (2017) lmer test package: tests in linear mixed. Effects Models J Stat Softw 82:1–26
41. Monod J (1949) The growth of bacterial cultures. Annu Rev Microbiol 3:371–394
42. Holm S (1979) A simple sequentially rejective multiple test procedure. Scand J Stat 6:65–70
43. Hewald S, Linne U, Scherer M, Marahiel MA, Kämper J, Bölker M (2006) Identification of a gene cluster for biosynthesis of mannosylerythritol lipids in the basidiomycetous fungus *Ustilago maydis*. Appl Environ Microbiol 72:5469–5477
44. Baune M-C, Lansing H, Fischer K, Meyer T, Charton L, Linka N, von Schaewen A (2020) The arabidopsis plastidial glucose-6-phosphate transporter GPT1 is dually targeted to peroxisomes via the endoplasmic reticulum[OPEN]. Plant Cell 32:1703–1726
45. Jansen RLM, Santana-Molina C, van den Noort M, Devos DP, van der Klei IJ (2021) Comparative genomics of peroxisome biogenesis proteins: making sense of the PEX proteins. Front Cell Dev Biol 9:654163

Check for updates

Chapter 35

Methods for Studying the Polyamines in *Ustilago maydis*

Fernando Pérez-Rodríguez, Claudia Geraldine Leon-Ramírez, José Ruiz-Herrera and Domingo Martínez-Soto

Abstract

Polyamines are small aliphatic molecules with two or more amino groups. They are present in almost all living organisms and regulate fundamental cellular processes through interactions with the DNA, RNA, and proteins. Although the mechanisms by which these micromolecules regulate cellular processes are not fully understood, the intracellular concentration of polyamines is crucial for the development of organisms. In fungi, the importance of polyamines lies in the cellular processes they regulate, since they are involved in cellular growth and differentiation and even in the virulence and pathogenesis of pathogenic fungi. *Ustilago maydis* is a basidiomycete, dimorphic fungus, and phytopathogen with a biotrophic lifestyle and synthesizes only two of the three most common polyamines. These features and advantages make this fungus a eukaryotic model for studying polyamines. The study of polyamines requires a suitable model that facilitates in vitro and in vivo experimentation and adequate methodologies for the appropriate quantification and determination of specific polyamines. This chapter presents the detailed protocols for the quantification of polyamines by liquid and gas chromatography, as well as the techniques to quantify the activity of enzymes involved in the biosynthesis of polyamines, and some relevant aspects to work with polyamine-auxotrophic strains of *U. maydis* as well. Collectively, the methods described here contribute to the study of polyamines in Ustilaginomycetes and even in evolutionarily distant fungi.

Keywords Determination of polyamines, *Ustilago maydis*, HPLC, Gas chromatography, Protein activity, Auxotrophic mutants

1 Introduction

Polyamines are small aliphatic molecules containing two or more amine groups in their chemical structure. Although a wide variety of molecules fulfill this general description, three polyamines are fundamental for cellular development and function of the cell: putrescine, a diamine; spermidine, a triamine; and spermine, a tetramine [1, 2]. These three molecules are the most widely distributed in living organisms, including most of the fungi. Putrescine is synthesized by decarboxylation of ornithine by the ornithine decarboxylase (Odc), and spermidine is produced by the enzyme spermidine synthase (Spe) by

An N. T. Phan et al. (eds.), *Ustilago maydis*, Methods in Molecular Biology, vol. 3050,
https://doi.org/10.1007/978-1-0716-5364-7_35,

adding a 3-aminopropyl from S-adenosylmethionine [3–7]. Interestingly, not all fungi synthesize spermine [8, 9], and some polyamines can be produced by the retrogradation. For example, spermidine can be transformed into putrescine [7, 10] (Fig. 1). This reconversion process is carried out by the spermidine N-acetyltransferase (Ssat) and polyamine oxidases (Pao) [2] (Fig. 1).

The distinctive feature of these molecules is that the amine groups are fully protonated at physiological pH, allowing them to interact with negatively charged molecules such as DNA, RNA, and proteins [11]. This characteristic confers the ability to specifically regulate various cellular processes such as cell reproduction, development, differentiation, growth, pathogenesis, etc. Therefore, the intracellular levels of polyamines must be subject to complex regulation involving synthesis, degradation, uptake, and excretion [6, 12]. Studies of polyamines in fungal models have facilitated our understanding of their functional roles in cellular processes. A particularly intriguing model

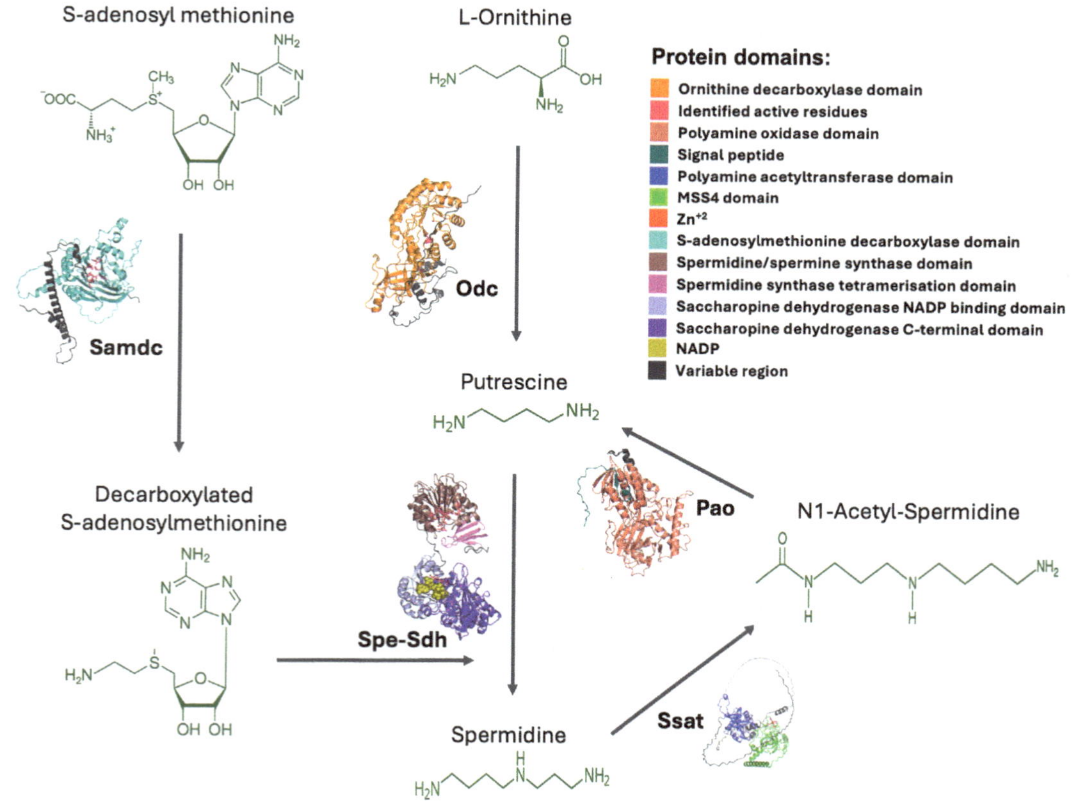

Fig. 1 Representative scheme of the biosynthesis pathway of polyamines in Ustilaginomycetes. S-adenosylmethionine decarboxylase (Samdc), ornithine decarboxylase (Odc), spermidine synthase-saccharopine dehydrogenase (Spe-Sdh), spermidine N-acetyltransferase (Ssat), polyamine oxidase (Pao). Colored boxes and legends at the upper right corner indicate domains in the proteins involved in the biosynthesis of polyamines. (Part of the figure was taken from [9])

for polyamine studies is the phytopathogenic fungus *Ustilago maydis*, responsible agent of common smut or "huitlacoche" in maize (*Zea mays* L.) (Fig. 2A) and its ancestor teozintle (*Zea mays* ssp. *parviglumis* and ssp. *mexicana*). This fungus offers experimental advantages for polyamine research. For example, it synthesizes only two polyamines, putrescine and spermidine [9]; there are well-established transformation techniques for generating auxotrophic mutant strains; there are rapid and straightforward microbiological methods for its propagation and maintenance; its genome is completely sequenced; and it possesses a life cycle with a parasitic and saprophytic phase. Additionally, *U. maydis* is a dimorphic fungus, growing like yeasts (Fig. 2B) or mycelial growth (Fig. 2C) depending on environmental conditions [13] and its pathogenic development [14], making this fungus an excellent model for studying the polyamine roles in fungal dimorphism and pathogenesis [8, 9, 15].

Since polyamines are small polycationic molecules with characteristics similar to certain amino acids, their detection by chromatography methods requires prior derivatization. In this process, the amino groups of polyamines are modified by functional groups that facilitate

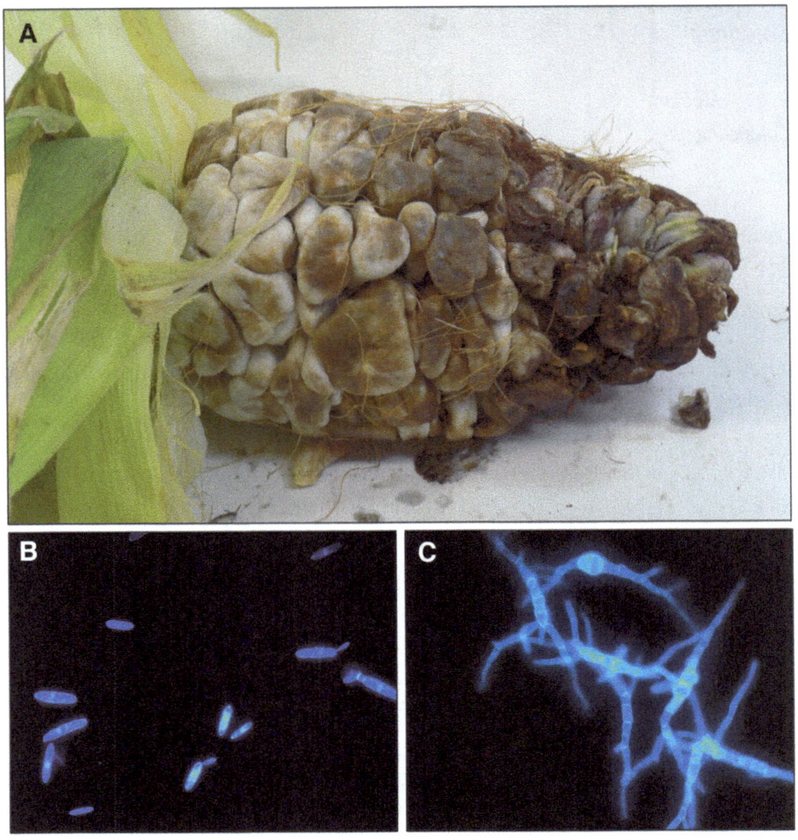

Fig. 2 Morphologies of *Ustilago maydis*. (**A**) Ear of corn plant showing sori filled with teliospores. (**B**) Yeast-like growth. (**C**) Mycelial growth

separation based on column affinity, enhancing sample preparation and facilitating detection by increasing intrinsic fluorescence and UV absorption when spectroscopic detection equipment is employed [16–18]. Several compounds are available for this process, such as benzoyl chloride, which reacts with amino groups, substituting a hydrogen atom with a benzoyl group [19].

Using the fungus *U. maydis* as a model, this work provides a detailed description of the methodologies employed for the detection and quantification of polyamines by liquid and gas chromatography. In addition, it thoroughly outlines the methods for determining the activity of proteins involved in polyamine biosynthesis. Finally, it describes the use of *U. maydis* mutants to analyze the role of polyamines in fungal development and pathogenesis. Collectively, these methods contribute to a deeper understanding of how these molecules regulate fungal growth and development, as well as their roles during fungal–plant interactions.

2 Materials

2.1 Reagents and Solutions

2.1.1 Medium Compositions

For the cultivation of *U. maydis* under laboratory conditions, commonly minimal medium (MM) and complete media (CM) [20, 21] are used.

Complete Medium (CM) (*see* Note 1)

Composition of the CM: 10 g glucose, 2 g hydrolyzed casein, 5 mL hydrolyzed nucleic acid [22], 10 mL vitamin solution [23], 1 g yeast extract, and 250 mL salt solution (formulation below). Adjust the final volume to 1 L with distilled deionized water (ddH_2O); adjust the pH to 7 prior to sterilization. Once sterile, CM can be stored at room temperature.

Minimal Medium (MM) (*see* Note 2)

Composition of the MM: 10 g carbon source (regularly glucose), 250 mL salt solution (formulation below) (*see* **Note 3**). Adjust to 1 L with ddH_2O. Once sterile, MM can be stored at room temperature, but it is recommended to use fresh.

Salt Solution

24 g potassium nitrate (KNO_3) for neutral pH, or ammonium nitrate (NH_4NO_3) for acidic pH, 16 g potassium dihydrogen phosphate (KH_2PO_4), 4 g sodium sulfate (Na_2SO_4), 8 g potassium chloride (KCl), 2 g magnesium sulfate heptahydrate ($MgSO_4 \cdot 7H_2O$), 1 g calcium chloride ($CaCl_2$), 8 mL trace element solution [21]. Adjust the volume to 2 L with ddH_2O; adjust the pH to 7 (MM7) prior to sterilization.

Solid Media

For a solid, complete, minimal, or supplemented minimal medium, add 1.5%–2% of bacteriological agar.

2.2 Solutions

2.2.1 Stock Standard Solutions

To quantify polyamines by high-performance liquid chromatography (HPLC), standard solutions should be prepared. Putrescine dihydrochloride (Sigma-Aldrich, 333-93-7), spermidine trihydrochloride (Sigma-Aldrich, 124-20-9), and spermine tetrahydrochloride (Sigma-Aldrich, 71-44-3) solutions at a concentration of 20 or 30 mM of each polyamine are used. Masses were calculated using the molarity formula. All standard solutions are prepared with ultrapure water and stored for up to 12 months under refrigeration (2–8°C) in a hermetic recipient, avoiding light.

2.2.2 Reagents and Solutions for Quantification of Polyamines by High-Performance Liquid Chromatography (HPLC)

- Benzoyl chloride (stored in its original flask under refrigeration (2–8°C), contained in a desiccant bottle) (*see* **Note 4**).
- Methanol HPLC-grade (stored in its original flask at room temperature, protected from heat and ignition sources).
- 6% (v/v) perchloric acid solution (storage at room temperature, in crystal or ceramic flask).
- 2 M NaOH solution (storage at room temperature, in a hermetic recipient).
- 2 M NaCl solution (storage at room temperature, in a hermetic recipient).

2.2.3 Reagents and Solutions for Quantification of Polyamines by Gas Chromatography

- Benzoyl chloride.
- Ethyl acetate HPLC grade (storage at room temperature, in a hermetic recipient).
- 1 mg/L 1,6-diaminohexane (DAH) (Sigma-Aldrich, 124-09-4) solution (store at 2–8°C, avoid light contact, do not use after 6 months of preparation).
- 5 M sodium hydroxide (NaOH) solution (storage at room temperature, in a hermetic recipient).

2.2.4 Reagents and Solutions for Determination of Polyamines in Maize Tumors

- Toluene (HPLC grade).
- Acetonitrile (HPLC grade).
- 5% (v/v) perchloric acid solution.
- Sodium carbonate (Na_2CO_3) saturated solution.
- 5 mg/mL dansyl chloride HPLC grade (Sigma-Aldrich, 605-65-2) solution diluted in acetone HPLC grade.
- 100 mM 1,7-diaminoheptane (HTD) (Sigma-Aldrich, 646-19-5) solution.

- 100 mg/mL proline solution.
- 12 N hydrochloric acid (HCl) solution.

2.2.5 Solution for Determination of the Ornithine Decarboxylase (ODC) Activity

Radioactive Method

FE buffer pH 7.5 (*see* **Note 5**, for solution pH adjustment): 20 mM monobasic potassium phosphate (KH_2PO_4), 100 mM ethylenediaminetetraacetic acid (EDTA). Prepare fresh before use or storage at 2–8°C for 2–6 weeks.

FEDP buffer pH 7.3 (*see* **Note 5**, for solution pH adjustment): 20 mM KH_2PO_4, 100 mM EDTA, 2 mM dithiothreitol (DTT) HPLC grade (Sigma-Aldrich, 3483-12-3), 0.2 mM pyridoxal phosphate (PLP) (Sigma-Aldrich, 853645-22-4). Prepare fresh and add DTT immediately before use.

Putrescine Detection

Homogenization buffer pH 8.0 (*see* **Note 5**, for solution pH adjustment): 0.1 mM sodium phosphate monobasic monohydrate ($NaH_2PO_4 \cdot H_2O$), 0.1 mM phenylmethylsulfonyl fluoride (PMSF) (Sigma-Aldrich, 329-98-6), 1 mM PLP, 5 mM DTT HPLC grade, 5 mM EDTA, 25 mM ascorbic acid, 1% polyvinylpolypyrrolidone (PVPP) (Supelco, 9003-39-8). Prepare fresh and add DTT immediately before use.

Reaction buffer pH 8.0: 10 mM 2-amino-2-methyl-1,3-propanediol (Tris-HCl) (Sigma-Aldrich, 115-69-5), 5 mM EDTA, 50 μM PLP, 5 mM DTT. Prepare fresh and maintain at room temperature.

Substrate solution pH 8 (*see* **Note 5**, for solution pH adjustment): 25 mM L-ornithine. Store at 4°C until use.

2.3 Equipment

2.3.1 Analytical Equipment

1. High-performance liquid chromatograph.
2. Spectrophotometer.
3. Gas chromatograph.
4. Scintillation counter.

2.3.2 Equipment for Fungal Culture and Derivatization (see Note 6)

1. Erlenmeyer flask for culture.
2. Petri dishes.
3. Centrifuge tubes (1.5, 15, and 50 mL).
4. Micropipettes (with capacities of 1–10 μL, 20–200 μL, and 100–1000 μL)
5. Micropipette tips (various capacities).

2.3.3 Complementary Materials

1. Analytical scale.
2. Centrifuge 15 and 50 mL tubes.
3. Microcentrifuge.
4. Incubator with orbital agitation at 25°C.
5. Dry oven at 80°C.
6. Water bath.

3 Methods

3.1 Polyamine Determination in Yeast/Mycelial Biomass of Ustilago maydis

This protocol was established to determine and quantify the concentrations of intracellular polyamines in *U. maydis*. Intracellular extractions contain many compounds with similar chemical structures and natures, which can be a problem when analyzing the concentration of a specific polyamine. To solve these experimental issues, Jin et al. [24] proposed a method using a derivatization process and then the identification and quantification of polyamines by HPLC, which allows the separation of molecules according to their mass (Fig. 3).

1. Collect 50 mL of fungal culture growing under the experimental design and desired conditions.
2. Centrifuge to obtain a cell pellet (*see* **Note 7**) and wash the cells with 2 mL of ddH_2O; repeat the centrifugation process two to three times.
3. Add 2 mL of 6% $HClO_4$ to the cell pellet and transfer the cells to a 15 mL tube and incubate for 3 h at room temperature in a fume hood.
4. Centrifuge at 4193 × *g* for 10 min, collect the supernatant, and transfer it to a new 15 mL tube.
5. Dry the supernatant in an oven at 80°C.
6. Resuspend the dry product in 1 mL of 6% $HClO_4$ and filter the resuspended suspension through a 0.45 μm membrane (MF-Millipore, HAWP04700F1).
7. Transfer 500 μL of the sample to a 15 mL tube, add 1 mL of 2 M NaOH and 10 μL of benzoyl chloride.
8. Vortex and incubate for 20 min at room temperature.
9. Add 2 mL of 2 M NaCl solution (to stop the reaction) and vortex.
10. Add 4 mL of diethyl ether and keep vortexing.
11. Recover the organic phase with a 1000 μL micropipette and add it to a new 1.5 mL microcentrifuge tube.
12. Evaporate the organic phase in a water bath until dry.

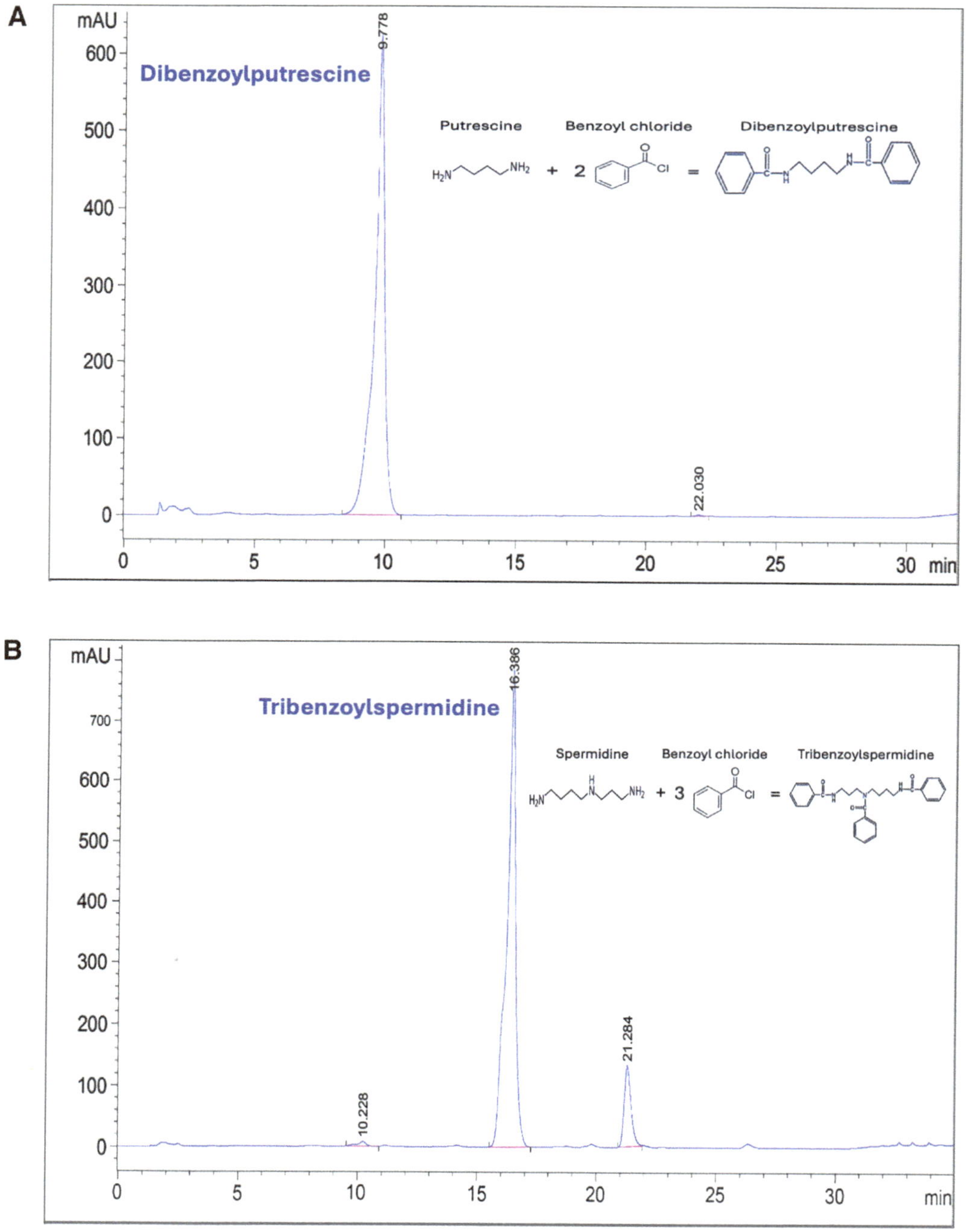

Fig. 3 Representative chromatograms from the determination of the two polyamines synthetized by *U. maydis*. After the derivatization process, dibenzoyl putrescine and tribenzyl spermidine are identified at 9.7 and 16.3 min of retention (chromatographic settings: mobile phase water-methanol, starting at 40%, increasing to 65% after 20 min, reaching 100% after 30 min, at 40°C, 121 bars, flow rate 1 mL/min)

13. Resuspend the sample by adding 200 μL of HPLC-grade methanol.
14. Analyze by HPLC using a C-18 column. The chromatography conditions recommended (*see* **Note 8**) utilized a water–methanol gradient as the mobile phase, starting at 40% at time zero, increasing to 65% after 20 min, and reaching 100% after 30 min; the temperature was maintained at 40°C, with a pressure of 121 bars and a flow rate of 1 mL/min. Detection was performed at 254 nm.

To calculate the concentration of the sample, it is necessary to obtain the equation of the line ($f1$) after running the standard solutions (20 and 30 mM standards, Fig. 3; *see* **Note 9**) and graphing the mAU*s (y-axis) vs. concentration (x-axis). These calculations can be performed in spreadsheet software or manually using the following equation, where it is necessary to isolate "X" once you know the mAU*s (Y) of the sample ($f2$).

$$f1.\ m = \frac{Y_2 - Y_1}{X_2 - X_1} \quad f2.\ (Y - Y_1) = m\ (X - X_1)$$

Polyamine concentrations calculated must be divided by the mg of biomass protein sample and expressing the results in nmol/mg protein.

3.1.1 Derivatization for Gas Chromatography

For determination and quantification of polyamines by gas chromatography, Peña et al. [25] used the extraction method of polyamines mentioned above (steps 1–6), but they proposed a different protocol for sample derivatization.

(Please follow the steps from 1 to 5 described in Sect. 3.1.)

6. 100 μL of sample solution and 900 μL of UHQ (ultra-high purity) water.
7. Add 100 μL of DAH solution, 1 mL of 5 M NaOH, and 5.0 μL of benzoyl chloride in a fume hood.
8. Seal the tube and mix for 60 s at room temperature (*see* **Note 9**).
9. Add 1.0 mL of ethyl acetate and stir for an additional 30 s.
10. Centrifuge at 1509 × g for 5 min at room temperature.
11. Carefully transfer 100 μL of the upper organic layer to a 1.0 mL GC vial containing a 300 μL insert.
12. Analyze by GC using an HP-5MS Ultra Inert GC Column (19091S-433UI, Agilent J&W), using helium as a carrier gas to a flow rate of 1.5 mL/min; the recommended oven program used by the authors initiates at 100–300°C (held for 5 min), increasing by 45°C/min. Total chromatographic running time: 9.46 min [25]. Similarly, as was previously

described for quantitation of polyamines by HPLC, apply the first line equation described at the end of the Sect. 3.1 (see above).

3.2 Determination of Free and Conjugated Polyamines in Maize Tumors Caused by U. maydis Plant Infection

The experimental advantage of *U. maydis* as a model fungus is that its pathogenic cycle is readily reproducible in maize plants (its natural host) or model plants under in vitro conditions [26, 27]. The concentration of polyamines can be determined under the *U. maydis*-plant host interaction. Although this method does not allow for discerning the polyamines produced by the fungus or the plant, it helps estimate the differential concentration of polyamines produced by those tissues colonized by the fungus [28].

This protocol permits the determination of free polyamines (Part A) and total polyamines after acid hydrolysis to release conjugated polyamines (Part B). The difference between total and free polyamines represents the concentration of conjugated polyamines.

Part A: Free Polyamines Extraction and Analysis

1. Collect 300 mg of fresh infected and mock-infected plant material.
2. Keep the sample overnight at 4°C with 1 mL of 5% perchloric acid.
3. Centrifuge and collect the supernatant.
4. Transfer 200 μL of the supernatant to a new 1.5 mL microcentrifuge tube and add 100 μL of saturated Na_2CO_3, 200 μL of dansylchloride solution, and 5 μL of 100 mM 1,7-diaminoheptane (HTD) as an internal standard.
5. Incubate the dansylation mixture overnight in darkness at room temperature.
6. Stop the reaction by adding 100 μL proline (100 mg/mL).
7. Extract the dansylated polyamines with 400 μL toluene.
8. Recover the organic phase and evaporate under vacuum.
9. Dissolve the residues in 100 μL of acetonitrile.
10. Analyze by reverse-phase HPLC.

Part B: Soluble Conjugated Polyamines Determination

11. Take 200 μL of the perchloric acid extract from step 3.
12. Add 200 μL of 12 N HCl and incubate for 18 h at 110°C.
13. Dissolve the dried product in 200 μL of 5% (v/v) perchloric acid.
14. Determine the liberated polyamines following the same procedure described in steps 4–10 above.

3.3 Determination of ODC Activity

Activity quantification of those enzymes initiating the biosynthesis of polyamines in maize plants colonized by *U. maydis* is crucial to understanding the role of the polyamines in the plant–host interaction and the manipulation of the synthesis of polyamines by this fungus in the maize plant host. For example, the quantification of the activity of the ornithine decarboxylase (ODC) and arginine decarboxylase (ADC) (Fig. 4), two of the initial reactions for polyamine synthesis, which are independent of degradative routes involving the oxidation of polyamine by the enzyme polyamine oxidase (PAO). The protocol proposed by Rodríguez-Kessler et al. [28] consists of the detection of $^{14}CO_2$, a product of decarboxylation of L-[1-14C]-ornithine to putrescine by the ODC, through a scintillation counter (Fig. 4). On the other hand, the quantification of the ODC activity by a non-radioactive method is based on the measurement of putrescine produced by a crude extract after adding ornithine, permitting the ODC enzymatic action and the measurement of putrescine by spectrophotometry. The enzymatic activity is calculated by comparing the putrescine concentration of the extract sample at the initial and the end of the enzymatic reaction.

3.3.1 Determination of ODC Activity by Radioactivity

1. Take aliquots of 20–100 mL of cultivated cells, previously exposed to experimental conditions; harvest the cells by centrifugation at 300 revolutions per minute for 5 min.
2. Wash the cells twice with FE buffer and resuspend in 1 mL of FEDP buffer.
3. In a reaction flask, place a 1 × 1.5 cm filter paper in the compartment and moisten it with 35 μL of 20% sodium hydroxide to trap CO_2.

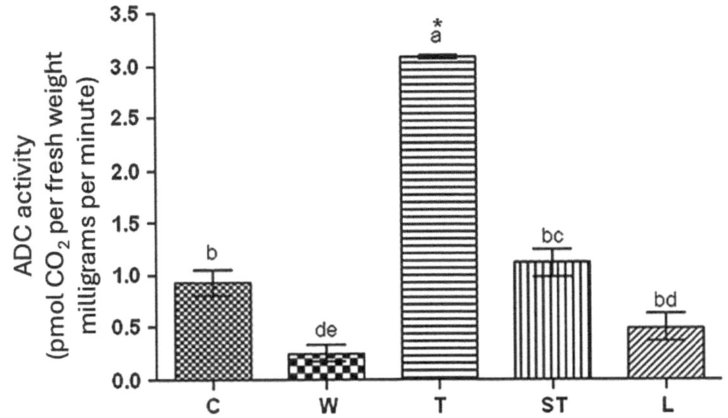

Fig. 4 Determination of the Odc activity in plant tissues colonized by *Ustilago maydis*. Notice a significative increasing of Odc activity in the tumors. Activity is expressed as pmol CO_2/mg FW/min. (This figure was taken from Rodríguez-Kessler et al. [28])

4. Place 200 µL of cell suspension and 190 µL of FEDP buffer in the outer compartment.
5. Initiate the reaction by adding 10 µL of radioactive ornithine 8.4 mM to the outer compartment.
6. Immediately seal the flasks with rubber stoppers, mix by rotation, and incubate at 37°C for 1 h.
7. Stop the reaction by injecting 100 µL of 2 N HCl through the rubber stopper and depositing it in the outer compartment.
8. Dry the filter papers at 60°C for 15–30 min and transfer them to vials containing approximately 7 mL of non-aqueous scintillation liquid.
9. Record the counts per minute for calculation of the specific activity of ODC expressed as nmol of CO_2 released/min*mg of protein.

3.3.2 Non-Radioactive Protocol of ODC Activity by Putrescine Detection [29]

1. Collect fresh fungal biomass and weigh exactly 2 g using an analytical balance; keep it cold during all processing.
2. Add 5 mL of cold homogenization buffer to 2 g of mycelium; homogenize thoroughly until complete tissue disruption.
3. Centrifuge homogenate at 12,000 × *g* for 20 min at 4°C.
4. Carefully collect supernatant (crude enzyme extract) and keep enzyme extract on ice until use.
5. For each reaction assay, add 1.5 mL of reaction buffer and 0.3 mL of crude enzyme extract, and mix gently by inversion.
6. Add 0.2 mL of 25 mmol/L L-ornithine (substrate solution) and mix.
7. Incubate the reaction mixture at 37°C for exactly 1 h.
8. After incubation, immediately add phosphoric acid solution and mix to end the reaction.
9. Centrifuge at 5000 × g for 10 min and collect clear supernatant, avoiding disturbing the pellet during collection.
10. Measure absorbance of the supernatant at 254 nm.
11. Calculate enzyme activity (expressed as U/g) using the formula:

1 unit (U) = 0.01 change in absorbance per minute per gram of mycelium.

3.4 Use of Polyamine Auxotrophic Mutant Strains

Another experimental advantage of *U. maydis* is that it is possible to obtain mutants affected in target genes. Moreover, the sequence of its whole genome provides valuable information for the functional analysis of genes of interest. For example, there are *U. maydis* strains affected in genes involved in the biosynthesis of its polyamines. However, its inability to synthesize polyamines makes it necessary

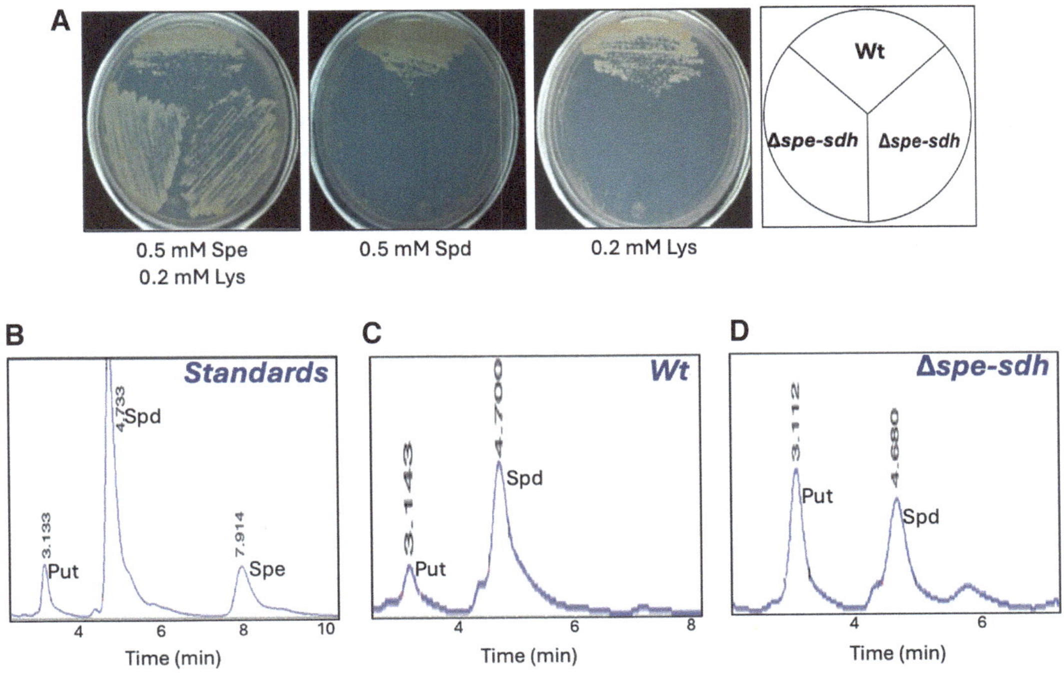

Fig. 5 *Ustilago maydis* mutants affected in genes encoding proteins involved in the biosynthesis of polyamines. (**A**) Growth of the mutant strain *Δspe-sdh* in minimal medium supplemented with polyamines (spermidine) and their precursors (lysine). Determination of polyamines in the standard solutions (**B**), wild type strain (**C**), and *Δspe-sdh* mutant strain (**D**). Chromatographic settings: mobile phase water-methanol 60% (v/v), at 40°C, 121 bars, flow rate 1 mL/min. (This figure was taken from Valdés-Santiago et al. [9])

to manipulate those strains adequately because they require the supplementation of polyamines in the media for growth (auxotrophic mutant strains) (Fig. 5). Lack of intracellular polyamines inhibits the growth of auxotrophic strains, and the intracellular polyamine concentration is related to the fungus's capacity to perform cellular processes such as dimorphism and mating [30, 31], highlighting the critical role of polyamines and their regulation during fungal development. For example, the *U. maydis* spermidine auxotrophic strain (LV54:a2b2) requires 0.4 mM of spermidine for the yeast-like vegetative growth and 5 mM for the dimorphic transition to mycelium [8]. The steps below describe the method for preparing the culture media for polyamine auxotrophic mutants, which can be used for further experiments according to the research interests.

1. Get mutant strains of *U. maydis* affected in genes involved in the synthesis of polyamines. Making mutations in this fungus is relatively easy due to the susceptibility of *U. maydis* to gene mutation by homologous recombination, by disrupting the target gene with an antimicrobial resistance gene (hygromycin, HygB is the main one used), and transformation of the fungus

by protoplasts [32]. The interference construction can be made using the method of double-joint PCR [33]. Recently, gene editing of *U. maydis* by the CRISPR-Cas system has also been reported [34]. Several *U. maydis* mutants with deletions of genes related to the biosynthesis of polyamines have been reported and analyzed, e.g., Δ*odc* [30], Δ*spe* (Fig. 5) [8], or Δ*odc-pao* [35].

2. For propagation of polyamine auxotrophic mutants, inoculate a preinoculum into complete medium (CM) at 25°C for 24–48 h with agitation at 250 rpm in an incubator with orbital shaker.
3. Harvest 1–5 mL of cells from the previous step.
4. Wash three times with sterile ddH_2O, then inoculate these cells with MM, and incubate at 25°C for 24 h with agitation at 250 rpm in an incubator with orbital shaker.
5. Repeat step 2 until no growth is observed (*see* **Note 10**).
6. Inoculate 10^6 cells/mL from the final growth culture into MM supplemented with an adequate concentration of the required polyamine.
7. Carry out experiments using the mutant strains under the required conditions. For example, follow the polyamine concentration in different stages of the pathogenic processes of *U. maydis* in maize plants [28]; use *U. maydis* double mutants such as Δ*odc/pao* for determination of the polyamines required for the fungal survival [35]; identify and quantify the polyamine(s) needed in the *U. maydis* dimorphic transition [30], for mating of compatible teliospores [31], and for the virulence of *U. maydis* in the maize host plant [28]. Moreover, the quantification of polyamines may be useful as an alternative mutant confirmation. For instance, in the double mutant Δ*odc/pao*, the absence of putrescine confirms that this polyamine is not essential for the *U. maydis* growth [35].

4 Notes and Special Considerations

1. Complete medium can be replaced by any other medium for the general growth of microorganisms, such as YPD, nutrient medium, or even LB (Luria–Bertani) medium.
2. Solutions added to prepare the minimum media culture are regularly prepared 2× (double concentrated) and separately. On one side, a carbon source solution plus bacteriological agar (in the case of solid media), and on the other hand, the salt solution. This technique reduces contamination and facilitates

non-scheduled experiments, modifying supplementation, conditions, or any other variable in the experiment.

3. CAUTION: Benzoyl chloride is corrosive and lachrymatory. Handle with complete security equipment: lab coat, nitrile gloves, and safety mask. Manipulate concentrated benzoyl chloride and its solutions in a fume hood.

4. For pH adjustment, it is recommended to use 1 N NaOH for solution alkalization and 1 N HCl for solution acidification. Alternatively, 1 N potassium hydroxide (KOH) and 1 N phosphoric acid (H_3PO_4) can be used. Always use a properly calibrated pH meter and adjust the pH before completing the desired volume of the solution. Once the desired pH is achieved, fill the solution up with ddH_2O. The volume of the prepared solutions should be related to the quantity of samples to be processed.

5. It is recommended that all the material for derivatization be new. Suppose you do not possess new crystal volumetric glassware material, such as flasks, beakers, or solution bottles. In that case, it is recommended to rinse them with concentrated sulfuric acid and then wash out with ultrapure water to eliminate residual organic material.

6. Cultivating 100 mL of cells in the desired conditions is recommended so that the remaining medium and cells can be used to measure biomass or protein amount. Polyamine concentration is reported as nmol/mg protein or nmol/mg biomass.

7. Chromatographic program protocols could be changed to the specification of the available chromatograph, column, or technician.

8. For the standard samples, take 1 mL of the known concentration standard and initiate the protocol for detection of the polyamines by HPLC in step 9.

9. Fungi can accumulate polyamines in some organelles, such as the vacuole. To control the intracellular polyamine contraction, it is necessary to empty polyamine pools; for this, two or three growth passes without polyamine supplementation are required.

General Note: It is essential to always include a negative and positive control for the quality control, reproducibility, and validation of the experiments. All assays will be performed in triplicate, including technical replicates within each experiment. Take special care in times of incubation in the different protocols; standardize temperatures and protocol conditions. Also, it is essential to use consistent mycelium/yeast age and growth conditions. Include ultrapure water as a blank in all experiments.

References

1. Tabor CW, Tabor H (1984) Polyamines. Annu Rev Biochem 53:749–790. https://doi.org/10.1146/annurev.bi.53.070184.003533
2. Salvi D, Tavladoraki P (2020) The tree of life of polyamine oxidases. Sci Rep 10: 17858. https://doi.org/10.1038/s41598-020-74708-3
3. Montañez R, Sánchez-Jiménez F, Aldana-Montes JF et al. (2007) Polyamines: metabolism to systems biology and beyond. Amino Acids 33:283–289. https://doi.org/10.1007/s00726-007-0521-4
4. Hoyt MA, Davis RH (2004) Polyamines in fungi. In: Brambl R, Marzluf GA (eds) Biochemistry and molecular biology. Springer, Berlin, Heilderberg, pp 335–347
5. Valdés-Santiago L, Cervantes-Chávez JA, León-Ramírez CG et al. (2012) Polyamine metabolism in fungi with emphasis on phytopathogenic species. J Amino Acids 2012:1–13. https://doi.org/10.1155/2012/837932
6. Valdés-Santiago L, Ruiz-Herrera J (2014) Stress and polyamine metabolism in fungi. Front Chem 1. https://doi.org/10.3389/fchem.2013.00042
7. Wuddineh W, Minocha R, Minocha SC (2017) Polyamines in the context of metabolic networks. Methods Mole Biol 1694:1–23. https://doi.org/10.1007/978-1-4939-7398-9_1
8. Valdés-Santiago L, Cervantes-Chávez JA, Ruiz-Herrera J (2009) *Ustilago maydis* spermidine synthase is encoded by a chimeric gene, required for morphogenesis, and indispensable for survival in the host. FEMS Yeast Res 9:923–935. https://doi.org/10.1111/j.1567-1364.2009.00539.x
9. Martínez-Soto D, Hernández-Rojas AJ, Valdés-Santiago L et al (2025) Conservation of the polyamines pathway in Ustilaginomycetes A genomic and experimental approach. J Basic Microbiol. 65. e2400561: 1-14 https://doi.org/10.1002/jobm.202400561
10. Stewart TM, Dunston TT, Woster PM et al. (2018) Polyamine catabolism and oxidative damage. J Biol Chem 293:18736–18745. https://doi.org/10.1074/jbc.tm118.003337
11. Kusano T, Berberich T, Tateda C et al. (2008) Polyamines: essential factors for growth and survival. Planta 228:367–381. https://doi.org/10.1007/s00425-008-0772-7
12. Rocha RO, Wilson RA (2018) Essential, deadly, enigmatic: polyamine metabolism and roles in fungal cells. Fungal Biol Rev 33:47–57. https://doi.org/10.1016/j.fbr.2018.07.003
13. Martínez-Soto D, Ruiz-Herrera J (2013) Transcriptomic analysis of the dimorphic transition of *Ustilago maydis* induced in vitro by a change in pH. Fungal Genet Biol 58–59:116–125. https://doi.org/10.1016/j.fgb.2013.08.011
14. Lanver D, Müller AN, Happel P et al. (2018) The biotrophic development of *Ustilago maydis* studied by RNA-Seq analysis. Plant Cell 30:300–323. https://doi.org/10.1105/tpc.17.00764
15. León-Ramírez CG, Sánchez-Arreguín JA, Ruiz-Herrera J (2014) *Ustilago maydis*, a delicacy of the Aztec cuisine and a model for research. Nat Resour 5:256–267. https://doi.org/10.4236/nr.2014.56024
16. Liu S-J, Xu J-J, Ma C-L et al. (2018) A comparative analysis of derivatization strategies for the determination of biogenic amines in sausage and cheese by HPLC. Food Chem 266:275–283. https://doi.org/10.1016/j.foodchem.2018.06.001
17. Özdestan Ö, Üren A (2009) A method for benzoyl chloride derivatization of biogenic amines for high performance liquid chromatography. Talanta 78:1321–1326. https://doi.org/10.1016/j.talanta.2009.02.001
18. David V, Moldoveanu SC, Galaon T (2020) Derivatization procedures and their analytical performances for HPLC determination in bioanalysis. Biomed Chromatogr 35:e5008. https://doi.org/10.1002/bmc.5008
19. Bito T, Okamoto N, Otsuka K et al. (2019) Involvement of spermidine in the reduced lifespan of *Caenorhabditis elegans* during vitamin B12 deficiency. Metabolites 9:192. https://doi.org/10.3390/metabo9090192
20. Holliday R (1961) The genetics of *Ustilago maydis*. Genet Res 2:204–230. https://doi.org/10.1017/s0016672300000719
21. Holliday R (1974) *Ustilago maydis* . In: King RC (Ed.)Eds. Handbook of genetics, bacteria, bacteriophages, and fungi. Springer, Boston, MA, 575–59
22. Pontecorvo G, Roper JA, Chemmons LM et al (1953) The genetics of Aspergillus nidulans. Adv Genet. 5:141–238. https://doi.org/10.1016/s0065-2660(08)60408-3
23. Beadle GW, Tatum EL (1945) *Neurospora*. II. Methods of producing and detecting mutations concerned with nutritional requirements. Am J Bot 32:678. https://doi.org/10.2307/2437625
24. Jin Y, Bok JW, Guzman-de-Peña D et al. (2002) Requirement of spermidine for developmental transitions in *Aspergillus nidulans*. Mol Microbiol 46:801–812. https://doi.org/10.1046/j.1365-2958.2002.03201.x
25. Peña J, Laespada MEF, Pinto CG et al. (2021) Development of a method for the determination of polyamines including N-acetylated forms in

human saliva via benzoylation and gas chromatography-mass spectrometry. J Chromatogr A 1651:462278. https://doi.org/10.1016/j.chroma.2021.462278
26. León-Ramírez CG, Cabrera-Ponce JL, Martínez-Espinoza AD et al. (2004) Infection of alternative host plant species by *Ustilago maydis*. New Phytol 164:337–346. https://doi.org/10.1111/j.1469-8137.2004.01171.x
27. Martínez-Soto D, Robledo-Briones AM, Estrada-Luna AA et al. (2013) Transcriptomic analysis of *Ustilago maydis* infecting *Arabidopsis* reveals important aspects of the fungus pathogenic mechanisms. Plant Signal Behav 8:e25059. https://doi.org/10.4161/psb.25059
28. Rodríguez-Kessler M, Ruiz OA, Maiale S et al. (2008) Polyamine metabolism in maize tumors induced by *Ustilago maydis*. Plant Physiol Biochem 46:805–814. https://doi.org/10.1016/j.plaphy.2008.05.012
29. Wang D, Meng L, Zhang H et al (2024) Exogenous L-arginine enhances pathogenicity of *Alternaria alternata* on kiwifruit by regulating metabolisms of nitric oxide, polyamines, reactive oxygen species (ROS), and cell wall modification. J Fungi 10:801. https://doi.org/10.3390/jof10110801
30. Guevara-Olvera L, Xoconostle-Cázares B, Ruiz-Herrera J (1997) Cloning and disruption of the ornithine decarboxylase gene of *Ustilago maydis*: evidence for a role of polyamines in its dimorphic transition. Microbiology 143:2237–2245. https://doi.org/10.1099/00221287-143-7-2237
31. Pérez-Rodríguez F, Valdés-Santiago L, García-Chávez JN et al. (2023) Analysis of gene expression related to polyamine concentration and dimorphism induced in ornithine decarboxylase (odc) and spermidine synthase (spd) *Ustilago maydis* mutants. Fungal Genet Biol 166:103792. https://doi.org/10.1016/j.fgb.2023.103792
32. Tsukuda T, Carleton S, Fotheringham S et al. (1988) Isolation and characterization of an autonomously replicating sequence from *Ustilago maydis*. Mol Cell Biol 8:3703–3709. https://doi.org/10.1128/mcb.8.9.3703-3709.1988
33. Yu J-H, Hamari Z, Han K-H et al. (2004) Double-joint PCR: a PCR-based molecular tool for gene manipulations in filamentous fungi. Fungal Genet Biol 41:973–981. https://doi.org/10.1016/j.fgb.2004.08.001
34. Schuster M, Schweizer G, Reissmann S et al. (2015) Genome editing in Ustilago maydis using the CRISPR-Cas system. Fungal Genet Biol 89:3–9. https://doi.org/10.1016/j.fgb.2015.09.001
35. Valdés-Santiago L, Guzmán-De-Peña D, Ruiz-Herrera J (2010) Life without putrescine: disruption of the gene-encoding polyamine oxidase in *Ustilago maydis* odc mutants. FEMS Yeast Res 10:928–940. https://doi.org/10.1111/j.1567-1364.2010.00675.x

Check for updates

Chapter 36

Extraction of Total Fatty Acids from *Ustilago maydis*

Selina Granderath-Miegel, Lars M. Blank and An N. T. Phan

Abstract

The utilization of microbial single-cell oils has emerged as a promising alternative to traditional plant oils, offering the potential to enhance sustainability and land efficiency in oil production. The natural accumulation of lipid droplets in *Ustilago maydis* suggests its potential as a suitable host for lipid production. Here, a refined method for the total fatty acid extraction from *U. maydis* is described. The method provides efficient cell disruption and high-purity fatty acid extracts suitable for lipid profiling via gas chromatography.

Keywords *Ustilago maydis*, *Ustilaginaceae*, lipids, fatty acids, GC–MS, GC-FID, triacylglycerides, single cell oil, applied microbiology, metabolic engineering

1 Introduction

Oleaginicity is a trait of certain microbial species that accumulate lipids to a large amount of up to 70% (w/w) of their cell dry weight (CDW) in specialized organelles, referred to as lipid droplets or lipid bodies [1]. Scientific records of oleaginous microorganisms date back to 1878 and industrial-scale productions were first attempted in 1915 with the fungus *Endomyces vernalis*, while the first standardization of lipid extraction protocols was only realized in 1957 and 1959 [2]. To date, the scientific study of oleaginous microorganisms remains a vibrant and active field of research with a significant increase in publications since the early 2000s. The industrial use of microbial oils in the food or biodiesel sector could replace the growing demand for vegetable fats, such as palm oil, whose production cause land conflicts in a growing world population and often compete with natural habitats [3–5].

Lipid droplet accumulation in *U. maydis* has been described in multiple studies [6–9]. According to the terminology of oleaginous organisms—which only includes those with a lipid content above 20% (w/w) of the CDW – wildtype *U. maydis* strains are not considered oleaginous [8]. However, multiple studies describe cultivation

An N. T. Phan et al. (eds.), *Ustilago maydis*, Methods in Molecular Biology, vol. 3050,
https://doi.org/10.1007/978-1-0716-5364-7_36,

techniques [7, 8] and engineered strains [9, 10] yielding significantly higher lipid contents.

The first standardized lipid extraction protocols were developed for the extraction from brain [11] or fish [12] tissue. Basing on these methods, protocols were refined for the extraction from different organisms, including microalgae and fungi [13]. Modified extraction protocols include different pretreatment methods as well as the replacement of the classical methanol/chloroform-based extraction solvents. The latter, however, remains the gold standard in broad-based, untargeted lipidomic studies due to the high extraction efficiency and reproducibility.

Here, a two-step extraction-transesterification method for the total fatty acid extraction from *U. maydis* is described. Cell disruption is realized by a combination of bead beating and ultrasonication, which is scalable, rapid and high-yielding. Extraction and transesterification are carried out in two distinct steps, thereby ensuring a high purity of the extract for subsequent analysis. Further, the protocol provides robust methods for fatty acid identification and quantification using gas chromatography.

2 Materials

2.1 Equipment

a. Freezer (−80°C and −20°C).

b. Incubation chamber (30°C and 37°C).

c. Shaker (shaking diameter: 50 mm; humidity control: 70%).

d. Spectrophotometer for OD_{600}.

e. Autoclave for media and waste sterilization.

f. Centrifuge for 15 mL Falcon tubes.

g. Centrifuge for 1.5 mL and 2 mL reaction tubes with cooling unit (4°C).

h. Ice bath.

i. Fume hood.

j. Heat chamber (60°C).

k. Precision scale.

l. Sonicator.

m. Bead-beating device.

n. Cooling chamber at 4°C.

o. Multi-vortex for 1.5 ml and 2 ml reaction tubes.

p. Vortex.

q. Crimping pliers.

r. Heating block.

s. Vacuum concentrator.

t. Gas chromatography–mass spectrometer (GC–MS) (*see* **Note 1**).

u. GC connected to a flame ionization detector (GC-FID) (*see* **Note 1**).

v. Organic solvent waste opportunities.

w. Microscope (63×).

2.2 Cultivation and Strain Storage (see Notes 2 and 3)

a. Cryogenic stocks: 80 % (v/v) glycerol solution.

b. YEPS agar plates: 15 g/L Agar; 10 g/L D-Saccharose; 10 g/L Peptone; 10 g/L Yeast extract.

c. YEPS medium: 10 g/L D-Saccharose; 10 g/L Peptone;10 g/L Yeast extract.

d. Modified Tabuchi medium (MTM) [14]: Stock solutions of the individual medium components: 200 g/L D-Glucose stock (*see* **Note 4**); 500 mM 2-(N-morpholino)ethanesulfonic acid (MES) buffer stock, adjusted to pH 6.5 with concentrated NaOH; 100 g/L NH_4Cl stock; 100 g/L $MgSO_4{\bullet}7H_2O$ stock; 10 g/L $FeSO_4{\bullet}7H_2O$ stock (*see* **Note 5**); 100 g/L KH_2PO_4; 1000× vitamin solution (0.1 g/L D-biotin; 1 g/L D-calcium-panthothenate; 1 g/L nicotinic acid; 25 g/L myo-Inositol; 1 g/L thiamine hydrochloride; 1 g/L pyridoxal hydrochloride; 0.2 g/L paraaminobenzoic acid) and 1000× trace element solution (1.5 g/L $EDTA{\bullet}Na_2{\bullet}2\ H_2O$; 0.45 g/L $ZnSO_4{\bullet}7\ H_2O$; 0.1 g/L $MnCl_2$; 0.03 g/L $CoCl_2{\bullet}6\ H_2O$; 0.03 g/L $CuSO_4{\bullet}5\ H_2O$; 0.04 g/L $Na_2MoO_4{\bullet}2\ H_2O$; 0.45 g/L $CaCl_2{\bullet}2\ H_2O$; 0.3 g/L $FeSO_4{\bullet}7\ H_2O$; 0.1 g/L H_3BO_3; 0.1 g/L KI) (*see* **Note 6**). Final concentrations of components in the medium (*see* **Note** 7): 50 g/L D-Glucose; 100 mM MES buffer; 0.8 g/L NH_4Cl; 0.2 g/L $MgSO_4{\bullet}7H_2O$; 0.01 g/L $FeSO_4{\bullet}7H_2O$; 0.5 g/L KH_2PO_4; 1 mL/L vitamin and 1 mL/L trace element solution. Adjust to the required final volume with aqua bidest.

e. Washing buffer: 0.9 % NaCl (w/v).

2.3 Fatty Acid Extraction (see Note 8)

a. Internal standard (IS) stock solution: 10 mM tridecanoic acid (CAS: 638-53-9) in methanol.

b. Extraction solution I: Methanol:Trichloromethane (TCM):H_2O (5:2:2) (*see* **Note 9**). Add tridecanoic acid to a final concentration of 100 μM from IS stock.

c. Esterification solution: Methanol:HCl (36 %):TCM (10:1:1), prepared on ice (*see* **Note 10**).

d. Extraction solution II: Hexane:TCM (4:1).

e. Additional solutions: TCM with 100 μM tridecanoic acid from IS stock; pure TCM, pure methanol, and pure hexane.

2.4 GC–MS Measurement

a. Fatty acid methyl esters (FAME) standard: Analytical grade FAME standard mixture (C4-C24), diluted to a concentration of 100 μM in hexane.

3 Methods

3.1 Cryogenic Cultures of *U. maydis* (see Note 11)

1. Inoculate YEPS medium with a single colony and incubate for 18–20 h until reaching an OD_{600} of 20 (*see* **Note 12**).
2. Aliquot 700 μL glycerol stock in 2 mL cryogenic tubes and mix well with 300 μL culture broth. Store at −80°C.

3.2 Cultivation of *U. maydis*

1. Streak a YEPS agar plate with the cryogenic culture using an inoculation loop or sterile pipette tip and incubate at 30°C for 24 h (*see* **Note 13**).
2. For the first preculture, inoculate YEPS medium with a single colony and incubate for 18–20 h until reaching an OD_{600} of 20.
3. Inoculate the second preculture with an OD_{600} of 0.5 in MTM and incubate for 18 h until reaching an OD_{600} of 20.
4. Inoculate the main culture with an OD_{600} of 0.5 in MTM and incubate for 72 h (*see* **Note 14**).

3.3 Cell Harvest and Estimation of CDW (see Note 15)

1. Transfer 5 mL of culture broth to a 15 mL Falcon tube. Centrifuge at 20,000 × *g* for 10 min (*see* **Note 16**).
2. Discard the supernatant and wash the cells 3× with 0.9 % NaCl.
3. Resuspend the cells in 1 mL aqua bidest. and transfer to a 2 mL screw cap reaction tube. For CDW estimation, the screw cap reaction tubes should be pre-labeled, dried at 60°C for at least 24 h and pre-weighed on a precision scale.
4. Centrifuge the cell suspension at 16,000 × *g* for 5 min, discard the supernatant and freeze the cell pellets at −80°C for 24 h.
5. Lyophilize the cells in a vacuum concentrator at 60 × *g* for 24 h (*see* **Note 17**).
6. Weigh the reaction tubes including the dried cells on a precision scale (*see* **Note 18**).
7. Calculate the weight difference to obtain the CDW.

3.4 Total Fatty Acid Extraction

1. Add 25 mg glass or ceramic beads (0.25–0.5 mm) to each reaction tube containing the freeze-dried cells (*see* **Note 19**).
2. Add 900 μL extraction solution I per tube. Process three controls containing no biological sample together with the samples.

3. Screw the tubes tightly and shake them at 4 m/s for 30 s in a bead-beating device. Repeat this step until the cell pellets are fully dispersed, cool 30 s on ice in between shaking.
4. Sonicate the cell suspension for 10 min and shake in a multi-vortex at full speed overnight at 4°C.
5. Add 400 μL of TCM including the IS to the solution. Vortex for 20 s or shake at 4 m/s for 30 s in a bead-beating device.
6. Centrifuge at 4°C and 14,000 × *g* for 5 min (*see* **Note 20**).
7. Gently push the cell pellet to the side and transfer the lower organic phase to a new 2 mL reaction tube using a pipette tip pre-wetted with TCM (*see* **Note 21**). 500–600 μL extract should be recoverable in the first extraction step.
8. Re-extract with 600 μL TCM including the IS.
9. Vortex at 4°C for 1 h and centrifuge at 4°C and 14,000 × *g* for 5 min.
10. Recover the lower organic phase (600–700 μL) and combine it with the first extraction fraction (*see* **Note 22**).
11. Transfer the extract to a 1.5 mL glass crimp top vial. For the total fatty acid estimation, the glass vials should be clean, pre-labeled, dried at 60°C for 24 h, and pre-weighed on a precision scale.
12. Evaporate the solvent in a vacuum concentrator at 40 × *g* for 2–3 h.
13. Weigh the glass vials on a precision scale post solvent evaporation (*see* **Note 23**).

3.5 Methanolic-HCl Transesterification [15]

1. Add 500 μL esterification solution to the vials and close them tightly with crimp caps.
2. Incubate the vials in a heating block at 90°C for 120 min and optionally 500 rpm shaking (*see* **Note 24**). Let the samples cool down on ice.
3. Transfer the liquid from the vials to a 2 mL reaction tube with a pipette tip pre-wetted with methanol.
4. Add 500 μL aqua bidest. to the glass vials and transfer it to the sample in the reaction tube.
5. Add 600 μL extraction solution II first to the glass vial, then transfer to the reaction tube.
6. Vortex the solution for 60 s and centrifuge at 4°C and 14000 × *g* for 5 min.

7. Transfer the upper phase—approximately 500–600 μL—to a new 2 mL reaction tube with a pipette tip prewetted in hexane.
8. Re-extract by adding 600 μL extraction solution II first to the glass vial, then to the tube containing the lower phase.
9. Vortex for 60 s and centrifuge at 4°C and 14,000 × *g* for 5 min.
10. Recover the upper layer (600–700 μL) and combine it with the first extraction fraction. Short-term storage of the extracts can be carried out at −20°C (*see* **Note 25**).

3.6 Quantification of Total Fatty Acid Content

1. Calculate the total amount of extracted fatty acids by the weight difference between the dried vials and the vials containing the lipids (*see* **Note 26**).
2. Use the control vials to estimate the quality of the measurement (expected weight difference of the control: ~0–1 mg). The total fatty acid content can be used to both calculate the fatty acid proportion in the CDW and the fatty acid yield per consumed carbon source.

3.7 Analysis of Fatty Acid Compositions with Gas Chromatography

GC–MS is used to identify the composition of total fatty acid extract.

1. Aliquot 200 μL of the extracts as well as the diluted FAMES standard into micro-vials or glass vials containing inserts, suitable for the used GC sampler.
2. Analyze the samples using the following method: Injection temperature: 250°C; injection volume: 1 μL; Split ratio: 20; ion source temperature: 300°C; carrier gas flow (He): 1.0 mL/min; column oven temperature ramp: 100°C, hold 2 min; ramp 1: 15°C/min to 185°C; ramp 2: 5°C/min to 205°C, hold 3 min; ramp 3: 0.8°C/min to 215°C; ramp 4: 15°C/min to 300°C, hold 5 min.
3. First, identify the retention time of the IS (tridecanoic acid methyl ester) from the controls using a spectral database.
4. Subsequently, identify the fatty acids contained in the biological samples (*see* **Notes 27 and 28**).
5. Estimate the quality of the identifications by the correct compound group (FAME); the library matching score obtained from the spectral database (≥70%); similarity of fatty acid profiles among triplicates and identity in retention times of saturated, even-numbered fatty acids in samples and FAME standard.
6. For the quantification, re-analyze the samples using a GC-FID. Export the peak area of the identified fatty acid compounds.

7. Normalize the peak areas to that of the IS for each sample.
8. Calculate the sum of the normalized peak areas and express the single compound proportions as percentage of the total fatty acids by diving through the calculated sum.

4 Notes

1. Equipped with a wax column (e.g., Zebron ZB-WAX (30 m length, 0.25 mm inner diameter, 0.25 μm film thickness); Phenomenex; USA).
2. Double-distilled water (aqua bidest.) should be used for all media and reagents.
3. Unless stated otherwise, all media and solutions are sterilized by autoclaving and can be stored at 4°C prior to use.
4. Sterilize by filtration to prohibit sugar caramelization during autoclaving at high glucose concentrations.
5. Sterilized by filtration and prepared freshly within the day of use to avoid oxidation, visible by yellow color of the solution. Alternatively, the stock can be acidified with 1 M HCl to pH 2 for long-term storage.
6. Both the vitamin and trace element solution should be sterilized by filtration.
7. The medium should be prepared freshly on the day of cultivation.
8. All solutions containing organic solvents should be prepared freshly on the day of use under the fume hood wearing solvent-resistant gloves. Using plasticware should be avoided. If the use of certain plastic equipment is unavoidable, first check whether the plastic type is compatible with the used solvents.
9. The ratio of the components has to be met in order to obtain a clear one-phased solution. Should the extraction solution turn cloudy, it should be discarded and redone. Usage of glass pipettes might aid with preparing the solution correctly. Due to the low surface tension of the organic solvents, the solvent should first be drawn and ejected once (pre-wetting). The formed liquid film will prevent dripping from the pipettes. When using a measuring cylinder, the glassware should first be flushed with the respective solvent.
10. Acidifying methanol generates heat, which might cause the solution to boil when insufficiently cooled. For preparing this solution, choose a shot flask with at least double the volume of the solution and wear acid- and solvent-resistant gloves. Precool the methanol at −20°C.

11. Preparation of media and inoculations should be carried out on a clean bench. All cultivations in liquid media are carried out with 10 % filling volume of the used flasks, at 30°C and 200 rpm shaking. Cultivation should be performed using three biological replicates. Microscopy of the cells should be carried out at the end of each cultivation to check for contamination.
12. Cell numbers of *U. maydis* can be correctly estimated photometrically up to an OD_{600} of 0.8. Dilute the culture broth 100× with aqua bidest. and mix well for OD_{600} measurements.
13. The cultivation on standard YPD agar plates is possible but prolongs the cultivation to 2 days.
14. The optimal timepoint for a maximum lipid content is shortly after glucose depletion and can vary based on the used strain. Preliminary studies supported by optical or respiratory online growth monitoring and determination of glucose consumption via time-course sampling using glucose test strips or HPLC analytics might aid in finding the proper cultivation duration. In a broad screening of multiple strains, 72 h is a good estimation to harvest lipids for fatty acid profiling.
15. Unless stated otherwise, all steps are carried out at room temperature (RT). Work with organic solvents should be carried out under a fume hood using solvent-resistant gloves and solvent-resistant plastic equipment. When pipetting organic solvents, pre-wet the pipette tips with the respective solvent to avoid dripping.
16. Since some strains of *U. maydis* grow filamentous and secrete extracellular polysaccharides turning the culture broth highly viscous, insufficient separation of cells and supernatant can occur after 10 min of centrifugation. In these cases, either dilute the culture broth 1:1 with 0.9 % NaCl or prolong the centrifugation time.
17. Freeze-drying of *U. maydis* cells might lead to rapid expansion of the cell pellet in the reaction tube and sample loss. To avoid this, prepare an additional set of screw caps and perforate the caps with at least three holes using a needle (Ø 1.2 mm).
18. Humidity can cause considerable deviations in weighing results. Re-cap the tubes immediately after the freeze-drying process with the non-perforated, pre-labeled screw caps and weigh them. For large sample sets, weigh the samples batchwise and use a desiccator for sample storage in between.
19. To shorten sample preparation time, weigh the beads in a 200 μL PCR reaction tube once and mark the filling volume on the tube. Attach the tube to a spatula and use it as a trowel for the beads.

20. If the upper layer is still milky, centrifuge again until you see clear separation between two phases and the upper layer is completely clear.
21. Be careful not to transfer cell debris from the interphase or glass beads from the bottom of the tube. The extract should be clear with a yellow color.
22. In case the extract looks cloudy it likely contains aqueous residues from the upper layer. Centrifuge the extract at 4°C and 14,000 × *g* for 5 min and remove the upper, aqueous layer. In case the extract contains cell debris from the interphase, filter with a solvent-proof filter (Ø 0.2 μm).
23. The extracted fatty acids oxidize rapidly. Work fast while weighing. If the samples are not esterified directly after extraction, evacuate the vials with argon gas and close with crimp caps.
24. Operate the heating block under the fume hood. Control the temperature with a thermometer and ensure it does not exceed 90°C. Using regular screw cap vials might cause sample loss due to leakage of solvent vapor. Temperatures greater than 90°C will cause the septa in the crimp caps to rupture, resulting in sample loss.
25. The extract can be stored maximally up to 1 month.
26. Alternatively, the total fatty acid content can be estimated from the weight difference of the CDW prior and post extraction. For this approach, evaporate the solvent from the reaction tubes with CDW post extraction using a vacuum concentrator at 60 × g for 2–3 h. Close the tubes with perforated lids. This method is however more error-prone than the one described.
27. In the methanolic transesterification, fatty acids are derivatized to fatty acid methyl esters. A C16:0 fatty acid compound with the chemical formula $C_{16}H_{32}O_2$ will thus be identified as $C_{17}H_{34}O_2$.
28. In case the peak intensities are close to the detection limit, concentrate the samples 5× by evaporating the solvent in a vacuum concentrator at 40 × *g* for 1–2 h. Resuspend in pure hexane. In case the concentrated samples precipitate, heat them in a water bath at 40°C for 30 min. Let the samples cool down to RT and check that the precipitation does not recur. Precipitated samples cannot be used for injection. Instead, lower the concentration to 2× or 3×.

Acknowledgments

This work was funded by the Deutsche Forschungsgemeinschaft (DFG, German Research Foundation) SFB1535—Project ID 458090666 and under Germany's Excellence Strategy—Cluster of Excellence 2186 "The Fuel Science Center"—Project ID: 390 919 832.

References

1. Papanikolaou S, Aggelis G (2011) Lipids of oleaginous yeasts. Part I: biochemistry of single cell oil production. Eur J Lipid Sci Technol 113:1031–1051. https://doi.org/10.1002/ejlt.201100014
2. Abeln F, Chuck CJ (2021) The history, state of the art and future prospects for oleaginous yeast research. Microb Cell Fact 20:221. https://doi.org/10.1186/s12934-021-01712-1
3. Parsons S, Raikova S, Chuck CJ (2020) The viability and desirability of replacing palm oil. Nat Sustain 3:412–418. https://doi.org/10.1038/s41893-020-0487-8
4. Carlson KM, Garrett RD (2018) Environmental impacts of tropical soybean and palm oil crops. In: Carlson KM, Garrett RD (Eds.)Eds. Oxford research encyclopedia of environmental science. Oxford University Press
5. Berenschot W, Dhiaulhaq A, Hospes O et al. (2024) Corporate contentious politics: palm oil companies and land conflicts in Indonesia. Political Geogr 114:103166. https://doi.org/10.1016/j.polgeo.2024.103166
6. Maassen N, Panakova M, Wierckx N et al. (2014) Influence of carbon and nitrogen concentration on itaconic acid production by the smut fungus *Ustilago maydis*. Eng Life Sci 14:129–134. https://doi.org/10.1002/elsc.201300043
7. Zavala-Moreno A, Arreguin-Espinosa R, Pardo JP et al. (2014) Nitrogen source affects glycolipid production and lipid accumulation in the phytopathogen fungus *Ustilago maydis*. AiM 04:934–944. https://doi.org/10.4236/aim.2014.413104
8. Aguilar LR, Pardo JP, Lomelí MM et al. (2017) Lipid droplets accumulation and other biochemical changes induced in the fungal pathogen *Ustilago maydis* under nitrogen-starvation. Arch Microbiol 199:1195–1209. https://doi.org/10.1007/s00203-017-1388-8
9. Richter P, Panchalingam J, Miebach K et al. (2024) Studying microbial triglyceride production from corn stover saccharides unveils insights into the galactose metabolism of *Ustilago maydis*. Microb Cell Fact 23:204. https://doi.org/10.1186/s12934-024-02483-1
10. Richter P, Maßjosthusmann M, Seidel T et al. (2025) Tailoring the fatty acid profile of microbial triglycerides in *Ustilago maydis* by adapting the cultivation conditions. Bioresour Technol Rep 30:102119. https://doi.org/10.1016/j.biteb.2025.102119
11. Folch J, Lees M, Sloane-Stanley GH (1957) A simple method for the isolation and purification of total lipides from animal tissues. J Biol Chem 226:497–509
12. Bligh EG, Dyer WJ (1959) A rapid method of total lipid extraction and purification. Can J Biochem Physiol 37:911–917. https://doi.org/10.1139/o59-099
13. Saini RK, Prasad P, Shang X et al (2021) Advances in lipid extraction methods – a review. Int J Mol Sci 22:13643. https://doi.org/10.3390/ijms222413643
14. Geiser E, Wiebach V, Wierckx N et al. (2014) Prospecting the biodiversity of the fungal family *Ustilaginaceae* for the production of value-added chemicals. Fungal Biol Biotechnol 1: 2. https://doi.org/10.1186/s40694-014-0002-y
15. Lewis T, Nichols PD, McMeekin TA (2000) Evaluation of extraction methods for recovery of fatty acids from lipid-producing microheterotrophs. J Microbiol Methods 43:107–116. https://doi.org/10.1016/S0167-7012(00)00217-7

Check for updates

Chapter 37

Metabolomic Study in Ustilaginaceae with GC–MS/MS

An N. T. Phan, Philipp Westhoff, Stefan Krüger and Lars M. Blank

Abstract

Ustilago maydis, the causative agent of corn smut in maize, is a well-established model organism for fundamental biological research. In addition, this fungus has growing biotechnological importance, as it proliferates in a haploid, non-filamentous form and efficiently converts renewable substrates into a range of industrially valuable products. To complement the extensive molecular and genetic tools already available, metabolomics provides essential insights into cellular physiology and supports the exploitation of its production potential. Here, a GC–MS-based workflow for the absolute quantification of intracellular metabolites in *U. maydis* is presented. This protocol enables systematic analysis of metabolic network operation and can be readily extended to metabolomics studies in related fungi in the Ustilaginaceae family.

Keywords *Ustilago maydis*, Ustilaginaceae, GC–MS, Metabolomics, Sample preparation, Central carbon metabolism, Metabolic engineering

1 Introduction

Ustilago maydis has been extensively used as a model organism for fundamental biological research, particularly in the study of pathogen–host interactions, fungal mating, DNA recombination, and DNA repair [1–4]. Genome sequencing efforts have revealed that approximately 10% of *U. maydis* proteins share high conservation with their human counterparts, whereas many of these proteins show little or no similarity to those in *Saccharomyces cerevisiae* [5, 6]. This distinct genomic composition positions *U. maydis* as a powerful experimental system for addressing cellular processes that cannot be adequately studied in yeast, such as endocytosis, long-distance mRNA transport, cell signaling, microtubule organization, and polarized growth [7–9].

In addition to its role in basic research, *U. maydis* has gained biotechnological importance. The organism grows in a non-filamentous haploid form, and can utilize sustainable substrates to produce a variety of industrially relevant chemicals, including mannitol, erythritol, mannosylerythritol lipids (MELs), ustilagic acid, itaconic acid,

An N. T. Phan et al. (eds.), *Ustilago maydis*, Methods in Molecular Biology, vol. 3050,
https://doi.org/10.1007/978-1-0716-5364-7_37,

malic acid, and hydroxyparaconic acid [10, 11]. To fully exploit these capabilities, a broad set of molecular, genetic, bioinformatic, and cell biological tools has been established for this species [7, 12, 13].

Comprehensive "omics" approaches—including genomics, transcriptomics, proteomics, and metabolomics—provide powerful means to gain insights into the physiology of *U. maydis*. The methodologies for genome analysis [14, 15], transcriptomics [16, 17], proteomics [18, 19], and metabolomics [20, 21] are already well established. Here, a gas chromatography–mass spectrometry (GC–MS)-based workflow for the absolute quantification of intracellular metabolites in *U. maydis* is presented. This protocol enables detailed investigation of metabolic network operation and can also be applied to metabolomics studies of other fungi within the Ustilaginaceae family.

2 Materials

2.1 Cell Cultivation

1. Modified Tabuchi medium (MTM) is used according to Geiser et al. [10] with 50 g L^{-1} glucose, 100 mM 2-(*N*-morpholino) ethanesulfonic acid (MES), pH 6.5, 0.2 g L^{-1} $MgSO_4 \cdot 7H_2O$, 10 mg L^{-1} $FeSO_4 \cdot 7H_2O$, 0.5 g L^{-1} KH_2PO_4, 0.8 g L^{-1} NH_4Cl, 1 mL L^{-1} vitamin solution, and 1 mL L^{-1} trace elements solution. The vitamin solution contains 0.05 g/L d-biotin, 1 g/L d-calcium pantothenate, 1 g/L nicotinic acid, 25 g/L myo-inositol, 1 g/L thiamine hydrochloride, 1 g/L pyridoxol hydrochloride, and 0.2 g/L *para*-aminobenzoic acid. The trace element solution contains 1.5 g/L EDTA, 0.45 g/L of $ZnSO_4 \cdot 7H_2O$, 0.10 g/L of $MnCl_2 \cdot 4H_2O$, 0.03 g/L of $CoCl_2 \cdot 6H_2O$, 0.03 g/L of $CuSO_4 \cdot 5H_2O$, 0.04 g/L of $Na_2MoO_4 \cdot 2H_2O$, 0.45 g/L of $CaCl_2 \cdot 2H_2O$, 0.3 g/L of $FeSO_4 \cdot 7H_2O$, 0.10 g/L of H_3BO_3, and 0.01 g/L of KI.
2. Cryogenic stocks: cell culture mix with 80% (v/v) glycerol with the ratio 1:1.
3. YEPS agar plates: 15 g/L Agar; 10 g/L D-Saccharose; 10 g/L Peptone; 10 g/L Yeast extract.

2.2 Sample Preparation

1. Quenching solution: Saline Buffer (SB) 0.9% NaCl at 4°C.
2. Extraction solution: 80% Ethanol with 20 μM ribitol as an internal standard. The solution is kept at −20°C.
3. Metabolite derivatization: methoxyamine hydrochloride (MeOX) (20 mg/mL in pyridine) and *N*-methyl-*N*-(trimethylsilyl) trifluoroacetamide (MSTFA) (*see* **Note 1**).

2.3 Equipments

1. Shake flasks and a shaker (speed: 200 rpm; shaking diameter: 50 mm; humidity control: 70%).

2. Fridge (4°C) and freezer (−80°C and −20°C) for cells and chemicals storage.
3. Incubation chamber (30°C).
4. Spectrophotometer for OD_{600}.
5. Autoclave for media and waste sterilization.
6. Precision scale to determine cell dry weight (CDW).
7. Centrifuges for 1.5 mL and 2 mL reaction tubes with cooling unit (4°C).
8. Multi-vortex for 1.5 ml and 2 ml reaction tubes.
9. Ice bath to store samples during extraction.
10. Fume hood.
11. Heating block for 1.5 mL and 2 mL reaction tubes (30°C and 37°C).
12. Vacuum concentrator with cooling unit (4°C).
13. GC–MS/MS: Trace GC Ultra coupled with TSQ 8000 mass spectrometry (*Thermo Fisher Scientific*). A VF-5ms capillary column (30 m × 250 μm i.d., 0.25 μm film thickness) with 10 m EZ-Guard (*Agilent*) is used.
14. Authentic standards of chemicals mentioned in Table 1 are of analytical grade.

3 Method

3.1 Cell Cultivation

1. Streak cells from cryo stock culture onto a YEPS agar plate and incubate at 30°C for 24–36 h until single colonies are clearly visible.
2. For each biological replicate, inoculate a single colony into MTM medium in a shake flask with a filling volume of 10%.
3. Cultivate the cells in a shaker at 30°C for 20–24 h until the cell culture reaches an OD_{600} of 15–20.
4. Inoculate from precultures to start main cultures at OD_{600} of 0.5 and monitor OD_{600} over time to determine the collection point.
5. At an OD_{600} of 20, corresponding to the mid-exponential phase of cell growth, collect a volume equal to 10 OD_{600} unit for intracellular metabolite measurement, followed by a quenching step (*see* **Note 2**).
6. Transfer the same volume of samples to pre-weighed dry Eppendorf tubes to determine CDW. Wash the samples twice with SB and then dry them in an oven at 70°C until a constant weight is achieved.

Table 1
Selected reaction monitoring (SRM) transitions for GC–MS/MS analysis

Name	Retention time (min)	Precursor ion *m/z*	Product Ion *m/z*	Collision energy
Pyruvate	4.25	115.1	74.1	6
Pyruvate	4.25	158.1	130.1	6
Pyruvate	4.25	174.1	115.1	10
Lactate	4.3	191.1	147.1	8
Lactate	4.3	219.1	147.1	10
Lactate	4.3	219.1	191.1	6
Alanine	4.66	116.1	73.1	10
Alanine	4.66	190.1	147.1	10
Valine	5.65	144.1	73.1	12
Valine	5.65	218.1	100.1	6
Glycerol	6.1	117	89.1	8
Glycerol	6.1	205	147.1	6
Glycerol	6.1	218	191.2	6
Leucine	6.14	158.1	102.1	6
Leucine	6.14	232.2	147.1	10
Isoleucine	6.34	158	102.1	6
Isoleucine	6.34	218.1	73.1	14
Isoleucine	6.34	232.2	147.1	10
Proline	6.44	142	73.1	12
Proline	6.44	216.1	147.1	12
Glycine	6.49	248.1	147.1	12
Glycine	6.49	276.1	248.2	8
Succinate	6.54	129.1	75	10
Succinate	6.54	172.1	113.1	8
Succinate	6.54	247.1	147.1	6
Uracil	6.77	241.1	126.1	8
Uracil	6.77	255.1	185	16
Uracil	6.77	255.1	239.1	10
Itaconate	6.79	131	115	10
Itaconate	6.79	259.1	131	24
Itaconate	6.79	259.1	147.1	8
Fumarate	6.85	143	75.1	12

Table 1
(continued)

Name	Retention time (min)	Precursor ion *m/z*	Product Ion *m/z*	Collision energy
Fumarate	6.85	245.1	147.1	10
Fumarate	6.85	245.1	217.1	6
Serine	6.89	188.1	73.1	12
Serine	6.89	204.1	188.2	6
Serine	6.89	218.1	73.1	14
Threonine	7.12	117	73.1	8
Threonine	7.12	218.1	128.1	8
Threonine	7.12	291.2	129.1	6
Beta-alanine	7.54	248.1	147.1	14
Beta-alanine	7.54	290.1	248.2	8
Malate	7.99	233.1	189.2	6
Malate	7.99	233.1	191.1	6
Malate	7.99	245.1	147.1	10
Meso-Erythritol	8.1	205.1	117.1	8
Meso-Erythritol	8.1	217.1	129.1	8
Meso-Erythritol	8.1	307.2	217.2	6
Aspatic acid	8.26	202.1	73.1	16
Aspatic acid	8.26	218.1	100.1	8
Aspatic acid	8.26	232.1	188.2	6
Methionine	8.33	128.1	73.1	10
Methionine	8.33	176.1	128.1	8
Methionine	8.33	250.1	202.2	6
Pyroglutamate	8.38	156.1	73.1	6
Pyroglutamate	8.38	230.1	147.1	10
Pyroglutamate	8.38	258.1	230.2	8
a-KG	8.73	198.1	170.1	6
a-KG	8.73	288.1	198.1	6
a-KG	8.73	304.1	156.1	6
PEP	8.91	243.1	211.1	8
PEP	8.91	369.1	217.1	8
PEP	8.91	369.1	299.1	8
Glutamate	9.1	156.1	73.1	10

(continued)

Table 1
(continued)

Name	Retention time (min)	Precursor ion *m/z*	Product Ion *m/z*	Collision energy
Glutamate	9.1	246.1	128.1	10
Glutamate	9.1	246.1	156.1	6
Phenylalanine	9.28	192.1	73.1	14
Phenylalanine	9.28	218.1	100.1	8
Xylose	9.35	103.1	73.1	6
Xylose	9.35	217.1	129.1	8
Xylose	9.35	307.2	217.2	6
Arabinose	9.41	103	73.1	6
Arabinose	9.41	217.1	129.1	8
Arabinose	9.41	307.2	217.2	6
Ribose	9.56	277.1	189.1	6
Ribose	9.56	307.1	129.1	12
Ribose	9.56	307.1	277.2	6
Xylitol	9.84	277.2	189.1	6
Xylitol	9.84	307.2	217.2	6
Xylitol	9.84	319.2	229.2	8
Ribitol	10.06	205.1	117.1	8
Ribitol	10.06	217.1	129.1	8
Ribitol	10.06	319.2	129.1	10
DHAP	10.31	400.1	299.2	12
DHAP	10.31	400.1	315.2	12
DHAP	10.31	400.1	369.2	8
Aconitate	10.39	229.1	147.1	10
Aconitate	10.39	285.1	147.1	10
Aconitate	10.39	375.1	285.1	6
G3P	10.48	328.1	298.2	10
G3P	10.48	328.1	313.2	8
G3P	10.48	357.1	341.2	8
G3P	10.48	445.1	299.2	8
2-PG	10.7	387.1	299.1	14
2-PG	10.7	459.2	299.2	10
3-PG	11.07	357.1	299.1	10

Table 1
(continued)

Name	Retention time (min)	Precursor ion *m/z*	Product Ion *m/z*	Collision energy
3-PG	11.07	387.1	299.1	14
3-PG	11.07	459.2	299.1	10
Iso/citrate	11.26	273.1	183.1	8
Iso/citrate	11.26	347.1	257.1	8
Iso/citrate	11.26	363.1	273.2	6
Iso/citrate	11.26	465.2	347.2	8
Fructose	11.99	103.1	73.1	6
Fructose	11.99	217.1	129.1	8
Fructose	11.99	307.1	217.2	6
Mannose	12.24	160.1	73.1	10
Mannose	12.24	205.1	73.1	16
Mannose	12.24	319.1	129.1	8
Adenine	12.34	192.1	165.1	12
Adenine	12.34	264.1	192.2	12
Adenine	12.34	279.1	264.2	8
Glucose	12.44	160.1	73.1	12
Glucose	12.44	205.1	117.1	8
Glucose	12.44	319.1	129.1	8
E4P	12.718	211	181	10
E4P	12.718	341.1	225.1	8
E4P	12.718	357.1	341.1	10
Mannitol	12.97	205.1	117.1	8
Mannitol	12.97	307.1	217.1	6
Mannitol	12.97	319.1	129.1	8
Sorbitol	13.08	205.1	117.1	8
Sorbitol	13.08	307.2	217.2	6
Sorbitol	13.08	319.1	129.1	8
Tyrosine	13.28	179.1	73.1	12
Tyrosine	13.28	218.1	100.1	10
Tyrosine	13.28	280.1	192.2	8
Gluconate	13.916	292.2	130.1	12
Gluconate	13.916	319.2	129.1	8

(continued)

Table 1
(continued)

Name	Retention time (min)	Precursor ion *m/z*	Product Ion *m/z*	Collision energy
Gluconate	13.916	333.2	171.1	8
Pantothenate	13.93	247.2	157.2	6
Pantothenate	13.93	291.1	201.2	6
Pantothenate	13.93	420.2	159.1	10
Myo-Inositol	14.88	217.1	143.1	10
Myo-Inositol	14.88	305.1	217.1	10
Myo-Inositol	14.88	318.1	215.2	10
Ri5P	14.971	315.1	299.1	10
Ri5P	14.971	403.1	299.1	10
Ri5P	14.971	459.2	315.2	8
Ru5P	15.026	341.1	225.2	10
Ru5P	15.026	357.1	299.2	10
Ru5P	15.026	387.2	299.2	12
Tryptophan	15.96	202.1	73.1	14
Tryptophan	15.96	291.2	101.1	16
Tryptophan	15.96	291.2	129.1	10
F6P	16.45	217.1	129.1	8
F6P	16.45	357.1	299.2	10
F6P	16.45	459.2	315.2	6
G6P	16.53	357.1	299.2	10
G6P	16.53	387.1	299.2	22
G6P	16.53	471.2	387.3	8
6PG	17.13	333.2	143.1	14
6PG	17.13	387.1	299.1	20
6PG	17.13	471.2	387.2	8
S7P	17.75	315.1	299.1	10
S7P	17.75	357.1	299.2	10
S7P	17.75	471.2	387.2	10
Sucrose	18.11	217.1	129.1	10
Sucrose	18.11	361.1	169.1	10
Sucrose	18.11	361.1	243.2	8
Trehalose	18.57	204.1	189.1	8

Table 1
(continued)

Name	Retention time (min)	Precursor ion *m/z*	Product Ion *m/z*	Collision energy
Trehalose	18.57	217.1	129.1	6
Trehalose	18.57	271.1	243.2	6
Trehalose	18.57	361.1	271.2	6

Abbreviations: *α-KG* alpha-ketoglutarate, *PEP* phosphoenolpyruvate, *DHAP* dihydroxyacetone phosphate, *G3P* glycerol-3-phosphate, *2-PG* 2-phosphoglycerate, *3-PG* 3-phosphoglycerate, *E4P* erythrose-4-phosphate, *Ri5P* ribose-5-phosphate, *Ru5P* ribulose-5-phosphate, *F6P* fructose-6-phosphate, *G6P* glucose-6-phosphate, *6PG* 6-phosphogluconate, *S7P* sedoheptulose-7-phosphate.

3.2 Quenching and Washing

1. Cast the broth culture into the precooled SB quenching solution at a ratio of 1:6 and mix thoroughly (*see* **Note 3**).
2. Centrifuge the mixture at 10,000 × *g* for 30 s (*see* **Note 4**).
3. Discard the supernatant.
4. Wash the pellet one more time with 1.5 mL of cold SB (*see* **Note 5**).
5. Retain the cell pellet and discard the supernatant (*see* **Note 6**).

3.3 Metabolite Extraction

1. Add 900 μL of extract solution to each sample (*see* **Note** 7).
2. Vortex at 4°C for 1 h.
3. Centrifuge at 10,000 g for 5 min to collect the supernatant.
4. Recover the analytes (remove the solvent) using a cold centrifugal concentrator at 4°C (*see* **Note 8**).

3.4 Metabolite Derivatization

1. Add 50 μL of MeOX to each sample for methoximation (*see* **Note 9**).
2. Incubate at 37°C for 90 min in a heat block with gentle shaking.
3. Adding 50 μL of MSTFA for silylation.
4. Incubate at 37°C for 30 min in a heat block with gentle shaking.

3.5 GC–MS/MS Method

1. Set the GC–MS/MS parameters as follows: injection temperature (250°C); helium carrier gas flow (1 mL/min); transfer line (280°C) and ion source temperatures (300°C). Ions were generated by electron ionization (EI) at 70 eV and analyzed in positive mode. Argon was utilized as a collision gas.
2. Column temperature: 100°C for 2 min, increased by 15°C/min to 200°C (held for 4 min) and then increased by 20°C/min to 325°C (held for 5 min).
3. Perform tuning and calibration before analysis.

4. The injection volume is 1 μL. Each sample was injected twice with two different split modes, including 20:1 (v/v) for low concentration metabolites and 100:1 (v/v) for high concentration metabolites (*see* **Note 10**).
5. All metabolites are detected using the Selected Reaction Monitoring (SRM) transitions (Table 1).
6. For quantification, analyze different concentrations of metabolites (Table 1) in the same manner as samples to build the calibration curves (*see* **Note 11**).
7. Normalize metabolite peak areas with the ribitol peak area.
8. Normalize the concentration of metabolites with the CDW of each sample.

4 Notes

1. Prepare the solution in a capped glass vial. If MeOX does not dissolve well in pyridine, place the vial in a sonication water bath until the solution becomes clear. Do not shake the vial, as MeOX may adhere to the walls or cap, leading to inaccurate concentrations.
2. From the moment cells are harvested, perform all steps rapidly and keep samples on ice to minimize metabolite turnover.
3. Cold SB is the optimal quenching buffer for *Ustilago maydis*, as the cell culture is highly viscous due to extracellular carbohydrate production. Avoid using organic solvents for cell quenching, as they will cause cell accumulation. The culture viscosity also limits the use of fast filtering methods.
4. The cells need to be washed one more time. Avoid prolonged centrifugation, which makes cell resuspension difficult. Applying a strong force might cause metabolite leakage.
5. Washing more than two times increase metabolite leakage. [20]
6. If the samples are not extracted immediately, lyophilize the samples and store at −80°C.
7. This extraction method is recommended when you are just interested in low molecular weight hydrophilic compounds. If you are interested in harvesting lipids from the same samples, use the extraction method with Methanol: Chloroform: Water [20]
8. If samples are not derivatized immediately, lyophilize them and store at −80°C. After derivatization, analyze samples by GC–MS/MS without delay, as prolonged storage may compromise metabolite stability.

9. Ensure that all samples are completely free of water, as residual moisture interferes with derivatization and GC–MS/MS analysis.
10. Depending on the sample and the metabolites of interest, different split ratios can be applied. However, this method requires high gas consumption, but less sample preparation. Analyzing the same sample with different concentrations is another option to save gas.
11. The concentrations of metabolites depend on both the samples and the GC–MS/MS system. Proper testing prior to the main experiment is recommended.

Acknowledgments

This work was partially funded by the Deutsche Forschungsgemeinschaft (DFG, German Research Foundation) SFB1535 – Project ID 458090666 and under Germany's Excellence Strategy—Cluster of Excellence 2186 "The Fuel Science Center" – Project ID: 390 919 832. This work was also partially funded by the Bundesministerium für Bildung und Forschung (BMBF), Germany in the project "UpRePP- Upcycling of Regional Waste Steams for the Production of Plattform Chemicals" in the project cluster "Modellregion, BioRevierPLUS: InBio, Innovationscluster Integrierte Bioraffinerie, TP2" (Grant Number 031B1135B/031B1135BX).

References

1. Brefort T, Doehlemann G, Mendoza-Mendoza A et al. (2009) *Ustilago maydis* as a pathogen. Annu Rev Phytopathol 47(1):423–445. https://doi.org/10.1146/annurev-phyto-080508-081923
2. Bölker M (2001) *Ustilago maydis* – a valuable model system for the study of fungal dimorphism and virulence. Microbio 147(6):1395–1401. https://doi.org/10.1099/00221287-147-6-1395
3. Martinez-Espinoza AD, Garcia-Pedrajas MD, Gold SE (2002) The Ustilaginales as plant pests and model systems. Fungal Genet Biol 35(1):1–20. https://doi.org/10.1006/fgbi.2001.1301
4. Matei A, Doehlemann G (2016) Cell biology of corn smut disease—*Ustilago maydis* as a model for biotrophic interactions. Curr Opin Microbiol 34:60–66. https://doi.org/10.1016/j.mib.2016.07.020
5. Kämper J, Kahmann R, Bölker M et al. (2006) Insights from the genome of the biotrophic fungal plant pathogen *Ustilago maydis*. Nature 444(7115):97–101. https://doi.org/10.1038/nature05248
6. Münsterkötter M, Steinberg G (2007) The fungus *Ustilago maydis* and humans share disease-related proteins that are not found in *Saccharomyces cerevisiae*. BMC Genomics 8(1):473. https://doi.org/10.1186/1471-2164-8-473
7. Steinberg G, Perez-Martin J (2008) *Ustilago maydis*, a new fungal model system for cell biology. Trends Cell Biol 18(2):61–67. https://doi.org/10.1016/j.tcb.2007.11.008
8. Etxebeste O, Espeso EA (2016) Neurons show the path: tip-to-nucleus communication in filamentous fungal development and pathogenesis. FEMS Microbiol Rev 40(5):610–624. https://doi.org/10.1093/femsre/fuw021
9. Haag C, Klein T, Feldbrügge M (2019) ESCRT mutant analysis and imaging of ESCRT components in the model fungus *Ustilago maydis*. Methods Mol Biol 1998:251–272
10. Geiser E, Wiebach V, Wierckx N, Blank LM (2014) Prospecting the biodiversity of the

fungal family Ustilaginaceae for the production of value-added chemicals. Fungal Biol and Biotechnol 1(1):2. https://doi.org/10.1186/s40694-014-0002-y
11. Becker J, Liebal UW, Phan ANT et al. (2023) Renewable carbon sources to biochemicals and -fuels: contributions of the smut fungi Ustilaginaceae. Curr Opin Microbiol 79:102849. https://doi.org/10.1016/j.copbio.2022.102849
12. Khrunyk Y, Münch K, Schipper K et al. (2010) The use of FLP-mediated recombination for the functional analysis of an effector gene family in the biotrophic smut fungus *Ustilago maydis*. New Phytol 187(4):957–968. https://doi.org/10.1111/j.1469-8137.2010.03413.x
13. Schuster M, Schweizer G, Reissmann S et al. (2016) Genome editing in *Ustilago maydis* using the CRISPR–Cas system. Fungal Genet Biol 89:3–9. https://doi.org/10.1016/j.fgb.2015.09.001
14. Schirawski J, Mannhaupt G, Münch K et al. (2010) Pathogenicity determinants in smut fungi revealed by genome comparison. Science 330(6010):1546–1548. https://doi.org/10.1126/science.1195330
15. Laurie JD, Ali S, Linning R et al. (2012) Genome comparison of barley and maize smut fungi reveals targeted loss of RNA silencing components and species-specific presence of transposable elements. Plant Cell 24(5):1733–1745. https://doi.org/10.1105/tpc.112.097261
16. Doyle CE, Donaldson ME, Morrison EN et al. (2011) *Ustilago maydis* transcript features identified through full-length cDNA analysis. Mol Genet Genomics 286(2):143. https://doi.org/10.1007/s00438-011-0634-z
17. Donaldson ME, Ostrowski LA, Goulet KM et al. (2017) Transcriptome analysis of smut fungi reveals widespread intergenic transcription and conserved antisense transcript expression. BMC Genomics 18(1):340. https://doi.org/10.1186/s12864-017-3720-8
18. Böhmer M, Colby T, Böhmer C et al. (2007) Proteomic analysis of dimorphic transition in the phytopathogenic fungus *Ustilago maydis*. Proteomics 7(5):675–685. https://doi.org/10.1002/pmic.200600900
19. Martínez-Salgado JL, León-Ramírez CG, Pacheco AB et al. (2013) Analysis of the regulation of the *Ustilago maydis* proteome by dimorphism, pH or MAPK and GCN5 genes. J Proteomics 79:251–262. https://doi.org/10.1016/j.jprot.2012.12.022
20. Phan ANT, Blank LM (2020) GC-MS-based metabolomics for the smut fungus *Ustilago maydis*: a comprehensive method optimization to quantify intracellular metabolites. Front Mol Biosci 7:211. https://doi.org/10.3389/fmolb.2020.00211
21. Phan ANT, Prigolovkin L, Blank LM (2024) Unlocking the potentials of *Ustilago trichophora* for up-cycling polyurethane-derived monomer 1, 4-butanediol. Microb Biotechnol 17(3):e14384. https://doi.org/10.1111/1751-7915.14384

Check for updates

Chapter 38

ER and Cell Wall Stress Assay of *Ustilago maydis* Strains

Nora Marie Kühne and Kai Heimel

Abstract

Analysis of stress resistance of *Ustilago maydis* strains using drop plates is a standard method for the characterization of generated mutant strains. Inclusion of appropriate control strains, different stressor concentrations, and proper handling allows to perform these crucial analyses within less than a week.

Keywords *Ustilago maydis* , ER-stress, Cell wall-stress, Unfolded-protein response, Cell-wall integrity, Drop-plate assay

1 Introduction

During the pathogenic development of *U. maydis*, various stress response pathways are activated [1, 2]. The unfolded protein response (UPR) ensures ER homeostasis and thereby enables effector secretion. Thus, intact ER stress resistance is important for the virulence of *U. maydis* and can be examined in a simple drop plate assay. Here, tunicamycin (TM) or dithiothreitol (DTT) is routinely used to induce ER stress. TM inhibits the first step of N-glycosylation and thus leads to the accumulation of dysfunctional glycoproteins in the ER lumen, while DTT is a reducing agent inhibiting disulfide bond formation, which results in the accumulation of un-/misfolded proteins in both the ER and the cytosol. The cell wall integrity (CWI) pathway constantly monitors the proper functioning and dynamics of cell wall assembly and disassembly during fungal growth. Besides a plethora of alternatives, Congo Red and Calcofluor White are most commonly used to assay cell wall stress resistance. Both of them interfere with proper cell wall formation by binding to chitin and/or ß-glucans, inducing activation of the cell wall integrity (CWI) pathway.

An N. T. Phan et al. (eds.), *Ustilago maydis*, Methods in Molecular Biology, vol. 3050,
https://doi.org/10.1007/978-1-0716-5364-7_38,

2 Materials

2.1 Preparation of Stress Plates

1. YNB solid medium: 1.7 g Difco™ Yeast Nitrogen Base w/o Amino Acids and Ammonium Sulfate (BD) in 1 L dH_2O, adjust pH to 5.6 using 5 M NaOH and add 18 g Bacto™ Agar. Autoclave at 121 °C for 10 min. Store at RT.
2. 50% glucose in dH_2O. Sterile filtrate. Store at RT.
3. 20% ammonium sulfate in dH_2O. Sterile filtrate. Store at RT.
4. 1 M dithiothreitol (DTT) in dH_2O. Sterile filtrate. Store at −20°C.
5. 5 mg/mL tunicamycin (TM) in DMSO. Sterile filtrate. Store at −20°C.
6. 10 mM Congo Red in dH_2O. Sterile filtrate. Store at 4°C.
7. 5 mM Calcofluor White in dH_2O. Sterile filtrate. Store at 4°C.

2.2 Cultivation of *U. maydis* Cells

1. $YEPS_{light}$ liquid medium [3]: 10 g yeast extract, 4 g Bacto Peptone, 4 g sucrose in 1 L dH_2O. Autoclave at 121 °C for 10 min. Store at RT.
2. Sterile 1.5 mL microcentrifuge tubes and 96-well plate.
3. Sterile dH_2O.

2.3 Incubation and Documentation

1. Incubator for flasks and plates set to 28°C.
2. Camera.

3 Methods

3.1 Preparation of Stress Plates

1. Heat YNB solid medium in water bath or microwave until it is fully re-liquified and add:
   - glucose to a final concentration (v/v) of 1% (2 mL of 50% glucose/100 mL medium).
   - ammonium sulfate to a final concentration (v/v) of 0.2% (1 mL of 20% ammonium sulfate/100 mL medium) to obtain YNB +/+ solid medium (*see* **Note 1**).
2. Prepare 50 mL Falcon tubes with the appropriate amounts of TM/DTT/Congo Red/Calcofluor White (*see* Table 1 and **Note 2**) and add 25 mL or 50 mL of YNB +/+ solid medium to pour one round plate or one squared plate, respectively (*see* **Note 3**). Remember to pour a control plate of YNB +/+ without added stressor.

Table 1
Amounts and concentrations of stressors to be added to the medium per plate

Stressor	Stock	Final concentration	Volume for 25 mL YNB [µL]	Volume for 50 mL YNB [µL]
DTT	1 M	0.5 mM	12.5	25
		0.75 mM	18.75	37.5
		1.0 mM	25	50
TM	5 mg/mL	0.5 µg/mL	2.5	5
		1.0 µg/mL	5.0	10
		2.0 µg/mL	10	20
Congo Red	10 mM	100 µM	250	500
Calcofluor White	5 mM	50 µM	250	500

3.2 Preparation of *U. maydis* Cell Suspensions

1. Grow strains overnight in 10 mL $YEPS_{light}$ liquid medium at 28 °C under constant shaking. Include appropriate WT and Δ *cib1* [1, 4] derivatives as control.
2. In the next morning, dilute cultures in 10 mL $YEPS_{light}$ to an $OD_{600}=0.25$ and incubate for another 4 h at 28°C under constant shaking.
3. Harvest cells by centrifugation at 1200 × *g* in a microcentrifuge for 5 min in 1.5 mL microcentrifuge tubes. Carefully discard supernatant and adjust to an $OD_{600}=1$ in sterile water.
4. Prepare a 10× dilution series using sterile water in a 96-well plate resulting in five cell suspensions per strain with $OD_{600}=1$, 0.1, 0.01, 0.001, and 0.0001, respectively (*see* **Note 4**).

3.3 Drop Assay

1. Pipette 3.5 µL of each cell suspension onto each stress plate. Ideally, this is performed using a multichannel pipette and a dropping template beneath the plate to achieve an even distance between the drops. Let the drops dry, seal plates, and incubate at 28°C for 2 days.
2. Take pictures of the plates after incubation to compare growth of your mutant strains to WT and Δ*cib1* strains (*see* **Note 5**).

4 Notes

1. The plates should be prepared the same day they are being used. Usually, this is done during the 4 h incubation time of the diluted overnight culture.

2. The ER stress-inducing potency of TM and DTT often varies between different batches. If growth of the WT is fully or not at all inhibited in comparison to the YNB+/+ control plate, try to adjust concentrations of stressors accordingly. Use only freshly thawed 1 M DTT stock solution. Do not reuse after thawing.
3. On a round plate (ø 9.2 cm) up to six strains can be assayed, while on a square plate (12 × 12 cm) there is space for up to 16 strains.
4. Usually, the dilution series is prepared in a total volume of 200 μL to ensure smooth pipetting with the multichannel pipette.
5. If your mutant strains show altered ER stress susceptibility, analysis of UPR target gene expression in the presence/absence of ER stressors (e.g., *bip1*, *lhs1*, *cne1*, *ero1*) [1] is advisable, as this is often connected to constantly active UPR signaling.

References

1. Heimel K, Freitag J, Hampel M et al (2013) Crosstalk between the unfolded protein response and pathways that regulate pathogenic development in *Ustilago maydis*. Plant Cell 25(10):4262–4277. https://doi.org/10.1105/tpc.113.115899
2. Brefort T, Doehlemann G, Mendoza-Mendoza A et al (2009) *Ustilago maydis* as a pathogen. Annu Rev Phytopathol 47:423–445. https://doi.org/10.1146/annurev-phyto-080508-081923
3. Tsukuda T, Carleton S, Fotheringham S et al (1988) Isolation and characterization of an autonomously replicating sequence from *Ustilago maydis*. Mol Cell Biol 8(9):3703–3709
4. Heimel K, Scherer M, Schuler D et al (2010) The *Ustilago maydis* Clp1 protein orchestrates pheromone and b-dependent signaling pathways to coordinate the cell cycle and pathogenic development. Plant Cell 22(8):2908–2922. https://doi.org/10.1105/tpc.110.076265

Chapter 39

Simultaneous Saccharification and Fermentation with *Ustilago maydis* and Analytical Methods for Cell and Product Quantification

Wolfgang Laudensack, Jens Weiermüller and Roland Ulber

Abstract

Ustilago maydis is an emerging microbial host for the sustainable production of itaconic acid, an important platform chemical. Its robustness toward impurities and compatibility with renewable resources make it an attractive alternative to the industrial producer *Aspergillus terreus.* Here, a detailed protocol for converting brewer's spent grains (BSG), an exemplary lignocellulosic feedstock, into itaconic acid using *U. maydis* is presented. The workflow describes an extended valorization route, integrating pretreatment, fermentation, and analysis. The solid fraction of BSG undergoes organosolv pretreatment to enhance enzymatic accessibility, while the liquid fraction from pretreatment is recovered and detoxified for potential fermentation use. Nitrogen availability is assessed to account for its key role in triggering secondary metabolism. Simultaneous saccharification and fermentation (SSF) is then applied, enabling continuous sugar release and conversion into itaconic acid. The protocol further details analytical methods, including HPLC for substrates, products, and inhibitors, as well as fluorescence microscopy–based cell counting for biomass monitoring.

Keywords *Ustilago maydis*, Itaconic acid, Lignocellulosic biomass, Organosolv pretreatment, SSF, Biomass valorization, Nitrogen limitation, Fermentation monitoring, Cell quantification

1 Introduction

Besides its role as a model organism for plant pathogenicity, *Ustilago maydis* has gained attention for its ability to produce a variety of interesting compounds [1]. Among these, three products stand out for their high industrial potential: itaconic acid, cellobiose lipids, and mannosylerythritol lipids. All three are linked to secondary metabolism and typically triggered under nitrogen limitation [2, 3], but their production conditions differ. Mannosylerythritol lipids require a hydrophobic carbon source, while itaconic acid and cellobiose lipids are produced from water-soluble carbon sources. In addition, itaconic

An N. T. Phan et al. (eds.), *Ustilago maydis*, Methods in Molecular Biology, vol. 3050,
https://doi.org/10.1007/978-1-0716-5364-7_39,

acid production is favored at near-neutral pH, whereas cellobiose lipid formation is stimulated under acidic conditions.

Of these three products, itaconic acid production is the most developed in *U. maydis*, with processes demonstrated at scales up to 150 L [4]. By contrast, only a few reports of stirred tank fermentations exist for the production of cellobiose lipids, while no reports exist so far for mannosylerythritol lipids [3]. Industrially, itaconic acid is produced by *Aspergillus terreus*, but this filamentous fungus is highly sensitive to shear stress, local oxygen limitation, and impurities in the medium [5]. Recent efforts in genetic modifications have improved *U. maydis* competitiveness [6], and its better compatibility with media based on renewable resources offers an additional advantage over *Aspergillus terreus* [7].

A promising route for more sustainable chemical production is the use of lignocellulosic biomass as a carbon source. Such resources can be derived from agricultural residues, forestry byproducts, and processing wastes such as brewer's spent grains (BSG) [7], and offer a sustainable alternative to pure sugars. Simultaneous saccharification and fermentation (SSF) reduces product inhibition and investment cost by using a single vessel [8], which also reduces processing time and contamination risk. *U. maydis* is particularly well-suited for such processes due to a close pH optimum of hydrolysis enzymes and itaconic acid production.

In this protocol, an extended valorization route for lignocellulosic biomass, exemplary on BSG, into itaconic acid by *U. maydis*, presenting a holistic usage of the generated fractions during pretreatment of the solid fraction, is described (Fig. 1). The solid fraction undergoes organosolv pretreatment to enhance enzymatic accessibility, while the soluble supernatant generated during pretreatment is recovered and detoxified for potential fermentation use. Nitrogen availability of the combined streams is assessed, with addition considered where necessary. SSF is then applied to continuously convert both soluble and polymeric carbohydrate fractions into itaconic acid. In addition to the cultivation workflow, the protocol details the accompanying

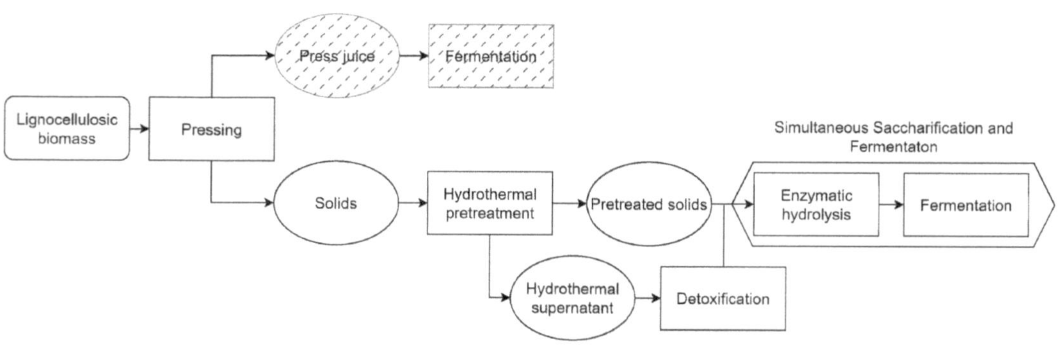

Fig. 1 Overview of workflow from raw lignocellulosic biomass to SSF

analytical methods, including HPLC for substrates, products, and inhibitors, as well as fluorescence-based cell counting for biomass quantification. While comprehensive, the individual methods and notes are broadly applicable and can be adapted to more streamlined processes or to other renewable resource-based feedstocks.

2 Materials

2.1 Processing of Lignocellulosic Biomass

1. SSF medium: 52 g BSG dry weight equivalent per 1 L of medium [7].
2. Screw press: Angel Juicer 8500 S (Angel Juicer Deutschland LUBA GmbH, Bad Homburg v.d.H., Germany) for liquid/solid separation.
3. Ethanol, technical grade.
4. High-pressure reactor: BR-500 (Berghof Products + Instruments GmbH, Eningen, Germany) for hydrothermal pretreatment.

2.2 Detoxification of Organosolv Supernatant

1. Rotary evaporator.
2. Anion-exchange resin: Purolite® A400, convert to OH^- form by washing three times with 5 M NaOH base, dry in a desiccator.
3. Zeolite Socony Mobil-5 ZSM-5 NH_4^+, dry in an incubator at 50°C.

2.3 Analysis of Organosolv Supernatant

1. Furan aldehyde HPLC: HPLC system with UV or photodiode array detector, Eurospher II 100–5 C18 column (250 × 3 mm) (Knauer GmbH, Berlin, Germany).
2. Eluent A: 0.1% (v/v) formic acid, prepared by adding 1 mL ≥99% formic acid to 1 L of double-deionized water, mixing thoroughly.
3. Eluent B: 0.1% (v/v) formic acid in methanol, prepared by adding 1 mL ≥99% formic acid to 1 L of methanol (HPLC grade), mixing thoroughly.
4. HMF and Furfural stock solutions: 500 $mg \cdot L^{-1}$ each (analytical grade), dilute working standards to 1–50 $mg \cdot L^{-1}$ for calibration, store stock solutions at −20°C, prepare working standards fresh.

2.4 Nitrogen Assessment

1. Ammonia assay: UV/Vis spectrometer.
2. Salicylate reagent: 0.5 g sodium nitroprusside dihydrate, 32 g sodium salicylate, 40 g trisodium phosphate dodecahydrate, fill up to 1 L with double-deionized water.

3. Hypochlorite reagent: dilute sodium hypochlorite solution to 0.25% (w/v) with double-deionized water, prepare fresh before use.
4. NH_4Cl stock solution: 1.07 g·L^{-1} NH_4Cl (20 mM NH_4^+), dilute working standards to 0.2–2 mM for calibration, store stock solutions at −20°C, prepare working standards fresh.
5. Amino acid HPLC: HPLC system with photodiode array detector, Resolve C18 column (5 μm, 150 × 3.9 mm) (Waters Corporation, Milford, USA), SecurityGuard C18 cartridge, 4 × 3.0 mm ID (Phenomenex, Torrance, USA).
6. Borate buffer: 24.72 g boric acid, adjust to pH 9.5 with 10 M NaOH, fill to 1 L with deionized water.
7. OPA reagent: dissolve 5.4 g ortho-phthaldialdehyde (OPA) in 100 mL ethanol, add 4 mL of 2–mercaptoethanol, fill to 1 L with 0.4 M borate buffer, allow to stand for 24 h before use (*see* **Note 1**).
8. Eluent A: 2.05 g sodium acetate, 3.45 g $NaH_2PO_4 \cdot H_2O$, adjust to pH 7 with 10 M NaOH, fill to 1 L with double-deionized water, add 21 mL tetrahydrofuran and 21 mL methanol (HPLC grade), mix thoroughly.
9. Eluent B: 50% (v/v) methanol (HPLC grade), mix thoroughly.
10. Sample preparation agent: ice-cold methanol.
11. Amino acid stock solution: amino acid standard AAS 18 and amino acid standard A 9781 (Sigma-Aldrich Inc., St. Louis, USA), dilute working standards to 0.125–1.25 mM for calibration, store stock solutions at −20°C, prepare working standards fresh.

2.5 Strain and Media Preparation

1. Strain: *Ustilago maydis* MB215 Δcyp3 ΔP_{ria1}::P_{etef}, kindly provided by Prof. Blank (RWTH Aachen, Germany) [9] (*see* **Note 2**).
2. Cryo cultures: stored at −80°C with 40% (v/v) glycerol.
3. Preculture media (YEPG): 10 g yeast extract, 20 g peptone, 20 g glucose, 167 mg streptomycin sulfate, fill to 1 L with deionized water; autoclave yeast extract + peptone and glucose solutions separately, sterile-filter streptomycin sulfate solution.
4. Nutrient additions: salt solution stock based on modified Tabuchi medium [10], aiming for a final concentration of 0.5 g KH_2PO_4, 0.2 g $MgSO_4 \cdot 7H_2O$, and 0.01 g $FeSO_4 \cdot 7H_2O$ per L, with adjusted NH_4Cl concentration in the fermentation media (*see* **Note 3**).
5. Vitamin solution: 0.05 g D-biotin, 1 g D-calcium pantothenate, 1 g nicotinic acid, 25 g myo-inositol, 1 g thiamine hydrochloride, 1 g pyridoxol hydrochloride, 0.2 g para-aminobenzoic

acid, fill to 1 L with deionized water [10]; sterile-filter and store at −20°C.

6. Trace element solution: 1.5 g EDTA, 0.45 g $ZnSO_4 \cdot 7H_2O$, 0.10 g $MnCl_2 \cdot 4H_2O$, 0.03 g $CoCl_2 \cdot 6H_2O$, 0.03 g $CuSO_4 \cdot 5H_2O$, 0.04 g $Na_2MoO_4 \cdot 2H_2O$, 0.45 g $CaCl_2 \cdot 2H_2O$, 0.3 g $FeSO_4 \cdot 7H_2O$, 0.10 g H_3BO_3, 0.01 g KI, fill to 1 L with deionized water [10]; sterile-filter and store at −20°C.

2.6 Reactor Equipment

1. Bioreactor system: BIOSTAT® Qplus controlled bioreactor system (Sartorius AG, Göttingen, Germany) with pH and DO sensor.
2. Process control solutions: 0.5 M NaOH and Rhodorsil Oils 47; sterilize by autoclaving.
3. Enzyme for SSF: Ultraflo® Max (Novozymes A/S, Bagsværd, Denmark) (*see* **Note 4**); sterile-filter before use.

2.7 Cell Counting

1. Counting chamber: Neubauer improved cell counting chamber.
2. Confocal imaging microscope: Leica TCS SP5 II (Leica Microsystems GmbH, Wetzlar, Germany).
3. Nile red stock solution: 0.025 g Nile red, dissolved in 50 mL acetone.

2.8 Analytics for Substrate and Product Concentrations

1. Substrate-product HPLC: HPLC system with RI detector and Bio-Rad Aminex HPX-87H column (300 × 7.8 mm, Hercules, USA).
2. 2.5 mM sulfuric acid eluent: 0.14 mL sulfuric acid (≥95% purity), fill to 1 L with double deionized water, mix thoroughly.
3. Substrate and product stock solutions: 50 $g \cdot L^{-1}$ of glucose, xylose, arabinose, itaconic acid, ethanol, acetic acid, and formic acid (analytical grade) in double-deionized water; mix stocks to create working standards of 0.5–10 $g \cdot L^{-1}$ (*see* **Note 5**); store stock solutions at −20°C, prepare working standards fresh.

3 Methods

3.1 Processing of Lignocellulosic Biomass

1. Wash the lignocellulosic biomass three times by adding deionized water at threefold the weight of the biomass (*see* **Note 6**).
2. Separate solids and liquid by pressing with a screw press (*see* **Note 7**). Store pressed solids at −20°C until use.
3. Add 50% (w/w) ethanol solvent to the solids at a solid-liquid ratio of 1:10 (w/w) (*see* **Note 8**). Perform organosolv

pretreatment in a high-pressure reactor under 5 bar N_2 atmosphere for a holding time of 15 min at 180°C and 600 rpm (*see* **Note 9**).

4. Separate solids and hydrothermal supernatant by centrifugation (10 min 4200 g) (*see* **Note 10**). Wash solids twice with 50% (w/w) ethanol and twice with deionized water (*see* **Note 11**).

3.2 Detoxification of the Hydrothermal Supernatant

1. Concentrate hydrothermal supernatant in a rotary evaporator at 40°C and 60 mbar for 180 min (*see* **Note 12**). Separate concentrated liquid from precipitated solids.
2. Analyze concentrated supernatant for furfural and HMF (furan aldehyde HPLC), and formic acid, acetic acid, and ethanol (substrate-product HPLC) to determine required detoxification steps (*see* **Note 13**).
3. For acid removal: Add 100 g of dried Purolite® A400 per 1 L of concentrated supernatant, mix for 30 min, and remove resin by filtration.
4. For furan aldehydes removal: Add 100 g of dried zeolite ZSM-5 per 1 L of concentrated supernatant, mix for 30 min, and remove zeolite by filtration.
5. Adjust pH to 5.0.

3.3 Furan Aldehyde HPLC

1. Filter samples through a 0.22 μm PTFE filter.
2. Set column temperature to 30°C and flow to 0.5 mL·min^{-1}. Apply gradient flow: 0–10 min, 20% B (isocratic), 10–11 min, 20 -> 90% B (linear), 11–17 min, 90% B (isocratic), 17–18 min 90 -> 20% B (linear), 18–25 min, 20% B (isocratic).
3. Inject 10 μL and detect at 280 nm. Dilute samples with double-deionized water if necessary (*see* **Note 14**).

3.4 Preparation of Media and Nitrogen Assessment

1. Combine solids with supernatant. Take a sample for ammonia assay and amino acid HPLC (*see* **Note 15**).
2. Ammonia assay: Add 50 μL sample to 1000 μL salicylate reagent and 250 μL hypochlorite reagent. Mix thoroughly and wait 20 min. Measure absorbance at 685 nm (*see* **Note 16**). Dilute samples with double-deionized water if necessary.
3. Amino acid HPLC: Precipitate proteins by adding four volumes of ice-cold methanol, store overnight at −20°C. Centrifuge the sample for 10 min at 8600 g.
4. Mix the protein-free supernatant with an equal volume of borate buffer. Dilute samples further with borate buffer if necessary (*see* **Note 17**).

5. Set column temperature to 30°C and flow to 1 mL·min^{-1}. Apply gradient flow: 0–50 min, 0 -> 100% B (linear), 50–55 min, 100% B (isocratic), 55–60 min, 100 -> 0% B (linear), 60–67 min, 0% B.
6. Place 80 μL sample in HPLC vial. Program an addition of 50 μL of OPA reagent to the sample, a mixing of three times with a volume of 50 μL, and an injection of 40 μL after 1 min. Detect amino acids at 230 nm.
7. Evaluate accessible nitrogen and determine if an additional nitrogen source is needed (*see* **Note 18**).

3.5 Cultivation Preparation

1. Autoclave a 500 mL baffled Erlenmeyer flask and add 150 mL of YEPG media.
2. Inoculate with 1.5 mL cryo culture.
3. Incubate at 120 rpm and 30°C overnight.
4. Prepare inoculum for main culture by centrifugation (2 min, 4200 g) to separate cells from media.

3.6 Simultaneous Saccharification and Fermentation (SSF) in Reactor

1. Adjust slurry to 52 g·L^{-1} solids. Fill slightly below the target volume to allow for later additions. Add the slurry to the reactor and autoclave.
2. Add additional nutrients (*see* **Note 19**) and 0.1% (v/v) vitamin and trace elements solutions.
3. Heat reactor to 50°C, stir at 400 rpm, adjust pH to 5.0, then add 0.1 $g_{enzymes} \cdot g_{solids}^{-1}$ of Ultraflo® Max.
4. Let enzymatic hydrolysis run for 24 h.
5. Adjust the temperature to 30°C, calibrate the oxygen sensor (*see* **Note 20**), set aeration to 1 vvm, and adjust pH to 6.3.
6. Resuspend the preculture cell pellet in 10 mL SSF media and inoculate the reactor at an OD_{600} of 0.5.
7. Regulate stirring rate between 400 and 800 rpm to keep pO_2 const. at 80% (*see* **Note 21**). Add antifoam as needed (*see* **Note 22**).
8. Sample using a 2.10 × 100 mm syringe (*see* **Note 23**). Save aliquots for cell counting, centrifuge the rest at 16,873 g for 2 min, and filter through a 0.22 μm PA filter. Store samples at −20°C before analysis.
9. Stop fermentation when base addition and agitation speed decrease (*see* **Note 24**).

3.7 Cell Counting in Solid-Loaded Media (Fluorescence)

1. Centrifuge samples, wash pellets three times with 0.9% NaCl, and resuspend in 0.9% NaCl. Stain with 0.2 μL of Nile red stock per 1 mL sample. Wash three more times and resuspend in 0.9% NaCl + 50% (v/v) glycerol [11].

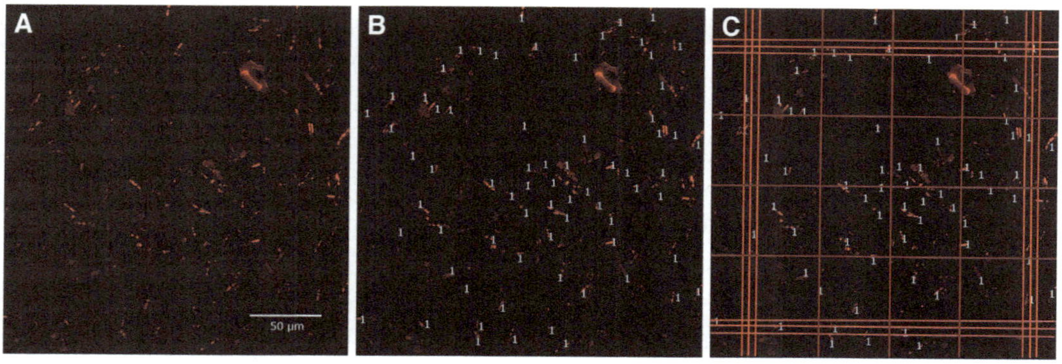

Fig. 2 Cell counting of fluorescence microscopy images from SSF with *U. maydis*. Original image (**A**), marked cells (**B**), and help grid over marked cells (**C**)

2. Add 10 µL to the Neubauer improved counting chamber (*see* **Note 25**).
3. Perform CLSM imaging using a water immersion objective (63x, 0.9 NA). Excite with an argon laser at 488 nm (15% laser power, 15% AOTF), detect at 570 and 630 nm. Mark and count cells on images according to standard counting rules (*see* **Note 26**) (Fig. 2).

3.8 Substrate-Product HPLC

1. Set column temperature to 80°C and flow to 0.6 mL·min^{-1}.
2. Inject 20 µL and run analysis for 25 min (*see* **Note 27**). Dilute samples with double-deionized water if necessary (*see* **Note 28**).

4 Notes

1. OPA reagent can be stored for one week. To prolong storage up to four weeks, add 20 µL of 2-mercaptoethanol weekly.
2. Genetically modified strains of *U. maydis* specialized for the production of a single compound are generally recommended over the wild type. In several strains, the formation of filamentous cells has been knocked out [9].
3. Either sterile filter a solution containing all components or autoclave separate solutions of KH_2PO_4, $MgSO_4{\cdot}7H_2O$, and NH_4Cl. Solutions of $FeSO_4{\cdot}7H_2O$ always need to be sterilized by filtration.
4. Optimized conditions for enzymatic hydrolysis of other types of lignocellulosic biomass might vary depending on biomass type and enzyme.

5. Acids can be combined into one stock solution, and the same applies to sugars. Formic and acetic acid are typical products of hydrothermal pretreatment and can be quantified with this HPLC method. Additional organic acids that may be present due to by-product formation during fermentation are malic acid, succinic acid, and aconitic acid. These acids can also be quantified with this HPLC method.
6. Washing removes residual soluble sugars from the solids, thereby reducing the formation of hydrothermal degradation products during pretreatment. It is only necessary if residual sugars are present, which can be verified by HPLC analysis of the wash water. If the biomass contains a sugar-rich juice with fermentation potential, pressing the biomass before washing can recover this fraction separately [12].
7. Other types of presses can be used. A screw press has the advantage of mechanically breaking up the biomass, thereby increasing the accessible surface and improving enzymatic hydrolysis yields [13].
8. When calculating ethanol addition, account for the water content in the wet biomass.
9. The goal of hydrothermal pretreatment is to disrupt the complex lignocellulosic structure to enhance enzymatic digestibility and to remove lignin [14]. Compared to liquid hot water pretreatment, organosolv achieves higher lignin solubilization and produces fewer hydrothermal degradation products [12].
10. The supernatant for the organosolv pretreatment contains parts of the hemicellulose, residual starch, and proteins [13], which can be exploited fermentatively. However, the benefit of using this fraction should be assessed carefully and may be optional.
11. Washing with ethanol removes residual hydrothermal degradation products from the solids, while washing with water removes residual ethanol. Perform sufficient washing steps to ensure complete ethanol removal.
12. Concentration in a rotary evaporator removes most of the ethanol and volatile hydrothermal degradation products and also precipitates dissolved lignin.
13. Detoxification of hydrothermal degradation products is recommended if inhibitory compounds are present at high levels [7]. Aim for a combined concentration of formic and acetic acid below $1\ g{\cdot}L^{-1}$ and furan aldehydes below $0.05\ g{\cdot}L^{-1}$ in the final medium. Tolerance may vary with different strains of *Ustilago maydis* and other *Ustilaginaceae* [15]. If concentrations are already below these thresholds,

detoxification may not be necessary. Analysis after each step is recommended.

14. Due to method sensitivity, high dilution factors may be required. Prepare serial dilutions in tenfold steps and analyze all to ensure accurate quantification.
15. Nitrogen-limited conditions are necessary to trigger itaconic acid production in *U. maydis* [2], but a nitrogen source is also required for biomass formation. Therefore, an assessment of whether a nitrogen source should be added to a renewable resource-based medium is needed. An exact nitrogen threshold for triggering itaconic acid production has not been reported and may differ depending on the type and biological value of the nitrogen source. Analysis of ammonia and free amino acids in the hydrolysate can provide valuable guidance. If both of these readily available nitrogen sources are absent or only present in trace amounts ($< 0.1\ g{\cdot}L^{-1}$), supplementation with ammonium chloride is likely required. Conversely, if significant levels of ammonia or amino acids are present ($> 1\ g{\cdot}L^{-1}$), additional nitrogen might prevent triggering of itaconic acid production. Testing growth and itaconic acid formation in a hydrolysate with and without added nitrogen prior to SSF is recommended.
16. Alternatively, the ammonia assay can be performed on a plate reader using one-quarter of the volumes for sample, salicylate reagent, and hypochlorite reagent. Measuring absorbance at 595 nm is also possible.
17. Samples should be adjusted to at least pH 10 before adding the OPA reagent.
18. If ammonium chloride supplementation is considered necessary, adjust the concentration to the expected sugar content from enzymatic hydrolysis. Following the ratio used in the modified Tabuchi medium of 62.5 g sugar per g NH_4Cl is recommended.
19. For shake flask fermentations, MES buffer should be added to media based on renewable resources, as only small amounts of itaconic acid may be formed before all carbon sources are consumed. In reactors with pH control, MES buffer is not strictly necessary, but its addition can serve as an internal standard on HPLC to track volume changes due to base addition or aeration-induced evaporation. Addition of streptomycin can also be considered to prevent contamination when using renewable resource-based media, where longer lag-phases of *U. maydis* are observed.
20. Determination of the k_La value under the employed reactor conditions before inoculation is recommended. Evaporation losses due to aeration can be minimized by passing inlet air

through a water-filled gas washer or by using an aeration rate of 1 vvm.

21. Oxygen availability strongly influences itaconic acid production in *U. maydis*. Most studies perform fermentations at high aeration rates and dissolved oxygen tensions between 30%–80%. In contrast, a recent study showed that fermenting at a constantly low agitation speed in modified Tabuchi medium increased itaconic acid yield [16], suggesting that reduced oxygen availability can be favorable. However, in renewable resource–based media, oxygen transfer may already be lower due to higher viscosity or solids content. In such cases, agitation speed should not be set too low, as insufficient oxygen may inhibit growth.
22. Renewable resource-based media tend to foam more than defined media. Early addition of antifoam and lower initial agitation speeds can minimize biomass losses due to foaming. Lower agitation speeds may also delay the formation of filamentous cells.
23. When fermenting with high solid loads, use needles with the maximal available diameter for sampling. Pointing the needle towards the reactor wall can help to avoid solid accumulation on the tip. Clogging should improve over time as solids degrade.
24. Fermentation can be considered finished if base addition has stopped for at least 3 h.
25. Solids elevate OD readings. At very high solid loadings, as in SSF, these effects might not be constant due to inhomogeneous sampling of solids and are subject to change due to solid degradation. This also interferes with cell dry weight determination, making offsets unreliable. Cell counting can be used as an alternative to assess biomass formation. Distinguishing *U. maydis* cells and solids can be enhanced by fluorescence staining and microscopy.
26. At optimal contrast for distinguishing cells from background biomass, the counting chamber rulings may not be visible. To overcome this, apply 10 μL of Nile Red stock solution to the chamber and adjust contrast until the rulings become visible. Capture an image of the rulings and construct a transparent help grid in graphical software. This overlay can then be placed above fluorescence images to facilitate accurate cell counting.
27. If high quantities of furan aldehydes are present, the HPLC run time may need to be increased.
28. In complex renewable resource-based media, several compounds may co-elute with itaconic acid. In such cases,

confirmatory analysis by HPLC–MS is recommended. Samples can be analyzed under similar chromatographic conditions as for substrate-product HPLC (BioRad Aminex HPX-87 H, eluent: 0.1% formic acid, 0.6 mL·min^{-1}, 80°C column temperature) with a PDA detector in parallel with MS detection. The MS is operated in negative electrospray ionization (ESI–) mode. Itaconic acid is identified by its precursor ion [M–H] at m/z 129 and the characteristic MRM transition m/z 129 → 85 (collision energy 12 eV, cone voltage 20 V).

References

1. Wierckx N, Miebach K, Ihling N et al (2021) Perspectives for the application of Ustilaginaceae as biotech cell factories. Essays Biochem 65:365–379. https://doi.org/10.1042/EBC20200141
2. Klement T, Milker S, Jäger G et al (2012) Biomass pretreatment affects Ustilago maydis in producing itaconic acid. Microb Cell Fact 11:43. https://doi.org/10.1186/1475-2859-11-43
3. Münßinger S, Beck A, Oraby A et al (2024) Past, present and future of glycolipids from Ustilaginaceae – a review on cellobiose lipids and mannosylerythritol lipids. J Surfact & Detergents 27:647–689. https://doi.org/10.1002/jsde.12764
4. Helm T, Stausberg T, Previati M et al (2024) Itaconate production from crude substrates with U. maydis: scale-up of an industrially relevant bioprocess. Microb Cell Fact 23:29. https://doi.org/10.1186/s12934-024-02295-3
5. Gyamerah M (1995) Factors affecting the growth form of Aspergillus terreus NRRL 1960 in relation to itaconic acid fermentation. Appl Microbiol Biotechnol 44:356–361. https://doi.org/10.1007/BF00169929
6. Becker J, Hosseinpour Tehrani H, Ernst P et al (2021) An Optimized Ustilago maydis for Itaconic Acid Production at Maximal Theoretical Yield. JoF 7:20. https://doi.org/10.3390/jof7010020
7. Weiermüller J, Akermann A, Laudensack W et al (2021) Brewers' spent grain as carbon source for itaconate production with engineered Ustilago maydis. Bioresour Technol 336:125262. https://doi.org/10.1016/j.biortech.2021.125262
8. Olofsson K, Bertilsson M, Lidén G (2008) A short review on SSF - an interesting process option for ethanol production from lignocellulosic feedstocks. Biotechnol Biofuels 1:7. https://doi.org/10.1186/1754-6834-1-7
9. Hosseinpour Tehrani H, Becker J, Bator I et al (2019) Integrated strain- and process design enable production of 220 g L-1 itaconic acid with Ustilago maydis. Biotechnol Biofuels 12:263. https://doi.org/10.1186/s13068-019-1605-6
10. Geiser E, Wiebach V, Wierckx N et al (2014) Prospecting the biodiversity of the fungal family Ustilaginaceae for the production of value-added chemicals. Fungal Biol Biotechnol 1:2. https://doi.org/10.1186/s40694-014-0002-y
11. Maassen N, Panakova M, Wierckx N et al (2014) Influence of carbon and nitrogen concentration on itaconic acid production by the smut fungus Ustilago maydis. Eng Life Sci 14:129–134. https://doi.org/10.1002/elsc.201300043
12. Akermann A, Weiermüller J, Ulber R (2019) Aufbau eines Bioraffineriekonzeptes für Biertreber mit vorgelagertem Pressschritt. Chemie Ingenieur Technik 91:1606–1614. https://doi.org/10.1002/cite.201900017
13. Weiermüller J, Akermann A, Sieker T et al (2020) Bioraffinerien auf basis schwach verholzter biomasse. Chemie Ingenieur Technik 92:1711–1722. https://doi.org/10.1002/cite.202000070
14. Mansfield SD, Mooney C, Saddler JN (1999) Substrate and enzyme characteristics that limit cellulose hydrolysis. Biotechnol Prog 15:804–816. https://doi.org/10.1021/bp9900864
15. Ullmann L, Phan ANT, Kaplan DKP et al (2021) Ustilaginaceae biocatalyst for co-metabolism of CO2-derived substrates toward carbon-neutral itaconate production. J Fungi (Basel) 7(2):98. https://doi.org/10.3390/jof7020098
16. Volkmar M, Laudensack W, Bartzack F et al (2025) Low oxygen availability increases itaconate production by *Ustilago maydis*. Biotechnol Bioeng 122(11):3007–3017. https://doi.org/10.1002/bit.70035

Check for updates

Chapter 40

BN-PAGE of *Ustilago maydis* Respiratory Complexes

Lucero Romero-Aguilar, Guadalupe Guerra Sánchez, Héctor Vázquez Meza and Juan Pablo Pardo

Abstract

Isolated mitochondria are valuable for the study of the respiratory chain, protein complex assembly, and the activities of the Krebs cycle enzymes, among others. Here, we present a protocol for the isolation of functional *U. maydis* mitochondria, which involves mechanically disrupting the cell wall with glass beads and performing two centrifugations. The first step is carried out at low speed to remove intact cells and cell debris, followed by a high-speed centrifugation to obtain mitochondria. For the separation of respiratory complexes and supercomplexes, we describe the solubilization of protein complexes from mitochondria with digitonin and their separation by blue native polyacrylamide gel electrophoresis (BN-PAGE), followed by the in-gel activity of respiratory complexes.

Keywords Respiratory complexes, Mitochondria isolation, Basidiomycete, Blue native PAGE

1 Introduction

1.1 The Mitochondrial Respiratory Chain of Ustilago maydis

The inner mitochondrial membrane of *U. maydis* hosts the electron transport chain (ETC), a series of multi-subunit protein complexes that sequentially transfer electrons from NADH or succinate to molecular oxygen, generating a proton motive force that is used for the synthesis of ATP [1]. *U. maydis* mitochondria contain the four classic respiratory complexes (I, II, III, IV), the ATP synthase (complex V), and other membrane-bound oxidoreductases such as the alternative oxidase, at least two type 2 NADH dehydrogenases, and the glycerol-3-phosphate dehydrogenase [2–6]. Complex I (NADH:ubiquinone oxidoreductase) serves as the primary entry point for electrons derived from the oxidation of matrix NADH, a major product of the tricarboxylic acid (TCA) cycle. In *U. maydis*, complex I is a large, proton-pumping enzyme, contributing significantly to the electrochemical gradient across the inner membrane [7, 8]. Complex II (succinate dehydrogenase) plays a dual role as both an

An N. T. Phan et al. (eds.), *Ustilago maydis*, Methods in Molecular Biology, vol. 3050,
https://doi.org/10.1007/978-1-0716-5364-7_40,

integral component of the ETC and a key enzyme of the TCA cycle, catalyzing the oxidation of succinate to fumarate. Electrons from this reaction are passed directly to ubiquinone in the membrane [9, 10]. Complex III (ubiquinol:cytochrome c oxidoreductase), also known as the cytochrome bc1 complex, accepts electrons from the ubiquinol pool and transfers them to cytochrome c [11]. This complex is a major proton pump, contributing to the proton motive force through the Q-cycle mechanism [11, 12]. In *U. maydis*, Complex III is a critical component of the classical respiratory pathway, and its inhibition by fungicides (e.g., carboxin) highlights its vulnerability and importance as a target for disease control [13]. Complex IV (cytochrome c oxidase) is the terminal enzyme of the classical ETC, receiving electrons from cytochrome c and reducing molecular oxygen to water. This complex also pumps protons, further contributing to the proton gradient. Complex V (ATP synthase), or F1Fo-ATP synthase, is the molecular engine that converts the energy stored in the proton motive force into ATP. Protons flowing back into the matrix through the Fo portion drive the rotation of the F1 portion, leading to ATP synthesis.

A distinguishing feature of fungal mitochondria, including those of *U. maydis*, is the presence of alternative respiratory pathways that provide metabolic flexibility, particularly under stressful conditions or when the classical cytochrome pathway is compromised [2]. The alternative NADH dehydrogenases (NDHs or type 2) in *U. maydis* represent a family of non-proton-pumping enzymes that oxidize NADH (and sometimes NADPH) and transfer electrons directly to ubiquinone, bypassing complex I. While they do not contribute to the proton gradient, they are important for the re-oxidation of cytosolic (glycolytic) or matrix (TCA) NADH [14–16]. The alternative oxidase (AOX) directly oxidizes ubiquinol and reduces oxygen to water, bypassing both complexes III and IV [17]. AOX does not pump protons, meaning that electron flow through this pathway generates less ATP per electron compared to the classical cytochrome pathway [18]. It has been speculated that by providing an alternative electron sink, AOX prevents the over-reduction of the ubiquinone pool and subsequent ROS production.

1.2 Structure and Function of Mitochondrial Respiratory Supercomplexes

While the individual complexes are capable of independent function, evidence over the past two decades has revealed their remarkable capacity to associate into larger, higher-order structures known as respiratory supercomplexes (SCs).

These supercomplexes might confer several advantages: (1) One of the most compelling proposed functions is the concept of "substrate channeling." By bringing CI and $CIII_2$ into proximity, the supercomplex could create a localized pool of ubiquinone, facilitating direct transfer of electrons from CI to $CIII_2$ without the need for extensive diffusion of ubiquinone within the membrane [19]. This "channeling" could reduce the diffusion distance for both ubiquinone and cytochrome *c* (between CIII and CIV), thereby accelerating the overall electron transfer rate and increasing the efficiency of OXPHOS. Studies have shown that supercomplex formation can indeed decrease cytochrome *c* diffusion distance and enhance electron transfer [20–22]. (2) It has also been proposed that the tightly organized architecture of supercomplexes minimizes electron leakage by ensuring a more efficient and directed electron flow, thereby reducing reactive oxygen species generation [21, 23]. (3) Supercomplex formation has been implicated in enhancing the stability of the individual respiratory complexes, particularly Complex I [24]. (4) The large size and specific interactions of supercomplexes and the ATP synthase dimer have been proposed to play a role in shaping the intricate folds of the inner mitochondrial membrane, known as cristae [25].

1.3 Blue Native Polyacrylamide Gel Electrophoresis (BN-PAGE)

BN-PAGE is a non-denaturing electrophoretic technique used to separate intact protein complexes based on their size, shape, and charge [26]. Unlike denaturing methods such as SDS-PAGE, BN-PAGE preserves the native structure and functionality of protein complexes, making it invaluable for studying protein–protein interactions, complex assembly, and enzymatic activities [27]. A key distinguishing feature of BN-PAGE is the use of Coomassie brilliant blue G-250 dye [26, 28]. This anionic dye non-covalently binds to the surface of proteins, imparting a uniform negative charge to the protein complexes [26, 28]. This negative charge is essential for their migration toward the anode (positive electrode) in the electric field.

Next, we will describe a protocol that facilitates the rapid isolation of *U. maydis* mitochondria based on glass beads rupture of the cell wall, followed by the separation of respiratory complexes and supercomplexes by BN-PAGE.

2 Materials

2.1 Solutions for Mitochondria Isolation

1. **Washing buffer, pH 7.4:** dissolve 113 g sucrose, 0.74 g EDTA, 0.38 g EGTA, 2.72 g KH_2PO_4, and 1.21 g Trizma base-HCl, in 900 mL ddH_2O. Adjust the pH to 7.4 and bring the volume to 1 L with ddH_2O.
2. **Lysis buffer, pH 7.4:** dissolve 113 g sucrose, 0.74 g EDTA, 0.38 g EGTA, 2.72 g KH_2PO_4, and 1.21 g Trizma base-HCl, in 900 mL ddH_2O. Adjust the pH to 7.4 and bring the volume to 1 L with ddH_2O. Before use, add 2 g BSA, 35.7 μL 2-mercaptoethanol 14 M, and protease inhibitors cocktail.
3. **1 M PMSF (protease inhibitor):** dissolve 0.174 g of phenylmethylsulfonyl fluoride in 1 mL of dimethyl sulfoxide.

4. **Protease inhibitor cocktail** (Sigma-Aldrich S8820-SIGMAFAST™, St. Louis, MO, USA): prepare a 50× stock solution by dissolving one tablet in 2 mL of water.

2.2 Solutions for Mitochondrial Membrane Solubilization

1. **Digitonin 10%:** dissolve 0.1 g in 1 mL (final volume) of ddH_2O.
2. **Solubilization buffer:** weigh 9.8 g of aminocaproic acid, 1 g of Bis-Tris/HCl, and dissolve in 80 mL of ddH_2O. Adjust the pH to 7.0 and bring the volume to 100 mL. Store at 4°C.

2.3 Solutions for BN-PAGE

1. **30% acrylamide:bisacrylamide (29/1):** weigh 29 g acrylamide and 1 g bisacrylamide. Dissolve in 80 mL of ddH_2O and complete the volume to 100 mL (*see* **Notes 1 and 2**).
2. **10% ammonium persulfate (APS):** dissolve 0.1 g ammonium persulfate in 1 mL of ddH_2O.
3. **1 M aminocaproic acid, pH 7.0:** dissolve 13.1 g aminocaproic acid in 80 mL of ddH_2O. Adjust the pH to 7.0 with NaOH and complete the volume to 100 mL.
4. **1 M Bis-Tris, pH 7.0:** dissolve 20.9 g Bis-Tris in 80 mL of ddH_2O, adjust the pH to 7.0 with HCl, and complete the volume to 100 mL.
5. **4% stacking gel:** mix 1 mL 30% acrylamide/bisacrylamide, 2.4 mL ddH_2O, 3.75 mL 1 M aminocaproic acid, pH 7.0, and 0.375 mL 1 M Bis-Tris, pH 7.0. To start the polymerization, add 40 μL 10% APS and 4 μL TEMED (N,N,N′,N′-tetramethylethylenediamine).
6. **Buffer solutions for a 4%–10% acrylamide gradient:**
    1. **4% resolving gel:** mix 2 mL 30% acrylamide/bisacrylamide, 4.75 mL ddH_2O, 7.5 mL 1 M aminocaproic acid, pH 7.0, and 0.75 mL 1 M Bis-Tris, pH 7.0. Add 80 μL 10% APS and 8 μL TEMED to start the polymerization.
    2. **10% resolving gel:** mix 5 mL 30% acrylamide/bisacrylamide, 1.75 mL ddH_2O, 7.5 mL 1 M aminocaproic acid, pH 7.0, and 0.75 mL 1 M Bis-Tris, pH 7.0. Add 80 μL 10% APS and 8 μL TEMED to start the polymerization.
7. **Cathode buffer:** dissolve 9 g Tricine and 3.14 g Bis-Tris/HCl in 800 mL ddH_2O. Adjust the pH to 7.0. Add 0.2 or 0.02 g Coomassie blue G250 and complete the volume to 1 L with ddH_2O.
8. **Anode buffer:** dissolve 10.5 g Bis-Tris/HCl in 800 mL ddH_2O. Adjust the pH to 7.0 and complete the volume to 1 L with ddH_2O.

9. **Loading buffer:** mix 1 mL glycerol, 200 μL of 1 M 6-aminocaproic acid solution, and 0.02 g Coomassie brilliant blue G-250 in 7 mL and complete the volume to 10 mL with ddH_2O.

2.4 In-Gel Activity Assays

1. **0.5 M Tris–HCl buffer:** dissolve 6.1 g of Tris-HCl in 80 mL ddH_2O, adjust the pH to 7.4 with HCl. Make up to 100 mL with ddH_2O.
2. **0.25 M phenazine methosulfate(PMS):** weigh 76.6 mg of PMS and dissolve in 1 mL of DMSO. Protect from light and store at 20°C.
3. **Complex I solution (0.2 mM NADH, 1.2 mM MTT):** dissolve 1.4 mg NADH, 5 mg of MTT, 200 μL of 0.5 M Tris-HCl, pH 7.4, in 9.8 mL of ddH_2O. Prepare fresh before use and protect from light.
4. **Complex II solution (0.2 mM PMS, 20 mM succinate, 2.4 mM MTT):** mix 10 mg MTT, 32.4 mg succinate, 8 μL of 0.25 M PMS, 200 μL of 0.5 M Tris-HCl, pH 7.4. Bring the volume to 10 mL with ddH_2O. Prepare fresh before use and protect from light.
5. **Complex IV solution:** mix 10 mg 3,3′-diaminobenzidine tetrahydrochloride hydrate (DAB), 10 mg cytochrome c, 750 mg sucrose, and 240 U of catalase in 8 mL of 10 mM K2HPO4 pH 7.4. Make up to 10 mL with the same buffer. Prepare fresh before use and protect from light exposure.

3 Methods

3.1 Cell Culture

Culture *U. maydis* in 1 L of YPD medium (0.5% yeast extract, 0.5% bactopeptone, and 0.5% glucose) for 24 h at 28°C and 180 rpm. This condition yields approximately 8 g of wet weight (*see* **Note 3**).

3.2 Cell Harvesting

1. Harvest the cells by centrifugation at 1000 × *g* for 5 min using a preweighed tube.
2. Rinse the cells twice with the washing buffer. Repeat step 1. Discard the supernatant.
3. Suspend the pellet with 1 mL of lysis buffer per gram of wet weight (1 mL:1 g).
4. Add 0.2% BSA, 1 mM PMSF, 5 mM 2-mercaptoethanol, and the protease inhibitors cocktail according to the manufacturer's instructions (*see* **Note 4**).

3.3 Cell Lysis and Mitochondria Isolation

1. Fill at least half of the bead beater tube with 0.5 μm glass beads and precool at −20°C. Keep at this temperature until use.
2. Add the cell suspension to fill the prepared tube until it is almost full. Keep it at 4°C (*see* **Note 5**).
3. Perform the cell disruption in the bead beater for 1 min at a frequency of 2000 oscillations/min, followed by 2 min of incubation in an ice-water bath. Cycle this bead-beating/cooling sequence for a total runtime of 4 min. At the end, separate the homogenate from the beads by decantation using a 50 mL centrifuge tube.
4. Centrifuge the homogenate at 3000 × *g* for 10 min at 4°C. Transfer the supernatant to a new 50 mL centrifuge tube and repeat the centrifugation. Discard the pellet.
5. Centrifuge the supernatant at 10,000 × *g* for 10 min. Discard the supernatant. *The pellet contains crude mitochondria.*
6. Resuspend the mitochondrial pellet with 500 μL of lysis buffer (free of BSA and 2-mercaptoethanol). Quantify the protein according to Lowry's method [29]. Keep the mitochondrial suspension at 4°C in an ice-water bath until used (*see* **Note 6**).

3.4 Gradient Gel Casting

1. Place the gradient mixer on the magnetic stirrer, and add a stir bar to the mixing chamber (destined for the 10% solution). Make sure that the inter-chamber valve is closed. Connect the outlet tubing of the mixing chamber to the preassembled gel cassette.
2. Add the 4% solution into the reservoir (non-mixing) chamber and the 10% solution in the mixing chamber.
3. **Begin mixing:** initiate magnetic stirring (200–300 rpm) in the mixing chamber. Add the 10% APS and TEMED to both chambers.
4. **Initiate gradient formation:** simultaneously open the inter-chamber valve and the outlet tubing leading to the gel cassette. Overlay the acrylamide solution with ddH_2O or isopropanol. Let the acrylamide polymerize for 30–60 min.
5. Remove the water/isopropanol from the top of the gel and rinse the gel surface. Mix the 4% stacking gel solution with the 10% APS and TEMED in a 50 mL conical tube, and immediately pipette the solution onto the polymerized gradient gel. Insert the comb and allow polymerization for 30–60 min. Keep at 4°C until use.

3.5 Solubilization of Respiratory Complexes with Digitonin and BN-PAGE

1. Pipette 400 μg of mitochondrial protein in four 1.5 mL microcentrifuge tubes and centrifuge at 10,000 × *g* for 5 min.
2. Discard the supernatant and resuspend each pellet with 40 μL of the solubilization buffer.

3. Titrate individual samples by adding digitonin dropwise with gentle stirring. Achieve final digitonin-to-protein ratios across a range of 0.5–3 (Fig. 1). Incubate samples in an ice-water bath for 30 min.
4. Centrifuge the sample at 13,000 × *g* for 30 min at 4°C.
5. Rssssecover the supernatant and mix it with 20 μL of the loading buffer.
6. Apply 10–30 μL of the prepared sample to a linear polyacrylamide gradient gel (4%–10%).
7. Assemble the electrophoresis unit and fill the anode and cathode chambers with their respective buffers. Make sure that the cathode buffer contains 0.02% of the anionic Coomassie dye.
8. Perform the BN-PAGE at 35 V for 1–3 h at 4°C (*see* **Note 7**).

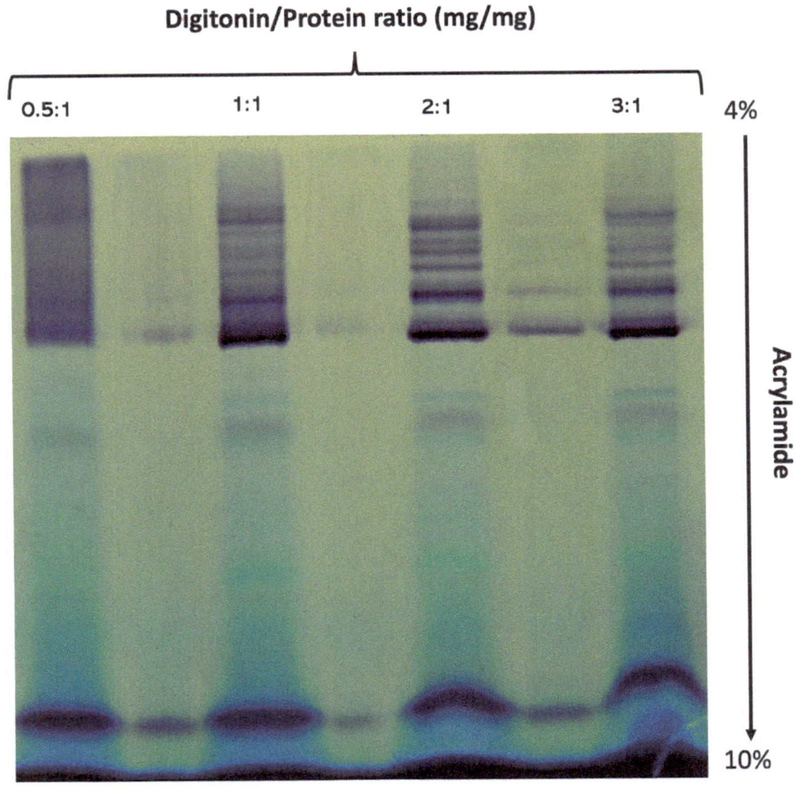

Fig. 1 Solubilization of respiratory complexes using varying detergent/protein ratios. Mitochondria were isolated by differential centrifugation. Then, the respiratory complexes were solubilized using digitonin-to-protein ratios of 0.5, 1, 2, and 3. The efficiency of solubilization for each condition was assessed by separating the complexes via BN-PAGE and evaluating the integrity of complex I through its NADH:MTT oxidoreductase activity

9. Replace the cathode buffer with the same buffer containing 0.002% Coomassie dye and continue the electrophoresis for an additional 7–8 h.
10. After determining the optimal detergent-to-protein ratio for the solubilization of the respiratory complexes, this ratio is used in subsequent experiments (Fig. 2).

3.6 Visualization of Respiratory Complexes by BN-PAGE

3.6.1 Complex I. NADH Dehydrogenase

1. Place the gel strip in 10 mL of the complex I solution.
2. Incubate for 10–30 min at 50 rpm on an orbital shaker at 28°C. The incubation time depends on the catalytic capacity of complex I in the gel. Visualize the purple formazan precipitate at 5-min intervals.

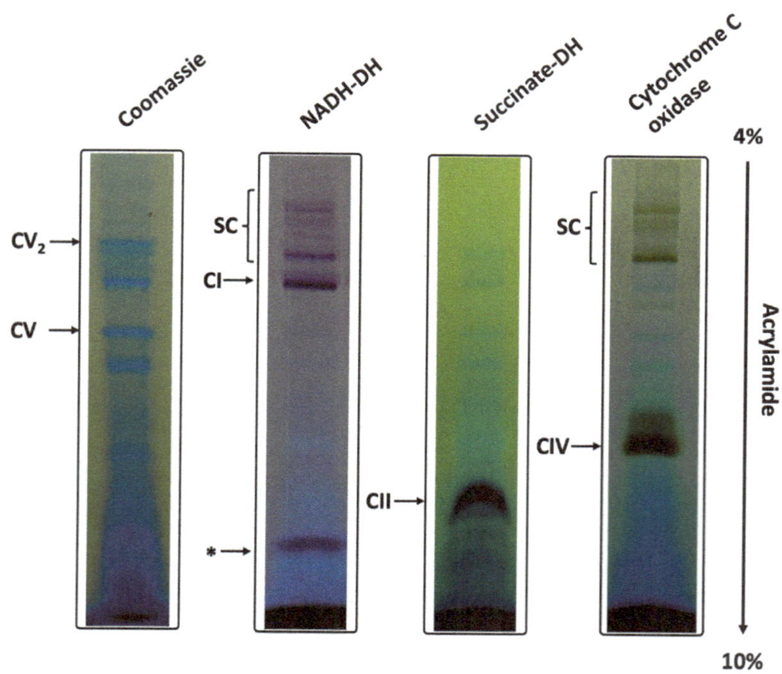

Fig. 2 In-gel activity assays of mitochondrial respiratory complexes. Mitochondrial complexes, solubilized at a digitonin-to-protein ratio of 2, were separated by BN-PAGE. The resulting gel was either stained with Coomassie blue to visualize total protein [1] or assayed for specific enzymatic activities: NADH dehydrogenase [2], succinate dehydrogenase [3], and cytochrome c oxidase [4]. Key complexes are indicated, including the positions of the monomer (CV) and dimer (CV_2) of complex V, and complexes I, II, and IV, the alternative NADH dehydrogenase (*), and the different supercomplexes (SC)

3. Stop the reaction by removing the reaction buffer. Rinse the gel strip with distilled water.

3.6.2 Complex II. Succinate Dehydrogenase

1. Place the gel strip in 10 mL of the complex II solution.
2. Incubate for 10–30 min at 50 rpm on an orbital shaker at 28°C. Visualize the purple formazan precipitate at 5-min intervals.
3. Stop the reaction by removing the reaction buffer. Rinse the gel with distilled water.

3.6.3 Complex IV. Cytochrome c Oxidase

1. Immerse the gel in 10 mL of the complex IV solution.
2. Incubate for 10–30 min with gentle agitation at 50 rpm on an orbital shaker at 28°C. Monitor continuously for the development of brown color (DAB oxidation product).
3. After the bands reveal the enzymatic activity, rinse the gel with distilled water.

4 Notes

1. Different acrylamide/bisacrylamide ratios are employed in the preparation of a 30% solution. Specifically, ratios of 29:1 (29 g acrylamide and 1 g bisacrylamide in 100 mL solution) and 37.5:1 (29.22 g acrylamide and 0.78 g bisacrylamide in 100 mL solution) are used to modulate the pore size of the resulting polyacrylamide gel, which consequently influences the separation of molecules during electrophoresis. A ratio of 29:1, indicative of a higher bisacrylamide content relative to acrylamide, leads to increased cross-linking and a denser gel network characterized by smaller pores, which is suited for separating medium to large proteins. Conversely, a 37.5:1 ratio, denoting a lower bisacrylamide content, produces a more porous gel network with larger pores, making it appropriate for the separation of larger molecules.
2. Because unpolymerized polyacrylamide is a potent neurotoxin, gloves must be worn during gel preparation. Avoid skin contact.
3. *Ustilago maydis* undergoes accelerated senescence when maintained at 4°C on a Petri dish containing YPD medium, which results in significant cellular death within 7–10 days. This aging results in reduced cell biomass yield when subsequently cultured in liquid medium, which correlates with the duration of the initial Petri dish incubation. Therefore, to obtain high-quality mitochondria, a 24-h cell biomass of at least 7 g/L wet weight is required, regardless of the culture volume. Starting from a young inoculum (2–3 days), *U. maydis* biomass

yield is approximately 8 g wet weight per liter when cultured in 2 L baffled beakers containing 1 L of YPD medium at 28°C, 180 rpm, and an initial optical density of 0.04 at 600 nm.

4. 2-Mercaptoethanol, BSA, and protease inhibitors are added to the lysis solution before use.
5. To prevent sample leaks during cell lysis, ensure there are no beads on the threads of the microtubes' caps.
6. Mitochondrial preparations can be assessed by measuring oxygen consumption in the presence of the appropriate respiratory substrates, ADP, and uncoupling agents to obtain a measure of respiratory control and the activity of the electron transport chain complexes.
7. Solubilized bovine complexes can be used as standards.

Acknowledgments

The authors would like to thank the Universidad Nacional Autónoma de México (PAPIIT-UNAM) for the financial support through the Proyectos de Investigación e Innovación Tecnológica program, grant PAPIIT-IN202525.

References

1. Marques HM (2024) Electron transfer in biological systems. J Biol Inorg Chem 29:641–683
2. Cárdenas-Monroy CA, Pohlmann T, Piñón-Zárate G et al (2017) The mitochondrial alternative oxidase Aox1 is needed to cope with respiratory stress but dispensable for pathogenic development in *Ustilago maydis*. PloS One 12:e0173389
3. Juárez O, Guerra G, Martínez F et al (2004) The mitochondrial respiratory chain of *Ustilago maydis*. Biochim Biophys Acta 1658:244–251
4. Matuz-Mares D, Flores-Herrera O, Guerra-Sánchez G et al (2021) Carbon and nitrogen sources have no impact on the organization and composition of *Ustilago maydis* respiratory supercomplexes. J Fungi (Basel) 7:42
5. Matuz-Mares D, Matus-Ortega G, Cárdenas-Monroy C et al (2018) Expression of alternative NADH dehydrogenases (NDH-2) in the phytopathogenic fungus *Ustilago maydis*. FEBS Open Bio 8:1267–1279
6. Sierra-Campos E, Velázquez I, Matuz-Mares D et al (2009) Functional properties of the *Ustilago maydis* alternative oxidase under oxidative stress conditions. Mitochondrion 9:96–102
7. Kravchuk V, Petrova O, Kampjut D et al (2022) A universal coupling mechanism of respiratory complex I. Nature 609:808–814
8. Kühlbrandt W, Carreira LAM, Yildiz Ö (2025) Cryo-EM of mitochondrial complex I and ATP synthase. Annu Rev Biophys 54:209–226
9. Goetzman E, Gong Z, Zhang B et al (2023) Complex II biology in aging, health, and disease. Antioxidants (Basel) 12:1477
10. Hederstedt L, Rutberg L (1981) Succinate dehydrogenase--a comparative review. Microbiol Rev 45:542–555
11. Crofts AR (2004) The cytochrome bc1 complex: function in the context of structure. Annu Rev Physiol 66:689–733
12. Snyder CH, Merbitz-Zahradnik T, Link TA et al (1999) Role of the Rieske iron-sulfur protein midpoint potential in the protonmotive Q-cycle mechanism of the cytochrome bc1 complex. J Bioenerg Biomembr 31:235–242
13. Ulrich JT, Mathre DE (1972) Mode of action of oxathiin systemic fungicides. V. Effect on electron transport system of *Ustilago maydis* and *Saccharomyces cerevisiae*. J Bacteriol 110:628–632
14. Antos-Krzeminska N, Jarmuszkiewicz W (2019) Alternative type II NAD(P)H dehydrogenases in

the mitochondria of protists and fungi. Protist 170:21–37

15. Melo AMP, Bandeiras TM, Teixeira M (2004) New insights into type II NAD(P)H:quinone oxidoreductases. Microbiol Mol Biol Rev MMBR 68:603–616
16. Rasmusson AG, Soole KL, Elthon TE (2004) Alternative NAD(P)H dehydrogenases of plant mitochondria. Annu Rev Plant Biol 55:23–39
17. Sluse FE, Jarmuszkiewicz W (1998) Alternative oxidase in the branched mitochondrial respiratory network: an overview on structure, function, regulation, and role. Braz J Med Biol Res 31:733–747
18. Amthor JS (2023) ATP yield of plant respiration: potential, actual and unknown. Ann Bot 132:133–162
19. Guo R, Zong S, Wu M et al (2017) Architecture of human mitochondrial respiratory megacomplex $I_2III_2IV_2$. Cell 170:1247.e12–1257.e12
20. Moe A, Di Trani J, Rubinstein JL et al (2021) Cryo-EM structure and kinetics reveal electron transfer by 2D diffusion of cytochrome c in the yeast III-IV respiratory supercomplex. Proc Natl Acad Sci USA 118:e2021157118
21. Reyes-Galindo M, Suarez R, Esparza-Perusquía M et al (2019) Mitochondrial respirasome works as a single unit and the cross-talk between complexes I, III_2 and IV stimulates NADH dehydrogenase activity. Biochim Biophys Acta Bioenerg 1860:618–627
22. Stuchebrukhov A, Schäfer J, Berg J et al (2020) Kinetic advantage of forming respiratory supercomplexes. Biochim Biophys Acta Bioenerg 1861:148193
23. Maranzana E, Barbero G, Falasca AI et al (2013) Mitochondrial respiratory supercomplex association limits production of reactive oxygen species from complex I. Antioxid Redox Signal 19:1469–1480
24. Acín-Pérez R, Bayona-Bafaluy MP, Fernández-Silva P et al (2004) Respiratory complex III is required to maintain complex I in mammalian mitochondria. Mol Cell 13:805–815
25. Waltz F, Righetto RD, Lamm L et al (2025) In-cell architecture of the mitochondrial respiratory chain. Science 387:1296–1301
26. Schägger H, Jagow Gvon (1991) Blue native electrophoresis for isolation of membrane protein complexes in enzymatically active form. Anal Biochem 199:223–231
27. Schertl P, Braun H-P (2015) Activity measurements of mitochondrial enzymes in native gels. Methods Mol Biol 1305:131–138
28. Eubel H, Braun H-P, Millar AH (2005) Blue-native PAGE in plants: a tool in analysis of protein-protein interactions. Plant Methods 1:11
29. Lowry OH, Rosebrough NJ, Farr AL et al (1951) Protein measurement with the Folin phenol reagent. J Biol Chem 193:265–275

Chapter 41

Trehalose Determination in *Ustilago maydis* Under Stress Conditions

Laura Valdés-Santiago and José Antonio Cervantes-Chávez

Abstract

Trehalose is a disaccharide involved in cellular protection under stress and energy storage. In the maize pathogen *Ustilago maydis*, intracellular trehalose levels fluctuate in response to environmental stressors. This protocol describes the intracellular trehalose quantification by HPLC in wild-type and mutant strains of *U. maydis* exposed to various *in vitro* stress conditions, including oxidative, osmotic, acid, and cell wall stress. This method enables the assessment of trehalose's protective role and the functionality of trehalases in stress response.

Keywords Trehalose quantification, HPLC, Refractive index detector, Ethanol extraction, Oxidative stress, Osmotic stress

1 Introduction

Trehalose is a non-reducing disaccharide of two glucose units linked via an α, α-1, 1-glycosidic bond. In fungi, including the basidiomycete *Ustilago maydis*, trehalose functions as a critical reserve carbohydrate, a carbon source upon its hydrolysis by trehalases, and a cell stress protectant. The synthesis of trehalose in *U. maydis* is carried out by the Tps1 and Tps2 genes and involves two trehalases: acid (Ath1) and neutral (Nth1) [1]. This method quantifies trehalose accumulation under defined stress conditions, offering insights into stress adaptation mechanisms. The protocol involves fungal culturing under stress, extraction of intracellular trehalose, and quantification via HPLC [1, 2, 3].

An N. T. Phan et al. (eds.), *Ustilago maydis*, Methods in Molecular Biology, vol. 3050,
https://doi.org/10.1007/978-1-0716-5364-7_41,

2 Materials

1. Tris–HCl buffer: 30 mM Tris–HCl, pH 7.0 (adjust with HCl), make up to final volume with HPLC-grade water.
2. Hydrogen peroxide: (H_2O_2), 2.5 mM final concentration
3. HPLC-grade ethanol, HPLC-grade water.
4. HPLC acetonitrile:water: (50:50; v/v).
5. HPLC-grade mobile phase: acetonitrile:water (75:25; v,v).
6. Sterile distilled water.
7. D-(+)-Trehalose.
8. Liquid nitrogen.
9. 0.2 μm syringe filters.
10. Erlenmeyer flasks (50 mL).
11. Mortar and pestle (sterile).
12. Benchtop microcentrifuge for 1.5–2.0 mL tubes.
13. Benchtop centrifuge for 15 and 50 mL tubes.
14. Shaking incubator (28°C).
15. HPLC system.
16. Carbohydrate column (4.6 × 250 mm, 5 μm) with an amino-propyl guard column.
17. Refractive index detector (RID).
18. Volumetric flasks and micropipettes.

3 Methods

3.1 Fungal Cultivation and Stress Induction

1. To recover strains, inoculate from frozen stocks onto complete medium (CM) agar plates and incubate at 28°C for 48–72 h. For mutant strains, supplement the CM medium with the appropriate antibiotics: 200 μg·mL^{-1} hygromycin B for Δnth1 and Δmca1 strains; 3 μg·mL^{-1} carboxin for Δtps2 strains; both antibiotics for Δtps2::Δmca1 double mutants.
2. Once recovered, select single colonies and propagate them in CM broth or on fresh CM plates, maintaining antibiotic selection as appropriate to ensure plasmid or mutation retention.
3. Inoculate *U. maydis* strains in 50 mL of fresh liquid CM and incubate for 18 h at 28°C with constant shaking.
4. Harvest cells by centrifugation at ≈ 755 × *g* for 5 min.
5. Discard the supernatant and wash the cell pellet twice with sterile distilled water.

6. Resuspend the washed cells in 200 mL of modified minimal medium with pH adjusted to seven using Tris–HCl buffer.
7. Adjust the initial OD_{600} to 0.02 using the appropriate volume of cell suspension.
8. Incubate for 15 h at 28°C with shaking.
9. After 15 h, add stress agents to the culture medium (H_2O_2 2.5 mM; Menadione 0.8 mM; tert-Butyl hydroperoxide 1.3 mM; Bengal rose 120 μg/mL; NaCl 0.8 M; KCl 1 M; Congo red 13 μg/mL; SDS 0.0075%; Caffeine 3 mM; UV light 30,000 J/m^2; Acetic acid 3.5 mM; LiCl 12 mM; $CoCl_2$ 0.4 mM; $CdCl_2$ 0.6 mM; $CdSO_4$ 0.8 mM).
10. Continue incubation for 3 more hours under the same conditions.
11. For control samples, follow the same procedure without adding any stressor agent.
12. After incubation, centrifuge the cultures to pellet the cells at ≈ 755 × *g* for 5 min.
13. Discard the supernatant and store the cell pellet at −80°C.

3.2 Trehalose Extraction

1. Recover the pellet (keep it frozen) and grind it until a fine powder using liquid nitrogen and a sterile mortar and pestle (*see* **Note 1**).
2. Weigh (~ 500 mg) the powdered biomass and transfer it to a sterile 50 mL Erlenmeyer flask using a spatula.
3. Add 30 mL of HPLC-grade ethanol to the flask.
4. Boil the mixture in a water bath until complete evaporation of ethanol is achieved (80°C) (*see* **Note 2**).
5. Once dried, add 1 mL of HPLC-grade acetonitrile:water (50:50 v/v).
6. Mix thoroughly until the sample is fully homogenized.
7. Transfer the solution to a new microcentrifuge tube.
8. Centrifuge the extract at ≈$1.42 \times 10^4 \times g$ for 10 min.
9. Recover and filter the supernatant through a 0.2 μm pore-size membrane for further analysis (*see* **Note 3**).

3.3 Trehalose Quantification by HPLC with Refractive Index Detection

1. Set up the HPLC system equipped with a refractive index detector.
2. Install a carbohydrate column (4.6 × 250 mm, 5 μm) and connect it to an aminopropyl guard column (*see* **Note 4**).
3. Prepare the HPLC-grade mobile phase solution.
4. Set the flow rate of the mobile phase to 1.2 mL/min (*see* **Note 5**).

5. Prepare a series of D-(+)-Trehalose standards by dissolving the compound in distilled water to obtain the following concentrations: 5, 4, 3, 2, 1, 0.9, 0.8, 0.7, 0.6, and 0.5 mg/mL.
6. Inject each standard solution into the HPLC system under the conditions described above (*see* **Note 6**).
7. Record the peak area corresponding to each trehalose standard.
8. Construct the calibration curve by plotting concentration (mg/mL) on the x-axis versus peak area (detector response) on the y-axis (*see* **Note 7**).
9. Use the resulting calibration equation to determine the trehalose concentration in experimental samples.

4 Notes

1. Trehalose extraction efficiency may vary with biomass amount; it is advisable to normalize by dry weight. Biomass should be ground to a fine powder to ensure efficient extraction.
2. Ensure complete ethanol evaporation before resuspension to avoid interference with downstream analysis.
3. Samples can be stored at −20°C after filtration if immediate analysis is not possible.
4. Maintain all solvents and water as HPLC-grade to prevent contamination or column damage.
5. The HPLC system should be equilibrated for at least 30 min before injecting standards or samples.
6. Prepare fresh standard solutions regularly to ensure accuracy.
7. The calibration curve should be linear within the selected range, and its R^2 value should exceed 0.99.

Acknowledgments

L.V.S. thanks the "Investigadoras e Investigadores por México SECIHTI" program (Project No. 5175). J.A.C.Ch, thanks to Universidad Autónoma de Querétaro, funding program: "FNB202537".

References

1. Cervantes-Chávez JA et al (2016) Trehalose is required for stress resistance and virulence of the Basidiomycota plant pathogen *Ustilago maydis*. Microbiol 162:1009–1022. https://doi.org/10.1099/mic.0.000287
2. Ferreira JC, Paschoalin VMF, Panek AD et al (1997) Comparison of three different methods for trehalose determination in yeast extracts. Food Chem 60:251–254. https://doi.org/10.1016/S0308-8146(96)00330-5
3. López Cabrera A (2020) Estudio del catabolismo de la trehalosa de *Ustilago maydis* en estrés *in vitro* y la patogenicidad en plantas de maíz. Master thesis, Universidad Autónoma de Querétaro

Chapter 42

H_2O_2 Consumption and Production Assay for the Characterization of *U. maydis* Oxidoreductases

Roseline A. Yao and Bastien Bissaro

Abstract

This protocol describes a validated method to quantify hydrogen peroxide (H_2O_2) generated by oxidases and to measure H_2O_2 consumption by other enzymes or chemical scavengers. This method is a peroxidase-coupled assay using Amplex® Red, which is both fluorogenic and chromogenic and allows to monitor H_2O_2 formation in kinetic or end-point modes. The assay can be conducted in various volumes, and we describe here in detail a 96-well plate format. Typically, depending on the number of samples, the so-called Amplex Red assay can be carried out within 1–3 h. This document covers reagent preparation, calibration, kinetic measurements, data analysis, critical parameters, troubleshooting, and expected results.

Keywords Hydrogen peroxide, Oxidases, Amplex Red, Horseradish peroxidase (HRP)

1 Introduction

Many oxidases (e.g., glucose oxidase, amino acid oxidases, monoamine oxidase, glycolate oxidase, etc.) reduce O_2 to H_2O_2 as part of their catalytic cycle. Hydrogen peroxide (H_2O_2) is essential in a wide variety of biological processes, from acting as a signaling and defense molecule [1] to being a central substrate in enzymatic networks involved in plant biomass degradation [2].

Accurate and sensitive measurement of H_2O_2 is thus required to characterize H_2O_2-producing enzymes (referred to as oxidases) and to monitor scavengers or peroxidases that consume H_2O_2. Several assays exist, with their advantages and drawbacks, such as the "FOX assay"—a colorimetric method [3] that is fast but not so sensitive—or electrochemical sensors, which enable continuous monitoring in complex matrices (such as heterogeneous media) but require specific setup and expertise. Peroxidase-coupled fluorometric and/or chromogenic assays are widely used assays because they are relatively easy to implement and amenable in any laboratory having access to a spectrophotometer. The most used one is probably the well-known "ABTS/HRP assay,"

An N. T. Phan et al. (eds.), *Ustilago maydis*, Methods in Molecular Biology, vol. 3050,
https://doi.org/10.1007/978-1-0716-5364-7_42,

in which ABTS (colorless) is oxidized into the corresponding radical cation ABTS$^{\bullet+}$ (green color) by horseradish peroxidase (HRP) in the presence of H_2O_2, in a 2:1 stoichiometric ratio (two molecules of ABTS$^{\bullet+}$ formed for one molecule of H_2O_2 consumed) [4]. A notable disadvantage of this assay is the poor stability of ABTS$^{\bullet+}$, which can lead to underestimations. Alternatively, *N*-acetyl-3,7-dihydroxyphenoxazine ("Amplex® Red") can be used in place of the ABTS, in a so-called analogous "Amplex Red/HRP assay" (Fig. 1). This "Amplex Red/HRP assay" has notably been extensively used to characterize the oxidase activity of lytic polysaccharide monooxygenases [5, 6], cytochromes P450s [7], NADPH oxidases [8], and many other type of oxidases over the past 20 years [9].

Filamentous fungi possess a wide variety of H_2O_2-producing and H_2O_2-consuming enzymes [2], and we anticipate that many remain to be discovered and characterized. *Ustilago maydis* is no exception. Actually, only a handful of oxidoreductases from *U. maydis* have been biochemically characterized, including its glyoxal oxidase Glo1 (*Um*AA5_1A; UMAG_02411) [10] and chitin-active lytic polysaccharide monooxygenase (*Um*AA10; UMAG_05439) [11]. We also underscore that such assay is very handy to determine if the electrons extracted from an organic compound by an oxidoreductase are used to reduce O_2 to H_2O_2, to reduce other non-O_2 electron acceptors, or both. The "Amplex Red/HRP assay" was, for instance, key to show that *U. maydis* β-1,3/1,6-glucooligosaccharide oxidoreductase (*Um*AA3_2A; UMAG_03551) was not able to reduce O_2 and instead reduced organic electron acceptors (using a complementary chromogenic assay), allowing to define its activity as "β-1,3/1,6-glucooligosaccharide dehydrogenase" [12].

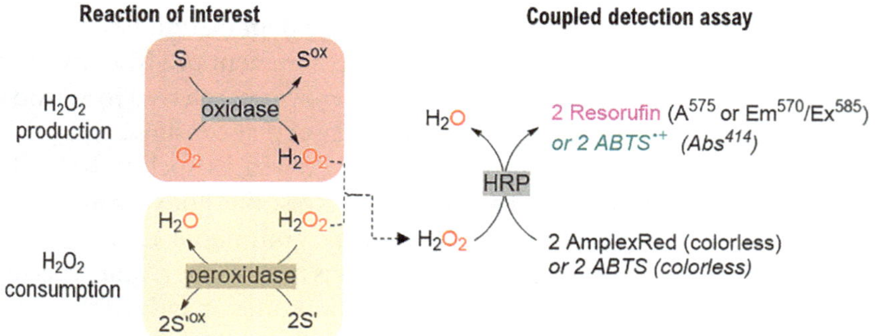

Fig. 1 Flow chart illustrating the detection assay for hydrogen peroxide (H_2O_2) relying on HRP-coupled reporter reactions. In the H_2O_2-production reaction, an oxidase enzyme converts its substrate (S) and molecular oxygen (O_2) into the corresponding oxidized substrate (S^{ox}) and H_2O_2. In the consuming reaction, a peroxidase converts H_2O_2 and its substrate (S') into water and the corresponding oxidized substrate (S'^{ox}). The available H_2O_2 reacts with horseradish peroxidase (HRP) to oxidize either two Amplex Red molecules into two Resorufin, or two ABTS molecules into two ABTS$^{\bullet+}$ molecules

Importantly, as the assay measures the apparent concentration of H_2O_2 at a given time, it actually reflects the outcome of antagonistic production and consumption fluxes (Fig. 1). It can thus be used to measure H_2O_2 production only (when the candidate enzyme is an oxidase), consumption only (when working with peroxidases/peroxygenases), or the apparent result of both fluxes (when working with mixtures of both types of activities).

2 Materials

Prepare all solutions using ultrapure water (prepared by purifying deionized water to attain a resistivity of 18.2 MΩ · cm at 25°C) and analytical grade reagents. Redox-sensitive compounds (such as H_2O_2 and reducing agents) are prepared in trace metal-free water (such as TraceSELECT™). Prepare all reagents at room temperature, make aliquots, and store at −20°C (unless indicated otherwise). Diligently follow all waste disposal regulations when disposing of waste materials. We do not add sodium azide to reagents.

2.1 Consumables

1. 96-well clear or black flat-bottom microplates (black for fluorescence) or cuvettes (for spectrophotometer), where appropriate.
2. Micropipettes and tips. Ideally, multichannel micropipettes to initiate simultaneously multiple reactions.
3. Ice bucket, vortex, microcentrifuge tubes.

2.2 Reagents and Buffers

4. Amplex Red reagent (10-acetyl-3,7-dihydroxyphenoxazine): dissolve the Amplex Red powder directly in the glass vial it was supplied with dimethylsulfoxide (DMSO) to reach a concentration of 10 mM. Protect from light. Make small aliquots (e.g., 50 μL) of this solution in PCR tubes, store at −20 °C (*see* **Note 1**).
5. Hydrogen peroxide (H_2O_2) stock solution, analytical grade (30% w/w, eq. ~9.8 M; *see* **Note 2** for accurate determination of concentration). Make serial dilutions in TraceSELECT™ water down to 10 mM, store at −20 °C (*see* **Note 3**).
6. Phosphate-buffered saline (PBS) or other assay buffer (commonly 0.5 M sodium phosphate, pH 6.0 or 7.0) and equilibrate to the desired assay temperature (e.g., 25 or 37°C). Store at 4°C or room temperature.
7. Horseradish peroxidase (HRP; 0.5 mg·mL^{-1}): dissolve 5 mg of HRP commercial powder (*see* **Note 4**) in 10 mL of sodium phosphate buffer (50 mM, pH 6.0). Filter sterilize. Store at 4 °C (*see* **Note 5**).

8. Candidate oxidoreductase enzyme(s) and positive control oxidoreductase (use, for instance, commercial glucose oxidase as benchmark).
9. Substrates to be oxidized (e.g., glucose for glucose oxidase or ascorbate for lytic polysaccharide monooxygenases). Dissolve in ultrapure water, aliquot, and store at −20°C.

2.3 Equipment

10. Fluorescence microplate reader (excitation = 540–560 nm; emission = 590–610 nm for resorufin). Alternatively, a standard microplate reader measuring absorbance can be used (absorbance of resorufin at 575 nm).
11. Cuvette spectrophotometer (for determination of H_2O_2 concentration; absorbance read at 240 nm).

3 Methods

Carry out all procedures at room temperature unless otherwise specified.

3.1 Preparation of Calibration and Master Mix

1. Prepare calibration standards of H_2O_2: (10×) e.g., 0, 0.1, 0.5, 1, 5, 10, 50 μM (for fluorescence measurements) or 0, 50, 100, 200, 300, 400 μM (for absorbance measurements). Prepare fresh immediately before assay by serial dilutions of a 10 mM stock of H_2O_2 (stored at −20°C, and the concentration of which should be verified spectrophotometrically; *see* **Note 2**) in ultrapure water. Keep on ice. Prepare extemporaneously a working solution of Amplex Red (1 mM final; 10% DMSO) by diluting the stock solution of Amplex Red (10 mM, in 100%DMSO) in ultrapure water. Keep on ice.
2. Prepare a master mix A containing all reagents that do not react together (volumes are given for one reaction; multiply all volumes by *n* + 1 for *n* reactions; concentrations given in parenthesis are that of solutions from which the indicated volumes should be pipetted): in a test tube, mix in the following order: 48 μL of ultrapure water, 10 μL of buffer (10×), 2 μL of candidate enzyme (0.5–50 μM; *see* **Notes 6 and 7**), 20 μL of HRP (0.5 mg·mL^{-1}), and 10 μL of Amplex Red (1 mM; 10% DMSO) (*see* **Note 8**). Prepare a master mix B that will be used for the H_2O_2 standard curve (so total volume for *n*, where *n* is the number of points of the standard curve) and that contains the same components as master mix A, in which the volume of candidate enzyme is replaced by the equivalent volume of enzyme storage buffer (**Note 9**).
3. Prepare working solutions of the substrate to be oxidized by the candidate oxidoreductase: make serial dilutions of the

stock solution in ultrapure water. For redox-sensitive substrates, prepare extemporaneously the working solution in TraceSELECT™ water: for instance, for ascorbate (500 µM working solution), add 25 µL of ascorbate (100 mM stock) in 4975 µL of TraceSELECT™ water. Keep on ice.

3.2 96-Well Plate Setup

4. In a 96-well plate (*see* **Note 10**), distribute 90 µL of the master mix A or B in each well (the final volume of the reaction will be 100 µL after all additions). Then, preincubate the plate at assay temperature for 2–5 min to equilibrate. Then, in wells containing master mix B, add 10 µL of H_2O_2 standard solution (10×). In wells containing master mix A, initiate the reaction by adding 10 µL of the substrate of the candidate enzyme (e.g., glucose for glucose oxidase or ascorbate for monitoring the oxidase activity of lytic polysaccharide monooxygenases). Immediately start kinetic acquisition (see Sect. 3.3).
Several controls must be included:
   - Control 1 or "no-enzyme" control: a reaction mixture in which the candidate enzyme is replaced by the equivalent volume of storage buffer (*see* **Note 11**).
   - Control 2 or "no-substrate" control: a reaction mixture in which the substrate is replaced by the equivalent volume of ultrapure water (*see* **Note 12**).
   - Control 3 or "HRP cross-reactivity" control: a standard curve in which the substrate of the candidate enzyme (e.g., glucose, ascorbate) is added (*see* **Note 13**).

3.3 Data Acquisition

5. Monitor the absorbance or the fluorescence, under constant shaking and at controlled temperature, every 15–60 s for 5–60 min depending on rate of the reaction (see example in Fig. 2). For an end-point experiment, incubate for a fixed time (e.g., 10–30 min) and read.
6. Fluorescence settings: Ex = 540–560 nm/Em = 590–610 nm, and set a fixed gain (*see* **Note 14**). Absorbance: 575 nm.

3.4 Converting RFU or Abs to [H_2O_2] and Derive a Rate

7. Subtract blank value (usually the data point corresponding to [H_2O_2] = 0 µM in the standard curve) from all raw fluorescence or absorbance values (*see* **Note 15** in case of low or no signal, and **Note 16** in case of too high background).
8. Use the calibration wells (i.e., with known [H_2O_2]) treated identically (Amplex Red + HRP), subtract blank (value corresponding to [H_2O_2] = 0 µM), and build a standard curve (RFU or Abs vs. µM H_2O_2) (*see* **Note 17**).
9. Apply linear equation to the linear zone of the time course of test sample (within limits of detection) to derive the slope (ΔRFU/min^{-1} or ΔAbs/min^{-1}). Convert this slope in initial

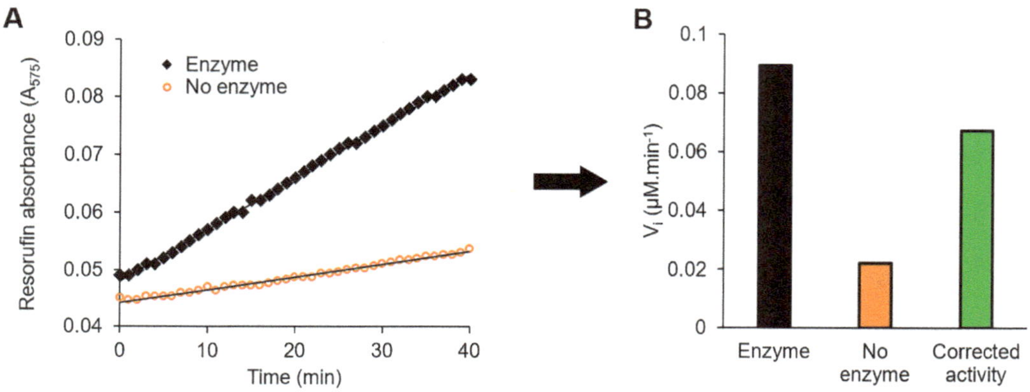

Fig. 2 Example of time-course production of resorufin and calculation of the initial velocity of H_2O_2 production. (**A**) Evolution of resorufin absorbance over time in the presence ("Enzyme") and absence ("No enzyme") of *Um*AA10, the LPMO from *Ustilago maydis*. Data taken from [11]. (**B**) Bar chart showing the initial rates of H_2O_2 production by *Um*AA10 before (in black) and after (in green) correction. Correction was made by subtracting the value of the control reaction without enzyme (in orange). For the sake of clarity, only one replicate is shown

rates V_i (μM H_2O_2/min^{-1}). If non-null, the rate of Amplex Red auto-oxidation measured in control reactions in the absence of enzyme should be subtracted from the initial rate determined for the candidate enzyme.

10. Calculate the specific activity V_i/[E] (in min^{-1} or s^{-1}) (where [E] is the concentration of candidate enzyme in μM; *see* **Note 18**) for each substrate concentration (*see* **Notes 19 and 20**) and plot Vi/[E] vs. [S] to derive the kinetic parameters (*see* **Note 21**).

3.5 Measuring H_2O_2 Consumption

11. To test consumption by an H_2O_2-consuming enzyme (or by a nonenzymatic scavenger), the test can be applied only in an offline mode, meaning that the reaction cannot be monitored online directly in the microplate well and, instead, needs to be carried out in two steps, as described below.
12. Step 1: H_2O_2-consumption reaction. Pregenerate a fixed H_2O_2 concentration (or add H_2O_2 standard) and add the candidate enzyme/compound under reaction conditions suited for the candidate enzyme (*see* **Note 22**).
13. Step 2: Quantification of residual H_2O_2. Sample 10 μL of the reaction of step 1 at fixed time intervals (*see* **Note 23**), predilute in ultrapure water if needed (depending on initial H_2O_2 concentration used in step 1) to fit within the standard curve range, and quickly mix the 10 μL (prediluted or not) to 90 μL of master mix B (described in Sect. 3) to reach a final volume of 100 μL. Monitor absorbance or fluorescence and convert in [H_2O_2] concentration, as described in Sect. 3.3 (*see* **Note 24**).

14. Repeating step 2 at fixed time intervals allows to monitor the decrease in [H_2O_2] concentration over time relative to a no-consumer control (e.g., without candidate enzyme).

4 Notes

1. The stock solution of Amplex Red might be slightly pink.
2. Always quantify the concentration spectrophotometrically if absolute accuracy is critical ($\varepsilon_{240} = 43.6\,M^{-1} \cdot cm^{-1}$ for H_2O_2).
3. Be careful with storage of too diluted solutions as they might prove unstable. All working dilutions below 10 mM should be discarded after use.
4. Note that protein content in HRP commercial powder varies from batches to batches and can be as low as 20% of protein only. We recommend evaluating protein content by Bradford assay.
5. HRP (0.5 mg · mL^{-1}) in sodium phosphate buffer (50 mM, pH 6.0) can be stable for several months at 4°C.
6. Dilute the enzyme in storage buffer to keep it stable. Do not keep diluted solution (<1 μM) and always work with freshly made diluted enzyme solutions.
7. Depending on the type of tested oxidoreductases, the H_2O_2 production rate is more or less elevated. Various enzyme concentrations should thus be tested to fit in the linear range and allow determination of the initial rate.
8. Avoid final DMSO concentrations >1% v/v in wells unless tested beforehand, as DMSO can affect enzymes.
9. As some tested enzymes might have H_2O_2-consuming activity, the candidate enzyme should not be included in the standard curve.
10. Use 96-well clear flat-bottom microplates for absorbance, or 96-well black flat-bottom microplates for fluorescence.
11. This "no-enzyme" control will show the enzyme-independent oxidation of Amplex Red, which should be very low if not null compared to the tested enzyme.
12. This "no-substrate" control allows to verify that the tested enzyme does not react with Amplex Red.
13. This "cross-reactivity" control allows to verify to what extent the HRP acts as a substrate peroxidase. For instance, the HRP can act as an ascorbate peroxidase. Therefore, the H_2O_2 produced by the tested oxidase will enter either the HRP-catalyzed Amplex Red oxidation pathway or the

HRP-catalyzed ascorbate oxidation pathway. The more pronounced the latter, the more underestimated will be the H_2O_2 concentration (as it is estimated on the basis of Amplex Red conversion).

14. Keep short reads to avoid photobleaching when using the fluorescence mode.
15. No or low signal can be due to several factors, among which:
    - Sodium azide and other reaction compounds can inhibit HRP. Run standard curves spiked with reaction components as controls.
    - Assay carried out at too low pH (pH < 6.0): the pH of the reaction mixture has a very strong impact on resorufin absorbance and fluorescence. The higher the pH, the higher the signal. For good enough sensitivity, we recommend using pH > 6.0. If you need to work at an acidic pH, we recommend to use ABTS instead of Amplex Red. Indeed, ABTS is oxidized into $ABTS^{\bullet+}$ cation radical by HRP in the presence of H_2O_2, and, in contrast to Resorufin, the extinction coefficient of $ABTS^{\bullet+}$ is not pH-sensitive. It can thus be used in a broad range of pH values. However, note that $ABTS^{\bullet+}$ is not very stable, and time courses of standard curves of H_2O_2 will show that at fixed H_2O_2 concentration the signal of $ABTS^{\bullet+}$ decreases over time. We thus suggest to use the extinction coefficient ($\varepsilon_{414} = 36{,}000\,M^{-1} \cdot cm^{-1}$) rather than a standard curve for converting the absorbance into concentrations (stoichiometry of two molecules of $ABTS^{\bullet+}$ formed for one molecule of H_2O_2 converted).
    - HRP is missing/inactive and/or Amplex Red degraded: check reagents, prepare fresh, minimize freeze-thaw cycles, and run positive H_2O_2 standard.
    - Candidate enzyme is too diluted or it is not the correct substrate for the candidate enzyme: use higher concentration of candidate enzyme. For typical oxidases, initial H_2O_2 formation rates should be measurable using between ca. 10 nM and 1 μM of candidate enzyme. The rate will also depend on substrate concentration.
16. High background signal could be due to a contaminated buffer or oxidation by light. Use fresh buffer and dark conditions.
17. Calibration can be nonlinear because of (i) limiting amount of Amplex Red compared to added H_2O_2 (check concentrations; *see* **Note 24**), (ii) improper mixing (mix well), (iii) photobleaching (protect from light), (iv) out-of-linear-range quantification (use linear portion of dilute samples, or use a higher concentration of Amplex Red; *see* **Note 24**).

18. Dividing the initial rate (in μM. min^{-1} of product) by [E] (in μM) leads to the standardized rate, with the assumption of a single active site per enzyme molecule (meaning that [E] really reflects the total number of active sites).
19. The tested substrate should not interfere with the HRP-catalyzed conversion of Amplex Red (which can, for instance, be the case when using ascorbate).
20. To verify the initial rate conditions, i.e., verify the steady-state assumption, V_i must be calculated during early phases of the reaction, before a significant amount of substrate is consumed (<10%).
21. The standard Michaelis–Menten equation can be used if no substrate/product inhibition is observed.
22. Mind to use buffer conditions in step 1 (test reaction) that do not affect the conditions used in Step 2 (H_2O_2 quantification assay). Dilution of the sample between steps 1 and 2 can be a mean to prevent such issue. In complement, carry out a control standard curve reaction (without enzyme and with different H_2O_2 concentrations) in step 1 conditions, and add 10 μL of this mixture to 90 μL of master mix B. Compare the values to a standard curve in which H_2O_2 was prepared in ultrapure water (as described in Sect. 3.1).
23. Mind that the reaction of step 1 is not stopped here. If the reaction needs to be stopped consider either of the following options: (i) predilution before mixing with master mix B will considerably slow down the reaction; (ii) boiling the sample will inactivate the candidate enzyme, but might affect H_2O_2 stability. A control with a known concentration of H_2O_2 needs to be done if such an option is retained; (iii) separation of the candidate enzyme from the bulk H_2O_2 by ultracentrifugation, as done by Munzone et al. [13].
24. Note that Amplex Red might become a limiting factor if too high concentrations of H_2O_2 are present. Indeed, given the stoichiometry of 2:1, 100 μM of Amplex Red will allow the detection of a maximum of 50 μM of H_2O_2. If higher concentrations of H_2O_2 are expected, either dilute further the sample or increase the Amplex Red concentration to 200 μM, while checking linearity of the method.

Acknowledgments

This work was supported by the French National Agency for Research ("Agence Nationale de la Recherche") through the ANR-JCJC "OXIWALL" project (ANR-23-CE44-0023) to B.B.

References

1. Sies H (2017) Hydrogen peroxide as a central redox signaling molecule in physiological oxidative stress: oxidative eustress. Redox Biol 11:613–619. https://doi.org/10.1016/j.redox.2016.12.035
2. Bissaro B, Várnai A, Røhr ÅK et al (2018) Oxidoreductases and reactive oxygen species in conversion of lignocellulosic biomass. Microbiol Mol Biol Rev 82:e00029-18. https://doi.org/10.1128/MMBR.00029-18
3. Deiana L, Carru C, Pes G et al (1999) Spectrophotometric measurement of hydroperoxides at increased sensitivity by oxidation of Fe^{2+} in the presence of xylenol orange. Free Radic Res 31:237–244. https://doi.org/10.1080/10715769900300801
4. Re R, Pellegrini N, Proteggente A et al (1999) Antioxidant activity applying an improved ABTS radical cation decolorization assay. Free Radic Biol Med 26:1231–1237. https://doi.org/10.1016/S0891-5849(98)00315-3
5. Kittl R, Kracher D, Burgstaller D et al (2012) Production of four *Neurospora crassa* lytic polysaccharide monooxygenases in *Pichia pastoris* monitored by a fluorimetric assay. Biotechnol Biofuels 5:79. https://doi.org/10.1186/1754-6834-5-79
6. Munzone A, Eijsink VGH, Berrin J-G et al (2024) Expanding the catalytic landscape of metalloenzymes with lytic polysaccharide monooxygenases. Nat Rev Chem 8:106–119. https://doi.org/10.1038/s41570-023-00565-z
7. Mishin V, Gray JP, Heck DE et al (2010) Application of the Amplex Red/horseradish peroxidase assay to measure hydrogen peroxide generation by recombinant microsomal enzymes. Free Radic Biol Med 48:1485–1491. https://doi.org/10.1016/j.freeradbiomed.2010.02.030
8. Mishin V, Heck DE, Laskin DL et al (2020) The Amplex Red/horseradish peroxidase assay requires superoxide dismutase to measure hydrogen peroxide in the presence of NAD(P)H. Free Radic Res 54:620–628. https://doi.org/10.1080/10715762.2020.1821883
9. Guang H, Du G (2006) High-throughput screening for monoamine oxidase-A and monoamine oxidase-B inhibitors using one-step fluorescence assay. Acta Pharmacol Sin 27:760–766. https://doi.org/10.1111/j.1745-7254.2006.00336.x
10. Leuthner B, Aichinger C, Oehmen E et al (2005) A H_2O_2-producing glyoxal oxidase is required for filamentous growth and pathogenicity in *Ustilago maydis*. Mol Genet Genomics 272:639–650. https://doi.org/10.1007/s00438-004-1085-6
11. Yao RA, Reyre J-L, Tamburrini KC et al (2023) The *Ustilago maydis* AA10 LPMO is active on fungal cell wall chitin. Appl Environ Microbiol 89:e00573-23. https://doi.org/10.1128/aem.00573-23
12. Reyre J-L, Grisel S, Haon M et al (2022) The maize pathogen *Ustilago maydis* secretes glycoside hydrolases and carbohydrate oxidases directed toward components of the fungal cell wall. Appl Environ Microbiol 88:e01581-22. https://doi.org/10.1128/aem.01581-22
13. Munzone A, Pujol M, Badjoudj M et al (2024) Design of plastic binding lytic polysaccharide monooxygenases via modular engineering. Chem Bio Eng 1:863–875. https://doi.org/10.1021/cbe.4c00125

Chapter 43

Online Monitoring of Oxygen Transfer Rate as a Versatile Tool for Fermentation Characterization and Optimization

Luca Antonia Grebe and Jørgen Barsett Magnus

Abstract

Microbial oxygen consumption is a reliable indicator of microbial metabolism in aerobic cultivations. By measuring the oxygen transfer rate (OTR) online, common characteristics of microbial cultivations such as biomass growth, nutrient limitation, oxygen limitation, and stress factors can easily be identified. This chapter presents a guideline to confidently interpret OTR data and to reliably identify typical phenomena.

Keywords Respiratory activity, Oxygen transfer rate, Oxygen consumption, Metabolic activity, Secondary substrate limitation, Oxygen limitation, Diauxic growth, Substrate inhibition

1 Introduction

Microbial metabolism is inherently linked to oxygen consumption in aerobic cultivations. Consequently, online monitoring of oxygen consumption can give valuable quantitative information on the physiological state of a microbial culture. Common phenomena such as biomass growth, nutrient limitation, oxygen limitation, or stress factors can thus be easily identified [1]. The oxygen transfer rate (OTR) can be regarded as equal to the oxygen uptake rate if the change in dissolved oxygen concentration in the liquid over time is sufficiently small [2]. Nowadays, commercial systems allow for the parallel measurement of the OTR in up to 16 shake flasks or 96-well microtiter plates in parallel [3–5]. These systems use a recurring cycle of aeration and measurement phases. During measurement, aeration is stopped, and the oxygen transfer rate can be calculated from the dynamic decrease of the oxygen partial pressure over time. Typical applications include strain screening [6, 7], medium optimization [8, 9], feedstock screening [6, 10], and process optimization [11]. In the following, a guideline to reliably identify typical phenomena from OTR data is presented.

An N. T. Phan et al. (eds.), *Ustilago maydis*, Methods in Molecular Biology, vol. 3050,
https://doi.org/10.1007/978-1-0716-5364-7_43,

2 Materials

The cultivations can be conducted using standard cultivation protocols for *U. maydis* while using a suitable device for online monitoring of the respiratory activity. Potential commercial systems are suggested below.

2.1 Shake Flask Scale

1. Respiration online monitoring system for shake flasks, e.g., Kuhner TOM (Adolf Kuhner AG, Birsfelden, Switzerland) or RAMOS (Hitec Zang GmbH, Herzogenrath, Germany).
2. Compatible shake flasks (depending on device).
3. Sterile barrier, e.g., cotton plugs, cigarette filters (depending on device).

2.2 Microtiter Scale

1. Respiration online monitoring system for microtiter plates, e.g., Kuhner microTOM (Adolf Kuhner AG, Birsfelden, Switzerland).
2. Compatible 24-/96-well microtiter plates (depending on device).
3. Self-adhesive sterile top layer.

3 Methods

The experiments can be conducted as usual while using a suitable device for online monitoring of the OTR. A selection of typical phenomena and their influence on the OTR is discussed in the following (*see* **Note 1**) (Fig. 1).

3.1 Unlimited Growth

1. Unlimited growth is characterized by an exponential increase in OTR.
2. The depletion of the carbon source (e.g., glucose) is indicated by an abrupt decrease in the OTR.
3. Metabolism of storage molecules, byproducts, or autophagy may lead to further respiratory activity afterward.
4. **How to be sure:**
   a. To reliably identify exponential growth, the OTR can either be plotted on a logarithmic scale or transformed with the natural logarithm [12]. For exponential growth, this results in a linear increase over time. The slope of the corresponding regression function can be used to calculate the growth rate (μ) according to Eq. 1 [13]:

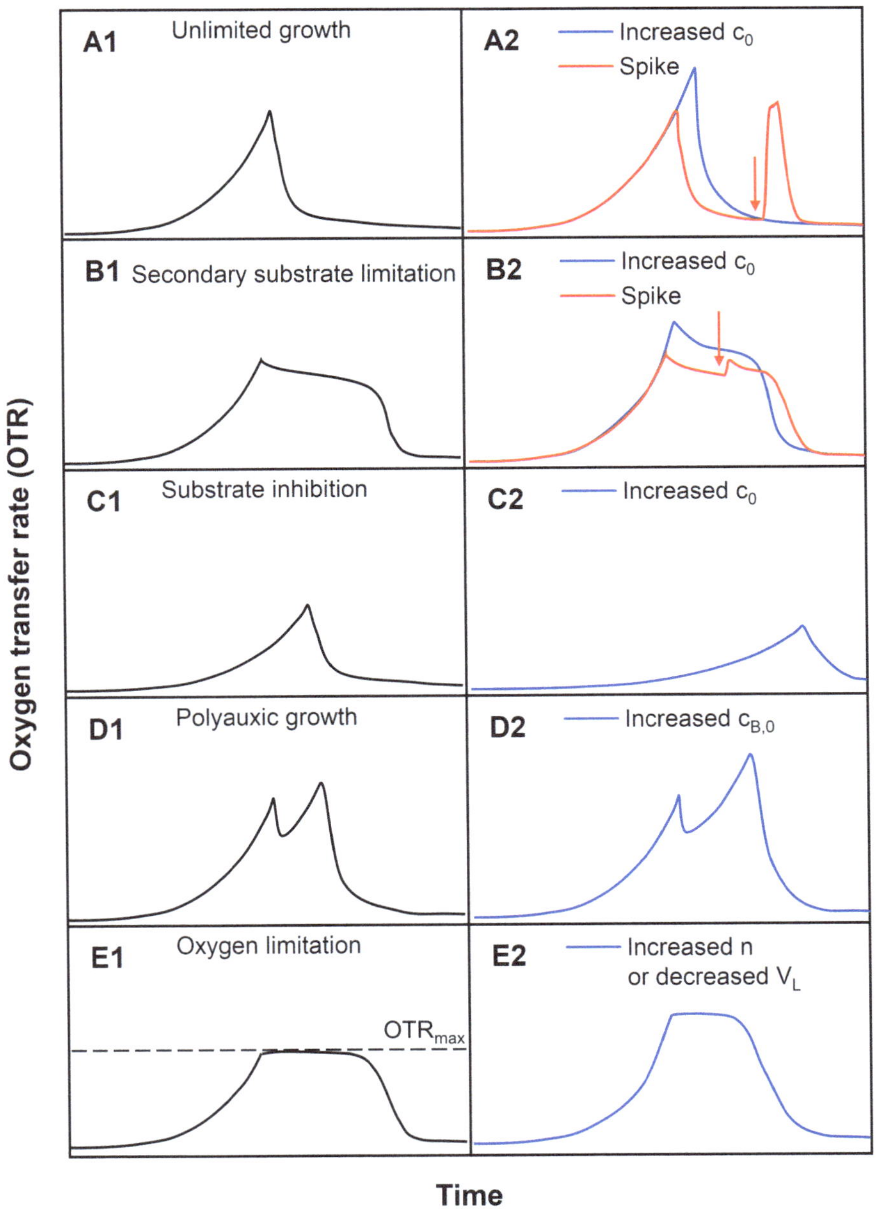

Fig. 1 Phenomena typically observed during *U. maydis* cultivations and their effect on the oxygen transfer rate. (**1**) depicts the characteristic OTR curve for the respective phenomenon, whereas (**2**) illustrates the influence of suitable approaches to verify the respective phenomenon. Arrows indicate time points of spikes. (**A1**) Unlimited growth, (**A2**) increased initial concentration of the respective carbon source and carbon spike. (**B1**) Secondary substrate limitation (here: nitrogen limitation), (**B2**) increased initial concentration of the nitrogen source and nitrogen spike. (**C1**) Substrate inhibition, (**C2**) increased initial substrate concentration. (**D1**) Polyauxic growth (here: diauxic growth on two carbon sources A+B), (**D2**) increased initial concentration of the carbon source B. (**E1**) Oxygen limitation with calculated OTR_{max}, (**E2**) increased shaking frequency or decreased filling volume

$$\mu = \frac{\ln(\mathrm{OTR(t)}) - \ln(\mathrm{OTR}(t_0))}{t - t_0} \quad (1)$$

b. Depletion of the carbon sources can be identified by either increasing the amount of carbon source in the medium or spiking the respective carbon source after depletion. This should result in either a higher maximal OTR or a resumption of growth, respectively [14].

3.2 Secondary Substrate Limitation

1. Secondary substrate limitation signifies that a nutrient other than the carbon source is depleted during the cultivation prior to carbon depletion. Generally, nitrogen limitation is introduced to induce organic acid or microbial lipid production and to redirect the carbon flow to product instead of biomass formation (*see* **Note 2**) [15–17].
2. When nitrogen is depleted, the OTR immediately displays a characteristic peak, followed by a decreasing plateau [17, 18].
3. **How to be sure:** Secondary substrate depletion can be identified by either increasing the initial concentration in the medium or by spiking the suspected nutrient during the OTR plateau. This should result in a prolonged growth phase or a resumption of growth, respectively [17, 18].

3.3 Substrate Inhibition

1. Substrate inhibition especially occurs when using alternative feedstocks, e.g., methanol and acetate from carbon capture, lignocellulosic hydrolysates containing furfural and HMF, or residue streams from the food industry with high osmolality such as molasses [19].
2. This may result in increased lag phases and decreased growth rates. If the substrate inhibition becomes too severe, the organism will not grow at all [17, 20, 21].
3. **How to be sure:** Compare cultivations with increasing amounts of the respective substrate while keeping all other conditions constant. With increasing concentration, the substrate inhibition will become more severe.

3.4 Polyauxic Growth

1. Renewable feedstocks typically contain various carbon sources simultaneously [22].
2. *U. maydis* exhibits polyauxic growth when presented with multiple carbon sources and consumes them in a distinct order according to its carbon source preference (*see* **Note 3**) [23].
3. Polyauxic growth is marked by multiphase exponential growth intermitted by lag phases of minimal growth [24].
4. As aerobic carbon metabolization and oxygen consumption are strictly correlated, this phenomenon is visible in the OTR.

The growth phases on each carbon source are consequently separated by short drops in the OTR, and the growth rates decrease with each phase.

5. **How to be sure:** Increase the initial concentration of each carbon source individually in parallel cultivations. The respective growth phase should be prolonged [23].

3.5 Oxygen Limitation

1. Oxygen availability is imperative for cultivations with *U. maydis*. However, inadequate cultivation conditions may easily lead to oxygen limitation in small-scale cultivations [18]. Oxygen limitation occurs when the oxygen demand of the microbial culture surpasses the maximal oxygen transfer capacity [25].
2. Oxygen limitation can be recognized by a plateau of constant OTR (*see* **Note 4**) [25].
3. Common sources of oxygen limitation are too high filling volumes and too low shaking frequencies.
4. **How to be sure:**
   a. The maximum oxygen transfer capacity (OTR_{max}) in shake flasks can be approximated using the following Eq. 2 [26]:

$$\mathrm{OTR}_{\max} = 3.72 \times 10^{-7} \times \mathrm{Osmol}^{0.05} \times n^{1.18-\frac{\mathrm{Osmol}}{10.1}} \times V_L^{-0.74} \times d_0^{0.33} \times d^{1.88} \times p_R \times y_{O2}^{*} \quad (2)$$

   Here, the OTR_{max} is dependent on the medium osmolality (Osmol in Osmol/kg), the shaking frequency (n in rpm), the filling volume (V_L in mL), the shaking diameter (d_0 in mm), the maximum shake flask diameter (d in mm), the pressure (p_R in bar), and the oxygen mole fraction in the gas phase (y_{O2}^* in mol/mol) (*see* **Note 5**) [26].

   b. The straightforward way to test for oxygen limitation is to repeat the experiment using a lower filling volume or higher shaking frequency to see if the maximal reached OTR increases (*see* **Note 6**).

4 Notes

1. If both the oxygen and the carbon dioxide transfer rate (CTR) are measured, the respiratory quotient (RQ) can be calculated according to Eq. 3:

$$\mathrm{RQ} = \frac{\mathrm{CTR}}{\mathrm{OTR}} \quad (3)$$

The RQ gives further insights into the metabolic activity of the organism. Biomass production on glucose results in an RQ of 1. An RQ above 1 indicates production of oxidized metabolites (e.g., itaconic acid) or consumption of more reduced compounds (e.g., ethanol). Vice versa, an RQ below 1 signifies consumption of oxidized metabolites or production of more reduced metabolites (e.g., erythritol, lipids) [4, 17].

2. The most common secondary substrate limitation for *U. maydis* is nitrogen limitation. However, other limitations such as phosphate limitation are superior when using a nitrogen-rich feedstock, e.g., sugar beet molasses. Contrary to nitrogen limitation, the metabolic activity of the culture increases even after phosphate depletion due to internal storage molecules. Polyphosphate, which is stored in the vacuoles, is used to buffer intracellular phosphate levels. As the internal storage decreases, the growth rate of the organism slowly decreases until the OTR displays the typical decreasing OTR plateau [18].
3. Polyauxic growth can also occur when using several nitrogen sources (e.g., ammonium nitrate) [27].
4. Oxygen-limited cultivations may also display a slightly rising OTR plateau. This occurs when the osmolality decreases over the cultivation time, e.g., due to substrate consumption, or if the filling volume decreases due to evaporation.
5. Common values for the calculation of the OTR_{max} are ambient pressure (1 bar), a gas phase concentration close to ambient air ($y_{O2}^{*}= 0.21$), and a maximal flask diameter of 83.3 mm (250 mL shake flask). The osmolality can be determined by an osmometer and is dependent on the medium composition. Typical osmolalities for standard Verduyn medium [28] start from 0.3 Osmol/kg (25 g/L glucose, 0.1 M MES buffer) but easily increase to 2 Osmol/kg with higher sugar, salt, and buffer concentrations (100 g/L molasses, 0.3 M MES buffer).
6. Although baffled shake flasks can increase the OTR_{max}, the results are often irreproducible due to inconsistent manufacturing, splashing, and out-of-phase conditions [29]. The use of unbaffled shake flasks is therefore recommended. Filling volume, shaking diameter, and shaking frequency should be adapted to circumvent oxygen limitation.

References

1. Anderlei T, Büchs J (2001) Device for sterile online measurement of the oxygen transfer rate in shaking flasks. Biochem Eng J 7:157–162. https://doi.org/10.1016/s1369-703x(00)00116-9
2. Ihling N, Munkler LP, Berg C et al (2021) Time-resolved monitoring of the oxygen transfer rate of Chinese hamster ovary cells provides insights into culture behavior in shake flasks. Front Bioeng Biotechnol 9:725498. https://doi.org/10.3389/fbioe.2021.725498
3. Dinger R, Lattermann C, Flitsch D et al (2022) Device for respiration activity measurement enables the determination of oxygen transfer rates of

microbial cultures in shaken 96-deepwell microtiter plates. Biotechnol Bioeng 119:881–894. https://doi.org/10.1002/bit.28022
4. Anderlei T, Zang W, Papaspyrou M et al (2004) Online respiration activity measurement (OTR, CTR, RQ) in shake flasks. Biochem Eng J 17:187–194. https://doi.org/10.1016/S1369-703X(03)00181-5
5. Schulte A, Brockmann J, Müller N et al (2025) A new approach to off-gas analysis for shaken bioreactors showing high CTR and RQ accuracy. J Biol Eng 19:11. https://doi.org/10.1186/s13036-025-00480-5
6. Niehoff P-J, Müller W, Pastoors J et al (2023) Development of an itaconic acid production process with Ustilaginaceae on alternative feedstocks. BMC Biotechnol 23:34. https://doi.org/10.1186/s12896-023-00802-9
7. Straaten SL, Zöllner M, Forsten E et al (2025) Primary high-throughput screening of engineered phytases by online monitoring of the oxygen transfer rate of Komagataella phaffii. Microb Cell Fact 24:180. https://doi.org/10.1186/s12934-025-02806-w
8. Battling S, Pastoors J, Deitert A et al (2022) Development of a novel defined minimal medium for Gluconobacter oxydans 621H by systematic investigation of metabolic demands. J Biol Eng 16:31. https://doi.org/10.1186/s13036-022-00310-y
9. Müller J, Beckers M, Mußmann N et al (2018) Elucidation of auxotrophic deficiencies of Bacillus pumilus DSM 18097 to develop a defined minimal medium. Microb Cell Fact 17:106. https://doi.org/10.1186/s12934-018-0956-1
10. Wahjudi SMW, Petrzik T, Oudenne F et al (2023) Unraveling the potential and constraints associated with corn steep liquor as a nutrient source for industrial fermentations. Biotechnol Prog 39:e3386. https://doi.org/10.1002/btpr.3386
11. Pastoors J, Baltin C, Bettmer J et al (2023) Respiration-based investigation of adsorbent-bioprocess compatibility. Biotechnol Biofuels Bioprod 16:49. https://doi.org/10.1186/s13068-023-02297-0
12. Hemmerich J, Wiechert W, Oldiges M (2017) Automated growth rate determination in high-throughput microbioreactor systems. BMC Res Notes 10:617. https://doi.org/10.1186/s13104-017-2945-6
13. Stöckmann C, Maier U, Anderlei T et al (2003) The oxygen transfer rate as key parameter for the characterization of Hansenula polymorpha screening cultures. J Ind Microbiol Biotechnol 30:613–622. https://doi.org/10.1007/s10295-003-0090-9
14. Steinhoff H, Finger M, Osthege M et al (2023) Experimental kS estimation: a comparison of methods for Corynebacterium glutamicum from lab to microfluidic scale. Biotechnol Bioeng 120:1288–1302. https://doi.org/10.1002/bit.28345
15. Zambanini T, Hartmann SK, Schmitz LM et al (2017) Promoters from the itaconate cluster of *Ustilago maydis* are induced by nitrogen depletion. Fungal Biol Biotechnol 4:11. https://doi.org/10.1186/s40694-017-0040-3
16. Wierckx N, Miebach K, Ihling N et al (2021) Perspectives for the application of Ustilaginaceae as biotech cell factories. Essays Biochem 65:365–379. https://doi.org/10.1042/EBC20200141
17. Klement T, Milker S, Jäger G et al (2012) Biomass pretreatment affects *Ustilago maydis* in producing itaconic acid. Microb Cell Fact 11:43. https://doi.org/10.1186/1475-2859-11-43
18. Grebe LA, Lichtenberg PG, Hürter K et al (2024) Phosphate limitation enhances malic acid production on nitrogen-rich molasses with Ustilago trichophora. Biotechnol Biofuels Bioprod 17:92. https://doi.org/10.1186/s13068-024-02543-z
19. Ghatak HR (2011) Biorefineries from the perspective of sustainability: feedstocks, products, and processes. Renew Sustain Energy Rev 15:4042–4052. https://doi.org/10.1016/j.rser.2011.07.034
20. Romero-Aguilar L, Hernández-Morfín KD, Guerra-Sánchez G, et al (2023) Metabolic changes and antioxidant response in Ustilago maydis grown in acetate. J Fungi (Basel) 9:749. https://doi.org/10.3390/jof9070749
21. Helm T, Niehoff P-J, Gätgens J et al (2023) Introducing molasses as an alternative feedstock into itaconate production using *Ustilago* sp. N Biotechnol 77:30–39. https://doi.org/10.1016/j.nbt.2023.06.003
22. Richter P, Panchalingam J, Miebach K et al (2024) Studying microbial triglyceride production from corn stover saccharides unveils insights into the galactose metabolism of Ustilago maydis. Microb Cell Fact 23:204. https://doi.org/10.1186/s12934-024-02483-1
23. Grebe LA, Richter P, Altenkirch T et al (2025) Sampling-free investigation of microbial carbon source preferences on renewable feedstocks via online monitoring of oxygen transfer rate. Bioprocess Biosyst Eng 48:413–425. https://doi.org/10.1007/s00449-024-03117-x
24. Monod J (1949) The growth of bacterial cultures. Annu Rev Microbiol 3:371–394. https://doi.org/10.1146/annurev.mi.03.100149.002103
25. Hansen S, Hariskos I, Luchterhand B et al (2012) Development of a modified respiration activity monitoring system for accurate and highly resolved measurement of respiration activity in shake flask fermentations. J Biol Eng 6:11. https://doi.org/10.1186/1754-1611-6-11

26. Meier K, Klöckner W, Bonhage B et al (2016) Correlation for the maximum oxygen transfer capacity in shake flasks for a wide range of operating conditions and for different culture media. Biochem Eng J 109:228–235. https://doi.org/10.1016/j.bej.2016.01.014
27. Horst RJ, Zeh C, Saur A et al (2012) The Ustilago maydis Nit2 homolog regulates nitrogen utilization and is required for efficient induction of filamentous growth. Eukaryot Cell 11:368–380. https://doi.org/10.1128/EC.05191-11
28. Pastoors J, Deitert A, Michel C et al (2023) In situ adsorption of itaconic acid from fermentations of *Ustilago cynodontis* improves bioprocess efficiency. Biotechnol Biofuels Bioprod 16:181. https://doi.org/10.1186/s13068-023-02433-w
29. Brauneck G, Engel D, Grebe LA et al (2025) Pitfalls in early bioprocess development using shake flask cultivations. Eng Life Sci 25:e70001. https://doi.org/10.1002/elsc.70001

Check for updates

Chapter 44

Online Monitoring of *U. maydis* Utilizing Fluorescence and Scattered Light Measurements in Microtiter Plate Cultivations

Paul Richter and Jørgen Barsett Magnus

Abstract

High-throughput cultivation at the microtiter scale enables rapid, parameter-rich screening of various organisms, including *Ustilago maydis*. This chapter presents a detailed protocol for online monitoring methods, which integrate scattered light (SL) and fluorescence readouts. This chapter outlines the preparation of microtiter plates and medium, inoculation, plate sealing, and software setup. It also covers the selection of SL wavelengths and appropriate fluorescence filter cassettes. In addition, robust normalization strategies using uninoculated controls or first-measurement baselines are presented. The chapter examines how SL measurements can be calibrated and used to resolve growth phases, detect nutrient limitations relevant to product formation, and point to rare events such as spore formation. Complementary fluorescence channels expand observability. Naturally metabolic fluorophores (e.g., NADH, FAD) support biomass tracking when SL is unsuitable, while tagged strains or dye-based approaches enable pathway- and product-level readouts. As an application case, a real-time monitoring of intracellular lipid accumulation by staining with BODIPY 493/503 is demonstrated. Step-by-step procedures are paired with data-interpretation guidelines, including control design, reference selection, and pitfalls when media composition or strain background alter optical properties.

Keywords High-throughput analysis, Microtiter plate cultivation, Scattered-light monitoring, Fluorescence monitoring, *Ustilago maydis* growth, Real-time monitoring of intracellular lipids

1 Introduction

High-throughput screening at the microtiter scale is a robust and flexible method to gain detailed insights into specific processes or for screening with a broad range of parameters. There are various methods for collecting online data in such experimental setups. One effective technique involves measurements based on light scattering or fluorescence. In 2005, Samorski et al. developed a device capable of combined measurement of scattered light and fluorescence [1]. Today, this device is commercially available from Beckman Coulter, Inc. under the name BioLector [2]. In this protocol, the experimental setup of this device and its application with *U. maydis* will be discussed.

An N. T. Phan et al. (eds.), *Ustilago maydis*, Methods in Molecular Biology, vol. 3050,
https://doi.org/10.1007/978-1-0716-5364-7_44,

2 Materials

2.1 Device for Fluorescence and Scattered Light Measurement

There are some devices available that can measure both scattered light and fluorescence in 48-well microtiter plates. While the specific configurations may differ, most devices operate on a similar principle:

1. A light source is guided, typically by glass fiber, to the transparent bottom of a microtiter plate well.
2. The wavelength of the light is selected either:
   a. using a spectrofluorometer with monochromators (commonly in prototypes), or
   b. with optical filters that block all but the desired wavelength (often used in commercial devices).
3. Light passes through the well bottom and interacts with the sample:
   a. it is either reflected (scattered light) or
   b. excites molecules, causing fluorescence.
4. The emitted or scattered light is collected, usually via the same glass fiber, and passes through a secondary filter or monochromator before reaching the detector.

2.2 Microtiter Plates

48-well, black, 0.3–2 mL working volume, with transparent flat bottom without optical irritations.

2.3 Sterile Seal

Sterile seal for 48-well plates, pierced for sufficient oxygen supply. Here the Art.No. 900371-T from HJ-BIOANALYTIK GmbH is recommended.

2.4 Cultivation Hood

Either the device fits in the hood, as with various prototypes, or the hood is integrated in the device, as in the BioLector platform.

3 Methods

3.1 Scattered Light (SL) Measurement

For scattered light (SL) measurements, it is necessary that the excitation and emission wavelengths are identical. When using the BioLector platform, ensure that the filter cassette for scattered light is used. Recommended wavelengths for SL measurements are 520, 600, or 700 nm, depending on the fluorescence activity of the sample and the filter cassettes available.

3.1.1 Procedure

1. Cultivation setup:
   a. Preparing the medium.

   b. Inoculation.
   c. Passaging into MTP wells.
   d. Sealing the plate.

2. Set up cultivation conditions in BioLector software.
3. Start the experiment + data logging.
4. End the experiment + export scattered light data + data interpretation.

3.1.2 Data Interpretation

1. Scattered light data are measured in arbitrary units and need a reference point to sufficiently compare it between experiments.
   - For experiments with the same inoculation density, the first measurement can serve as the reference point.
   - For experiments with different inoculation densities, a triplicate without inoculation should be used as the reference point.
2. When there are no solid media components and the organisms grow exponentially and do not produce optically active molecules, the SL data normalized to the reference point have a linear correlation with the cell density in the culture broth.
3. Through the analysis of these data, it is possible to observe the growth phases of *U. maydis*, identify potential nutrient limitations relevant to product formation, and detect rare events such as spore formation.

3.2 Fluorescence Measurement

Fluorescence measurement can be used to track a multitude of different properties of the organism's metabolism as well as for monitoring biomass. Fluorescence can be measured at wavelength combinations of certain used fluorescence tags (e.g., GFP or mCherry) or at wavelength combinations of different fluorescent metabolic products such as NADH (Ex = 355 nm, Em = 470 nm) or FAD (Ex = 370 or 450 nm, Em = 525 nm) [3]. The online monitoring of biomass can be achieved through the use of natural fluorescent metabolic products (NADH, FAD) or by employing fluorescence-tagged strains. These fluorescent tags have been found to be particularly useful in experimental setups where scattered light is not a suitable biomass indicator, such as in the presence of solid media components [4]. Furthermore, the application of fluorescence tagging to other cellular components or even the product itself enables real-time monitoring of processes such as product formation (See Notes 1-3). When using the commercially available BioLector, the appropriate filter cassette for the desired fluorescence must be purchased.

3.2.1 Procedure

The procedure is analogous to that of SL.

3.2.2 Data Interpretation

Fluorescence data are also measured in arbitrary units and need a reference point to sufficiently compare it between experiments.

- For experiments with the same fluorescence properties, e.g., the same media composition, the first measurement can serve as the reference point.
- For experiments with different media compositions or different strains, a triplicate without special fluorescence properties (deionized water) should be used as the reference point.

4 Notes

1. Since fluorescence data interpretation is heavily dependent on the monitored molecule and the specificity of its fluorescence, an example of microbial oil production is provided. In the process, the oil was stained with the fluorescent dye BODIPY 493/503 and could therefore be monitored with the BioLector. Detailed information about the staining technique can be found in Qiu and Simon [5]. The cultivation and analysis in the BioLector were already described in *Yarrowia lipolytica* by Back et al. [6, 7].
2. Figure 1 illustrates intracellular oil production in *U. maydis*, monitored using the fluorescent dye BODIPY, which starts fluorescing inside the accumulated oil and is quenched by potassium ions in the cultivation medium [7]. A clear correlation between nutrient limitation, indicated by a drop in the oxygen transfer rate (vertical dotted line) and the onset of oil production can be observed. Oil accumulation in *U. maydis* begins with nutrient limitation, regardless of whether nitrogen (standard), sulfur, or phosphate is limiting. Notably, different types of nutrient limitation result in distinct oil production patterns, all of which can be tracked using this fluorescence-based method. Interestingly, not only the production but also the metabolization of the accumulated oil under sulfur limitation is visible in the decreasing fluorescence data (blue) after around 43 h.
3. With the described fluorescence analysis and staining technique, it is possible to monitor product that typically shows no fluorescence itself in *U. maydis* cultivations online and compare different limitations with varying production kinetics.

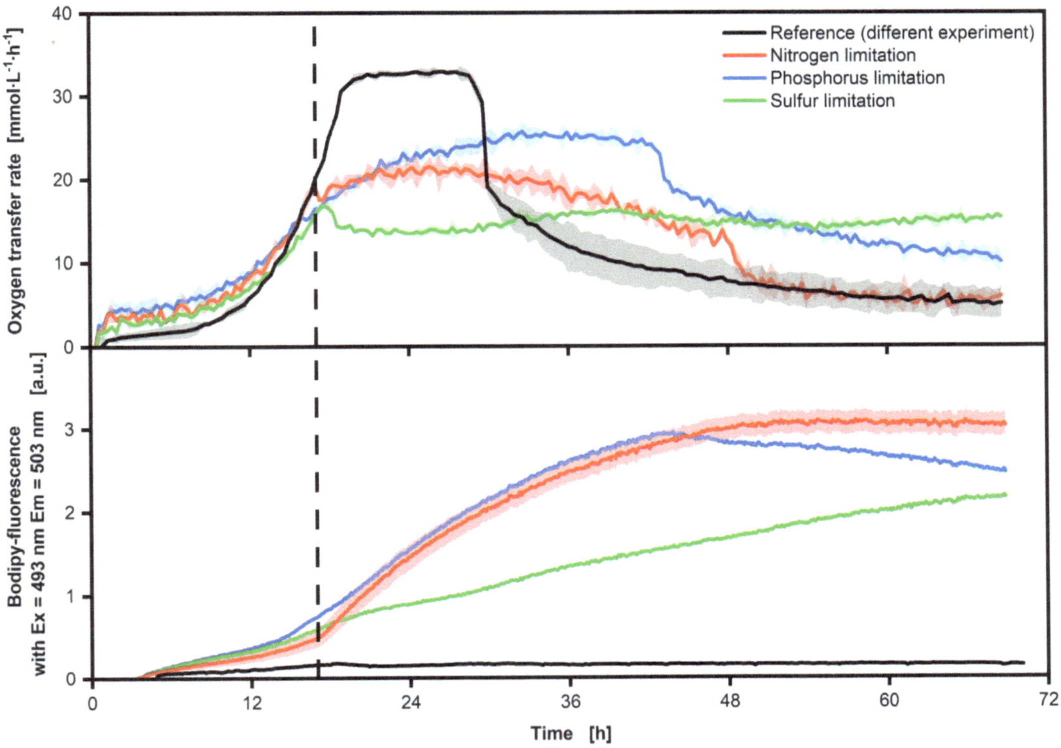

Fig. 1 Oxygen transfer rate and BODIPY 493/503 fluorescence intensity of *U. maydis* MB215ΔcyplΔemt1 cultivations with different secondary substrate limitations. Lines represent the averages of biological triplicates, and shadows indicate the standard deviation of these triplicates. The vertical dashed line represents the switch between growth and triglyceride production at the time of nutrient limitation. BODIPY fluorescence intensity was normalized at $t = 0$ h. Cultivation conditions: μRAMOS-BioLector combination, modified Verduyn medium, $OD_{start} = 0.1$, $V_L = 1$ mL, $T = 30°C$, $n = 1000$ rpm at 3 mm diameter, BODIPY 493/503 fluorescence measurement with excitation at 493 nm, emission at 500 nm, and an integration time of 600 ms

References

1. Samorski M, Müller-Newen G, Büchs J (2005) Quasi-continuous combined scattered light and fluorescence measurements: a novel measurement technique for shaken microtiter plates. Biotechnol Bioeng 92(1):61–68
2. Beckman Coulter Life Sciences (2025) Mikrobioreaktor. https://www.beckman.de/microbioreactor. Accessed 22 Jul 2025
3. Lukina MM, Shirmanova MV, Sergeeva TF et al (2016) Metabolic imaging in the study of oncological processes (review). Sovremennye tehnologii v medicine 8(4):113–126
4. Robertz S, Philipp M, Schipper K et al (2024) Monitoring corn stover processing by the fungus *Ustilago maydis*. Bioresour Bioprocess 11(1):87
5. Qiu B, Simon MC (2016) BODIPY 493/503 staining of neutral lipid droplets for microscopy and quantification by flow cytometry. Bio Protoc 6(17):e1912
6. Nicaud J-M, Cruz-Le Coq A-M, Rossignol T et al (2017) Protocols for monitoring growth and lipid accumulation in oleaginous yeasts. In: McGenity TJ, Timmis KN, Nogales B (Eds.) Hydrocarbon and lipid microbiology protocols. Springer Berlin Heidelberg, Berlin, Heidelberg, pp 153–169
7. Back A, Rossignol T, Krier F et al (2016) High-throughput fermentation screening for the yeast Yarrowia lipolytica with real-time monitoring of biomass and lipid production. Microb Cell Fact 15(1):147

Chapter 45

Integrating pH Value and Base Consumption in Online Monitoring of Organic Acid Production

Luca Antonia Grebe and Jørgen Barsett Magnus

Abstract

Microbial organic acid production is always associated with either a pH decrease or corresponding base titration. Here, we present two methods for online monitoring of organic acid production via these phenomena and give relevant interpretation suggestions. In shake flask and microtiter scales, the pH value can be measured online to determine qualitative information on growth and production phase. For pH-controlled cultivations, a linear regression can be derived to correlate organic acid production with base consumption and hence quantitatively monitor the organic acid production.

Keywords Organic acid, pH monitoring, pH adjuvant, Base titration, Product correlation, Soft sensor

1 Introduction

The pH value is a critical parameter in fermentation processes with *U. maydis*, as it directly influences the production profile. Especially organic acid production is highly pH-sensitive as low pH values are associated with weak organic acid stress. In small-scale cultivations, e.g., microtiter plates or shake flasks, soluble buffers are often used to mitigate the pH decrease caused by organic acid production [1, 2]. In pH-controlled fermentations, the pH value is maintained constant by automatic titration with a suitable base [2]. When monitored online, both the decrease in pH value and the base consumption are useful indicators of the cultivation performance [3, 4]. Moreover, the base consumption can be used for quantitative monitoring of organic acid formation [5, 6].

Organic acid production in *U. maydis* is commonly triggered by nitrogen starvation, which is ensured through deliberate selection of the medium composition [7]. Defined minimal media are especially suited because they enable controlled adjustments of the nitrogen source (e.g., ammonium salts) and the carbon source (e.g., glucose) [8, 9]. Growth and organic acid production are thus decoupled into two

An N. T. Phan et al. (eds.), *Ustilago maydis*, Methods in Molecular Biology, vol. 3050,
https://doi.org/10.1007/978-1-0716-5364-7_45,

consecutive phases. The pH change in the first phase is hence associated with biomass growth. After nitrogen depletion, organic acid formation starts. Depending on the cultivation systems, this either leads to a decrease in pH value or to the consumption of the pH adjuvant [3, 4].

2 Materials

The fermentations can be conducted using standard protocols for *U. maydis* while using a suitable device for online monitoring of the pH value or the base consumption. Potential commercial systems are suggested below (*see* **Note 1**).

2.1 pH Monitoring in Shake-Flask Scale

1. Transparent Erlenmeyer flasks.
2. Fiber-optic pH meter pH-1 SMA LG1 (PreSens, Regensburg, Germany).
3. Polymer optical fiber (PreSens, Regensburg, Germany).
4. Self-adhesive pH sensor spots SP-LG1-SA (PreSens, Regensburg, Germany).

2.2 pH Monitoring in Microtiter Scale

1. BioLector microbioreactor (Beckman Coulter Life Sciences, Indianapolis, USA).
2. Microtiter plates with pH optodes (Beckman Coulter Life Sciences, Indianapolis, USA).
3. Self-adhesive sterile top layer.

2.3 Online Monitoring of Base Consumption in Fermenter Scale

1. Fermenter with automatic pH control.
2. Base, e.g., 10 M NaOH.
3. Data logging
   a. Volumetric: Logging of consumed volume using a calibrated pump.
   b. Gravimetric: Logging of consumed mass using an electronic scale.
4. High-performance liquid chromatography (HPLC) for organic acid quantification.

3 Methods

3.1 Qualitative Evaluation of pH Decrease or Base Consumption

The change in pH value or the base consumption can be used for a qualitative monitoring of the process.

1. When monitoring the pH value: Prepare the pH sensor spots or microtiter plates according to the respective user manual. Depending on the device, calibration might be necessary.

2. When monitoring the base consumption: Log the base consumption volumetrically or gravimetrically.
3. Conduct the experiment as usual.
4. After the experiment is completed, plot the pH value or the base consumption over time.
5. In the following, a fermentation using a minimal medium containing glucose and ammonium is analyzed exemplarily. This prior knowledge is necessary for interpretation, as explained in **Note 2**. The different cultivation phases should be clearly distinguishable as depicted in Fig. 1. Base consumption and changes in the pH value exhibit an inverse relationship.

   a. The first phase of each cultivation is characterized by unlimited exponential growth, during which ammonium is consumed until nitrogen depletion. To maintain homeostasis, protons are exported equimolarly to ammonium import, resulting in a decrease in pH value or corresponding base consumption [10].

   b. When ammonium is depleted, the organism adapts its metabolism, resulting in a brief pause in both curves [7].

   c. Afterward, organic acid is produced until the carbon source is depleted, leading to a further decrease in pH value or base consumption.

   d. Depending on the strain, the produced organic acid is consumed afterward, which may be visible in an increase in pH value [3].

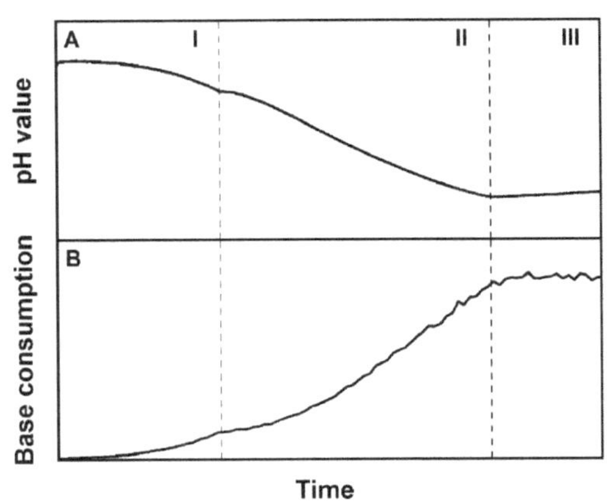

Fig. 1 Typical trajectory of (**A**) the pH value in cultivations without pH control and (**B**) cumulative base consumption in cultivations with pH control over time for organic acid production on ammonium and glucose. **I:** Unlimited exponential growth and ammonium consumption until depletion of the nitrogen source. **II:** Organic acid production under nitrogen limitation until glucose depletion. **III:** Organic acid consumption

3.2 Quantitative Correlation and Monitoring of Organic Acid Production via Base Consumption

The monitored base consumption can be used for the online quantification of the organic acid concentration (*see* **Note 3**).

1. A calibration dataset is first collected. A representative fermentation is frequently sampled, and the organic acid concentration is measured via HPLC.
2. For each sample, the measured organic acid concentration is then matched to the respective cumulative base consumption as depicted in Fig. 2.
3. In the beginning of the fermentation, base is consumed during the growth phase without organic acid production. Samples taken during the growth phase are plotted directly on the *y*-axis. The amount of base used for the growth phase can be determined when plotting the base consumption over time as depicted in Fig. 1. Alternatively, additional online signals such as respiratory activity can be used as an indicator of production start. The amount of base consumed during growth is dependent on the initial ammonium concentration of the medium and can thus vary when adjusting the medium composition.
4. After nitrogen depletion, base is consumed to counteract the pH decrease due to organic acid production. A linear relationship between produced organic acid and consumed base should be visible (*see* **Note 4**).
5. All data points during the production phase are subsequently used for a linear regression. The amount of base consumed during the growth phase is taken into account by the *y*-axis intercept. The slope of the regression function represents the

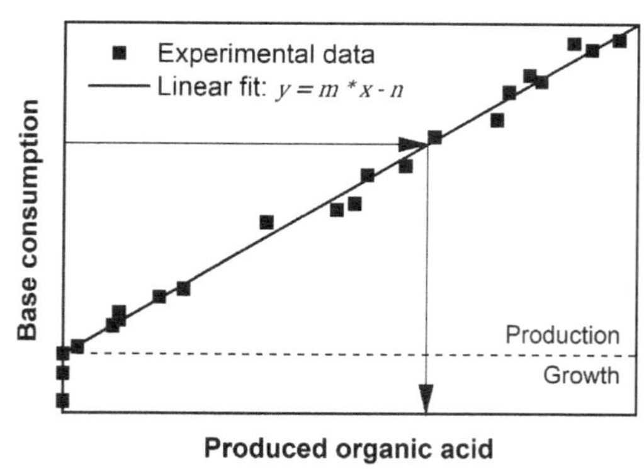

Fig. 2 Base consumption is linearly proportional to the produced organic acid. A linear fit can be derived from plotting the consumed base over the produced organic acid. All base consumption below the dashed line, as considered by the *y*-axis intercept, is consumed during growth and is therefore dependent on the ammonium concentration. The slope represents the proportionality factor and can be used to predict the organic acid concentration based on the online-monitored base consumption as indicated by the arrows

proportionality factor between organic acid production (g, kg, or mol) and base consumption (mL, L, g, kg, or mol). Both can be used to derive the following Eq. 1:

$$\text{Base consumption} = \text{Proportionality factor} \times \text{Produced organic acid} + \text{Base consumption during the growth phase} \quad (1)$$

6. This equation can now be used to calculate the current organic acid concentration from the base consumption monitored online (*see* **Notes 5 and 6**).

4 Notes

1. Benchtop fermenters equipped with a pH probe can also be used for pH monitoring if no small-scale pH monitoring is available.
2. Both pH monitoring and correlation of base consumption with organic acid production have to be conducted with care, as the change in the pH value is the sum of multiple factors. Changes in medium composition or cultivation conditions can greatly impact the observations described above. When using nitrate instead of ammonium as the nitrogen source, the pH value will increase instead of decreasing during the growth phase [11]. Consequently, the pH value will first decrease and then increase when using ammonium nitrate, as *U. maydis* prefers ammonium over nitrate [12]. The carbon source used will also have an effect. Complex feedstocks such as agricultural residue streams often contain organic acids themselves, which leads to an increase in pH value upon consumption.
3. The decrease in pH value in cultivations without pH control can, in theory, also be used to quantify organic acid production. It is, however, significantly more complex due to the interplay of the dissociation equilibria of both the product and the buffer in use.
4. The linear correlation between base consumption and organic acid production can be influenced by noticeable byproduct formation (e.g., when itaconic acid is the molecule of interest, but the organism also produces malic acid). If the main product and the byproduct are produced with a stable stoichiometry, this effect can be neglected.
5. When using an aqueous solution as pH adjuvant, such as NaOH, the amount of consumed base is inherently dependent on its concentration. Consequently, if the volume or mass of the aqueous solution is used in the correlation (e.g., 100 mL or

133 g of 10 M NaOH per 65 g organic acid), the concentration of the base has to be taken into account for each experiment. Alternatively, the amount of NaOH molecules could be used directly, so that the correlation is independent of the concentration of the base (e.g., 1 mol or 40 g NaOH per 65 g itaconic acid).

6. The proportionality factor between base consumption and organic acid formation depends on the acid's dissociation equilibrium and may change when the pH control setpoint is adjusted.

References

1. Niehoff P-J, Müller W, Pastoors J et al (2023) Development of an itaconic acid production process with Ustilaginaceae on alternative feedstocks. BMC Biotechnol 23:34. https://doi.org/10.1186/s12896-023-00802-9
2. Helm T, Niehoff P-J, Gätgens J et al (2023) Introducing molasses as an alternative feedstock into itaconate production using *Ustilago* sp. N Biotechnol 77:30–39. https://doi.org/10.1016/j.nbt.2023.06.003
3. Sarikaya B, Günster K, Grebe LA et al (2025) High-sensitivity real-time monitoring of pH and respiration activity unveils metabolic dynamics in shake flask cultures. Biotechnol Prog 41:e3525. https://doi.org/10.1002/btpr.3525
4. Pastoors J, Deitert A, Michel C et al (2023) In situ adsorption of itaconic acid from fermentations of *Ustilago cynodontis* improves bioprocess efficiency. Biotechnol Biofuels Bioprod 16:181. https://doi.org/10.1186/s13068-023-02433-w
5. Feitkenhauer H, Meyer U (2004) Software sensors based on titrimetric techniques for the monitoring and control of aerobic and anaerobic bioreactors. Biochem Eng J 17:147–151. https://doi.org/10.1016/S1369-703X(03)00150-5
6. Siano SA (1995) On the use of the pH control reagent addition rate for fermentation monitoring. Biotechnol Bioeng 47:651–665. https://doi.org/10.1002/bit.260470606
7. Zambanini T, Hartmann SK, Schmitz LM et al (2017) Promoters from the itaconate cluster of *Ustilago maydis* are induced by nitrogen depletion. Fungal Biol Biotechnol 4:11. https://doi.org/10.1186/s40694-017-0040-3
8. Klement T, Milker S, Jäger G et al (2012) Biomass pretreatment affects *Ustilago maydis* in producing itaconic acid. Microb Cell Fact 11:43. https://doi.org/10.1186/1475-2859-11-43
9. Grebe LA, Lichtenberg PG, Hürter K et al (2024) Phosphate limitation enhances malic acid production on nitrogen-rich molasses with Ustilago trichophora. Biotechnol Biofuels Bioprod 17:92. https://doi.org/10.1186/s13068-024-02543-z
10. Yang X, Yang Y, Huang J et al (2021) Comparisons of urea or ammonium on growth and fermentative metabolism of Saccharomyces cerevisiae in ethanol fermentation. World J Microbiol Biotechnol 37:98. https://doi.org/10.1007/s11274-021-03056-9
11. Galafassi S, Capusoni C, Moktaduzzaman M et al (2013) Utilization of nitrate abolishes the "Custers effect" in Dekkera bruxellensis and determines a different pattern of fermentation products. J Ind Microbiol Biotechnol 40:297–303. https://doi.org/10.1007/s10295-012-1229-3
12. Lewis CM, Fincham JRS (1970) Regulation of nitrate reductase in the basidiomycete Ustilago maydis. J Bacteriol 103:55–61. https://doi.org/10.1128/jb.103.1.55-61.1970

Chapter 46

Online Monitor the Viable Biomass Formation of *U. maydis* by Capacitance Measurement in Lab Scale Fermenter

Paul Richter and Jørgen Barsett Magnus

Abstract

Online capacitance (permittivity) measurements provide a simple, robust alternative to optical and gravimetric readouts for tracking viable biomass in *Ustilago maydis*. By applying an alternating electric field and measuring the dielectric response of intact cells, the signal reflects the volume of membrane-bound, living biomass and remains largely insensitive to turbidity, color, solids, or intracellular product accumulation. This is particularly beneficial during periods of lipid accumulation, where offline analytics can be influenced by changes in scattering or density. In contrast, capacitance maintains a closer correlation with the viable cell volume and enhances the clarity of growth phases and physiological shifts. This chapter introduces the measurement principle, practical probe handling with sterilizable, in situ sensors, and step-by-step setup in a 2 L lab-scale batch. Best practices for placement and autoclaving, baseline removal, logging intervals, and give guidance on sensor configuration, alongside straightforward calibration and interpretation strategies, will be discussed. Collectively, these procedures provide a reproducible workflow for continuous, non-invasive monitoring of viable biomass in *U. maydis*. This complements optical methods and enables more reliable kinetic analysis and process decisions during phases of poor offline analytics due to oil accumulation or solids within the system.

Keywords Capacitance, Permittivity, Cell monitoring, Intracellular product

1 Introduction

Monitoring the capacitance/permittivity of a microbial cultivation offers an easy, reliable route to online monitoring of viable biomass. By applying an alternating electric field and measuring the dielectric response of the cells, the method selectively tracks the volume of biomass with intact membranes, as the cells act as small capacitors inside the electric field. Unlike optical density or cell dry weight, which are sensitive to turbidity, color, non-cellular solids, and cell debris, capacitance is largely insensitive to optical artifacts and thus also remains robust when intracellular products alter cell appearance. This advantage is especially relevant for *U. maydis*, where intracellular lipid accumulation can distort OD and complicate gravimetric

An N. T. Phan et al. (eds.), *Ustilago maydis*, Methods in Molecular Biology, vol. 3050,
https://doi.org/10.1007/978-1-0716-5364-7_46,

methods. Lipid-filled cells scatter light differently and shed material that alters offline measurements, but their membranes still polarize in the alternating electric field. As a result, capacitance signals maintain a closer correspondence to viable cell volume during lipid build-up, supporting clearer interpretation of growth phases and physiological shifts (*see* **Notes 1-2**)[1, 2].

Practically, capacitance is measured with a sterilizable, in situ probe mounted directly in the fermenter. In a 2 L lab-scale batch setup, the probe delivers continuous, real-time data without sampling. This chapter introduces the principle of capacitance measurement with living cells, best practices for probe placement, and simple strategies to calibrate capacitance to biomass for *U. maydis*.

2 Materials

1. Capacitance probe: Probe that enables the measurement of capacitance in the bioreactor. There probes are available from a variety of companies.
2. Monitoring software: Probe setup, data logging, "Mark Zero" offset, scans, and reporting.
3. Bioreactor: Every standard bioreactor with a PG13.5 port is suitable.

3 Methods

3.1 Probe Installation and Sterilization

Mount the probe in a PG13.5 port of the lab-scale bioreactor, with the deepening at the power connector in the flow direction of the reactor. Autoclave the reactor filled with water, after protecting data/power connections of the probes. After autoclaving, empty the reactor and fill it with the cultivation medium. Connect the power cable and power supply.

3.2 Software Connection and Setup

Connect the sensor to the software. Enable sensor data logging. Create a new experiment, set a sampling time ≥ 6 s (recommended). Before inoculation, use Mark Zero in the sterilized medium to remove the background offset. The cell factor can be set at 1.0 initially (pF/cm) before conducting calibration with offline sampling. After calibration against viable cell density, the cell factor can be adjusted accordingly (*See* **Note 3**). Select the required cell type; for *U. maydis*, the yeast cell type is best suited. After adjusting the measurement settings, the experiment can be started. The inoculation timestamp can be used software wise to mark cell addition.

After the stop of cultivation, capacitance data can be exported via the software.

4 Notes

1. As an example for data interpretation, Fig. 1 shows exponential growth across all signals until the implemented nitrogen limitation at 15 h. This timepoint appears simultaneously in the culture's respiration activity and in viability-related measurements, such as capacitance and cell count, marking the physiological switch from growth to production.
2. Immediately after nitrogen depletion, the capacitance signal displays a brief transition phase: some cells still divide while biomass composition adapts and average cell size increases. Once this adjustment ends, capacitance stabilizes, indicating a viable but non-growing population even as intracellular oil synthesis proceeds. In contrast, OD and dry cell weight continue to rise beyond 15 h. This divergence from the flat capacitance profile reflects intracellular lipid accumulation, which inflates optical and gravimetric readouts without increasing viable cell volume.

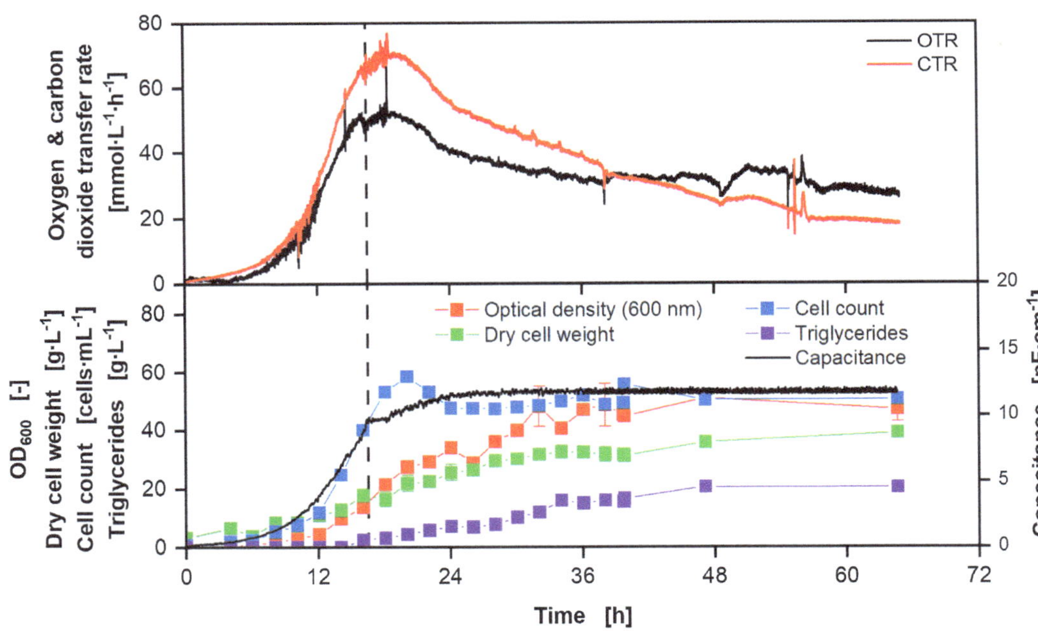

Fig. 1 Cultivation data of the triglyceride production process in *U. maydis* MB215Δcyp1Δemt1 at bioreactor scale, with closely timed sampling to evaluate different biomass measurements. OTR and CTR are presented as a reference for the culture's activity. All biomass measurements were conducted in technical triplicate. Cultivation conditions: modified Verduyn medium, pH = 6.5, 2 L New Brunswick bioreactor; $V_L = 1.2$ L, $T =$ 30°C, initial $OD_{600} = 0.1$, stirrer speed = varied between 300 and 1200 rpm to keep DO at 30%

3. Cell counting, based on principles closely related to capacitance, is well-suited for cultures with intracellular product formation. Use it to calibrate the capacitance probe against an independent viability metric and track size-distribution shifts during lipid accumulation, thereby reinforcing the interpretation of the post-limitation plateau in capacitance signal.

References

1. Maskow T, Röllich A, Fetzer I et al. (2008) On-line monitoring of lipid storage in yeasts using impedance spectroscopy. J Biotechnol 135:64–70. https://doi.org/10.1016/j.jbiotec.2008.02.014
2. Yardley JE, Kell DB, Barrett J et al. (2000) On-line, real-time measurements of cellular biomass using dielectric spectroscopy. Biotechnol Genet Eng Rev 17:3–35. https://doi.org/10.1080/02648725.2000.10647986

Subject Index

An N. T. Phan et al. (eds.), *Ustilago maydis*, Methods in Molecular Biology, vol. ,3050
https://doi.org/10.1007/978-1-0716-5364-7,

G

H

I

L

S

T

U

V

Y

Z

MIX
Papier | Fördert
gute Waldnutzung
FSC® C083411

Kolibri 360 GmbH
Ferdinand-Jühlke-Straße 7
99095 Erfurt, Deutschland
produktsicherheit@kolibri360.de